LABORATORY MANUAL

Cat Version

HOLE'S HUMAN ANATOMY & PHYSIOLOGY

TWELFTH EDITION

DAVID SHIER JACKIE BUTLER RICKI LEWIS

Terry R. Martin

Kishwaukee College

Higher Education

Boston Burr Ridge, IL Dubuque, IA New York San Francisco St. Louis
Bangkok Bogotá Caracas Kuala Lumpur Lisbon London Madrid Mexico City
Milan Montreal New Delhi Santiago Seoul Singapore Sydney Taipei Toronto

Higher Education

LABORATORY MANUAL TO ACCOMPANY HOLE'S HUMAN ANATOMY & PHYSIOLOGY, TWELFTH EDITION (CAT)

Published by McGraw-Hill, a business unit of The McGraw-Hill Companies, Inc., 1221 Avenue of the Americas, New York, NY 10020.

This book is printed on acid-free paper.

1 2 3 4 5 6 7 8 9 0 BAN/BAN 0 9

ISBN 978-0-07-331596-6
MHID 0-07-331596-6

Publisher: *Michelle Watnick*
Senior Sponsoring Editor: *James F. Connely*
Director of Development: *Kristine Tibbetts*
Developmental Editor: *Fran Schreiber*
Marketing Manager: *Lynn M. Breithaupt*
Senior Project Manager: *Jayne L. Klein*
Lead Production Supervisor: *Sandy Ludovissy*
Lead Media Project Manager: *Stacy A. Patch*
Designer: *Laurie B. Janssen*
Cover Designer: *Elise Lansdon*
(USE) Cover Image: © *The McGraw-Hill Companies, Inc./Gerald Wofford, photographer (left photo);* © *Design Pics/PunchStock (middle photo);* © *Getty Images (right photo)*
Senior Photo Research Coordinator: *John C. Leland*
Photo Research: *Mary Reeg*
Compositor: *Precision Graphics*
Typeface: *10/12 Slimbach Book*
Printer: *R. R. Donnelley Menasha, WI*

Some of the laboratory experiments included in this text may be hazardous if materials are handled improperly or if procedures are conducted incorrectly. Safety precautions are necessary when you are working with chemicals, glass test tubes, hot water baths, sharp instruments, and the like, or for any procedures that generally require caution. Your school may have set regulations regarding safety procedures that your instructor will explain to you. Should you have any problems with materials or procedures, please ask your instructor for help.

www.mhhe.com

Contents

Surface Anatomy

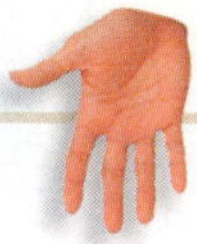

Nervous System

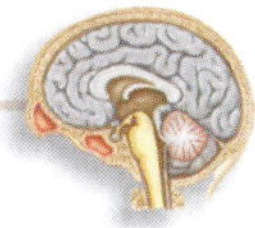

General and Special Senses

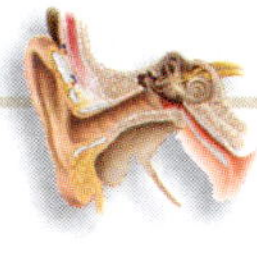

Endocrine System

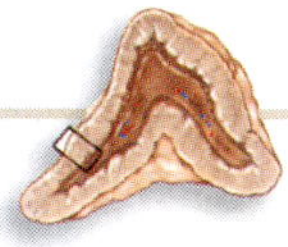

Cardiovascular System

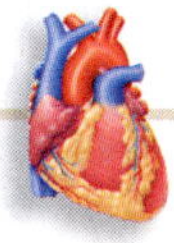

Lymphatic System

Digestive System

Respiratory System

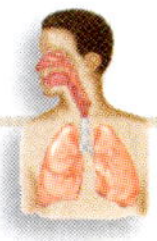

Urinary System

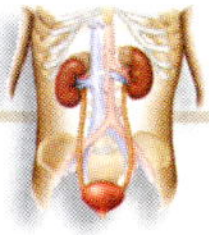

Reproductive System and Development

Cat Dissection Laboratory Exercises

Online Laboratory Exercises

*These laboratory exercises are available at www.mhhe.com/shier12 (Click on this laboratory manual cover).

Preface

This laboratory manual was prepared to be used with the textbook *Hole's Human Anatomy and Physiology,* 12th edition, by David Shier, Jackie Butler, and Ricki Lewis. As with the textbook, the laboratory manual is designed for students with minimal backgrounds in the physical and biological sciences who are pursuing careers in professional health fields.

The laboratory manual contains sixty-three laboratory exercises and reports, which are closely integrated with the chapters of the textbook. (Four of these exercises appear only online.) The exercises are planned to illustrate and review anatomical and physiological facts and principles presented in the textbook and to help students investigate some of these ideas in greater detail.

Often the laboratory exercises are short or are divided into several separate procedures. This allows an instructor to select those exercises or parts of exercises that will best meet the needs of a particular program. Also, exercises requiring a minimal amount of laboratory equipment have been included.

The laboratory exercises include a variety of special features that are designed to stimulate interest in the subject matter, to involve students in the learning process, and to guide them through the planned activities. These special features include the following:

Materials Needed

This section lists the laboratory materials that are required to complete the exercise and to perform the demonstrations and learning extensions.

Safety

A list of safety guidelines is included inside the front cover. Each lab session that requires special safety guidelines has a safety section following "Materials Needed." Your instructor might require some modifications of these guidelines.

Introduction

The introduction briefly describes the subject of the exercise or the ideas that will be investigated.

Purpose of the Exercise

The purpose provides a statement about the intent of the exercise—that is, what will be accomplished.

Learning Outcomes

The learning outcomes list what a student should be able to do after completing the exercise. Each learning outcome will have a matching assessment indicated by a corresponding icon 1 in the laboratory exercise or the laboratory report.

Procedure

The procedure provides a set of detailed instructions for accomplishing the planned laboratory activities. Each procedure heading includes an explore icon indicating ways to accomplish the learning outcomes. Usually these instructions are presented in outline form so that a student can proceed efficiently through the exercise in stepwise fashion. Often the student is referred to particular sections of a textbook for necessary background information or for review of subject matter presented previously.

The procedures include a wide variety of laboratory activities and, from time to time, direct the student to complete various tasks in the laboratory reports.

Laboratory Reports

A laboratory report to be completed by the student immediately follows each exercise. These reports include various types of review activities, spaces for sketches of microscopic objects, tables for recording observations and experimental results, and questions dealing with the analysis of such data.

As a result of these activities, students will develop a better understanding of the structural and functional characteristics of their bodies and will increase their skills in gathering information by observation and experimentation. Some of the exercises also include demonstrations, optional activities, and useful illustrations. By completing all of the assessments in the laboratory report, students will be able to determine if they were able to accomplish all of the learning outcomes.

Demonstrations

Demonstrations appear in separate boxes. They describe specimens, specialized laboratory equipment, or other materials of interest that an instructor may want to display to enrich the student's laboratory experience.

Learning Extensions

Learning extensions also appear in separate boxes. They encourage students to extend their laboratory experiences. Some of these activities are open-ended in that they suggest the student plan an investigation or experiment and carry it out after receiving approval from the laboratory instructor. Some of the figures are illustrated as line art or in grayscale. This will allow colored pencils to be used as a visual learning activity to distinguish various structures.

The Use of Animals in Biology Education*

The National Association of Biology Teachers (NABT) believes that the study of organisms, including nonhuman animals, is essential to the understanding of life on Earth. NABT recommends the prudent and responsible use of animals in the life science classroom. NABT believes that biology teachers should foster a respect for life. Biology teachers also should teach about the interrelationship and interdependency of all things.

Classroom experiences that involve nonhuman animals range from observation to dissection. NABT supports these experiences so long as they are conducted within the long-established guidelines of proper care and use of animals, as developed by the scientific and educational community.

As with any instructional activity, the use of nonhuman animals in the biology classroom must have sound educational objectives. Any use of animals, whether for observation or dissection, must convey substantive knowledge of biology. NABT believes that biology teachers are in the best position to make this determination for their students.

NABT acknowledges that no alternative can substitute for the actual experience of dissection or other use of animals and urges teachers to be aware of the limitations of alternatives. When the teacher determines that the most effective means to meet the objectives of the class do not require dissection, NABT accepts the use of alternatives to dissection, including models and the various forms of multimedia. The Association encourages teachers to be sensitive to substantive student objections to dissection and to consider providing appropriate lessons for those students where necessary.

To implement this policy, NABT endorses and adopts the "Principles and Guidelines for the Use of Animals in Precollege Education" of the Institute of Laboratory Animals Resources (National Research Council). Copies of the "Principles and Guidelines" may be obtained from the ILAR (2101 Constitution Avenue, NW, Washington, DC 20418; 202-334-2590).

Illustrations

Diagrams from the textbook and diagrams similar to those in the textbook often are used as aids for reviewing subject matter. Other illustrations provide visual instructions for performing steps in procedures or are used to identify parts of instruments or specimens. Micrographs are included to help students identify microscopic structures or to evaluate student understanding of tissues.

In some exercises, the figures include line drawings suitable for students to color with colored pencils. This activity may motivate students to observe the illustrations more carefully and help them to locate the special features represented in the figures. Students can check their work by referring to the corresponding full-color illustrations in the textbook.

Frequent variations exist in anatomical structures among humans. The illustrations in the textbook and the laboratory manual represent normal (normal means the most common variation) anatomy. Variations from normal anatomy do not represent abnormal anatomy unless some function is impaired.

Features of This Edition

Many of the changes in this new edition are a result of evaluations and suggestions from anatomy and physiology students. Many suggestions from users and reviewers of the eleventh edition have been incorporated. Some features include the following:

1. Learning outcomes with icons (1) have matching assessments with icons 1 so students can be sure they have accomplished all of the content of the laboratory exercise. Learning outcomes and assessments include all levels of learning skills: knowledge, comprehension, application, analysis, synthesis, and evaluation.
2. For the first time a surface anatomy lab has been added to the lab manual. Lurana Bain LMT, massage therapy program coordinator and instructor at Elgin Community College, was the contributing author of this laboratory exercise. Surface anatomy is important for students selecting careers in massage therapy, physical therapy, nursing, and emergency medical services.
3. The endocrine system lab contains more histology, diabetic physiology, and an expanded laboratory report.
4. The joints lab has been expanded to include additional material and figures on movements.
5. The blood cells lab and the blood typing lab have been combined into a single laboratory exercise.
6. The blood vessels lab and the arteries and veins lab have been combined into a single laboratory exercise.
7. A new more colorful design has been incorporated into this lab manual. This facilitates locating labs, laboratory reports, and studying tables.
8. The six cat dissection labs are together as laboratory exercises 54–59. Shaded portion of icon designates figure focus area.
9. To meet the need for clearer and more definite safety guidelines, a revised and updated safety list is located inside the front cover, and safety sections are found in appropriate labs.
10. A section called "Study Skills for Anatomy and Physiology" is located in the front material. This section was written by students enrolled in a Human Anatomy and Physiology course.

*Adopted by the Board of Directors in October 1995. This policy supersedes and replaces all previous NABT statements regarding animals in biology education.

11. "Critical Thinking Applications" are incorporated 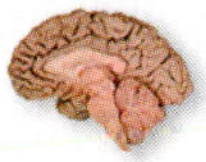 within most of the laboratory exercises to enhance valuable critical thinking skills that students will need throughout their lives.
12. The three labs using live frogs and the blood testing lab have been moved from the lab manual to the text website at www.mhhe.com/shier12.
13. Two assessment tools (rubrics) for laboratory reports are included in Appendix 2.
14. The *Instructor's Manual for the Laboratory Manual to Accompany Hole's Human Anatomy and Physiology* has been revised and updated. It contains a "Student Safety Contract" and a "Student Informed Consent Form" for possible inclusion in the class. The forms could be adapted for a specific institution based upon school policy. The instructor's manual is found on the text website for *Hole's Human Anatomy & Physiology,* Twelfth Edition, www.mhhe.com/shier12 and www.mhhe.com/biosci.ap/labcentral.

Acknowledgments

I value all the support and encouragement by the staff at McGraw-Hill Higher Education, including Michelle Watnick, Jim Connely, Fran Schreiber, Jayne Klein, and Cathy Schmitt. Special recognition is granted to Colin Wheatley for his insight, confidence, wisdom, warmth, and friendship.

I am grateful for the professional talent of David Shier, Jackie Butler, and Ricki Lewis for all of their coordination efforts and guidance with the laboratory manual. The deepest gratitude is extended to John W. Hole, Jr. for his years of dedicated effort in early editions of this classic work and for the opportunity to revise his established laboratory manual.

I am particularly thankful to Dr. Norman Jenkins and Dr. David Louis, retired presidents of Kishwaukee College, and Dr. Thomas Choice, president of Kishwaukee College, for their support, suggestions, and confidence in my endeavors. I am appreciative for the expertise of Womack Photography for numerous contributions. The professional reviews of the nursing procedures were provided by Kathy Schnier. I am also grateful to Laura Anderson, Rebecca Doty, Michele Dukes, Troy Hanke, Jenifer Holtzclaw, Stephen House, Shannon Johnson, Brian Jones, Marissa Kannheiser, Morgan Keen, Marcie Martin, Angele Myska, Sparkle Neal, Bonnie Overton, Susan Rieger, Eric Serna, Robert Stockley, Shatina Thompson, Nancy Valdivia, Marla Van Vickle, Jana Voorhis, Joyce Woo, and DeKalb Clinic Chartered for their special contributions. There have been valuable contributions from my students, who have supplied thoughtful suggestions and assisted in the clarification of details.

To my son Ross, I owe gratitude for his keen eye and creative suggestions. Foremost, I am appreciative to Sherrie Martin, my spouse and best friend, for advice, understanding, and devotion throughout the writing and revising.

Terry R. Martin
Kishwaukee College
21193 Malta Road
Malta, IL 60150

Reviewers

I would like to express my sincere gratitude to all users of the laboratory manual who provided suggestions for its improvement. I am especially grateful for the contributions of the reviewers who kindly reviewed the manual and examined the manuscript of the new edition. Their thoughtful comments and valuable suggestions are greatly appreciated. They include the following:

Steven Amdur, *Nassau Community College*
Mary Atchison, *St. Charles Community College*
Mary Bauer, *Chippewa Valley Technical College*
Moges Bizuneh, *Ivy Tech Community College*
Shawn Bjerke, *Minnesota State Community and Technical College*
Winnifred Bryant, *University of Wisconsin, Eau Claire*
Nathan L. Collie, *Texas Tech University*
Ethel R. Cornforth, *San Jacinto College South*
Mary Elizabeth Dawson, *Kingsborough Community College*
Clair B. Eckersell, *Brigham Young University–Idaho*
Clare Hays, *Metropolitan State College of Denver*
Andrew S. Lacerenza, *Nassau Community College*
Claudia I. Stanescu, *University of Arizona*
Rachel Venn Beecham, *Mississippi Valley State University*

About the Author

This twelfth edition is the sixth revision by Terry R. Martin of Kishwaukee College. Terry's teaching experience of over thirty years, his interest in students and love for college instruction, and his innovative attitude and use of technology-based learning enhance the solid tradition of John Hole's laboratory manual. Among Terry's awards are the 1972 Kishwaukee College Outstanding Educator, 1977 Phi Theta Kappa Outstanding Instructor Award, 1989 Kishwaukee College ICCTA Outstanding Educator Award, 1996 and 2004 Who's Who Among America's Teachers, 1996 Kishwaukee College Faculty Board of Trustees Award of Excellence, and 1998 Continued Excellence Award for Phi Theta Kappa Advisors. Terry's professional memberships include the National Association of Biology Teachers, Illinois Association of Community College Biologists, Human Anatomy and Physiology Society, former Chicago Area Anatomy and Physiology Society (founding member), Phi Theta Kappa (honorary member), and Nature Conservancy. In addition to writing many publications, he co-produced with Hassan Rastegar a videotape entitled *Introduction to the Human Cadaver and Prosection*, published by Wm. C. Brown Publishers in 1989. Terry revised the *Laboratory Manual to Accompany Hole's Essentials of Human Anatomy and Physiology*, 10th edition, and authored *Human Anatomy and Physiology Laboratory Manual, Fetal Pig Dissection*, 3rd edition, and *Laboratory Manual for Human Anatomy & Physiology*, cat, main, and fetal pig versions, first editions. A series of seven LabCam videos of anatomy and physiology laboratory processes were produced at Kishwaukee College, published by McGraw-Hill Higher Education in 2007. Cadaver dissection experiences have been provided for his students for over twenty years. Terry teaches portions of EMT and paramedic classes and serves as a Faculty Consultant for the Advanced Placement Biology examination readings. During 1994, Terry was a faculty exchange member in Ireland. The author locally supports historical preservation, natural areas, scouting, and scholarship. We are pleased to have Terry continue the tradition of John Hole's laboratory manual.

The Editor

To the Student

The exercises in this laboratory manual will provide you with opportunities to observe various anatomical parts and to investigate certain physiological phenomena. Such experiences should help you to relate specimens, models, microscope slides, and your body to what you have learned in the lecture and read about in the textbook.

The following list of suggestions and study skills, all placed together into one section, may help to make your laboratory activities more effective and profitable.

1. Prepare yourself before attending the laboratory session by reading the assigned exercise and reviewing the related sections of the textbook. It is important to have some understanding of what will be done in the laboratory before you come to class.
2. Bring your laboratory manual and textbook to each laboratory session. These books are closely integrated and will help you complete most of the exercises.
3. Be on time. During the first few minutes of the laboratory meeting, the instructor often will provide verbal instructions. Make special note of any changes in materials to be used or procedures to be followed. Also listen carefully for information about special techniques to be used and precautions to be taken.
4. Keep your work area clean and your materials neatly arranged so that you can locate needed items. This will enable you to efficiently proceed and will reduce the chances of making mistakes.
5. Pay particular attention to the purpose of the exercise, which states what you are to accomplish in general terms, and to the learning outcomes, which list what you should be able to do as a result of the laboratory experience. Then, before you leave the class, review the outcomes and make sure that you can meet all of the assessments.
6. Precisely follow the directions in the procedure and proceed only when you understand them clearly. Do not improvise procedures unless you have the approval of the laboratory instructor. Ask questions if you do not understand exactly what you are supposed to do and why you are doing it.
7. Handle all laboratory materials with care. These materials often are fragile and expensive to replace. Whenever you have questions about the proper treatment of equipment, ask the instructor.
8. Treat all living specimens humanely and try to minimize any discomfort they might experience.
9. Although at times you might work with a laboratory partner or a small group, try to remain independent when you are making observations, drawing conclusions, and completing the activities in the laboratory reports.
10. Record your observations immediately after making them. In most cases, such data can be entered in spaces provided in the laboratory reports.
11. Read the instructions for each section of the laboratory report before you begin to complete it. Think about the questions before you answer them. Your responses should be based on logical reasoning and phrased in clear and concise language.
12. At the end of each laboratory period, clean your work area and the instruments you have used. Return all materials to their proper places and dispose of wastes, including glassware or microscope slides that have become contaminated with human blood or body fluids, as directed by the laboratory instructor. Wash your hands thoroughly before leaving the laboratory.

Study Skills for Anatomy and Physiology

My students have found that certain study skills worked well for them while they were enrolled in Human Anatomy and Physiology. Although everyone has his or her learning style, there are techniques that work well for most students. Using some of the skills listed here could make your course more enjoyable and rewarding.

1. **Time management:** Prepare monthly, weekly, and daily schedules. Include dates of quizzes, exams, and projects on the calendar. On your daily schedule, budget several short study periods. Daily repetition alleviates cramming for exams. Prioritize your time so that you still have time for work and leisure activities. Find an appropriate study atmosphere with minimum distractions.
2. **Note taking:** Look for the main ideas and briefly express them in your own words. Organize, edit, and review your notes soon after the lecture. Add textbook information to your notes as you reorganize them. Underline or highlight with different colors the important points, major headings, and key terms. Study your notes daily, as they provide sequential building blocks of the course content.
3. **Chunking:** Organize information into logical groups or categories. Study and master one chunk of information at a time. For example, study the bones of the upper limb, lower limb, trunk, and head as separate study tasks.
4. **Mnemonic devices:** An *acrostic* is a combination of association and imagery to aid your memory. It is often in the form of a poem, rhyme, or jingle in which the first letter of each word corresponds to the first letters of the

words you need to remember. **S**o **L**ong **T**op **P**art, **H**ere **C**omes **T**he **T**humb is an example of such a mnemonic device for remembering the eight carpals in the correct sequence. *Acronyms* are words formed by the first letters of the items to remember. *IPMAT* is an example of this type of mnemonic device to help you remember the phases of the cell cycle in the correct sequence. Try to create some of your own.

5. **Note cards/flash cards:** Make your own. Add labels and colors to enhance the material. Keep them with you in your pocket or purse. Study them often and for short periods. Concentrate on a small number of cards at one time. Shuffle your cards and have someone quiz you on their content. As you become familiar with the material, you can set aside cards that don't require additional mastery.
6. **Recording and recitation:** An auditory learner can benefit by recording lectures and review sessions with a cassette recorder. Many students listen to the taped sessions as they drive or just before they go to bed. Reading your notes aloud can help also. Explain the material to anyone (even if there are no listeners). Talk about anatomy and physiology in everyday conversations.
7. **Study groups:** Small study groups that meet periodically to review course material and compare notes have helped and encouraged many students. However, keep the group on the task at hand. Work as a team and alternate leaders. This group often becomes a support group.

Practice sound study skills during your anatomy and physiology endeavor.

Correlation of Textbook Chapters and Laboratory Exercises

Textbook Chapters		Related Laboratory Exercises	
Chapter 1	Introduction to Human Anatomy and Physiology	Exercise 1	Scientific Method and Measurements
		Exercise 2	Body Organization and Terminology
Chapter 2	Chemical Basis of Life	Exercise 3	Chemistry of Life
Chapter 3	Cells	Exercise 4	Care and Use of the Microscope
		Exercise 5	Cell Structure and Function
		Exercise 6	Movements Through Membranes
		Exercise 7	Cell Cycle
Chapter 4	Cellular Metabolism		
Chapter 5	Tissues	Exercise 8	Epithelial Tissues
		Exercise 9	Connective Tissues
		Exercise 10	Muscle and Nervous Tissues
Chapter 6	Integumentary System	Exercise 11	Integumentary System
Chapter 7	Skeletal System	Exercise 12	Bone Structure and Classification
		Exercise 13	Organization of the Skeleton
		Exercise 14	Skull
		Exercise 15	Vertebral Column and Thoracic Cage
		Exercise 16	Pectoral Girdle and Upper Limb
		Exercise 17	Pelvic Girdle and Lower Limb
Chapter 8	Joints of the Skeletal System	Exercise 18	Joint Structure and Movements
Chapter 9	Muscular System	Exercise 19	Skeletal Muscle Structure
		Exercise 20	Muscles of the Face, Head, and Neck
		Exercise 21	Muscles of the Chest, Shoulder, and Upper Limb
		Exercise 22	Muscles of the Deep Back, Abdominal Wall, and Pelvic Outlet
		Exercise 23	Muscles of the Hip and Lower Limb
		Exercise 24	Surface Anatomy
		Exercise 54	Cat Dissection: Musculature
Chapter 10	Nervous System I: Basic Structure and Function	Exercise 25	Nervous Tissue and Nerves
Chapter 11	Nervous System II: Divisions of the Nervous System	Exercise 26	Spinal Cord and Meninges
		Exercise 27	Reflex Arc and Reflexes
		Exercise 28	Brain and Cranial Nerves
		Exercise 29	Dissection of the Sheep Brain
Chapter 12	Nervous System III: Senses	Exercise 30	Receptors and General Senses
		Exercise 31	Smell and Taste
		Exercise 32	Ear and Hearing
		Exercise 33	Equilibrium
		Exercise 34	Eye Structure
		Exercise 35	Visual Tests and Demonstrations
Chapter 13	Endocrine System	Exercise 36	Endocrine Histology and Diabetic Physiology
Chapter 14	Blood	Exercise 37	Blood Cells and Blood Typing
Chapter 15	Cardiovascular System	Exercise 38	Heart Structure
		Exercise 39	Cardiac Cycle
		Exercise 40	Blood Vessel Structure, Arteries, and Veins
		Exercise 41	Pulse Rate and Blood Pressure
		Exercise 55	Cat Dissection: Cardiovascular System
Chapter 16	Lymphatic System and Immunity	Exercise 42	Lymphatic System
Chapter 17	Digestive System	Exercise 43	Digestive Organs
		Exercise 44	Action of a Digestive Enzyme
		Exercise 56	Cat Dissection: Digestive System
Chapter 18	Nutrition and Metabolism		

Textbook Chapters		Related Laboratory Exercises	
Chapter 19	Respiratory System	Exercise 45	Respiratory Organs
		Exercise 46	Breathing and Respiratory Volumes
		Exercise 47	Control of Breathing
		Exercise 57	Cat Dissection: Respiratory System
Chapter 20	Urinary System	Exercise 48	Kidney Structure
		Exercise 49	Urinalysis
		Exercise 58	Cat Dissection: Urinary System
Chapter 21	Water, Electrolyte, and Acid-Base Balance		
Chapter 22	Reproductive Systems	Exercise 50	Male Reproductive System
		Exercise 51	Female Reproductive System
		Exercise 59	Cat Dissection: Reproductive Systems
Chapter 23	Pregnancy, Growth, and Development	Exercise 52	Fertilization and Early Development
Chapter 24	Genetics and Genomics	Exercise 53	Genetics

Correlation of Textbook Chapters and Online Laboratory Exercises

Textbook Chapters		Related Laboratory Exercises	
Chapter 9	Muscular System	*Exercise 60	Skeletal Muscle Contraction
Chapter 10	Nervous System I: Basic Structure and Function	*Exercise 61	Nerve Impulse Stimulation
Chapter 14	Blood	*Exercise 62	Blood Testing
Chapter 15	Cardiovascular System	*Exercise 63	Factors Affecting the Cardiac Cycle

*These laboratory exercises are available at www.mhhe.com/shier12 (click on this laboratory manual cover).

Laboratory Exercise 1

Scientific Method and Measurements

Materials Needed

Meterstick
Calculator
Human skeleton

Scientific investigation involves a series of logical steps to arrive at explanations for various biological phenomena. This technique, called the *scientific method,* is used in all disciplines of science. It allows scientists to draw logical and reliable conclusions about phenomena.

The scientific method begins with *observations* related to the topic under investigation. This step commonly involves the accumulation of previously acquired information and/or your observations of the phenomenon. These observations are used to formulate a tentative explanation known as the *hypothesis.* An important attribute of a hypothesis is that it must be testable. The testing of the hypothesis involves performing a carefully controlled *experiment* to obtain data that can be used to support, reject, or modify the hypothesis. An *analysis of data* is conducted using sufficient information collected during the experiment. Data analysis may include organization and presentation of data as tables, graphs, and drawings. From the interpretation of the data analysis, *conclusions* are drawn. (If the data do not support the hypothesis, you must reexamine the experimental design and the data, and if needed, develop a new hypothesis.) The final presentation of the information is made from the conclusions. Results and conclusions are presented to the scientific community for evaluation through peer reviews, presentations at professional meetings, and published articles. If many investigators working independently can validate the hypothesis by arriving at the same conclusions, the explanation becomes a **theory.** A theory verified continuously over time and accepted by the scientific community becomes known as a **scientific law** or **principle.** A scientific law serves as the standard explanation for an observation unless it is disproved by new information.

The five components of the scientific method are summarized as

Observations
↓
Hypothesis
↓
Experiment
↓
Analysis of data
↓
Conclusions

Metric measurements are characteristic tools of scientific investigations. The English system of measurements is often used in the United States, so the investigator must make conversions from the English system to the metric system. Use table 1.1 for the conversion of English units of measure to metric units for length, mass, volume, time, and temperature.

Purpose of the Exercise

To become familiar with the scientific method of investigation, to learn how to formulate sound conclusions, and to provide opportunities to use the metric system of measurements.

LEARNING OUTCOMES

After completing this exercise, you should be able to

1. Convert English measurements to the metric system, and vice versa.
2. Measure and record upper limb lengths and heights of ten subjects.
3. Apply the scientific method to test the validity of a hypothesis concerning the direct, linear relationship between human upper limb length and height.
4. Design an experiment, formulate a hypothesis, and test it using the scientific method.

Table 1.1 Metric Measurement System and Conversions

Measurement	Unit & Abbreviation	Metric Equivalent	Conversion Factor Metric to English (approximate)	Conversion Factor English to Metric (approximate)
Length	1 kilometer (km)	1,000 (10^3) m	1 km = 0.62 mile	1 mile = 1.61 km
	1 meter (m)	100 (10^2) cm 1,000 mm	1 m = 1.1 yards = 3.3 feet = 39.4 inches	1 yard = 0.9 m 1 foot = 0.3 m
	1 decimeter (dm)	0.1 (10^{-1}) m	1 dm = 3.94 inches	1 inch = 0.25 dm
	1 centimeter (cm)	0.01 (10^{-2}) m	1 cm = 0.4 inches	1 foot = 30.5 cm 1 inch = 2.54 cm
	1 millimeter (mm)	0.001 (10^{-3}) m 0.1 cm	1 mm = 0.04 inches	
	1 micrometer (µm)	0.000001 (10^{-6}) m 0.001 mm		
Mass	1 metric ton (t)	1,000 kg	1 t = 1.1 ton	1 ton = 0.91 t
	1 kilogram (kg)	1,000 g	1 kg = 2.2 pounds	1 pound = 0.45 kg
	1 gram (g)	1,000 mg	1 g = 0.04 ounce	1 pound = 454 g 1 ounce = 28.35 g
	1 milligram (mg)	0.001 g		
Volume (liquids and gases)	1 liter (L)	1,000 mL	1 L = 1.06 quarts	1 gallon = 3.78 L 1 quart = 0.95 L
	1 milliliter (mL)	0.001 L 1 cubic centimeter (cc or cm^3)	1 mL = 0.03 fluid ounce 1 mL = 1/4 teaspoon 1 mL = 15–16 drops	1 quart = 946 mL 1 fluid ounce = 29.6 mL 1 teaspoon = 5 mL
Time	1 second (s)	1/60 minute	same	same
	1 millisecond (ms)	0.001 s	same	same
Temperature	Degrees Celsius (°C)		°F = 9/5°C + 32	°C = 5/9 (°F – 32)

EXPLORE

Procedure A—Using the Steps of the Scientific Method

1. Many people have observed a correlation between the length of the upper and lower limbs and the height (stature) of an individual. For example, a person who has long upper limbs (the arm, forearm, and hand combined) tends to be tall. Make some visual observations of other people in your class to observe a possible correlation.
2. From such observations, the following hypothesis is formulated: The length of a person's upper limb is equal to 0.4 (40%) of the height of the person. Test this hypothesis by performing the following experiment.
3. In this experiment, use a meterstick (fig. 1.1) to measure an upper limb length of ten subjects. For each measurement, place the meterstick in the axilla (armpit) and record the length in centimeters to the end of the longest finger (fig. 1.2). Obtain the height of each person in centimeters by measuring them without shoes against a wall (fig. 1.3). The height of each person can be calculated by multiplying each individual's height in inches by 2.54 to obtain his/her height in centimeters. Record all your measurements in Part A of Laboratory Report 1.
4. The data collected from all of the measurements can now be analyzed. The expected (predicted) correlation between upper limb length and height is determined using the following equation:

   Height × 0.4 = expected upper limb length

   The observed (actual) correlation to be used to test the hypothesis is determined by

   Length of upper limb/height = actual % of height
5. A graph is an excellent way to display a visual representation of the data. Plot the subjects' data in

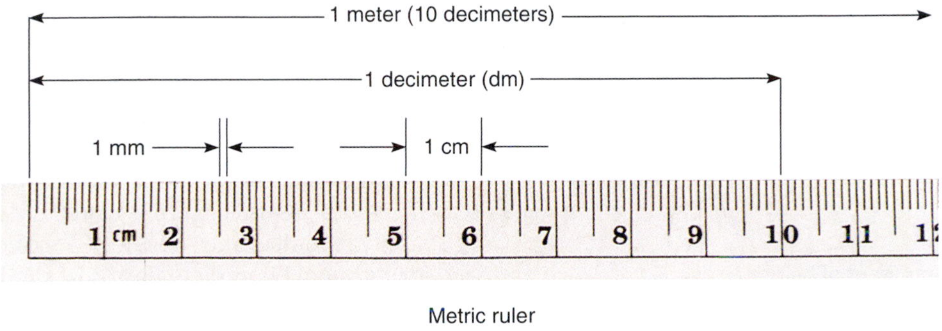

Figure 1.1 Metric ruler with metric lengths indicated. A meterstick length would be 100 centimeters. (The image size is approximately to scale.)

Figure 1.2 Measurement of upper limb length.

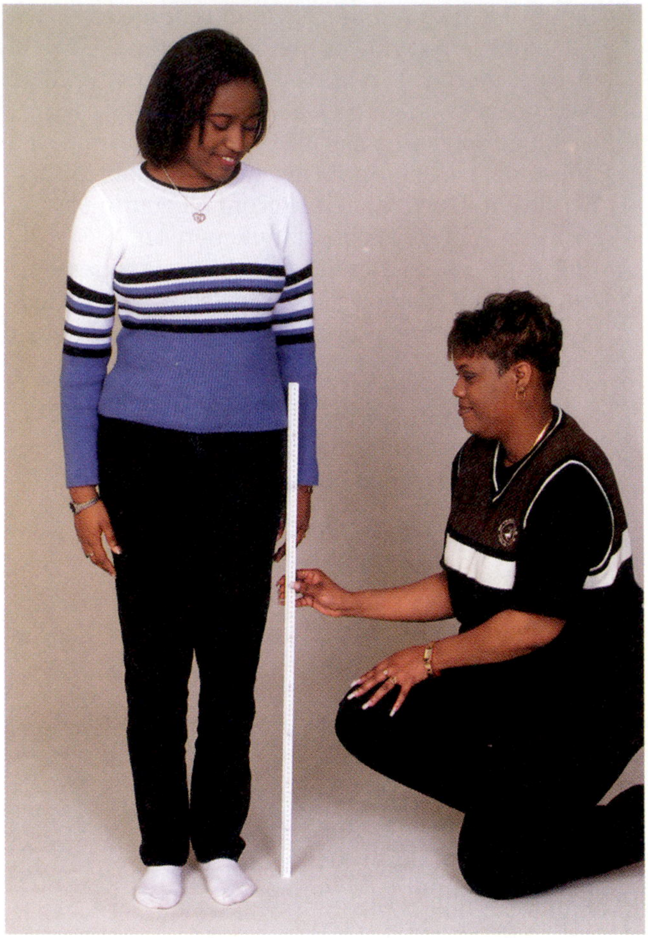

Figure 1.3 Measurement of height.

Part A of the laboratory report. Plot the upper limb length of each subject on the x-axis and the height of each person on the y-axis. A line is already located on the graph that represents a hypothetical relationship of 0.4 (40%) upper limb length compared to height. This is a graphic representation of the original hypothesis.

6. Compare the distribution of all of the points (actual height and upper limb length) that you placed on the graph with the distribution of the expected correlation represented by the hypothesis.
7. Complete Part A of the laboratory report.

EXPLORE

Procedure B—Design an Experiment

Critical Thinking Application

You have probably concluded that there is some correlation of the length of body parts to height. Often when a skeleton is found, it is not complete, especially when paleontologists discover a skeleton. It is occasionally feasible to use the length of a single bone to estimate the height of an individual. Observe human skeletons and locate the radius bone in the forearm. Use your observations to identify a mathematical relationship between the length of the radius and height. Formulate a hypothesis that can be tested. Make measurements, analyze data, and develop a conclusion from your experiment. Complete Part B of the laboratory report.

Name ______________________________

Date ______________________________

Section ______________________________

The [A] corresponds to the indicated Learning Outcome(s) listed at the beginning of the laboratory exercise.

Scientific Method and Measurements

Laboratory Report 1

Part A Assessments

1. Record measurements for the upper limb length and height of ten subjects. Use a calculator to determine the expected upper limb length and the actual percentage (as a decimal or a percentage) of the height for the ten subjects. Record your results in the following table. [2]

Subject	Measured Upper Limb Length (cm)	Height* (cm)	Height × 0.4 = Expected Upper Limb Length (cm)	Actual % of Height = Upper Limb Length (cm)/ Height (cm)
1.				
2.				
3.				
4.				
5.				
6.				
7.				
8.				
9.				
10.				

**The height of each person can be calculated by multiplying each individual's height in inches by 2.54 to obtain his/her height in centimeters.* [1]

2. Plot the distribution of data (upper limb length and height) collected for the ten subjects on the following graph. The line located on the graph represents the *expected* 0.4 (40%) ratio of upper limb length to measured height (the original hypothesis). (The x-axis represents upper limb length and the y-axis represents height.) Draw a line of *best fit* through the distribution of points of the plotted data of the ten subjects. Compare the two distributions (expected line and the distribution line drawn for the ten subjects). 3

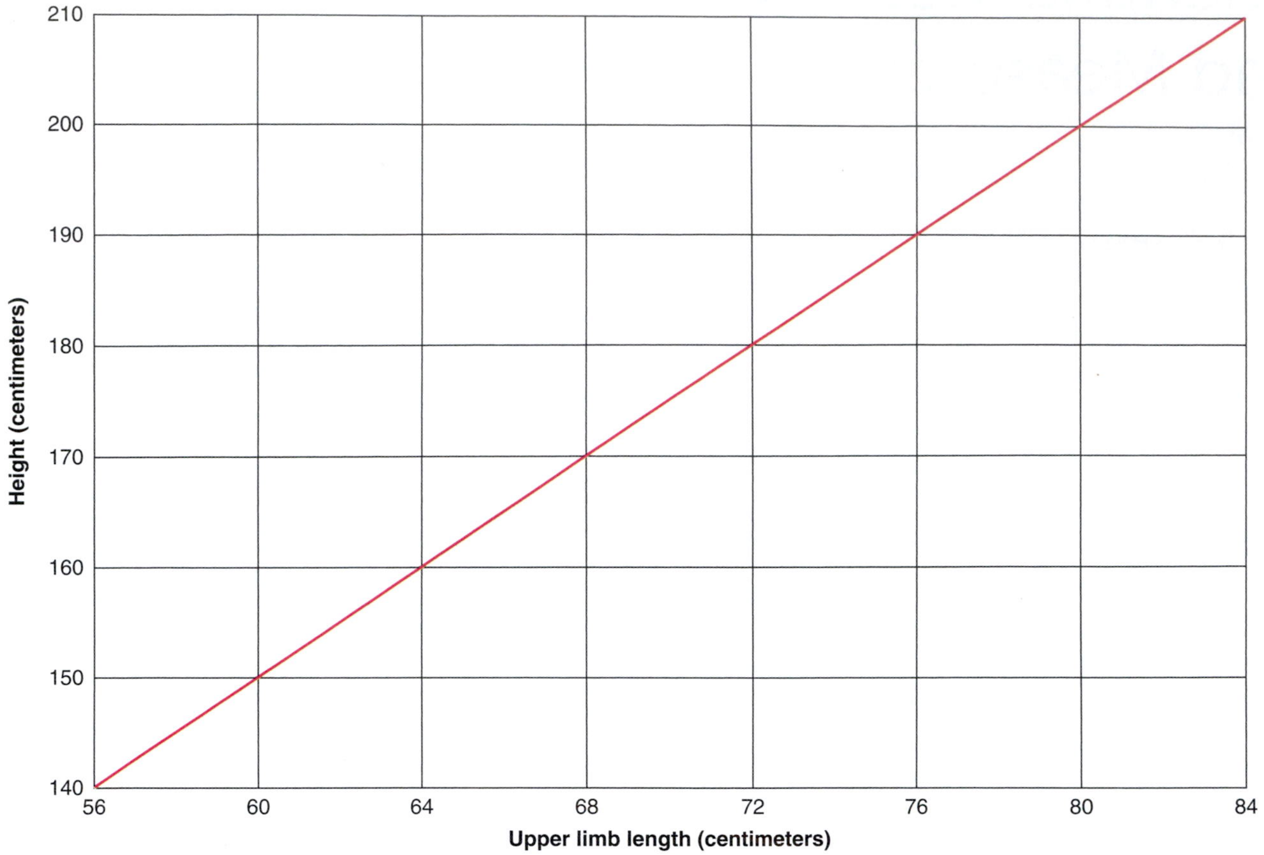

3. Does the distribution of the ten subjects' measured upper limb lengths support or reject the original hypothesis? ___________ Explain your answer. 3

Part B Assessments

1. Describe your observations of a possible correlation between the radius length and height. 4

2. Write a hypothesis based on your observations. 4

3. Describe the design of the experiment that you devised to test your hypothesis. 4

4. Place your analysis of the data in this space in the form of a table and a graph. 4

5. Based on an analysis of your data, what conclusions can you make? Do these conclusions confirm or refute your original hypothesis?

6. Discuss your results and conclusions with classmates. What common conclusion can the class formulate about the correlation between radius length and height?

Laboratory Exercise 2

Body Organization and Terminology

Materials Needed

Textbook
Dissectible human torso model (manikin)
Variety of specimens or models sectioned along various planes

For Learning Extension:
Colored pencils

The major features of the human body include certain cavities, a set of membranes associated with these cavities, and a group of organ systems composed of related organs. In order to communicate effectively with each other about the body, scientists have devised names to describe these body features. They also have developed terms to represent the relative positions of body parts, imaginary planes passing through these parts, and body regions.

Purpose of the Exercise

To review the organizational pattern of the human body, to review its organ systems and the organs included in each system, and to become acquainted with the terms used to describe the relative position of body parts, body sections, and body regions.

LEARNING OUTCOMES

After completing this exercise, you should be able to

1. Locate and name the major body cavities and identify the membranes associated with each cavity.
2. Differentiate the general functions of the organ systems of the human body.
3. Associate the organs included within each system and locate the organs in a dissectible human torso model.
4. Select the terms used to describe the relative positions of body parts.
5. Match the terms used to identify body sections and identify the plane along which a particular specimen is cut.
6. Label the body regions and associate the terms used to identify body regions.

EXPLORE

Procedure A—Body Cavities and Membranes

1. Review the sections entitled "Body Cavities" and "Thoracic and Abdominopelvic Membranes" in chapter 1 of the textbook.
2. As a review activity, label figures 2.1 and 2.2.
3. Locate the following features on the reference plates near the end of chapter 1 of the textbook and on the dissectible human torso model (fig. 2.3):

body cavities
- cranial cavity
- vertebral canal (spinal cavity)
- thoracic cavity
  - mediastinum (region between the lungs; includes pericardial cavity)
  - pleural cavities
- abdominopelvic cavity
  - abdominal cavity
  - pelvic cavity

diaphragm

smaller cavities within the head
- oral cavity
- nasal cavity with connected sinuses
- orbital cavity
- middle ear cavity

membranes and cavities
- pleural cavity
  - parietal pleura
  - visceral pleura

pericardial cavity
- parietal pericardium (covered by fibrous pericardium)
- visceral pericardium (epicardium)

peritoneal cavity
- parietal peritoneum
- visceral peritoneum

4. Complete Part A of Laboratory Report 2.

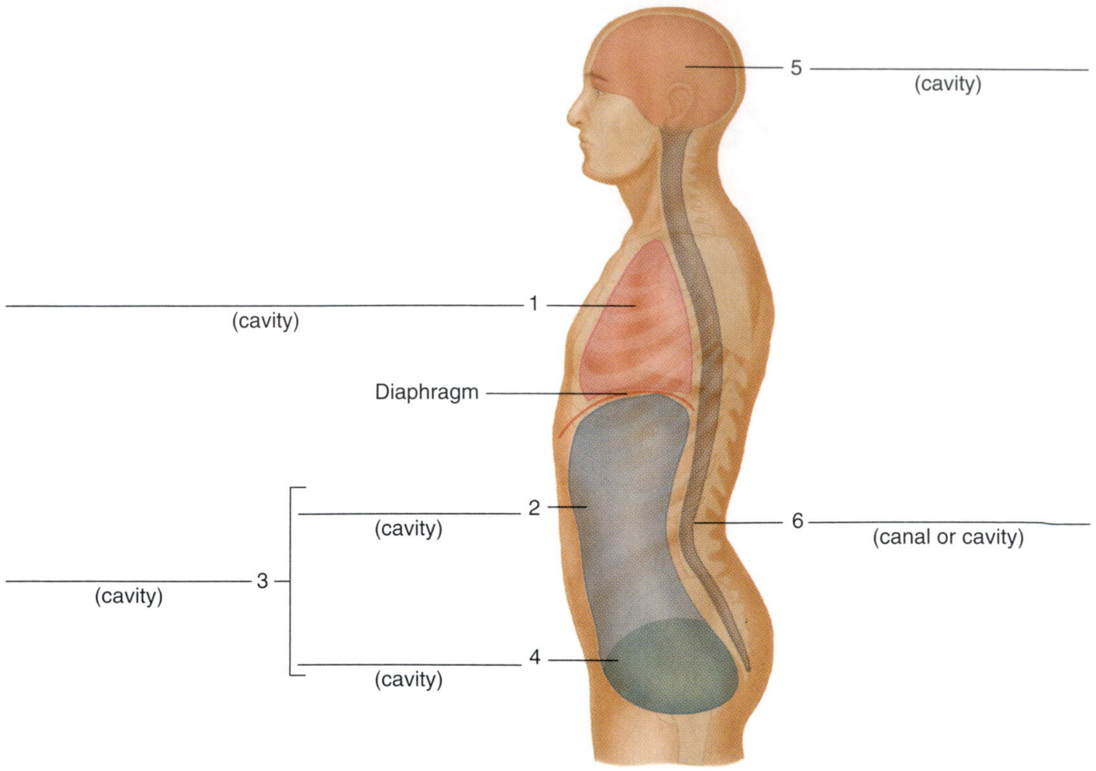

Figure 2.1 Label these body cavities. 1

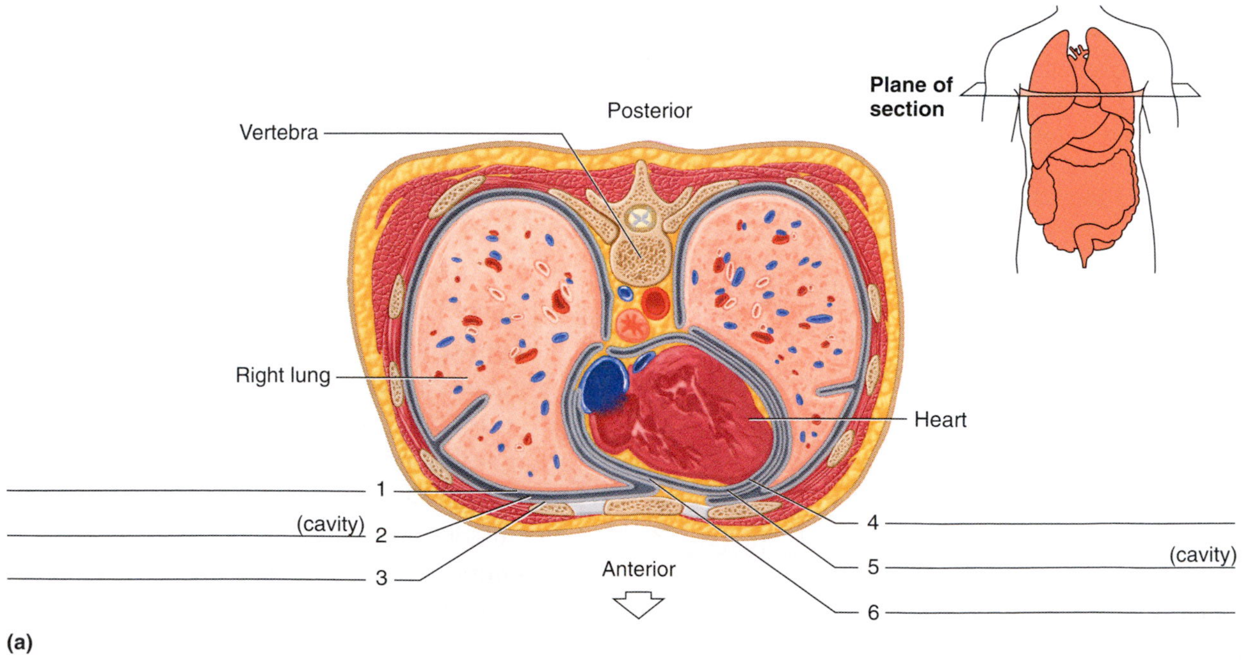

Figure 2.2 Label the thoracic membranes and cavities in (*a*) and the abdominopelvic membranes and cavity in (*b*) as shown in these superior views of transverse sections. 1

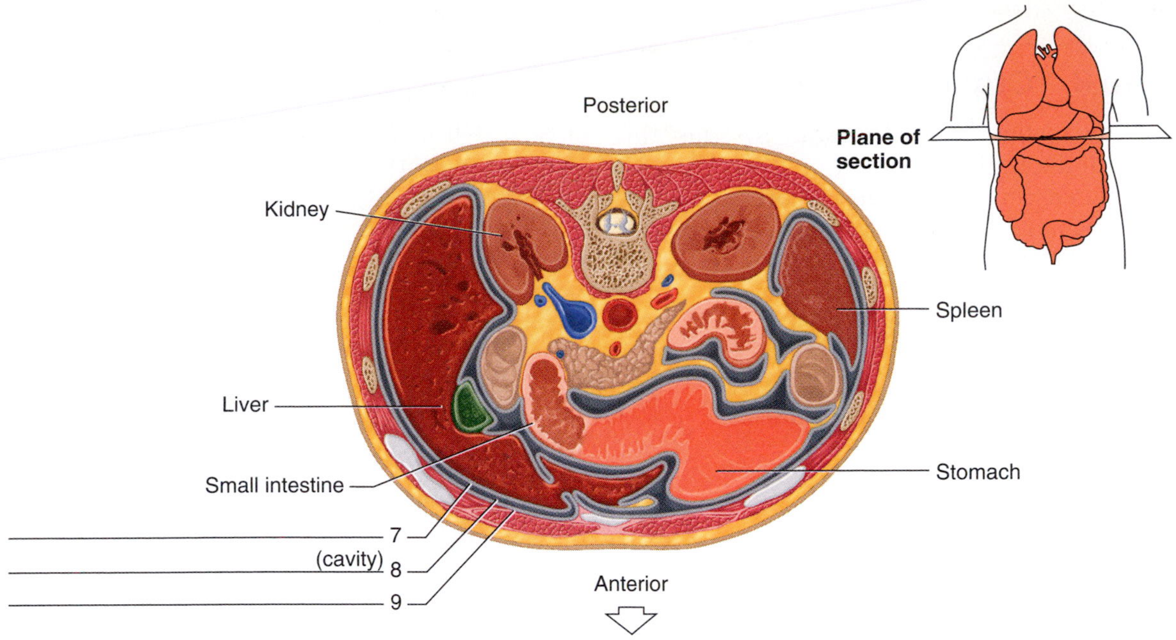

(b)

Figure 2.2 *Continued.*

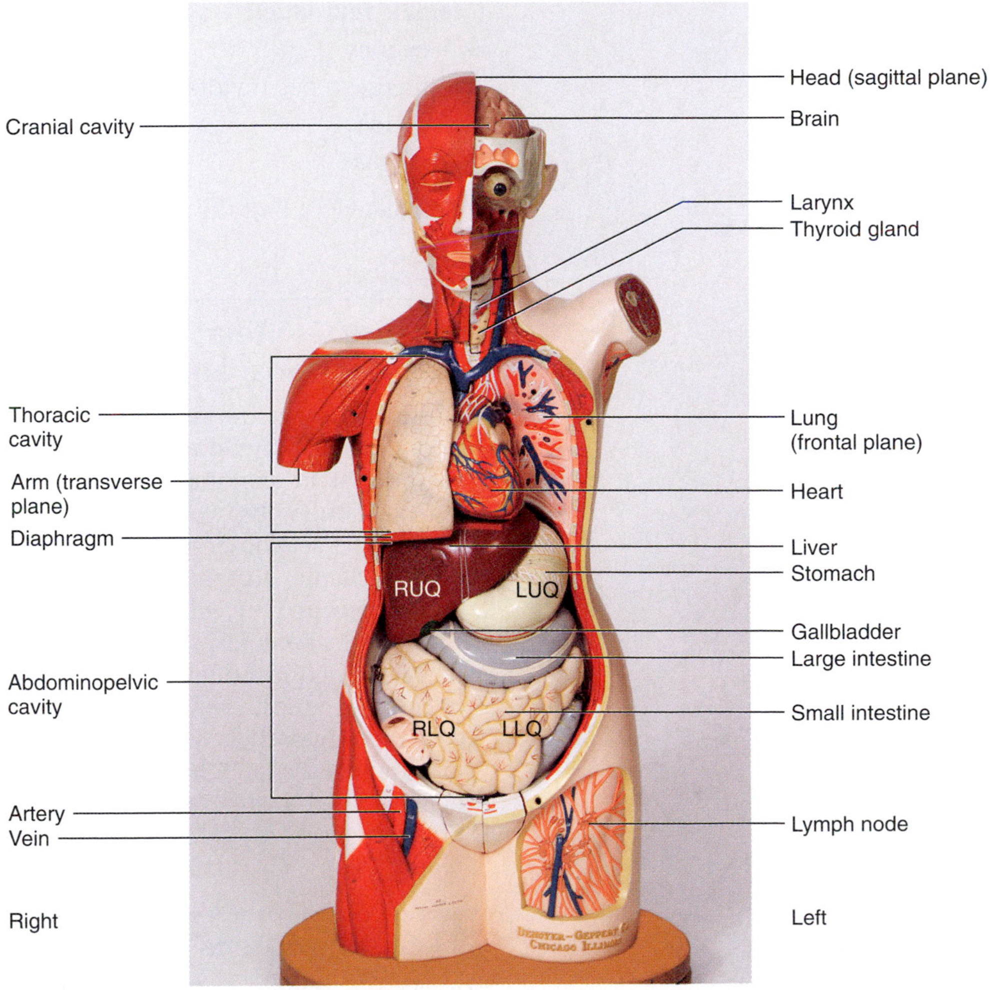

Figure 2.3 Dissectible human torso model with body cavities, abdominopelvic quadrants, body planes, and major organs indicated.

EXPLORE

Procedure B—Organ Systems

1. Review the section entitled "Organ Systems" in chapter 1 of the textbook.
2. Use the reference plates near the end of chapter 1 of the textbook and the dissectible human torso model (fig. 2.3) to locate the following systems and their major organs:

integumentary system
- skin
- accessory organs such as hair and nails

skeletal system
- bones
- ligaments

muscular system
- skeletal muscles
- tendons

nervous system
- brain
- spinal cord
- nerves

endocrine system
- pituitary gland
- thyroid gland
- parathyroid glands
- adrenal glands
- pancreas
- ovaries
- testes
- pineal gland
- thymus

cardiovascular system
- heart
- arteries
- veins

lymphatic system
- lymphatic vessels
- lymph nodes
- thymus
- spleen

digestive system
- mouth
- tongue
- teeth
- salivary glands
- pharynx
- esophagus
- stomach
- liver
- gallbladder
- pancreas
- small intestine
- large intestine

respiratory system
- nasal cavity
- pharynx
- larynx
- trachea
- bronchi
- lungs

urinary system
- kidneys
- ureters
- urinary bladder
- urethra

male reproductive system
- scrotum
- testes
- penis
- urethra

female reproductive system
- ovaries
- uterine tubes (oviducts; fallopian tubes)
- uterus
- vagina

3. Complete Part B of the laboratory report.

EXPLORE

Procedure C—Relative Positions, Planes, Sections, and Regions

1. Observe the person standing in anatomical position (fig. 2.4). Anatomical terminology assumes the body is in anatomical position even though a person is often observed differently.
2. Review the section entitled "Anatomical Terminology" in chapter 1 of the textbook.
3. As a review activity, label figures 2.5, 2.6, and 2.7.
4. Examine the sectioned specimens on the demonstration table and identify the plane along which each is cut. Cylindrical structures, such as a long bone or a blood vessel, may be cut in cross section, or longitudinal section. The same three sections can be demonstrated by three cuts of a banana (fig. 2.8).
5. Complete Parts C, D, E, and F of the laboratory report.

Learning Extension

Use different colored pencils to distinguish body regions in figure 2.7.

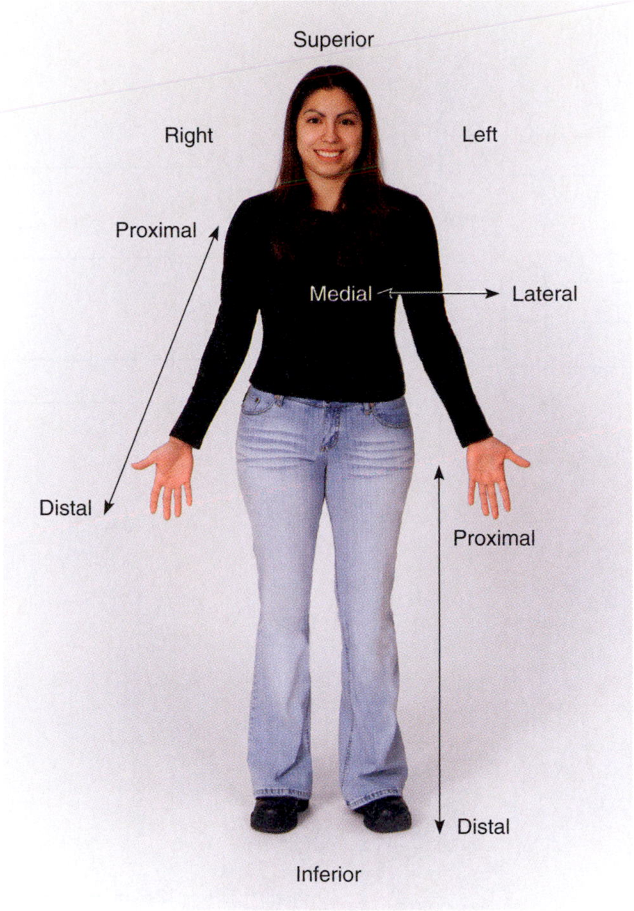

Figure 2.4 Anatomical position with directional terms indicated. The body is standing erect, face forward, with upper limbs at the sides and palms forward. This results in an anterior view of the body.

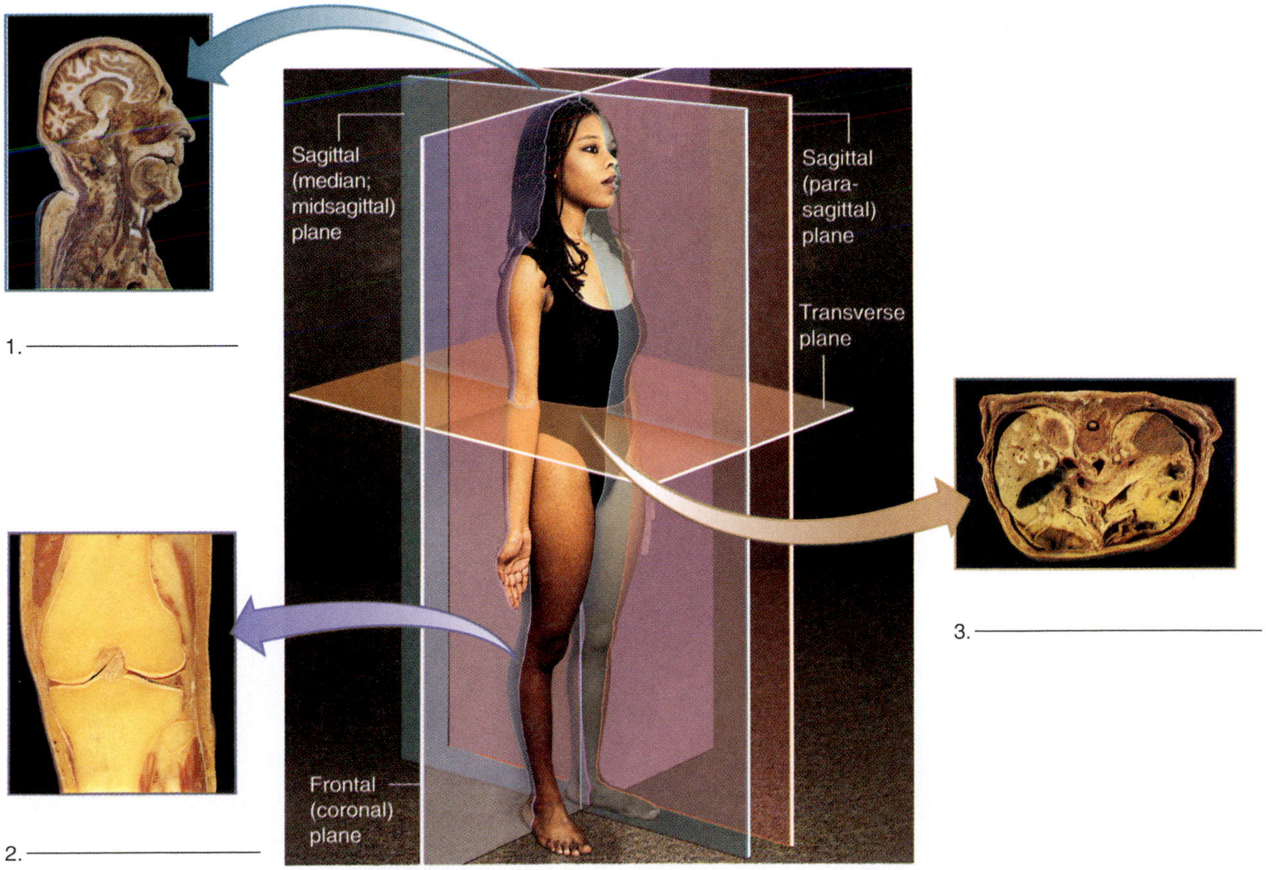

Figure 2.5 Label the planes represented in this illustration. 5

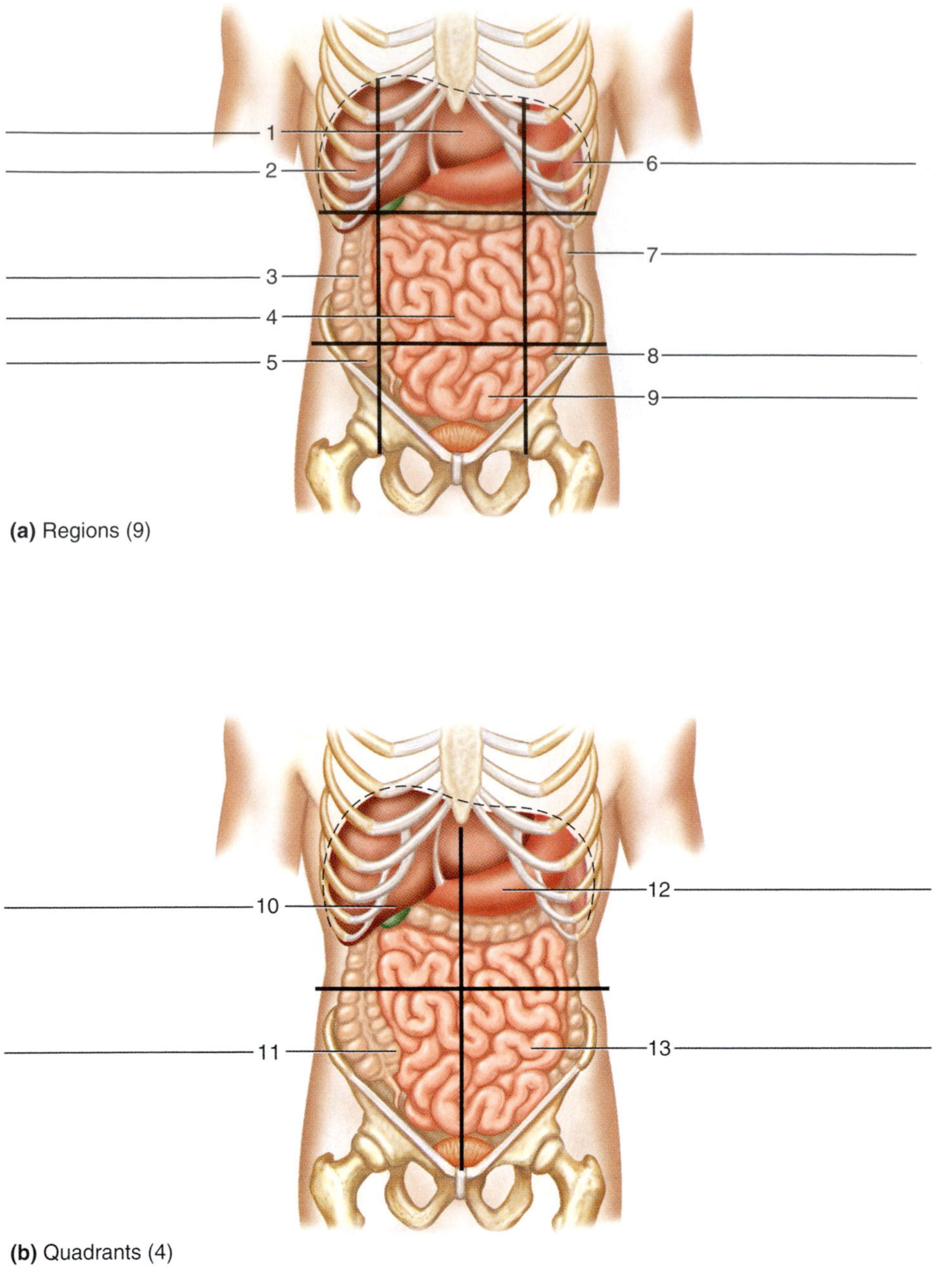

Figure 2.6 Label (*a*) the regions and (*b*) the quadrants of the abdominopelvic cavity. 6

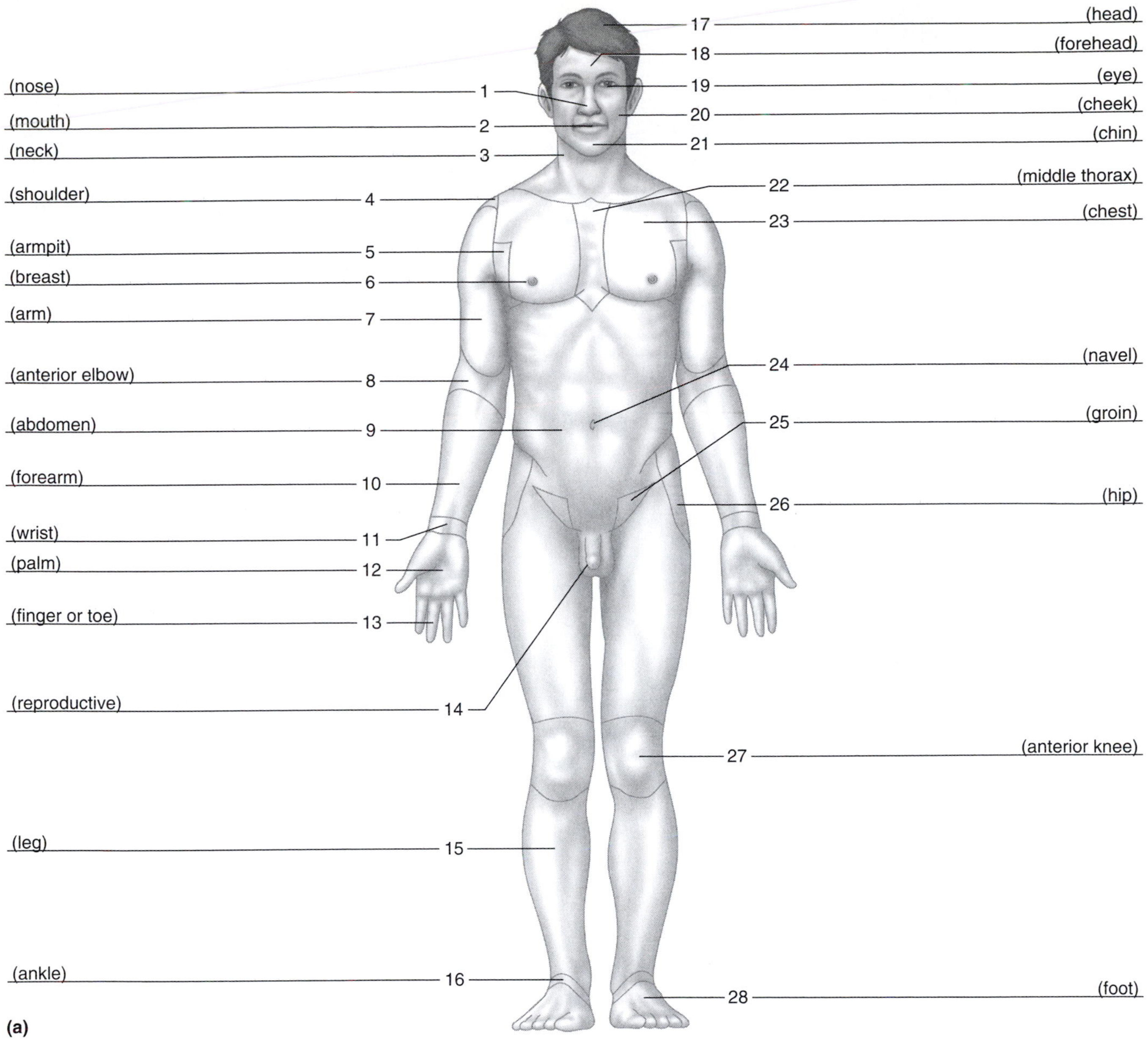

Figure 2.7 Label these diagrams with terms used to describe body regions: (*a*) anterior regions; (*b*) posterior regions. 6

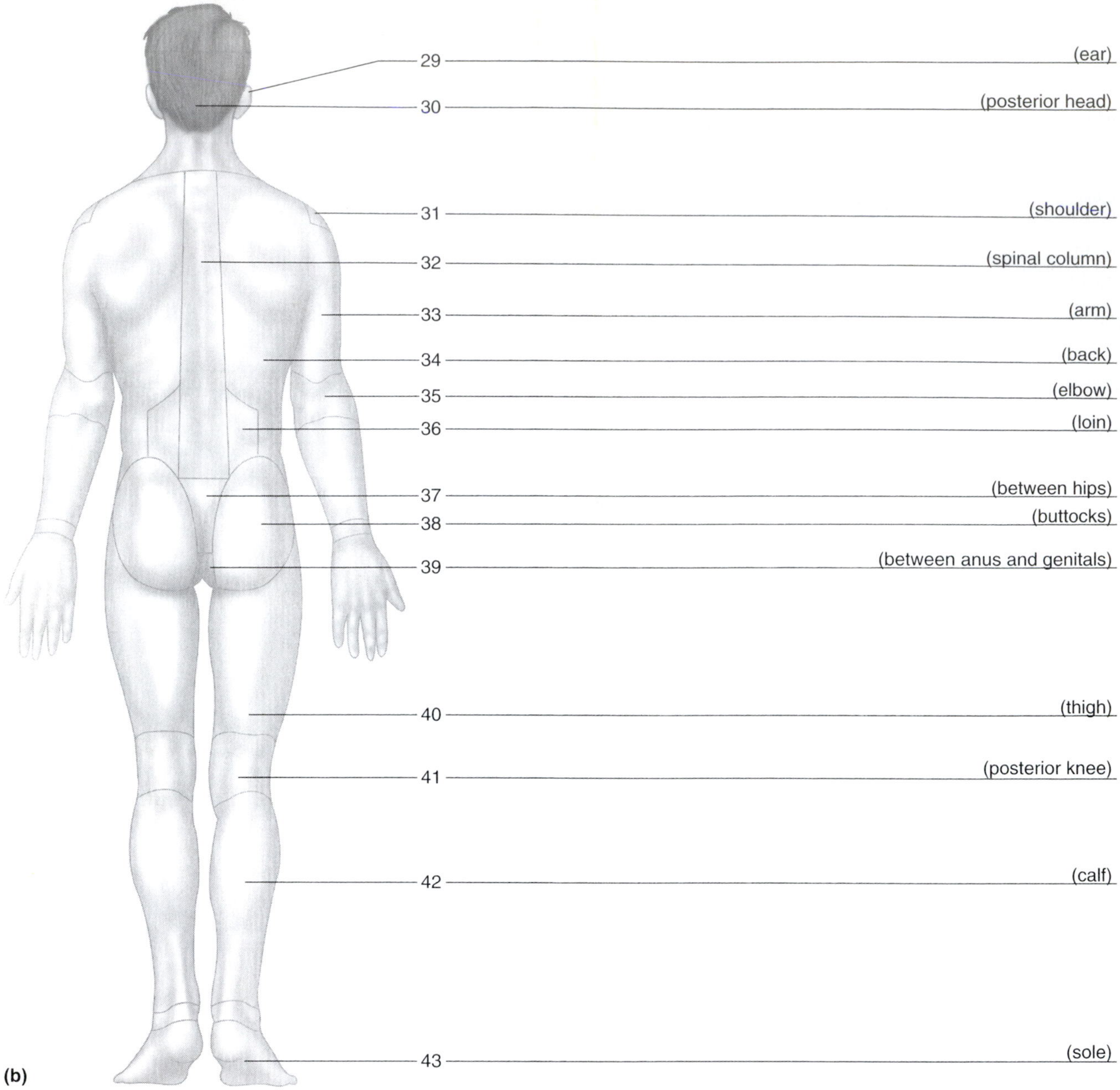

Figure 2.7 *Continued.*

Figure 2.8 Three possible cuts of a banana: (*a*) cross section; (*b*) oblique section; (*c*) longitudinal section. Sections through an organ, as a body tube, frequently produce views similar to the cut banana.

Name ______________________________

Date ______________________________

Section ______________________________

The [A] corresponds to the indicated Learning Outcome(s) listed at the beginning of the laboratory exercise.

Laboratory Report 2

Body Organization and Terminology

Part A Assessments

Match the body cavities in column A with the organs contained in the cavities in column B. Place the letter of your choice in the space provided. [1]

Column A

a. Abdominal cavity
b. Cranial cavity
c. Pelvic cavity
d. Thoracic cavity
e. Vertebral canal (spinal cavity)

Column B

_______ 1. Liver
_______ 2. Lungs
_______ 3. Spleen
_______ 4. Stomach
_______ 5. Brain
_______ 6. Internal reproductive organs
_______ 7. Gallbladder
_______ 8. Urinary bladder
_______ 9. Mediastinum
_______ 10. Spinal cord
_______ 11. Heart
_______ 12. Small intestine

Part B Assessments

Match the organ systems in column A with the general functions in column B. Place the letter of your choice in the space provided. [2]

Column A

a. Cardiovascular system
b. Digestive system
c. Endocrine system
d. Integumentary system
e. Lymphatic system
f. Muscular system
g. Nervous system
h. Reproductive system
i. Respiratory system
j. Skeletal system
k. Urinary system

Column B

_______ 1. Main system that secretes hormones
_______ 2. Provides an outer covering of the body
_______ 3. Produces gametes (eggs and sperm)
_______ 4. Stimulates muscles to contract and interprets information from sensory organs
_______ 5. Provides a framework and support for soft tissues and produces blood cells in red marrow
_______ 6. Exchanges gases between air and blood
_______ 7. Transports excess fluid from tissues to blood
_______ 8. Movements via contractions and creates most body heat
_______ 9. Removes liquid and wastes from blood and transports them to the outside of the body
_______ 10. Converts food molecules into absorbable forms
_______ 11. Transports nutrients, wastes, and gases throughout the body

Part C Assessments

Indicate whether each of the following sentences makes correct or incorrect usage of the word in boldface type (assume that the body is in the anatomical position as observed in fig. 2.4). If the sentence is incorrect, in the space provided supply a term that will make it correct. 4

1. The mouth is **superior** to the nose. ______________________
2. The stomach is **inferior** to the diaphragm. ______________________
3. The trachea is **anterior** to the spinal cord. ______________________
4. The larynx is **posterior** to the esophagus. ______________________
5. The heart is **medial** to the lungs. ______________________
6. The kidneys are **inferior** to the adrenal glands. ______________________
7. The hand is **proximal** to the elbow. ______________________
8. The knee is **proximal** to the ankle. ______________________
9. Blood in **deep** blood vessels gives color to the skin. ______________________
10. A **peripheral** nerve passes from the spinal cord into the limbs. ______________________
11. The spleen and gallbladder are **ipsilateral.** ______________________
12. The dermis is the **superficial** layer of the skin. ______________________

Part D Assessments

Match the body regions in column A with the locations in column B. Place the letter of your choice in the space provided. 6

Column A		Column B
a. Antecubital	______	**1.** Wrist
b. Axillary	______	**2.** Reproductive organs
c. Brachial	______	**3.** Armpit
d. Buccal	______	**4.** Elbow
e. Carpal	______	**5.** Buttocks
f. Cephalic	______	**6.** Back
g. Cervical	______	**7.** Neck
h. Crural	______	**8.** Arm
i. Cubital	______	**9.** Cheek
j. Dorsal	______	**10.** Leg
k. Genital	______	**11.** Head
l. Gluteal	______	**12.** Front of elbow

Part E Assessments

Match the body regions in column A with the locations in column B. Place the letter of your choice in the space provided. 6

Column A		Column B
a. Inguinal	______	1. Pelvis
b. Lumbar	______	2. Breasts
c. Mammary	______	3. Between anus and reproductive organs
d. Occipital	______	4. Sole
e. Palmar	______	5. Middle of thorax
f. Pectoral	______	6. Chest
g. Pedal	______	7. Back of knee
h. Pelvic	______	8. Foot
i. Perineal	______	9. Lower posterior region of head
j. Plantar	______	10. Abdominal wall near thigh (groin)
k. Popliteal	______	11. Lower back
l. Sternal	______	12. Palm

Part F Assessments

Critical Thinking Application

State the quadrant of the abdominopelvic cavity where the pain or sound would be located for each of the six conditions listed. In some cases, there may be more than one correct answer, and pain is sometimes referred to another region. This phenomenon, called *referred pain,* occurs when pain is interpreted as originating from some area other than the parts being stimulated. When referred pain is involved in the patient's interpretation of the pain location, the proper diagnosis of the ailment is more challenging. For the purpose of this exercise, assume the pain is interpreted as originating from the organ involved. 3

1. Stomach ulcer ______________________________
2. Appendicitis ______________________________
3. Bowel sounds ______________________________
4. Gallbladder attack ______________________________
5. Kidney stone in left ureter ______________________________
6. Ruptured spleen ______________________________

Notes

Laboratory Exercise 3

Chemistry of Life

Materials Needed

For pH Tests:
Chopped fresh red cabbage
Beaker (250 mL)
Distilled water
Tap water
Vinegar
Baking soda
Laboratory scoop for measuring
Pipets for measuring
Full-range pH test papers
7 assorted common liquids clearly labeled in closed bottles on a tray or in a tub
Droppers labeled for each liquid

For Organic Tests:
Test tubes
Test-tube rack
Test-tube clamps
China marker
Hot plate
Beaker for hot water bath (500 mL)
Pipets for measuring
Benedict's solution
Biuret reagent (or 10% NaOH and 1% $CuSO_4$)
Iodine-potassium-iodide (IKI) solution
Sudan IV dye
Egg albumin
10% glucose solution
Clear carbonated soft drink
10% starch solution
Potatoes for potato water
Distilled water
Vegetable oil
Brown paper
Numbered unknown organic samples

Safety

- Review all safety guidelines inside the front cover of your laboratory manual.
- Clean laboratory surfaces before and after laboratory procedures using soap and water.
- Use extreme caution when working with chemicals.
- Safety goggles must be worn at all times.
- Wear disposable gloves while working with the chemicals.
- Precautions should be taken to prevent chemicals from contacting your skin.
- Do not mix any of the chemicals together unless instructed to do so.
- Clean up any spills immediately and notify the instructor at once.
- Wash your hands before leaving the laboratory.

The complexities of the human body arise from the organization and interactions of chemicals. Organisms are made of matter, and the most basic unit of matter is the chemical element. The smallest unit of an element is an atom, and two or more of those can unite to form a molecule. All processes that occur within the body involve chemical reactions—interactions between atoms and molecules. We breathe to supply oxygen to our cells for energy. We eat and drink to bring chemicals into our bodies that our cells need. Water fills and bathes all of our cells and allows an amazing array of reactions to occur, all of which are designed to keep us alive. Chemistry forms the basis of life and thus forms the foundation of anatomy and physiology.

Purpose of the Exercise

To review the organization of atoms and molecules, types of chemical interactions, and basic categories of organic compounds, and to differentiate between types of organic compounds.

LEARNING OUTCOMES

After completing this exercise, you should be able to

1. Associate and illustrate the basic organization of atoms and molecules.
2. Measure pH values of various substances through testing methods.
3. Determine categories of organic compounds with basic colorimetric tests.
4. Discover the organic composition of an unknown solution.

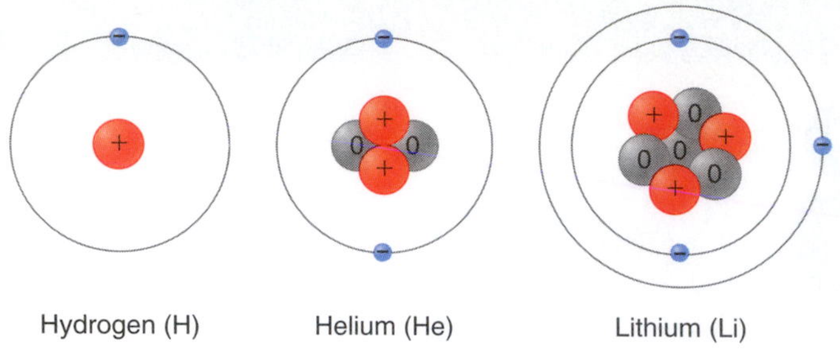

Figure 3.1 The single electron of a hydrogen atom is located in its first shell. The two electrons of a helium atom fill its first shell. Two of the three electrons of a lithium atom are in the first shell, and one is in the second shell.

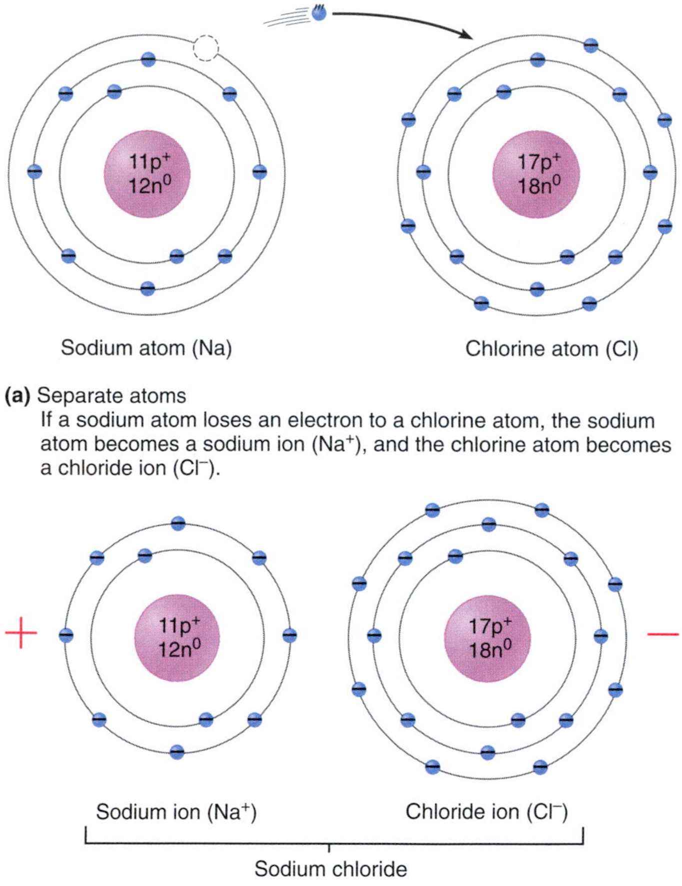

(a) Separate atoms
If a sodium atom loses an electron to a chlorine atom, the sodium atom becomes a sodium ion (Na^+), and the chlorine atom becomes a chloride ion (Cl^-).

(b) Bonded ions
These oppositely charged particles attract electrically and join by an ionic bond.

Figure 3.2 Formation of an ionic bond.

EXPLORE

Procedure A—Matter, Molecules, Bonding, and pH

1. Review the section entitled "Structure of Matter" in chapter 2 of the textbook. Use the Periodic Table of Elements, located inside the back cover of the laboratory manual, as a reference.
2. Review figures 3.1, 3.2, and 3.3, which show molecular diagrams, formation of an ionic bond, and formation of a covalent bond, respectively.
3. Complete Part A of Laboratory Report 3.

EXPLORE

Procedure B—The pH Scale

1. Review the section entitled "Acid and Base Concentrations" in chapter 2 of the textbook.
2. Review figure 3.4, which shows the pH scale. Note the range of the scale and the pH value considered neutral.
3. **Cabbage water tests.** Many tests can determine a pH value, but among the more interesting are colorimetric tests in which an indicator chemical changes color when it reacts. Many plant pigments, especially anthocyanins (which give plants color, from blue to red), can be used as colorimetric pH indicators. One that works well is the red pigment in red cabbage.
   - **a.** Prepare cabbage water to be used as a general pH indicator. Fill a 250 mL beaker to the 100 mL level with chopped red cabbage. Add water to make 150 mL. Place the beaker on a hot plate and simmer the mixture until the pigments come out of the cabbage and the water turns deep purple. Allow the water to cool. (You may proceed to step 4 while you wait for this to finish.)
   - **b.** Label three clean test tubes: one for water, one for vinegar, and one for baking soda.
   - **c.** Place 2 mL of cabbage water into each test tube.
   - **d.** To the first test tube, add 2 mL of distilled water and swirl the mixture. Record the color in Part B of the laboratory report.

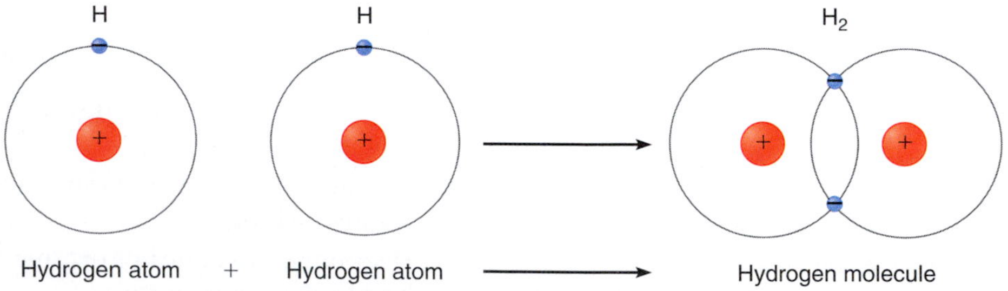

Figure 3.3 A hydrogen molecule forms when two hydrogen atoms share a pair of electrons and join by a covalent bond.

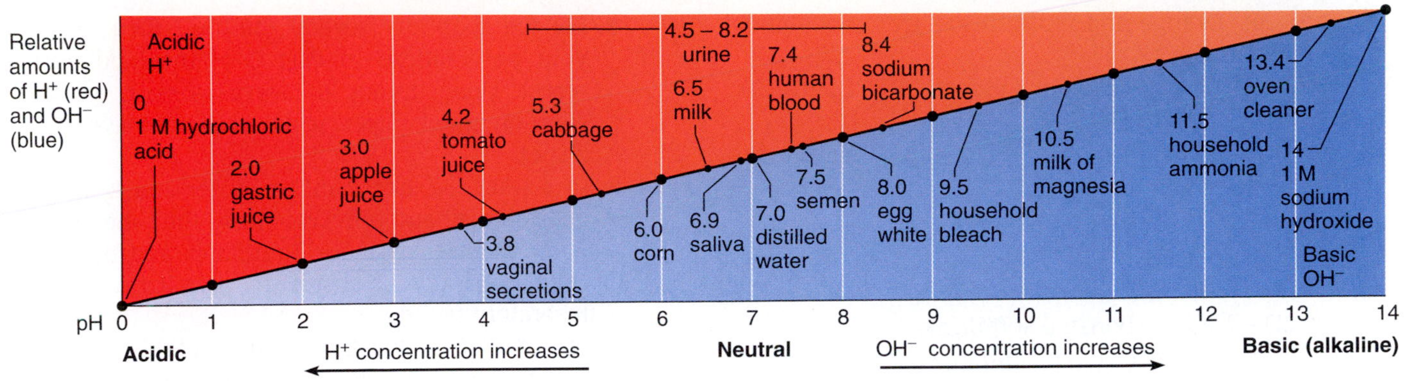

Figure 3.4 As the concentration of hydrogen ions (H^+) increases, a solution becomes more acidic, and the pH value decreases. As the concentration of ions that combine with hydrogen ions (such as hydroxide ions) increases, a solution becomes more basic (alkaline), and the pH value increases. The pH values of some common substances are shown.

e. Repeat this procedure for test tube 2, adding 2 mL of vinegar and swirling the mixture. Record the results.

f. Repeat this procedure for test tube 3, adding one laboratory scoop of baking soda. Swirl the mixture. Record the results.

4. **Testing with pH paper.** Many commercial pH indicators are available. A simple one to use is pH paper, which comes in small strips. Don gloves for this procedure so your skin secretions do not contaminate the paper, and to protect you from the chemicals you are testing.
   a. Test the pH of distilled water by dropping one or two drops of distilled water onto a strip of pH paper, and note the color. Compare the color to the color guide on the container. Record the pH value in Part B of the laboratory report.
   b. Repeat this procedure for tap water. Record the results.
   c. Individually test the various common substances found on your lab table. Record your results.
   d. Complete Part B of the laboratory report.

EXPLORE

Procedure C—Organic Molecules

In this section, you will perform some simple tests to check for the presence of the following categories of organic molecules (biomolecules): protein, sugar, starch, and lipid. Specific color changes will occur if the target compound is present. Please note the original color of the indicator being added so you can tell if the color really changed. For example, Biuret reagent and Benedict's solution are both initially blue, so an end color of blue would indicate no change. Use caution when working with these chemicals, and follow all directions. Carefully label the test tubes and avoid contaminating any of your samples. Only the Benedict's test requires heating and a time delay for accurate results. Do not heat any other tubes. To prepare for the Benedict's test, fill a 500 mL beaker half full with water and place it on the hot plate. Turn the hot plate on and bring the water to a boil. To save time, start the water bath before doing the Biuret test.

1. **Biuret test for protein.** In the presence of protein, Biuret reagent reacts with peptide bonds and changes to violet or purple. A pinkish color indicates that shorter polypeptides are present. The color intensity is proportional to the number of peptide bonds, thus the intensity reflects the length of polypeptides (amount of protein).
   a. Label six test tubes as follows: 1W, 1E, 1G, 1D, 1S, and 1P.
   b. To each test tube, add 2 mL of one of the samples as follows:
      1W—2 mL distilled water
      1E—2 mL egg albumin
      1G—2 mL 10% glucose solution
      1D—2 mL carbonated soft drink
      1S—2 mL 10% starch solution
      1P—2 mL potato juice
   c. To each, add 2 mL of Biuret reagent, swirl the tube to mix it, and note the final color. [*Note:* If Biuret reagent is not available, add 2 mL of 10% NaOH (sodium hydroxide) and about 10 drops of 1% CuSO4 (copper sulfate)]. **Be careful—NaOH is very caustic.**
   d. Record your results in Part C of the laboratory report.
   e. Mark a "+" on any tubes with positive results and retain them for comparison while testing your unknown sample. Put remaining test tubes aside so they will not get confused with future trials.

2. **Benedict's test for sugar (monosaccharides).** In the presence of sugar, Benedict's solution changes from its initial blue to a green, yellow, orange, or reddish color, depending on the amount of sugar

present. Orange and red indicate greater amounts of sugar.

a. Label six test tubes as follows: 2W, 2E, 2G, 2D, 2S, and 2P.
b. To each test tube, add 2 mL of one of the samples as follows:
   2W—2 mL distilled water
   2E—2 mL egg albumin
   2G—2 mL 10% glucose solution
   2D—2 mL carbonated soft drink
   2S—2 mL 10% starch solution
   2P—2 mL potato juice
c. To each, add 2 mL of Benedict's solution, swirl the tube to mix it, and place all tubes into the boiling water bath for 3 to 5 minutes.
d. Note the final color of each tube after heating and record the results in Part C of the laboratory report.
e. Mark a "+" on any tubes with positive results and retain them for comparison while testing your unknown sample. Put remaining test tubes aside so they will not get confused with future trials.

3. **Iodine test for starch.** In the presence of starch, iodine turns a dark purple or blue-black color. Starch is a long chain formed by many glucose units linked together side by side. This regular organization traps the iodine molecules and produces the dark color.
   a. Label six test tubes as follows: 3W, 3E, 3G, 3D, 3S, and 3P.
   b. To each test tube, add 2 mL of one of the samples as follows:
      3W—2 mL distilled water
      3E—2 mL egg albumin
      3G—2 mL 10% glucose solution
      3D—2 mL carbonated soft drink
      3S—2 mL 10% starch solution
      3P—2 mL potato juice
   c. To each, add 0.5 mL of IKI (iodine solution) and swirl the tube to mix it.
   d. Record the final color of each tube in Part C of the laboratory report.
   e. Mark a "+" on any tubes with positive results and retain them for comparison while testing your unknown sample. Put remaining test tubes aside so they will not get confused with future trials.
4. **Tests for lipids.**
   a. Label two separate areas of a piece of brown paper as "water" or "oil."
   b. Place a drop of water on the area marked "water" and a drop of vegetable oil on the area marked "oil."
   c. Let the spots dry several minutes, then record your observations in Part C of the laboratory report. Upon drying, oil leaves a stain (grease spot) on brown paper; water does not leave such a spot.
   d. In a test tube, add 2 mL of water and 2 mL of vegetable oil and observe. Shake the tube vigorously then let it sit for 5 minutes and observe again. Record your observations in Part C of the laboratory report.
   e. Sudan IV is a dye that is lipid-soluble but not water-soluble. If lipids are present, the Sudan IV will stain them pink or red. Add a small amount of Sudan IV to the test tube that contains the oil and water. Swirl it then let it sit for a few minutes. Record your observations in Part C of the laboratory report.

EXPLORE

Procedure D—Identifying Unknown Compounds

Now you will apply the information you gained with the tests in the previous section. You will retrieve an unknown sample that contains none, one, or any combination of the following types of organic compounds: protein, sugar, starch, or lipid. You will test your sample using each test from the previous section and record your results in Part D of the laboratory report.

1. Label four test tubes (1, 2, 3, 4).
2. Add 2 mL of your unknown sample to each test tube.
3. **Test for protein.** Add 2 mL of Biuret reagent to tube 1. Swirl the tube and observe the color. Record your observations in Part D of the laboratory report.
4. **Test for sugar.** Add 2 mL of Benedict's solution to tube 2. Swirl the tube and place it in a boiling water bath for 3 to 5 minutes. Record your observations in Part D of the laboratory report.
5. **Test for starch.** Add several drops of iodine solution to tube 3. Swirl the tube and observe the color. Record your observations in Part D of the laboratory report.
6. **Test for lipid.** Add 2 mL of water to tube 4. Swirl the tube and note if there is any separation.
7. Add a small amount of Sudan IV to test tube 4, swirl the tube and record your observations in Part D of the laboratory report.
8. Based on the results of these tests, determine if your unknown sample contains any organic compounds and, if so, what are they? Record and explain your identification in Part D of the laboratory report.

Name ______________________________

Date ______________________________

Section ______________________________

The [A] corresponds to the indicated Learning Outcome(s) listed at the beginning of the laboratory exercise.

Chemistry of Life

Laboratory Report 3

Part A Assessments

Match the terms in column A with the descriptions in column B. Place the letter of your choice in the space provided. [1]

Column A

- **a.** Atomic number
- **b.** Atomic weight (mass)
- **c.** Base
- **d.** Catalyst (enzyme)
- **e.** Compound
- **f.** Covalent bond
- **g.** Electrolyte
- **h.** Electron
- **i.** Elements
- **j.** Ion
- **k.** Isotopes
- **l.** Nucleus

Column B

_______ **1.** The number of protons plus the number of neutrons

_______ **2.** A combination of two or more atoms of different elements

_______ **3.** A small, negatively charged particle that orbits the nucleus

_______ **4.** An atom that has gained or lost electrons, and thus carries an electrical charge

_______ **5.** Atoms are held together by sharing electrons

_______ **6.** The most fundamental substances of matter

_______ **7.** A substance that combines with hydrogen ions

_______ **8.** A molecule that influences the rate of chemical reactions but is not consumed in the reaction

_______ **9.** Two atoms with the same atomic number but different atomic weights

_______ **10.** A substance that releases ions in water

_______ **11.** The number of protons in an atom

_______ **12.** The area of an atom where protons and neutrons are located

Molecules and Bonding

After reviewing figures 3.1, 3.2, and 3.3, complete the following:

1. The atomic number of hydrogen is ____________, and it has ____________ electron(s) in its outer shell. [1]
2. The atomic number of chlorine is ____________, and it has ____________ electron(s) in its outer shell. [1]
3. Using the style shown in figure 3.2, draw an atom of hydrogen and one of chlorine, clearly indicating the numbers and positions of the protons, neutrons, and electrons. [1]

Hydrogen

Chlorine

4. Is the hydrogen atom stable? 1 ______________________________

Is the chlorine atom stable? 1 ______________________________

5. What type of bond is likely to form the compound HCl (hydrochloric acid)? 1 ______________

6. Repeat your drawings below, but also show and explain how this bond of HCl would form. Clearly indicate the positions of protons, neutrons, and electrons. 1

Part B—The pH Scale Assessments

1. Results from cabbage water test: 2

Substance	Cabbage Water	Distilled Water	Vinegar	Baking Soda
Color				
Acid, base, or neutral?				

2. Results from pH paper tests: 2

Substance Tested	Distilled Water	Tap Water	Sample 1:_____	Sample 2:_____	Sample 3:_____	Sample 4:_____	Sample 5:_____	Sample 6:_____	Sample 7:_____
pH Value									

3. Are the pH values the same for distilled water and tap water? 2 ______________

4. If not, what might explain this difference? ______________________________

____

5. Draw the pH scale here, and indicate the following values: 0, 7 (neutral), and 14. Now label the scale with the names and indicate the location of the pH values for each substance you tested.

Part C—Testing for Organic Molecules (Biomolecules) Assessments

1. **Biuret test results for protein.** Enter your results from the Biuret test for protein. A color change to purple indicates that protein is present; pink indicates that short polypeptides are present. 3

Tube	Contents	Color	Protein Present (+) or Absent (–)
1W	Distilled water		
1E	Egg albumin		
1G	Glucose solution		
1D	Soft drink		
1S	Starch solution		
1P	Potato juice		

2. **Benedict's test results for most sugars (monosaccharides).** Enter your results from the Benedict's test for sugars. A color change to green, yellow, orange, or red indicates that sugar is present. Note the color after the mixture has been heated for 3 to 5 minutes. 3

Tube	Contents	Color	Sugar Present (+) or Absent (–)
2W	Distilled water		
2E	Egg albumin		
2G	Glucose solution		
2D	Soft drink		
2S	Starch solution		
2P	Potato juice		

3. **Iodine test results for starch.** Enter your results from the iodine test for starch. A color change to dark blue or black indicates that starch is present. 3

Tube	Contents	Color	Starch Present (+) or Absent (–)
3W	Distilled water		
3E	Egg albumin		
3G	Glucose solution		
3D	Soft drink		
3S	Starch solution		
3P	Potato juice		

4. **Lipid test results.** What did you observe when you allowed the drops of oil and water to dry on the brown paper? ______________________________

What did you observe when you mixed the oil and water together? ______________________________

____

What did you observe when you added the Sudan IV dye? 3 ______________________________

Critical Thinking Application

If a person were on a low-carbohydrate, high-protein diet, which of the six substances tested would the person want to increase?__________

Explain your answer.

____

____

____

____

Part D—Testing for Unknown Organic Compounds Assessments

1. What is the number of your sample? __________
2. Record the results of your tests on the unknown here: 4

Test Performed	Results
Biuret test for protein	
Benedict's test for sugars	
Iodine test for starch	
Water test for lipid	
Sudan IV test for lipid	

3. Based on these results, what organic compound(s) does your unknown sample contain? 4 __________

____

4. Explain your answer. 4

____

____

____

____

Laboratory Exercise 4

Care and Use of the Microscope

Materials Needed

Compound light microscope
Lens paper
Microscope slides
Coverslips
Transparent plastic millimeter ruler
Prepared slide of letter *e*
Slide of three colored threads
Medicine dropper
Dissecting needle (needle probe)
Specimen examples for wet mounts
Methylene blue (dilute) or iodine-potassium-iodide stain

For Demonstrations:

Micrometer scale
Stereomicroscope (dissecting microscope)

The human eye cannot perceive objects less than 0.1 mm in diameter, so a microscope is an essential tool for the study of small structures such as cells. The microscope usually used for this purpose is the *compound light microscope.* It is called compound because it uses two sets of lenses: an eyepiece or ocular lens system and an objective lens system. The eyepiece lens system magnifies, or compounds, the image reaching it after the image is magnified by the objective lens system. Such an instrument can magnify images of small objects up to about one thousand times.

Purpose of the Exercise

To become familiar with the major parts of a compound light microscope and their functions and to make use of the microscope to observe small objects.

LEARNING OUTCOMES

After completing this exercise, you should be able to

1. Locate and identify the major parts of a compound light microscope and differentiate the functions of these parts.
2. Calculate the total magnification produced by various combinations of eyepiece and objective lenses.
3. Demonstrate proper use of the microscope to observe and measure small objects.
4. Prepare a simple microscope slide and sketch the objects you observed.

EXPLORE

Procedure A—Microscope Basics

1. Familiarize yourself with the following list of rules for care of the microscope:
   a. Keep the microscope under its *dustcover* and in a cabinet when it is not being used.
   b. Handle the microscope with great care. It is an expensive and delicate instrument. To move it or carry it, hold it by its *arm* with one hand and support its *base* with the other hand (fig. 4.1).
   c. Always store the microscope with the scanning or lowest-power objective in place. Always start with this objective when using the microscope.
   d. To clean the lenses, rub them gently with *lens paper* or a high-quality cotton swab. If the lenses need additional cleaning, follow the directions in the "Lens-Cleaning Technique" section that follows.
   e. If the microscope has a substage lamp, be sure the electric cord does not hang off the laboratory table where someone might trip over it. The bulb life can be extended if the lamp is cool before the microscope is moved.
   f. Never drag the microscope across the laboratory table after you have placed it.
   g. Never remove parts of the microscope or try to disassemble the eyepiece or objective lenses.
   h. If your microscope is not functioning properly, report the problem to your laboratory instructor immediately.

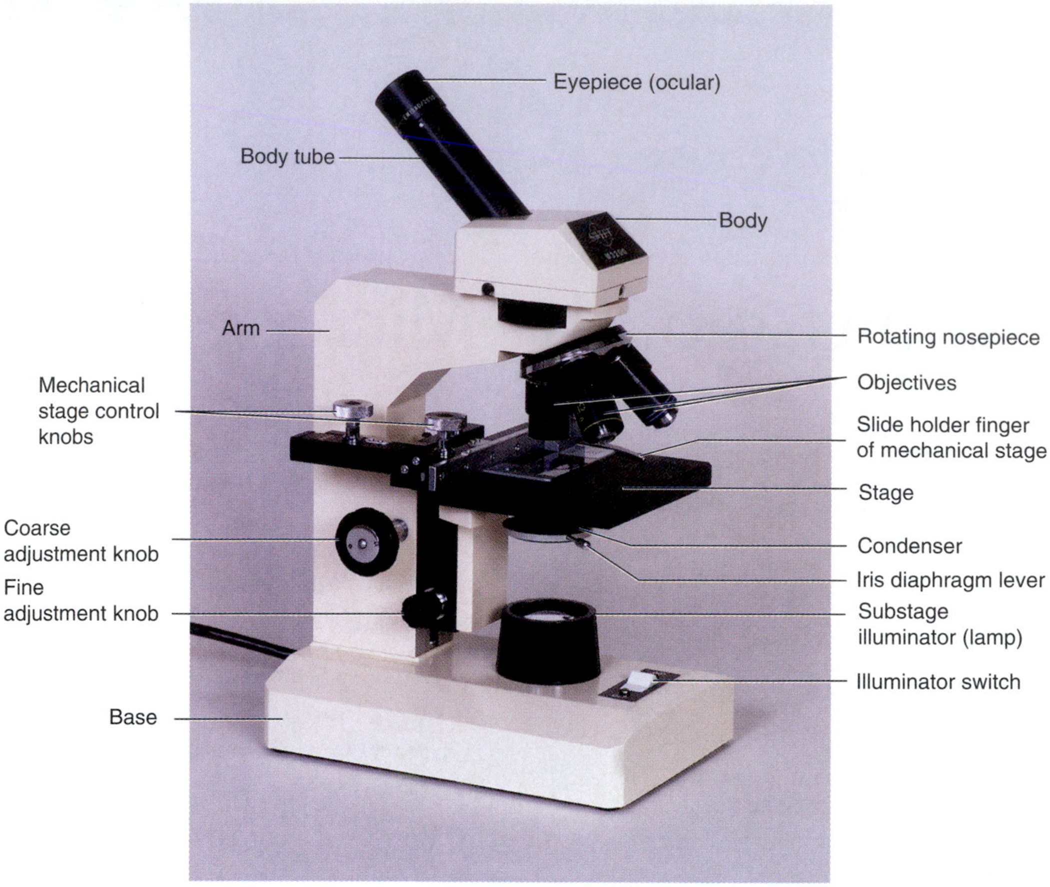

Figure 4.1 Major parts of a compound light microscope with a monocular body and a mechanical stage. Some microscopes are equipped with a binocular body.

Lens-Cleaning Technique

1. Moisten one end of a high-quality cotton swab with one drop of Kodak lens cleaner. Keep the other end dry.
2. Clean the optical surface with the wet end. Dry it with the other end, using a circular motion.
3. Use a hand aspirator to remove lingering dust particles.
4. Start with the scanning objective and work upward in magnification, using a new cotton swab for each objective.
5. When cleaning the eyepiece, do not open the lens unless it is absolutely necessary.
6. Use alcohol for difficult cleaning and only as a last resort use xylene. Regular use of xylene will destroy lens coatings.

2. Observe a compound light microscope and study figure 4.1 to learn the names of its major parts. The lens system of a compound microscope includes three parts—the condenser, objective lens, and eyepiece or ocular.

Light enters this system from a *substage illuminator (lamp)* or *mirror* and is concentrated and focused by a *condenser* onto a microscope slide or specimen placed on the *stage*. The condenser, which contains a set of lenses, usually is kept in its highest position possible.

The *iris diaphragm*, located between the light source and the condenser, can be used to increase or decrease the intensity of the light entering the condenser. Locate the lever that operates the iris diaphragm beneath the stage and move it back and forth. Note how this movement causes the size of the opening in the diaphragm to change. (Some microscopes have a revolving plate called a disc diaphragm beneath the stage instead of an iris diaphragm. Disc diaphragms have different-sized holes to admit varying amounts of light.) Which way do you move the diaphragm to increase the light intensity? ____________ Which way to decrease it? ____________

After light passes through a specimen mounted on a microscope slide, it enters an *objective lens system*. This lens projects the light upward into the *body tube*, where it produces a magnified image of the object being viewed.

(a)

(b)

Figure 4.2 The powers of this 10× eyepiece (*a*) and this 40× objective (*b*) are marked in the metal. DIN is an international optical standard on quality optics. The 0.65 on the 40× objective is the numerical aperture, a measure of the light-gathering capabilities.

The *eyepiece (ocular) lens* system then magnifies this image to produce another image, seen by the eye. Typically, the eyepiece lens magnifies the image ten times (10×). Look for the number in the metal of the eyepiece that indicates its power (fig. 4.2). What is the eyepiece power of your microscope? ____________________

The objective lenses are mounted in a *rotating nosepiece* so that different magnifications can be achieved by rotating any one of several objective lenses into position above the specimen. Commonly, this set of lenses includes a scanning objective (4×), a low-power objective (10×), and a high-power objective, also called a high-dry-power objective (about 40×). Sometimes an oil immersion objective (about 100×) is present. Look for the number printed on each objective that indicates its power. What are the objective lens powers of your microscope? ____________________

To calculate the *total magnification* achieved when using a particular objective, multiply the power of the eyepiece by the power of the objective used. Thus, the 10× eyepiece and the 40× objective produce a total magnification of 10 × 40, or 400×. See table 4.1.

Table 4.1 Microscope Lenses

Objective Lens Name	Common Objective Lens Magnification	Common Eyepiece Lens Magnification	Total Magnification
Scan	4×	10×	40×
Low power (LP)	10×	10×	100×
High power (HP)	40×	10×	400×
Oil immersion	100×	10×	1,000×

Note: If you wish to observe an object under LP, HP, or oil immersion, locate and then center and focus the object first under scan magnification.

3. Complete Part A of Laboratory Report 4.
4. Turn on the substage illuminator and look through the eyepiece. You will see a lighted circular area called the *field of view.*

   You can measure the diameter of this field of view by focusing the lenses on the millimeter scale of a transparent plastic ruler. To do this, follow these steps:

   a. Place the ruler on the microscope stage in the spring clamp of a slide holder finger on a mechanical stage or under the stage (slide) clips. (*Note:* If your microscope is equipped with a mechanical stage, it may be necessary to use a short section cut from a transparent plastic ruler. The section should be several millimeters long and can be mounted on a microscope slide for viewing.)

b. Center the millimeter scale in the beam of light coming up through the condenser, and rotate the scanning objective into position.

c. While you watch from the side to prevent the lens from touching anything, raise the stage until the objective is as close to the ruler as possible, using the *coarse adjustment knob* and then using the *fine adjustment knob* (fig. 4.3). (*Note:* The adjustment knobs on some microscopes move the body and objectives downward and upward for focusing.)

d. Look into the eyepiece and use the coarse adjustment knob to raise the stage until the lines of the millimeter scale come into sharp focus.

e. Adjust the light intensity by moving the *iris diaphragm lever* so that the field of view is brightly illuminated but comfortable to your eye. At the same time, take care not to overilluminate the field because transparent objects tend to disappear in bright light.

f. Position the millimeter ruler so that its scale crosses the greatest diameter of the field of view. Also, move the ruler so that one of the millimeter marks is against the edge of the field of view.

g. In millimeters, measure the distance across the field of view.

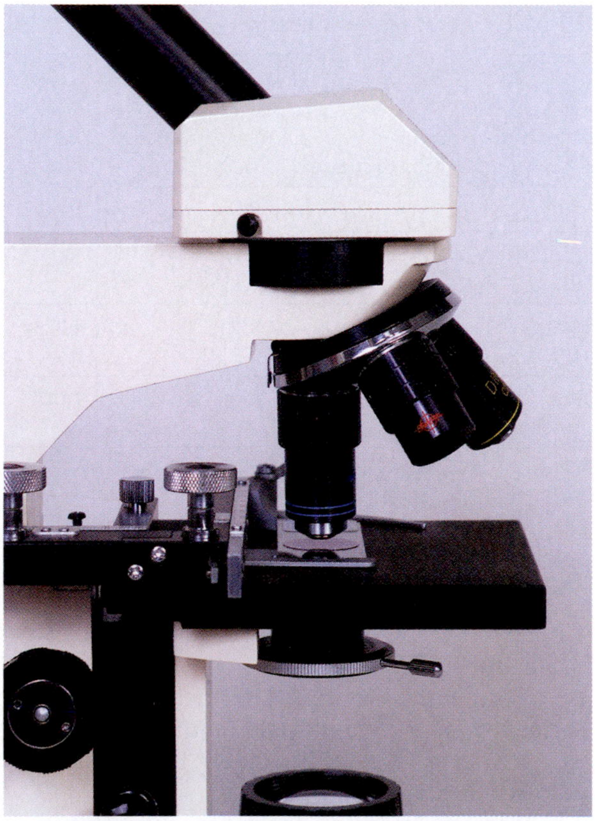

Figure 4.3 When you focus using a particular objective, you can prevent it from touching the specimen by watching from the side.

5. Complete Part B of the laboratory report.

6. Most microscopes are designed to be *parfocal.* This means that when a specimen is in focus with a lower-power objective, it will be in focus (or nearly so) when a higher-power objective is rotated into position. Always center the specimen in the field of view before changing to higher objectives.

Rotate the low-power objective into position, and then look at the millimeter scale of the transparent plastic ruler. If you need to move the low-power objective to sharpen the focus, use the *fine adjustment knob.*

Adjust the iris diaphragm so that the field of view is properly illuminated. Once again, adjust the millimeter ruler so that the scale crosses the field of view through its greatest diameter, and position the ruler so that a millimeter mark is against one edge of the field.

Try to measure the distance across the field of view in millimeters.

7. Rotate the high-power objective into position while you watch from the side, and then observe the millimeter scale on the plastic ruler. All focusing using high-power magnification should be done only with the fine adjustment knob. If you use the coarse adjustment knob with the high-power objective, you can accidently force the objective into the coverslip and break the slide. This is because the *working distance* (the distance from the objective lens to the slide on the stage) is much shorter when using higher magnifications.

Adjust the iris diaphragm for proper illumination. Usually more illumination when using higher magnifications will help you to view the objects more clearly. Try to measure the distance across the field of view in millimeters.

8. Locate the numeral 4 (or 9) on the plastic ruler and focus on it using the scanning objective. Note how the number appears in the field of view. Move the plastic ruler to the right and note which way the image moves. Slide the ruler away from you and again note how the image moves.

9. Observe a letter *e* slide in addition to the numerals on the plastic ruler. Note the orientation of the letter *e* using the scan, LP, and HP objectives. As you increase magnifications, note that the amount of the letter *e* shown is decreased. Move the slide to the left, and then away from you, and note the direction in which the observed image moves.

10. Examine the slide of the three colored threads using the low-power objective and then the high-power objective. Focus on the location where the three threads cross. By using the fine adjustment knob, determine the order from top to bottom by noting which color is in focus at different depths. The other colored threads will still be visible, but they will be blurred. Be sure to notice whether the stage or the body tube moves up and down with the adjustment knobs of the microscope being used for this depth determination. The vertical depth of the specimen clearly in focus is called the *depth of field* (*focus*). Whenever specimens are examined, continue to use the fine adjustment focusing knob to determine relative depths of structures clearly in focus within cells, giving a three-dimensional perspective. The depth of field is less at higher magnifications.

Critical Thinking Application

What was the sequence of the three colored threads from top to bottom? Explain how you came to that conclusion.

____

____

____

11. Complete Parts C and D of the laboratory report.

Demonstration

A compound light microscope is sometimes equipped with a micrometer scale mounted in the eyepiece. Such a scale is subdivided into fifty to one hundred equal divisions (fig. 4.4). These arbitrary divisions can be calibrated against the known divisions of a micrometer slide placed on the microscope stage. Once the values of the divisions are known, the length and width of a microscopic object can be measured by superimposing the scale over the magnified image of the object.

Observe the micrometer scale in the eyepiece of the demonstration microscope. Focus the low-power objective on the millimeter scale of a micrometer slide (or a plastic ruler) and measure the distance between the divisions on the micrometer scale in the eyepiece. What is the distance between the finest divisions of the scale in micrometers? ____________________________

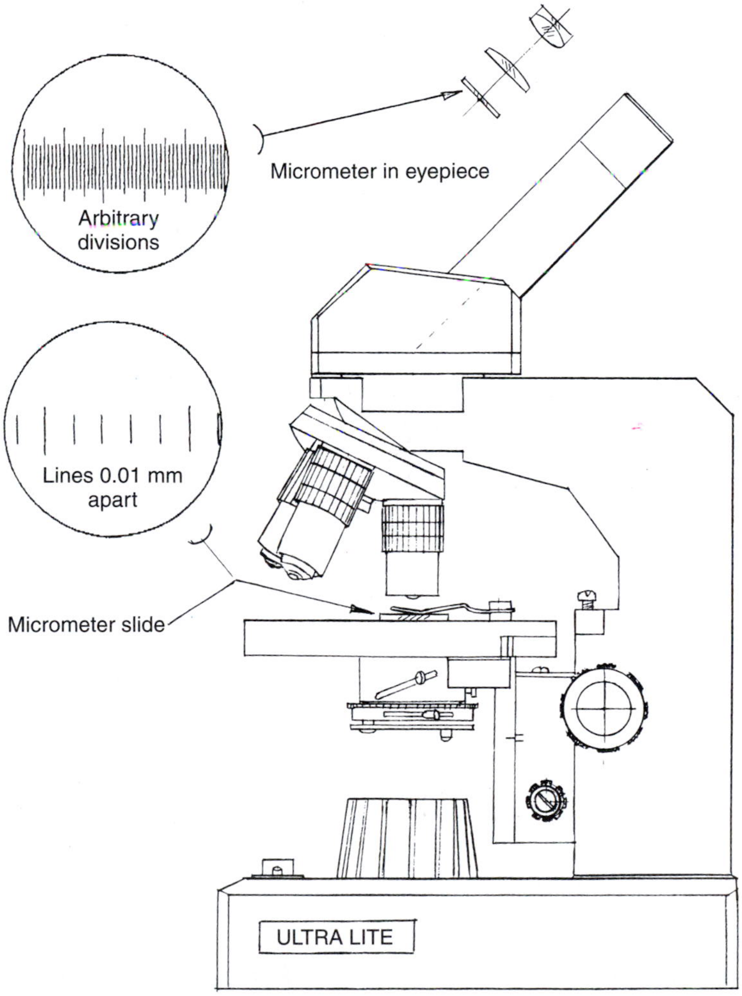

Figure 4.4 The divisions of a micrometer scale in an eyepiece can be calibrated against the known divisions of a micrometer slide. (Courtesy of Swift Instruments, Inc., San Jose, California)

EXPLORE

Procedure B—Slide Preparation

1. Prepare several temporary *wet mounts* using any small, transparent objects of interest, and examine the specimens using the low-power objective and then a high-power objective to observe their details. To prepare a wet mount, follow these steps (fig. 4.5):
   a. Obtain a precleaned microscope slide.
   b. Place a tiny, thin piece of the specimen you want to observe in the center of the slide, and use a medicine dropper to put a drop of water over it. Consult with your instructor if a drop of stain might enhance the image of any cellular structures of your specimen. If the specimen is solid, you might want to tease some of it apart with dissecting needles. In any case, the specimen must be thin enough so that light can pass through it. Why is it necessary for the specimen to be so thin?
   c. Cover the specimen with a coverslip. Try to avoid trapping bubbles of air beneath the coverslip by slowly lowering it at an angle into the drop of water.
   d. Remove any excess water from the edge of the coverslip with absorbent paper. If your microscope has an inclination joint, do not tilt the microscope while observing wet mounts because the fluid will flow.
   e. Place the slide under the stage (slide) clips or in the slide holder on a mechanical stage, and position the slide so that the specimen is centered in the light beam passing up through the condenser.
   f. Focus on the specimen using the scanning objective first. Next focus using the low-power objective, and then examine it with the high-power objective.
2. If an oil immersion objective is available, use it to examine the specimen. To use the oil immersion objective, follow these steps:
   a. Center the object you want to study under the high-power field of view.
   b. Rotate the high-power objective away from the microscope slide, place a small drop of immersion oil on the coverslip, and swing the oil immersion objective into position. To achieve sharp focus, use the fine adjustment knob only.
   c. You will need to open the iris diaphragm more fully for proper illumination. More light is

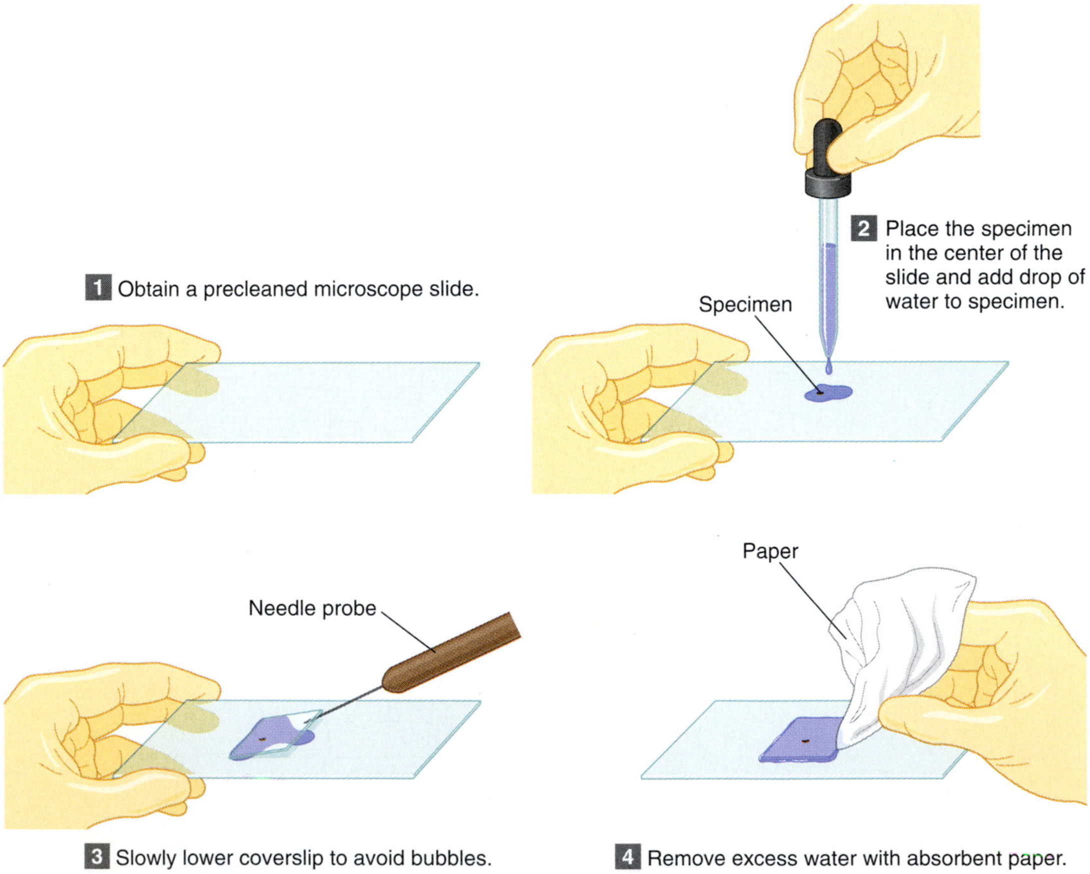

Figure 4.5 Steps in the preparation of a wet mount.

needed because the oil immersion objective covers a very small lighted area of the microscope slide.

d. The oil immersion objective must be very close to the coverslip to achieve sharp focus, so care must be taken to avoid breaking the coverslip or damaging the objective lens. For this reason, never lower the objective when you are looking into the eyepiece. Instead, always raise the objective to achieve focus, or prevent the objective from touching the coverslip by watching the microscope slide and coverslip from the side if the objective needs to be lowered. Usually when using the oil immersion objective, only the fine adjustment knob needs to be used for focusing.

3. When you have finished working with the microscope, remove the microscope slide from the stage and wipe any oil from the objective lens with lens paper or a high-quality cotton swab. Swing the scanning objective or the low-power objective into position. Wrap the electric cord around the base of the microscope and replace the dustcover.

4. Complete Part E of the laboratory report.

Demonstration

A stereomicroscope (dissecting microscope) (fig. 4.6) is useful for observing the details of relatively large, opaque specimens. Although this type of microscope achieves less magnification than a compound microscope, it has the advantage of producing a three-dimensional image rather than the flat, two-dimensional image of the compound light microscope. In addition, the image produced by the stereomicroscope is positioned in the same manner as the specimen, rather than being reversed and inverted as it is by the compound light microscope.

Observe the stereomicroscope. The eyepieces can be pushed apart or together to fit the distance between your eyes. Focus the microscope on the end of your finger. Which way does the image move when you move your finger to the right? ____________

When you move it away? ______________________

If the instrument has more than one objective, change the magnification to a higher power. Use the instrument to examine various small, opaque objects available in the laboratory.

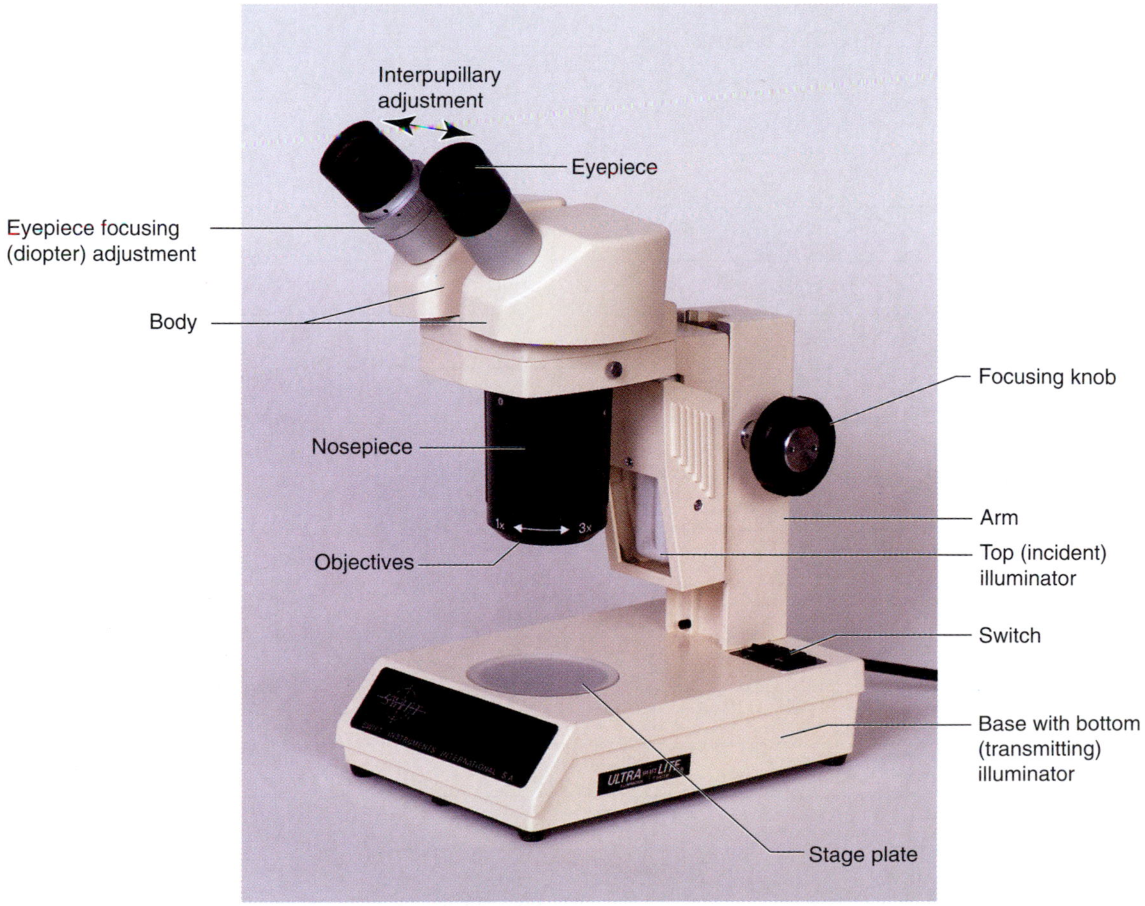

Figure 4.6 A stereomicroscope, also called a dissecting microscope.

Notes

Name ______________________________

Date ______________________________

Section ______________________________

The A corresponds to the indicated Learning Outcome(s) found at the beginning of the laboratory exercise.

Laboratory Report 4

Care and Use of the Microscope

Part A Assessments

Revisit Procedure A, number 2, then complete the following:

1. What total magnification will be achieved if the 10× eyepiece and the 10× objective are used? A2 __________
2. What total magnification will be achieved if the 10× eyepiece and the 100× objective are used? A2 __________

Part B Assessments

Revisit Procedure A, number 2, then complete the following:

1. Sketch the millimeter scale as it appears under the scanning objective magnification. (The circle represents the field of view through the microscope.)

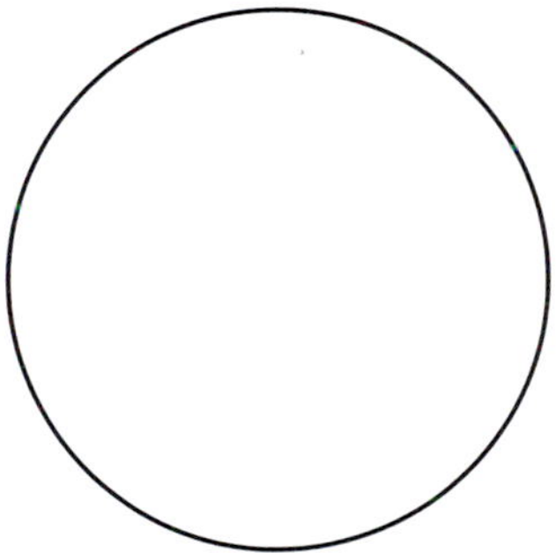

2. In millimeters, what is the diameter of the scanning field of view? A3 __________
3. Microscopic objects often are measured in *micrometers*. A micrometer equals 1/1,000 of a millimeter and is symbolized by µm. In micrometers, what is the diameter of the scanning power field of view? A3 __________
4. If a circular object or specimen extends halfway across the scanning field, what is its diameter in millimeters? A3 __________
5. In micrometers, what is its diameter? A3 __________

Part C Assessments

Complete the following:

1. Sketch the millimeter scale as it appears using the low-power objective.

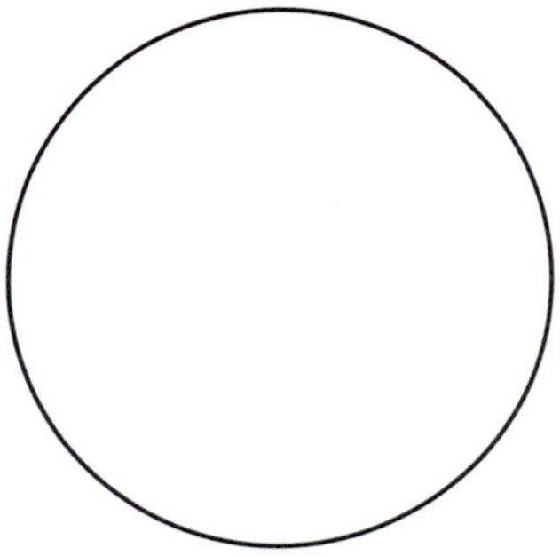

2. What do you estimate the diameter of this field of view to be in millimeters? A3 __________
3. How does the diameter of the scanning power field of view compare with that of the low-power field?

____

4. Why is it more difficult to measure the diameter of the high-power field of view than the low-power field?

____

____

5. What change occurred in the light intensity of the field of view when you exchanged the low-power objective for the high-power objective? ____

____

6. Sketch the numeral 4 (or 9) as it appears through the scanning objective of the compound microscope.

7. What has the lens system done to the image of the numeral? (Is it right side up, upside down, or what?) ______________________

8. When you moved the ruler to the right, which way did the image move? ______________________

9. When you moved the ruler away from you, which way did the image move? ______________________

Part D Assessments

Match the names of the microscope parts in column A with the descriptions in column B. Place the letter of your choice in the space provided.

Column A

- **a.** Adjustment knob
- **b.** Arm
- **c.** Condenser
- **d.** Eyepiece (ocular)
- **e.** Field of view
- **f.** Iris diaphragm
- **g.** Nosepiece
- **h.** Objective lens system
- **i.** Stage
- **j.** Stage (slide) clip

Column B

_______ **1.** Increases or decreases the light intensity

_______ **2.** Platform that supports a microscope slide

_______ **3.** Concentrates light onto the specimen

_______ **4.** Causes stage (or objective lens) to move upward or downward

_______ **5.** After light passes through the specimen, it next enters this lens system

_______ **6.** Holds a microscope slide in position

_______ **7.** Contains a lens at the top of the body tube

_______ **8.** Serves as a handle for carrying the microscope

_______ **9.** Part to which the objective lenses are attached

_______ **10.** Circular area seen through the eyepiece

Part E Assessments

Prepare sketches of the objects you observed using the microscope. For each sketch, include the name of the object, the magnification you used to observe it, and its estimated dimensions in millimeters and micrometers.

Laboratory Exercise 5

Cell Structure and Function

Materials Needed

Textbook
Animal cell model
Clean microscope slides
Coverslips
Flat toothpicks
Medicine dropper
Methylene blue (dilute) or iodine-potassium-iodide stain
Prepared microscope slides of human tissues
Compound light microscope

For Learning Extensions:

Single-edged razor blade
Plant materials such as leaves, soft stems, fruits, onion peel, and vegetables
Cultures of *Amoeba* and *Paramecium*

Safety

- Review all the safety guidelines inside the front cover.
- Clean laboratory surfaces before and after laboratory procedures.
- Wear disposable gloves for the wet-mount procedures of the cells lining the inside of the cheek.
- Work only with your own materials when preparing the slide of cheek cells. Observe the same precautions as with all body fluids.
- Dispose of laboratory gloves, slides, coverslips, and toothpicks as instructed.
- Precautions should be taken to prevent cellular stains from contacting your clothes and skin.
- Wash your hands before leaving the laboratory.

Cells are the "building blocks" from which all parts of the human body are formed. They account for the shape, organization, and construction of the body and are responsible for carrying on its life processes. Under the light microscope, with a properly applied stain to make structures visible, the **cell (plasma) membrane,** the **cytoplasm,** and a **nucleus** are easily seen. The cytoplasm is composed of a clear fluid, the *cytosol,* and numerous *cytoplasmic organelles* suspended in the cytosol.

The cell membrane, composed of lipids and proteins, forms the cell boundary and functions in various methods of membrane transport. The nucleus contains fine strands of DNA and protein called chromatin. Various cytoplasmic organelles, including mitochondria, endoplasmic reticulum, and Golgi apparatus, provide specialized metabolic functions.

Purpose of the Exercise

To review the structure and functions of major cellular components and to observe examples of human cells.

LEARNING OUTCOMES

After completing this exercise, you should be able to

1. Name and locate the components of a cell.
2. Differentiate the functions of cellular components.
3. Prepare a wet mount of cells lining the inside of the cheek; stain the cells; and identify the cell membrane, nucleus, and cytoplasm.
4. Examine cells on prepared slides of human tissues and identify their major components.

EXPLORE

Procedure—Cell Structure and Function

1. Review the section entitled "A Composite Cell" in chapter 3 of the textbook.
2. Observe the animal cell model and identify its structures.
3. As a review activity, label figure 5.1 and study figure 5.2.
4. Complete Part A of Laboratory Report 5.
5. Prepare a wet mount of cells lining the inside of the cheek. To do this, follow these steps:
   a. Gently scrape (force is not necessary and should be avoided) the inner lining of your cheek with the broad end of a flat toothpick.
   b. Stir the toothpick in a drop of water on a clean microscope slide and dispose of the toothpick as directed by your instructor.

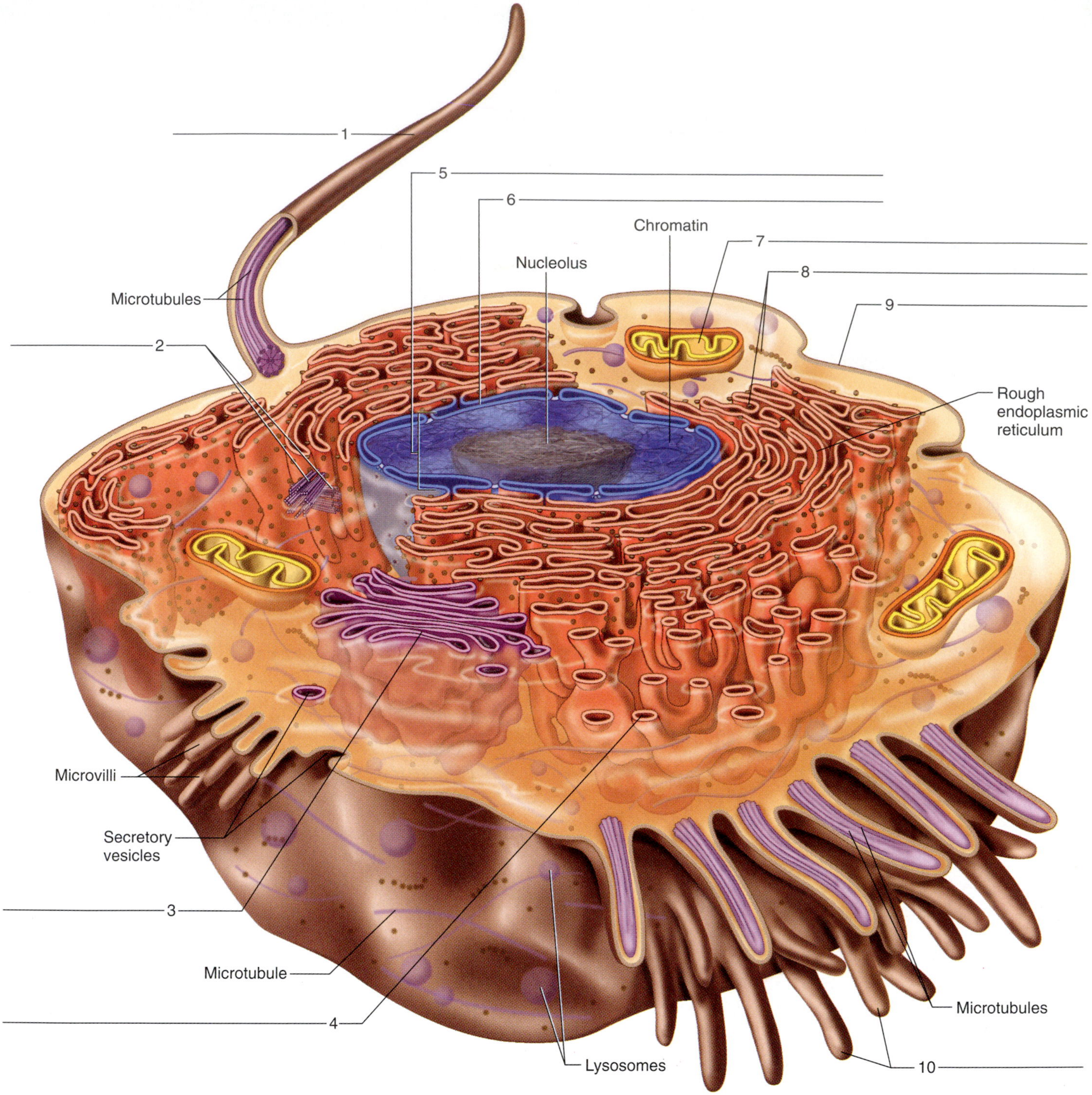

Figure 5.1 Label the structures of this composite cell. The structures are not drawn to scale. 1

c. Cover the drop with a coverslip.

d. Observe the cheek cells by using the microscope. Compare your image with figure 5.3. To report what you observe, sketch a single cell in the space provided in Part B of the laboratory report.

6. Prepare a second wet mount of cheek cells, but this time, add a drop of dilute methylene blue or iodine-potassium-iodide stain to the cells. Cover the liquid with a coverslip and observe the cells with the microscope. Add to your sketch any additional structures you observe in the stained cells.

7. Answer the questions in Part B of the laboratory report.

8. Using the microscope, observe each of the prepared slides of human tissues. To report what you observe, sketch a single cell of each type in the space provided in Part C of the laboratory report.

9. Complete Parts C and D of the laboratory report.

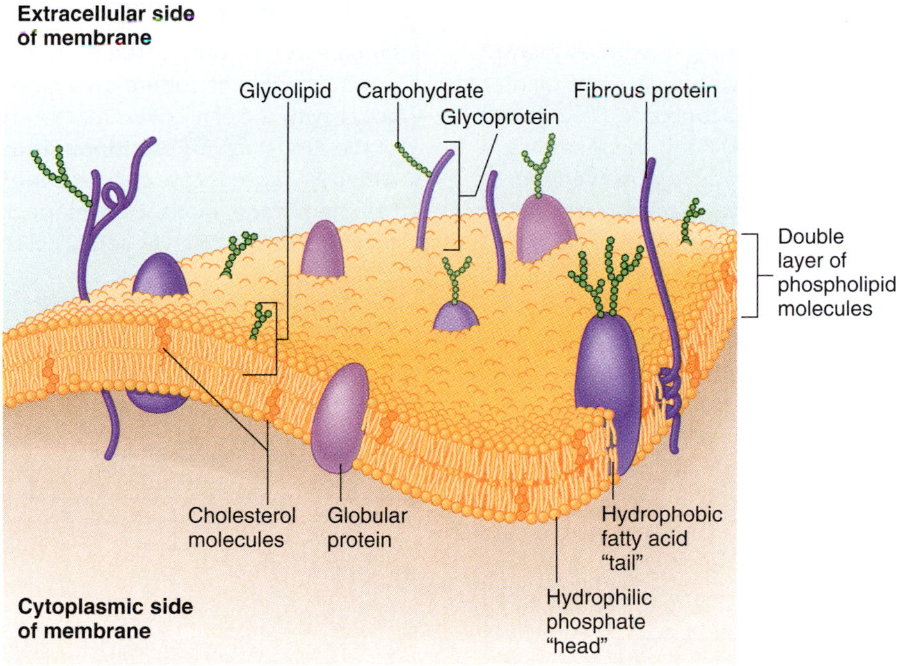

Figure 5.2 Structures of the cell membrane.

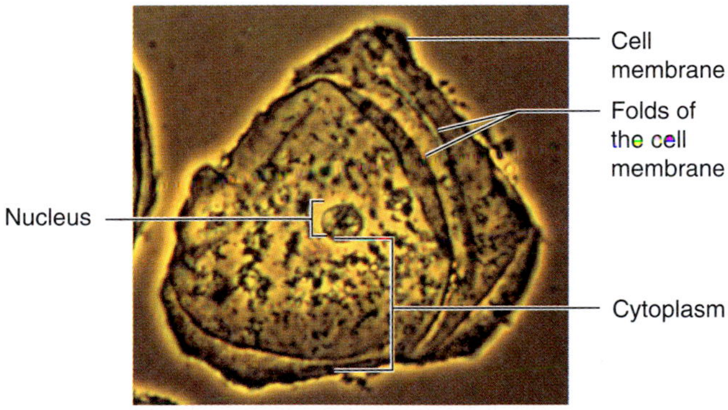

Figure 5.3 Stained cell lining the inside of the cheek as viewed through the compound light microscope using the high-power objective (400×).

Critical Thinking Application

The cells lining the inside of the cheek are frequently removed for making observations of basic cell structure. The cells are from stratified squamous epithelium. Explain why these cells are used instead of outer body surface tissue.

____

____

____

____

____

Learning Extension

Investigate the microscopic structure of various plant materials. To do this, prepare tiny, thin slices of plant specimens, using a single-edged razor blade. (*Take care not to injure yourself with the blade.*) Keep the slices in a container of water until you are ready to observe them. To observe a specimen, place it into a drop of water on a clean microscope slide and cover it with a coverslip. Use the microscope and view the specimen using low- and high-power magnifications. Observe near the edges where your section of tissue is most likely to be one cell thick. Add a drop of dilute methylene blue or iodine-potassium-iodide stain, and note if any additional structures become visible. How are the microscopic structures of the plant specimens similar to the human tissues you observed?

____

How are they different? ______________________________

____

Learning Extension

Prepare a wet mount of the *Amoeba* and *Paramecium* by putting a drop of culture on a clean glass slide. Gently cover with a clean coverslip. Observe the movements of the *Amoeba* with pseudopodia and the *Paramecium* with cilia. Try to locate cellular components such as the cell membrane, nuclear envelope, nucleus, mitochondria, and contractile vacuoles. Describe the movement of the *Amoeba*.

____

____

Describe the movement of the *Paramecium*.

____

____

Name ________________________________

Date ________________________________

Section ________________________________

The [A] corresponds to the indicated Learning Outcome(s) found at the beginning of the laboratory exercise.

Laboratory Report 5

Cell Structure and Function

Part A Assessments

Match the cellular components in column A with the descriptions in column B. Place the letter of your choice in the space provided. [2]

Column A		Column B
a. Chromatin	______	1. Loosely coiled fibers containing protein and DNA within nucleus
b. Cytoplasm	______	2. Location of ATP production from digested food molecules
c. Endoplasmic reticulum	______	3. Small RNA-containing particles for the synthesis of proteins
d. Golgi apparatus	______	4. Membranous sac formed by the pinching off of pieces of cell membrane
e. Lysosome	______	5. Dense body of RNA within the nucleus
f. Microtubule	______	6. Slender tubes that provide movement in cilia and flagella
g. Mitochondrion	______	7. Organelles composed of membrane-bound sacs, canals, and vesicles for tubular transport
h. Nuclear envelope	______	8. Occupies space between cell membrane and nucleus
i. Nucleolus	______	9. Flattened membranous sacs that package a secretion
j. Nucleus	______	10. Membranous sac that contains digestive enzymes
k. Ribosome	______	11. Separates nuclear contents from cytoplasm
l. Vesicle (vacuole)	______	12. Spherical organelle that contains chromatin and nucleolus

Part B Assessments

Complete the following:

1. Sketch a single cheek cell that has been stained. Label the cellular components you recognize. (The circle represents the field of view through the microscope.) [3]

Magnification ______ ×

2. After comparing the wet mount and the stained cheek cells, describe the advantage gained by staining cells.

____

____

3. Are the stained cheek cells nearly the same size and shape? ______________ Propose an explanation for your answer. ____

____

Part C Assessments

Complete the following:

1. Sketch a single cell of each type you observed in the prepared slides of human tissues. Name the tissue, indicate the magnification used, and label the cellular components you recognize. 4

______________ ×

Tissue ______________

______________ ×

Tissue ______________

______________ ×

Tissue ______________

______________ ×

Tissue ______________

2. What do the various types of cells in these tissues have in common? ________________________________

____

3. What are the main differences you observed among these cells? ________________________________

____

Part D Assessments

Electron micrographs represent extremely thin slices of cells. Each micrograph in figure 5.4 contains a section of a nucleus and some cytoplasm. Compare the organelles shown in these micrographs with organelles of the animal cell model and figure 5.1.

Identify the structures indicated by the arrows in figure 5.4. 1

1. ______________________
2. ______________________
3. ______________________
4. ______________________
5. ______________________
6. ______________________
7. ______________________
8. ______________________
9. ______________________
10. ______________________

Answer the following questions after observing the transmission electron micrographs in figure 5.4.

11. What cellular structures were visible in the transmission electron micrographs that were not apparent in the cells you observed using the microscope? ______________________

____

12. Before they can be observed by using a transmission electron microscope, cells are sliced into very thin sections. What disadvantage does this procedure present in the study of cellular parts? ______________________

____

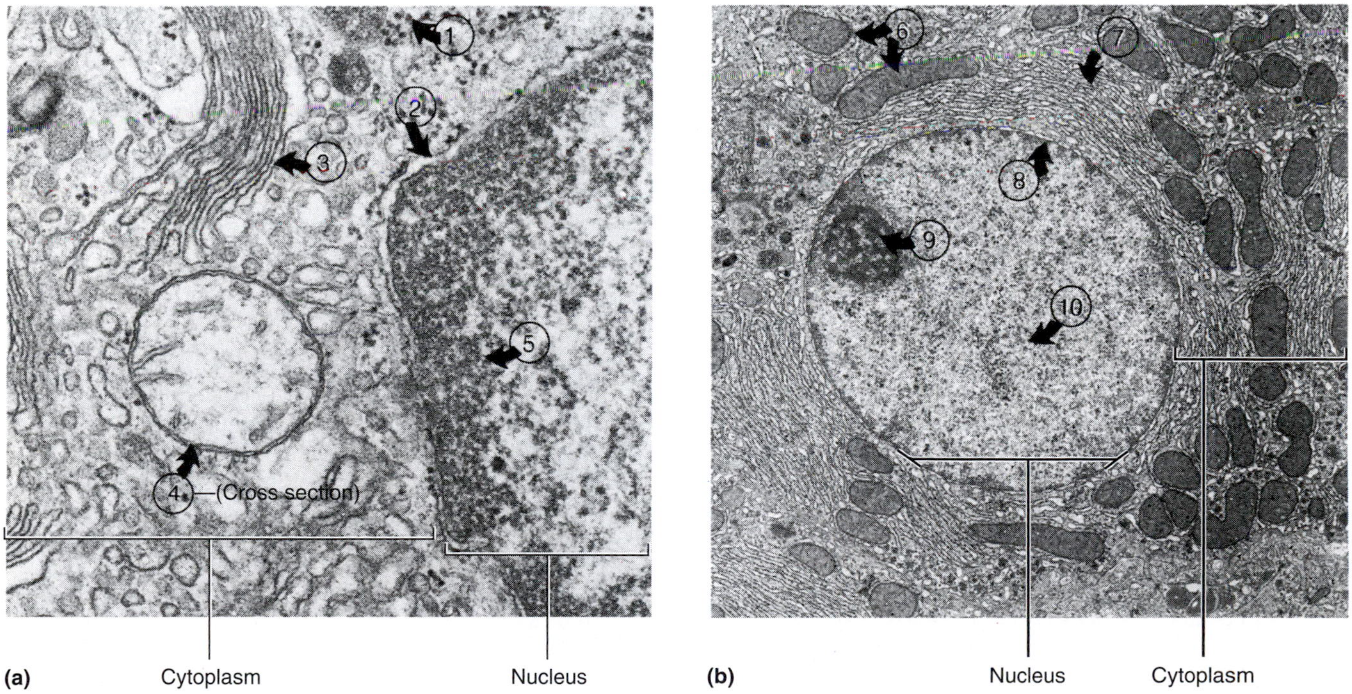

Terms:
Chromatin (use 2 times)
Endoplasmic reticulum
Golgi apparatus
Mitochondria
Mitochondrian (cross section)
Nuclear envelope (use 2 times)
Nucleolus
Ribosomes

Figure 5.4 Transmission electron micrographs of cellular components. The views are only portions of a cell. Magnifications: (*a*) 26,000×; (*b*)10,000×. Identify the numbered cellular structures, using the terms provided. 1

Notes

Laboratory Exercise 6

Movements Through Membranes

Materials Needed

For Procedure A—Diffusion:

Textbook
Petri dish
White paper
Forceps
Potassium permanganate crystals
Millimeter ruler (thin and transparent)

For Procedure B—Osmosis:

Thistle tube
Molasses (or Karo dark corn syrup)
Selectively permeable (semipermeable) membrane (presoaked dialysis tubing of 1 5/16″ or greater diameter)
Ring stand and clamp
Beaker
Rubber band
Millimeter ruler

For Procedure C—Hypertonic, Hypotonic, and Isotonic Solutions:

Test tubes
Marking pen
Test-tube rack
10 mL graduated cylinder
Medicine dropper
Uncoagulated animal blood
Distilled water
0.9% NaCl (aqueous solution)
3.0% NaCl (aqueous solution)
Clean microscope slides
Coverslips
Compound light microscope

For Procedure D—Filtration:

Glass funnel
Filter paper
Ring stand and ring
Beaker
Powdered charcoal or ground black pepper
1% glucose (aqueous solution)
1% starch (aqueous solution)
Test tubes
10 mL graduated cylinder
Water bath (boiling water)
Benedict's solution
Iodine-potassium-iodide solution
Medicine dropper

For Alternative Osmosis Activity:

Fresh chicken egg
Beaker
Laboratory balance
Spoon
Vinegar
Corn syrup (Karo)

Safety

- Clean laboratory surfaces before and after laboratory procedures.
- Wear disposable gloves when handling chemicals and animal blood.
- Wear safety glasses when using chemicals.
- Dispose of laboratory gloves and blood-contaminated items as instructed.
- Wash your hands before leaving the laboratory.

A cell membrane functions as a gateway through which chemical substances and small particles may enter or leave a cell. These substances move through the membrane by physical processes such as simple diffusion, osmosis, and filtration, or by physiological processes such as active transport, phagocytosis, or pinocytosis.

Purpose of the Exercise

To demonstrate some of the physical processes by which substances move through cell membranes.

LEARNING OUTCOMES

After completing this exercise, you should be able to

1. Demonstrate *simple diffusion* and identify examples of diffusion.
2. Interpret diffusion by preparing and explaining a graph.
3. Demonstrate *osmosis* and identify examples of osmosis.
4. Distinguish among hypertonic, hypotonic, and isotonic solutions and observe the effects of these solutions on animal cells.
5. Demonstrate *filtration* and identify examples of filtration.

EXPLORE

Procedure A—Simple Diffusion

1. Review the section entitled "Diffusion" in chapter 3 of the textbook.
2. To demonstrate simple diffusion, follow these steps:
   a. Place a petri dish, half filled with water, on a piece of white paper that has a millimeter ruler positioned on the paper. Wait until the water surface is still. Allow approximately 3 minutes. *Note:* The petri dish should remain level. A second millimeter ruler may be needed under the petri dish as a shim to obtain a level amount of the water inside the petri dish.
   b. Using forceps, place one crystal of potassium permanganate near the center of the petri dish and near the millimeter ruler (fig. 6.1).

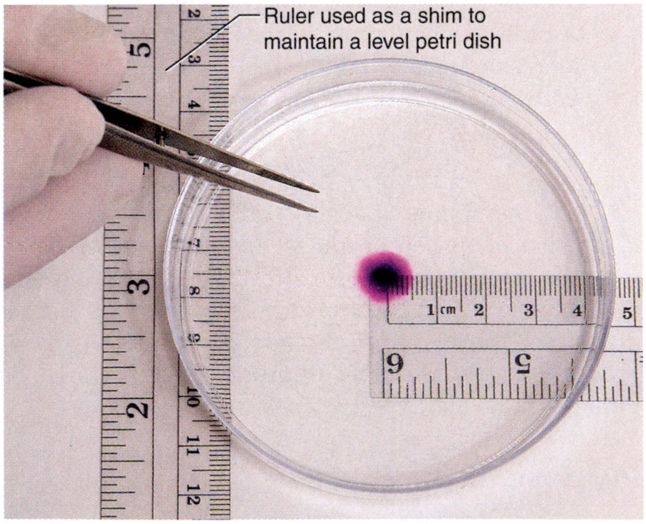

Figure 6.1 To demonstrate simple diffusion, place one crystal of potassium permanganate in the center of a petri dish containing water. Place the crystal near the millimeter ruler (positioned under the petri dish).

   c. Measure the radius of the purple circle at 1-minute intervals for 10 minutes and record the results in Part A of Laboratory Report 6.
3. Complete Part A of the laboratory report.

> **Learning Extension**
>
> Repeat the demonstration of simple diffusion using a petri dish filled with ice-cold water and a second dish filled with very hot water. At the same moment, add a crystal of potassium permanganate to each dish and observe the circle as before. What difference do you note in the rate of diffusion in the two dishes? How do you explain this difference?
>
> ____
>
> ____
>
> ____
>
> ____

EXPLORE

Procedure B—Osmosis

1. Review the section entitled "Osmosis" in chapter 3 of the textbook.
2. To demonstrate osmosis, refer to figure 6.2 as you follow these steps:
   a. One person plugs the tube end of a thistle tube with a finger.
   b. Another person then fills the bulb with molasses until it is about to overflow at the top of the bulb. Air remains trapped in the stem.
   c. Cover the bulb opening with a single-thickness piece of moist selectively permeable (semipermeable) membrane. Dialysis tubing that has been soaked for 30 minutes can easily be cut open because it becomes pliable.
   d. Tightly secure the membrane in place with several wrappings of a rubber band.
   e. Immerse the bulb end of the tube in a beaker of water. If leaks are noted, repeat the procedures.
   f. Support the upright portion of the tube with a clamp on a ring stand. Folded paper under the clamp will protect the thistle tube stem from breakage.
   g. Mark the meniscus level of the molasses in the tube. *Note:* The best results will occur if the mark of the molasses is a short distance up the stem of the thistle tube when the experiment starts.
   h. Measure the level changes after 10 minutes and 30 minutes and record the results in Part B of the laboratory report.
3. Complete Part B of the laboratory report.

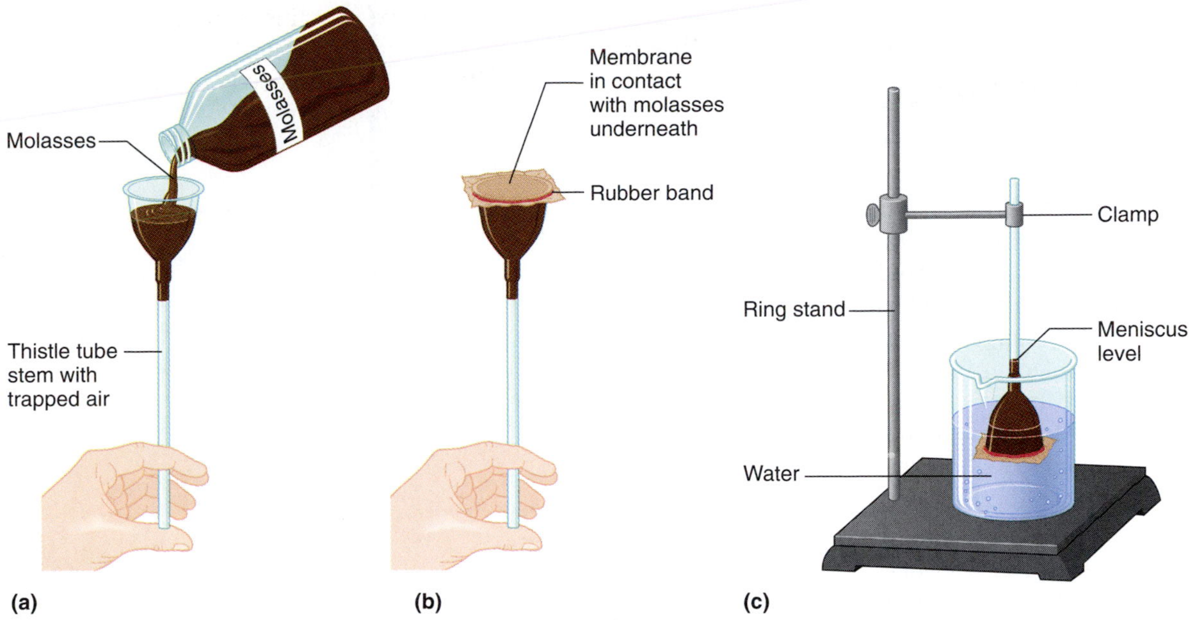

Figure 6.2 (*a*) Fill the bulb of the thistle tube with molasses; (*b*) tightly secure a piece of selectively permeable (semipermeable) membrane over the bulb opening; and (*c*) immerse the bulb in a beaker of water. *Note:* These procedures require the participation of two people.

Alternative Procedure

Eggshell membranes possess selectively permeable properties. To demonstrate osmosis using a natural membrane, soak a fresh chicken egg in vinegar for about 24 hours to remove the shell. Use a spoon to carefully handle the delicate egg. Place the egg in a hypertonic solution (corn syrup) for about 24 hours. Remove the egg, rinse it, and using a laboratory balance, weigh the egg to establish a baseline weight. Place the egg in a hypotonic solution (distilled water). Remove the egg and weigh it every 15 minutes for an elapsed time of 75 minutes. Explain any weight changes that were noted during this experiment.

____

____

____

____

____

EXPLORE

Procedure C—Hypertonic, Hypotonic, and Isotonic Solutions

1. Review the section entitled "Osmosis" in chapter 3 of the textbook.
2. To demonstrate the effect of hypertonic, hypotonic, and isotonic solutions on animal cells, follow these steps:
   a. Place three test tubes in a rack and mark them *tube 1, tube 2,* and *tube 3.* (*Note:* One set of tubes can be used to supply test samples for the entire class.)
   b. Using 10 mL graduated cylinders, add 3 mL of distilled water to tube 1; add 3 mL of 0.9% NaCl to tube 2; and add 3 mL of 3.0% NaCl to tube 3.
   c. Place three drops of fresh, uncoagulated animal blood into each of the tubes, and gently mix the blood with the solutions. Wait 5 minutes.
   d. Using three separate medicine droppers, remove a drop from each tube and place the drops on three separate microscope slides marked *1, 2,* and *3.*
   e. Cover the drops with coverslips and observe the blood cells, using the high power of the microscope.
3. Complete Part C of the laboratory report.

Alternative Procedure

Various substitutes for blood can be used for Procedure C. Onion, cucumber, or cells lining the inside of the cheek represent three possible options.

EXPLORE

Procedure D—Filtration

1. Review the section entitled "Filtration" in chapter 3 of the textbook.
2. To demonstrate filtration, follow these steps:
   a. Place a glass funnel in the ring of a ring stand over an empty beaker. Fold a piece of filter paper in half and then in half again. Open one thickness of the filter paper to form a cone. Wet the cone, and place it in the funnel. The filter paper is used to demonstrate how movement across membranes is limited by the size of the molecules, but it does not represent a working model of biological membranes.
   b. Prepare a mixture of 5 cc (approximately 1 teaspoon) powdered charcoal (or ground black pepper) and equal amounts of 1% glucose solution and 1% starch solution in a beaker. Pour some of the mixture into the funnel until it nearly reaches the top of the filter-paper cone. Care should be taken to prevent the mixture from spilling over the top of the filter paper. Collect the filtrate in the beaker below the funnel (fig. 6.3).
   c. Test some of the filtrate in the beaker for the presence of glucose. To do this, place 1 mL of filtrate in a clean test tube and add 1 mL of Benedict's solution. Place the test tube in a water bath of boiling water for 2 minutes and then allow the liquid to cool slowly. If the color of the solution changes to green, yellow, or red, glucose is present (fig. 6.4).
   d. Test some of the filtrate in the beaker for the presence of starch. To do this, place a few drops of filtrate in a test tube and add one drop of iodine-potassium-iodide solution. If the color of the solution changes to blue-black, starch is present.
   e. Observe any charcoal in the filtrate.
3. Complete Part D of the laboratory report.

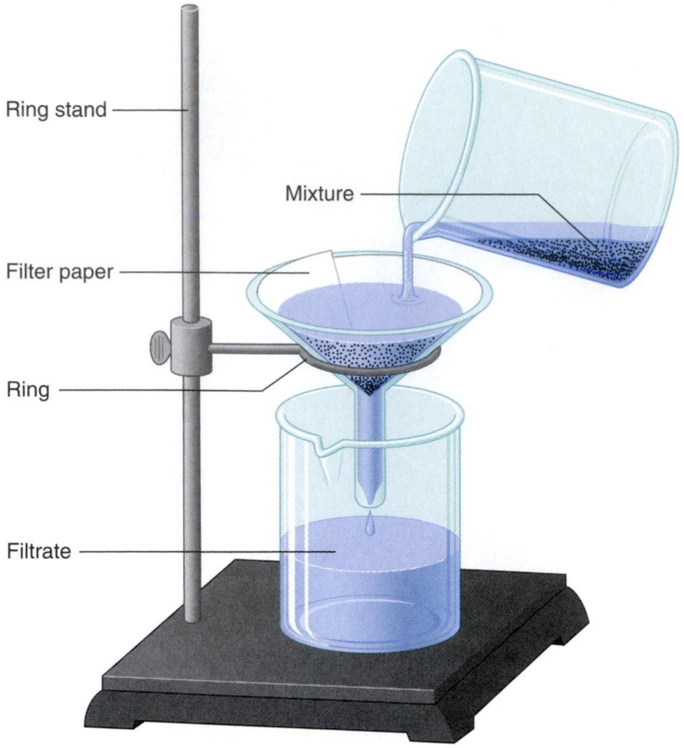

Figure 6.3 Apparatus used to illustrate filtration.

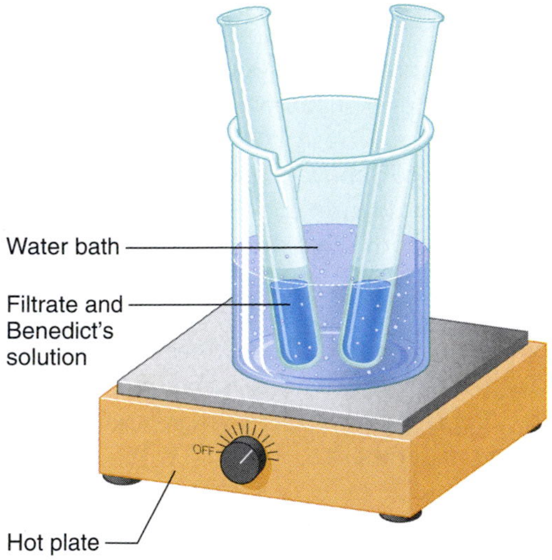

Figure 6.4 Heat the filtrate and Benedict's solution in a boiling water bath for 2 minutes.

Name ______________________

Date ______________________

Section ______________________

The A corresponds to the indicated Learning Outcome(s) listed at the beginning of the laboratory exercise.

Movements Through Membranes

Laboratory Report 6

Part A Assessments

Complete the following:

1. Enter data for changes in the movement of the potassium permanganate. 1

Elapsed Time	Radius of Purple Circle in Millimeters
Initial	______
1 minute	______
2 minutes	______
3 minutes	______
4 minutes	______
5 minutes	______
6 minutes	______
7 minutes	______
8 minutes	______
9 minutes	______
10 minutes	______

2. Prepare a graph that illustrates the diffusion distance of potassium permanganate in 10 minutes. 2

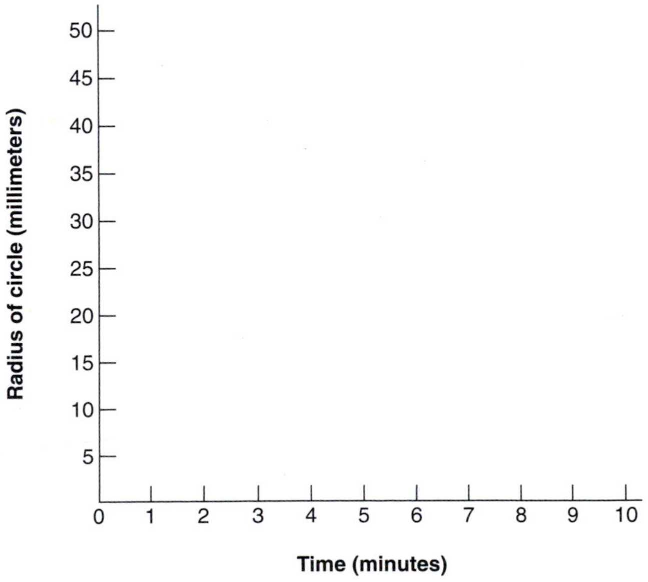

3. Explain your graph. 2 ____

____

4. Define *simple diffusion.* ____

____

Critical Thinking Application

By answering yes or no, indicate which of the following provides an example of simple diffusion. 1

1. A perfume bottle is opened, and soon the odor can be sensed in all parts of the room. ____________
2. A sugar cube is dropped into a cup of hot water, and, without being stirred, all of the liquid becomes sweet tasting. ____________
3. Water molecules move from a faucet through a garden hose when the faucet is turned on. ____________
4. A person blows air molecules into a balloon by forcefully exhaling. ____________
5. A crystal of blue copper sulfate is placed in a test tube of water. The next day, the solid is gone, but the water is evenly colored. ____________

Part B Assessments

Complete the following:

1. What was the change in the level of molasses in 10 minutes? 3 ______________________________
2. What was the change in the level of molasses in 30 minutes? 3 ______________________________
3. How do you explain this change? 3 ____

____

4. Define *osmosis.* ____

____

Critical Thinking Application

By answering yes or no, indicate which of the following involves osmosis. 3

1. A fresh potato is peeled, weighed, and soaked in a strong salt solution. The next day, it is discovered that the potato has lost weight. ____________
2. Garden grass wilts after being exposed to dry chemical fertilizer. ____________
3. Air molecules escape from a punctured tire as a result of high pressure inside. ____________
4. Plant seeds soaked in water swell and become several times as large as before soaking. ____________
5. When the bulb of a thistle tube filled with water is sealed by a selectively permeable membrane and submerged in a beaker of molasses, the water level in the tube falls. ____________

Part C Assessments

Complete the following:

1. In the spaces, sketch a few blood cells from each of the test tubes, and indicate the magnification. 4

Tube 1 (distilled water)	Tube 2 (0.9% NaCl)	Tube 3 (3.0% NaCl)
____________×	____________×	____________×

2. Based on your results, which tube contained a solution hypertonic to the blood cells? ______________________

   Give the reason for your answer. 4 ____

   ____

3. Which tube contained a solution hypotonic to the blood cells? ______________________

   Give the reason for your answer. 4 ____

   ____

4. Which tube contained a solution isotonic to the blood cells? ______________________

   Give the reason for your answer. 4 ____

   ____

Part D Assessments

Complete the following:

1. Which of the substances in the mixture you prepared passed through the filter paper into the filtrate? 5 ______

   ____

2. What evidence do you have for your answer to question 1? 5 ______________________

   ____

   ____

3. What force was responsible for the movement of substances through the filter paper? 5 ________________

____

4. What substances did not pass through the filter paper? 5 ________________________________

____

5. What factor prevented these substances from passing through? 5 ____________________________

____

____

6. Define *filtration*. ____

____

____

Critical Thinking Application

By answering yes or no, indicate which of the following involves filtration. 5

1. Oxygen molecules move into a cell and carbon dioxide molecules leave a cell because of differences in the concentrations of these substances on either side of the cell membrane. ____________
2. Blood pressure forces water molecules from the blood outward through the thin wall of a blood capillary. ____________
3. Urine is forced from the urinary bladder through the tubular urethra by muscular contractions. ____________
4. Air molecules enter the lungs through the airways when air pressure is greater outside these organs than inside. ____________
5. Coffee is made using a coffeemaker (not instant). ____________

Laboratory Exercise 7

Cell Cycle

Materials Needed

Textbook
Models of animal mitosis
Microscope slides of whitefish mitosis (blastula)
Compound light microscope

For Demonstration:

Microscope slide of human chromosomes from leukocytes in mitosis
Oil immersion objective

The cell cycle consists of the series of changes a cell undergoes from the time it is formed until it divides. Typically, a newly formed cell grows to a certain size and then divides to form two new cells (*daughter cells*). This cell division process involves two major steps: (1) division of the cell's nuclear parts, *mitosis* and (2) division of the cell's cytoplasm, *cytokinesis*. Before the cell divides, it must synthesize biochemicals and other contents. This period of preparation is called *interphase*. The extensive period of interphase is divided into three phases. The S phase, when DNA synthesis occurs, is between two gap phases (G_1 and G_2), when cell growth occurs and cytoplasmic organelles duplicate. Eventually, some specialized cells, such as skeletal muscle cells and most nerve cells, cease further cell division, but remain alive.

A special type of cell division, called *meiosis,* occurs in the reproductive system to produce gametes and is not included in this laboratory exercise.

Purpose of the Exercise

To review the stages in the cell cycle and to observe cells in various stages of their life cycles.

LEARNING OUTCOMES

After completing this exercise, you should be able to

1. Describe the cell cycle, and locate structures involved with the process.
2. Identify and sketch the stages in the life cycle of a particular cell.
3. Arrange into a correct sequence a set of models or drawings of cells in various stages of their life cycles.

EXPLORE

Procedure—Cell Cycle

1. Review the section entitled "The Cell Cycle" in chapter 3 of the textbook.
2. As a review activity, study the various stages of the cell's life cycle represented in figure 7.1 and label the structures indicated in figure 7.2.
3. Observe the animal mitosis models, and review the major events in a cell's life cycle represented by each of them. Be sure you can arrange these models in correct sequence if their positions are changed. The acronym IPMAT can help you arrange the correct order of phases in the cell cycle. This includes interphase followed by the four phases of mitosis. Cytokinesis overlaps anaphase and telophase.
4. Complete Part A of Laboratory Report 7.
5. Obtain a slide of the whitefish mitosis (blastula).
   a. Examine the slide using the high-power objective of a microscope. The tissue on this slide was obtained from a developing embryo (blastula) of a fish, and many of the embryonic cells are undergoing mitosis. The chromosomes of these dividing cells are darkly stained (fig. 7.3).
   b. Search the tissue for cells in various stages of cell division. There are several sections on the slide. If you cannot locate different stages in one section, examine the cells of another section because the stages occur randomly on the slide.
   c. Each time you locate a cell in a different stage, sketch it in an appropriate circle in Part B of the laboratory report.

Critical Thinking Application

Which stage (phase) of the cell cycle was the most numerous in the blastula? ____________________

Explain your answer. ____________________

____

____

6. Complete Parts C, D, and E of the laboratory report.

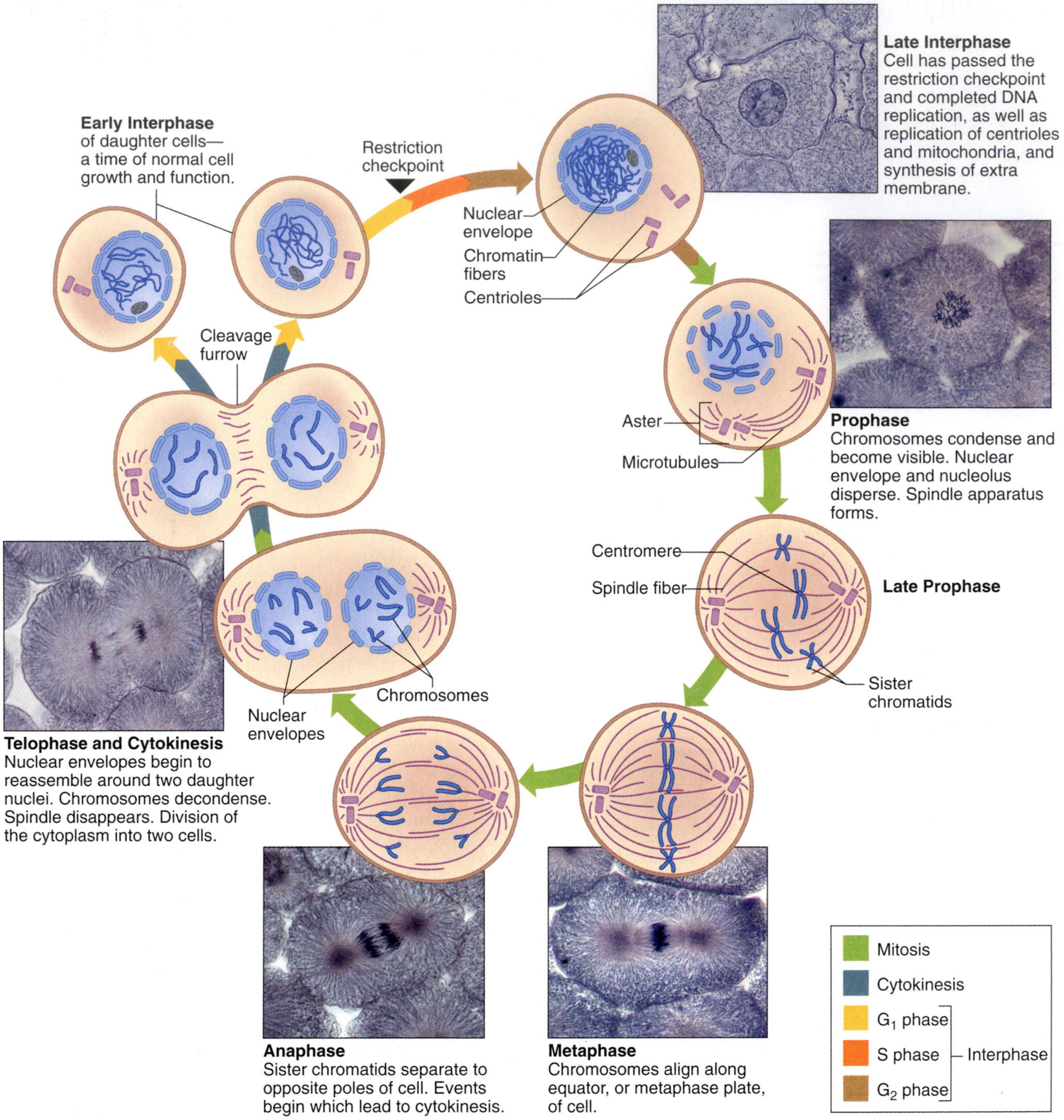

Figure 7.1 The cell cycle: interphase, mitosis, and cytokinesis.

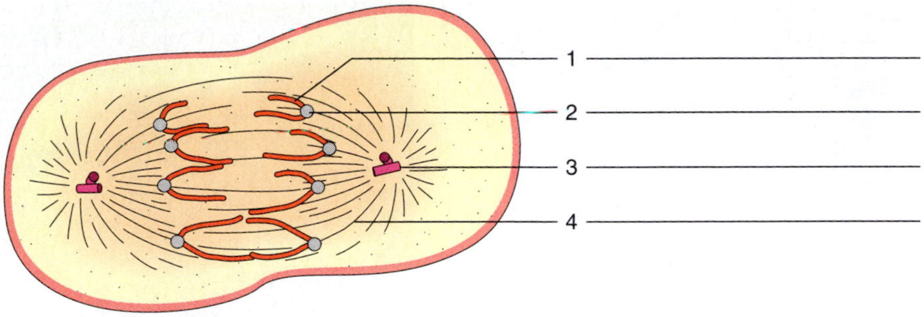

Figure 7.2 Label the structures indicated in the dividing cell. 1

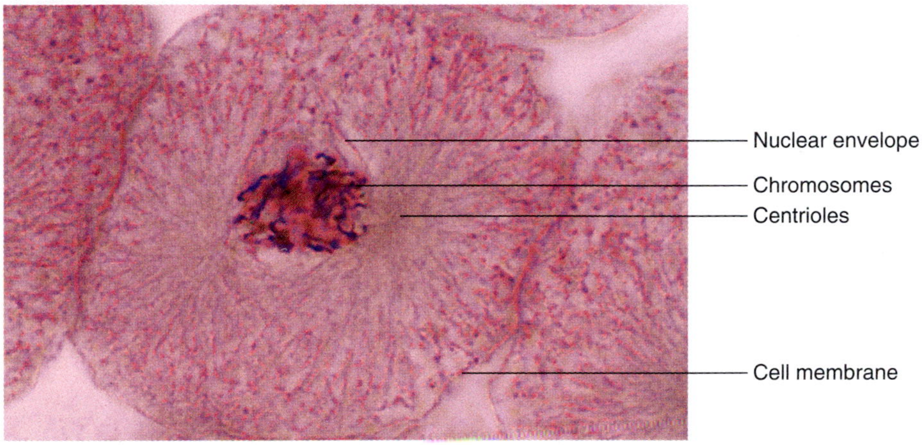

Figure 7.3 Cell in prophase (250× micrograph enlarged to 1,000×).

Demonstration

Using the oil immersion objective of a microscope, see if you can locate some human chromosomes by examining a prepared slide of human chromosomes from leukocytes. The cells on this slide were cultured in a special medium and were stimulated to undergo mitosis. The mitotic process was arrested in metaphase by exposing the cells to a chemical called colchicine, and the cells were caused to swell osmotically. As a result of this treatment, the chromosomes were spread apart. A complement of human chromosomes should be visible when they are magnified about 1,000×. Each chromosome is double-stranded and consists of two chromatids joined by a common centromere (fig. 7.4).

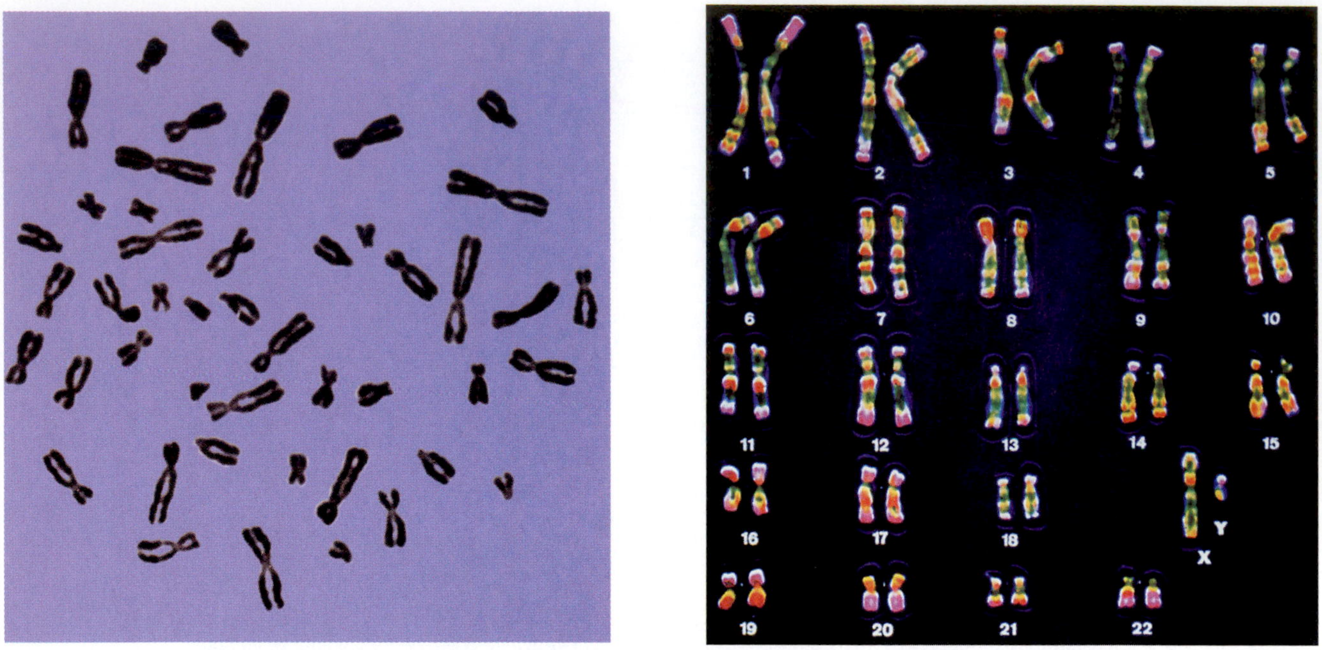

(a)

(b)

Figure 7.4 (*a*) A complement of human chromosomes (2,700×). (*b*) A *karyotype* can be constructed by arranging the homologous chromosome pairs together in a chart. The completed karyotype indicates a normal male.

Name ______________________________

Date ______________________________

Section ______________________________

The [A] corresponds to the indicated Learning Outcome(s) listed at the beginning of the laboratory exercise.

Laboratory Report 7

Cell Cycle

Part A Assessments

Complete the table. [1]

Stage	Major Events Occurring
Interphase (G_1, S, and G_2)	
Mitosis Prophase	
Metaphase	
Anaphase	
Telophase	
Cytokinesis	

Part B Assessments

Sketch an interphase cell and cells in different stages of mitosis to illustrate the whitefish cell's life cycle. Label the major cellular structures represented in the sketches and indicate cytokinesis locations. (The circles represent fields of view through the microscope.)

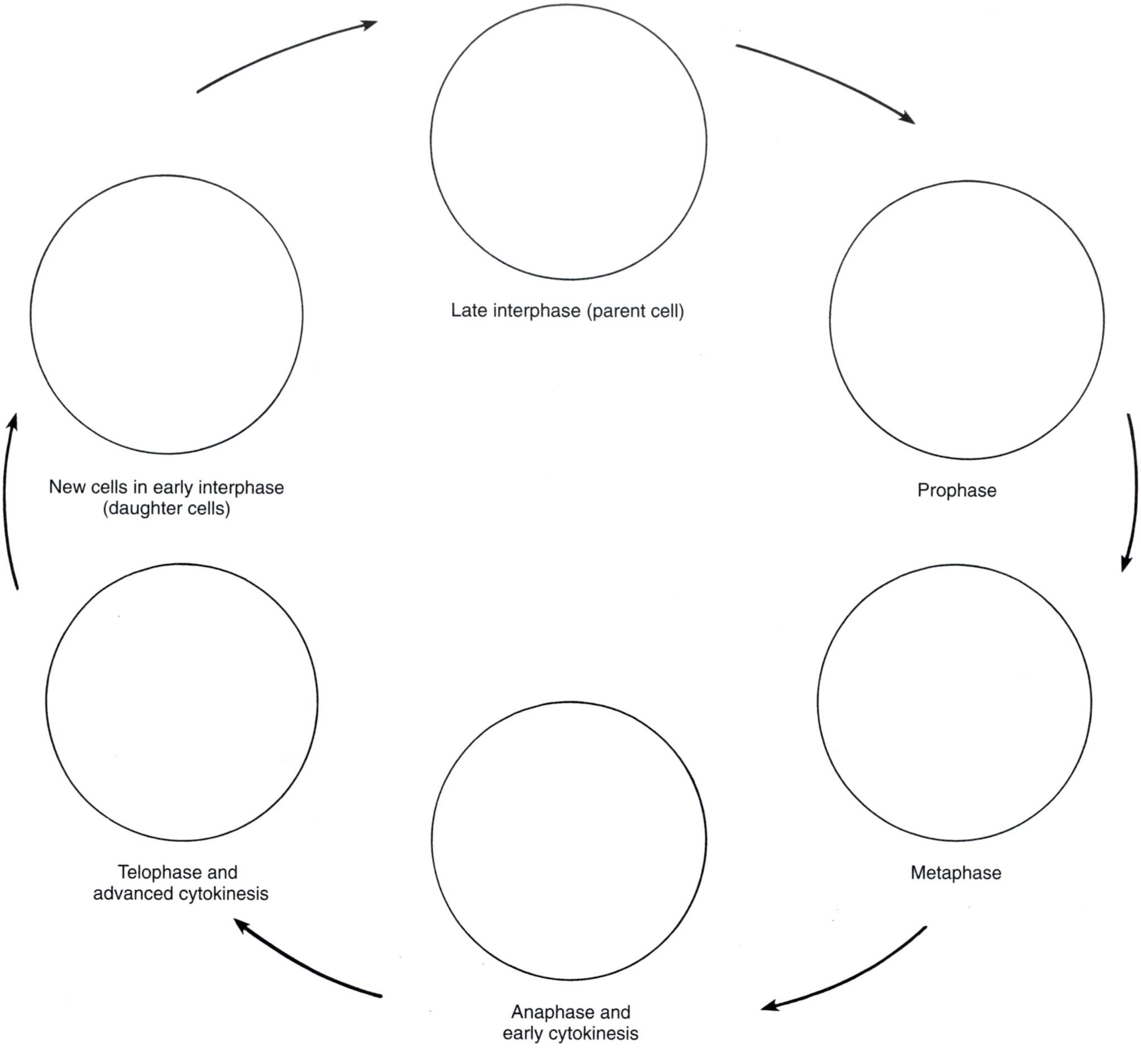

Magnification ____________ ×

Part C Assessments

Complete the following:

1. In what ways are the new cells (daughter cells), which result from a cell cycle, similar? ____________________

____

____

2. How do the new cells slightly differ? ____________________

____

____

3. Distinguish between mitosis and cytokinesis. 1 ____________________

____

____

Part D Assessments

Identify the mitotic stage represented by each of the micrographs in figure 7.5(*a–d*). 3

a. ____________________ c. ____________________

b. ____________________ d. ____________________

Part E Assessments

Identify the structures indicated by numbers in figure 7.5. 1

1. ____________________ 4. ____________________

2. ____________________ 5. ____________________

3. ____________________ 6. ____________________

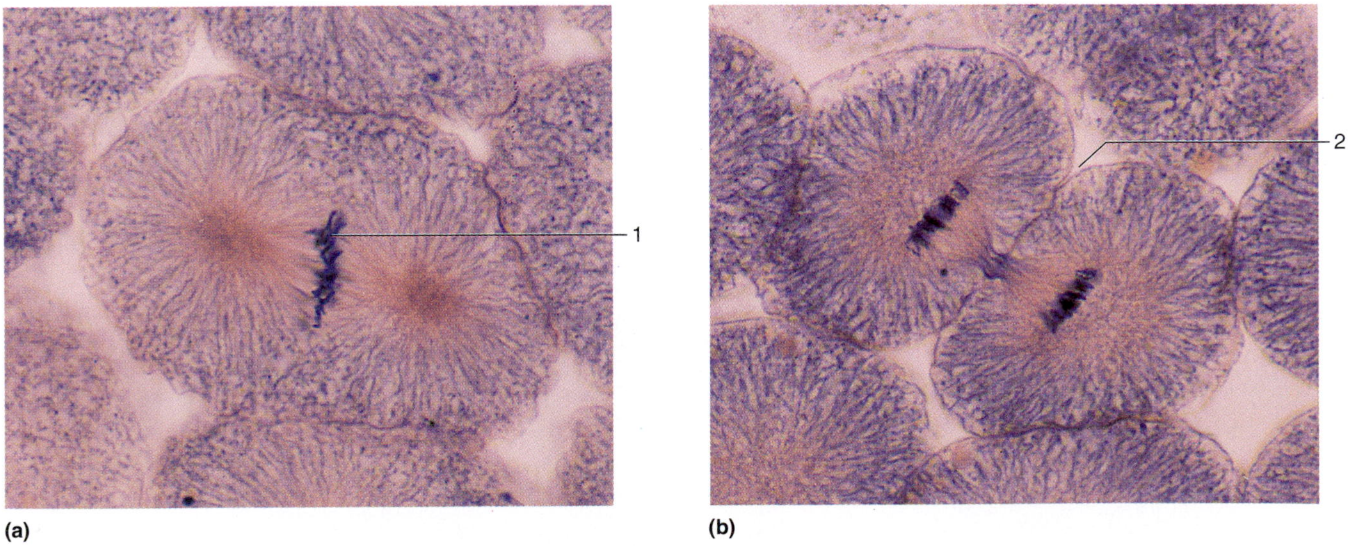

Figure 7.5 Identify the mitotic stage and structures of the cell in each of these micrographs of the whitefish blastula (250×).

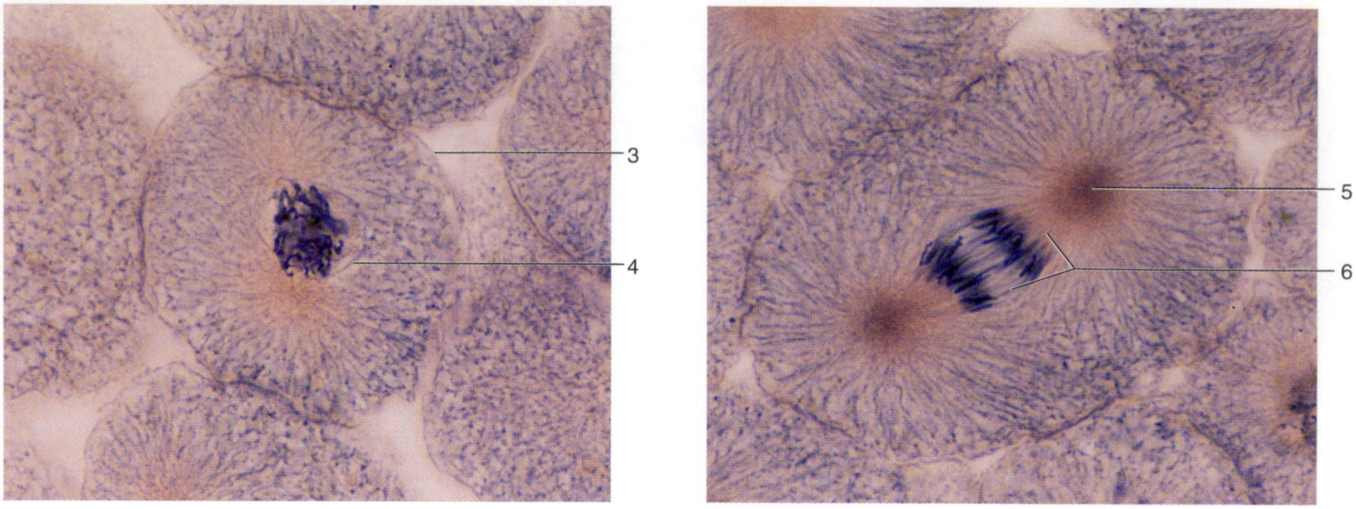

(c) (d)

Figure 7.5 *Continued.*

Laboratory Exercise 8

Epithelial Tissues

Materials Needed

Textbook
Compound light microscope
Prepared slides of the following epithelial tissues:
- Simple squamous epithelium (lung)
- Simple cuboidal epithelium (kidney)
- Simple columnar epithelium (small intestine)
- Pseudostratified (ciliated) columnar epithelium (trachea)
- Stratified squamous epithelium (esophagus)
- Transitional epithelium (urinary bladder)

For Learning Extension:
Colored pencils

A tissue is composed of a layer or group of cells similar in size, shape, and function. Within the human body, there are four major types of tissues: (1) epithelial, which cover the body's external and internal surfaces and most glands; (2) connective, which bind and support parts; (3) muscle, which make movement possible; and (4) nervous, which conduct impulses from one part of the body to another and help to control and coordinate body activities.

Epithelial tissues are tightly packed single (simple) to multiple (stratified) layers of cells that provide protective barriers. The underside of this tissue layer contains an acellular basement membrane layer composed of adhesives and collagen to which the epithelial cells anchor to underlying connective tissue. The cells readily divide and lack blood vessels. Epithelial cells always have a free (apical) surface exposed to the outside or to an open space internally and a basal surface that attaches to the basement membrane. Epithelial cell functions include protection, filtration, secretion, and absorption. Many shapes of the cells exist that are used to name and identify the variations. Many of the prepared slides contain more than the tissue to be studied, so be certain that your view matches the correct tissue. Also be aware that stained colors of all tissues might vary.

Purpose of the Exercise

To review the characteristics of epithelial tissues and to observe examples.

LEARNING OUTCOMES

After completing this exercise, you should be able to

1. Differentiate the special characteristics of each type of epithelial tissue.
2. Sketch and label the characteristics of epithelial tissues that you were able to observe.
3. Indicate a location and function of each type of epithelial tissue.
4. Identify the major types of epithelial tissues on microscope slides.

EXPLORE

Procedure—Epithelial Tissues

1. Review the section entitled "Epithelial Tissues" in chapter 5 of the textbook.
2. Complete Part A of Laboratory Report 8.
3. Use the microscope to observe the prepared slides of types of epithelial tissues. As you observe each tissue, look for its special distinguishing features as described in the textbook, such as cell size, shape, and arrangement. Compare your prepared slides of epithelial tissues to the micrographs in figure 8.1. As you observe each type of epithelial tissue, prepare a labeled sketch of a representative portion of the tissue in Part B of the laboratory report.
4. Complete Part B of the laboratory report.
5. Test your ability to recognize each type of epithelial tissue. To do this, have a laboratory partner select one of the prepared slides, cover its label, and focus the microscope on the tissue. Then see if you can correctly identify the tissue. 4

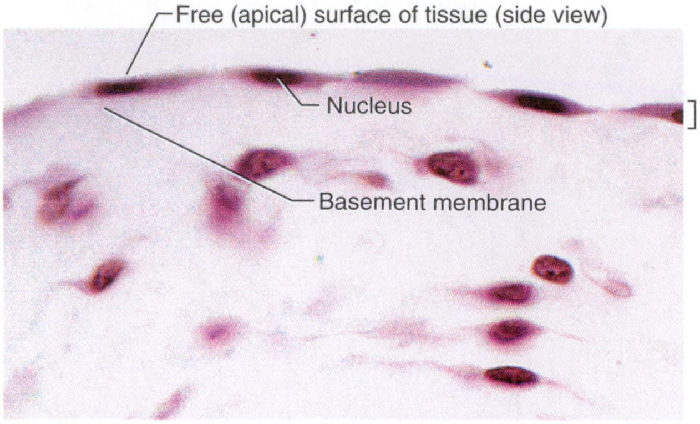

(a) Simple squamous epithelium (side view) (from lung)

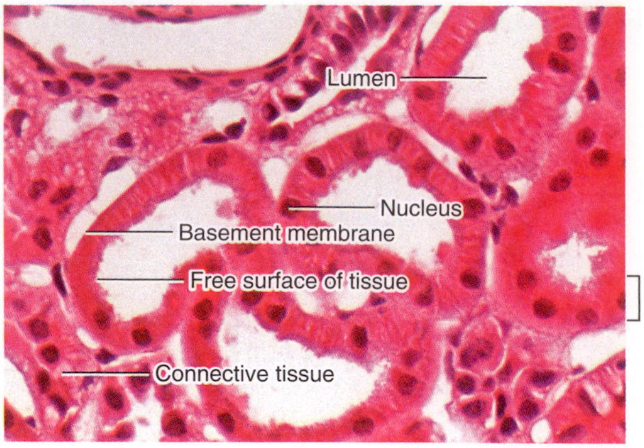

(b) Simple cuboidal epithelium (from kidney)

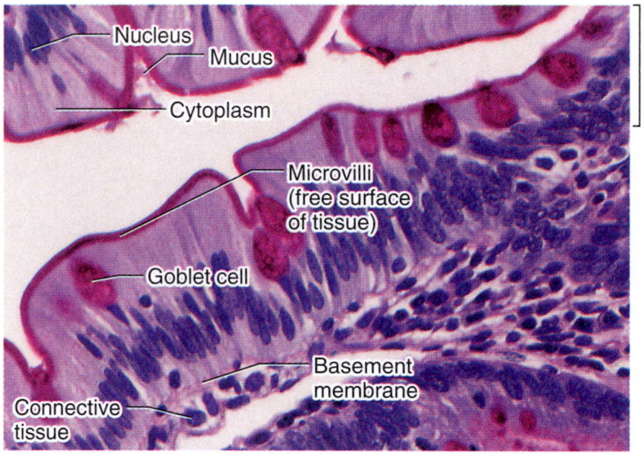

(c) Simple columnar epithelium (from intestine)

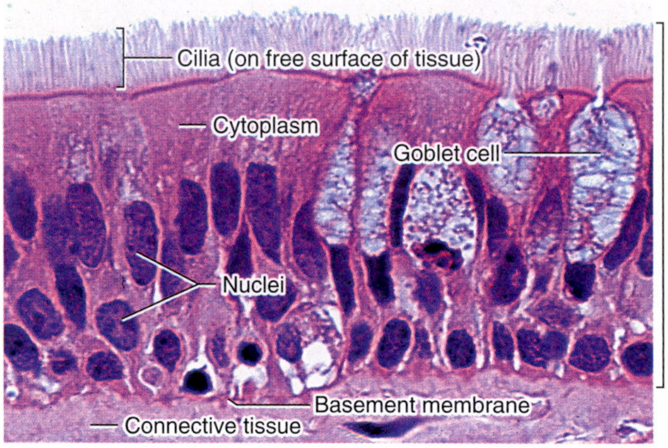

(d) Pseudostratified columnar epithelium with cilia (from trachea)

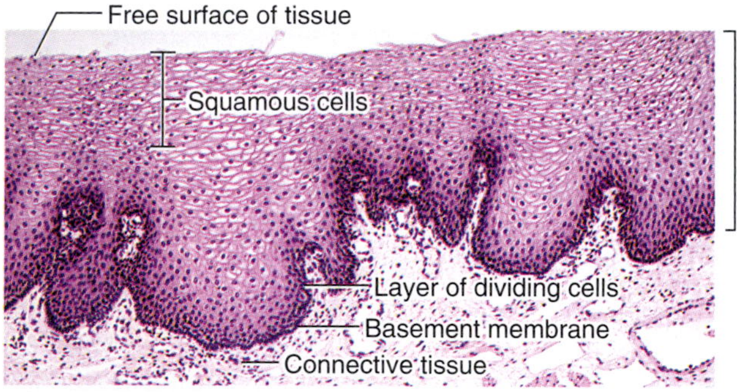

(e) Stratified squamous epithelium (nonkeratinized) (from esophagus)

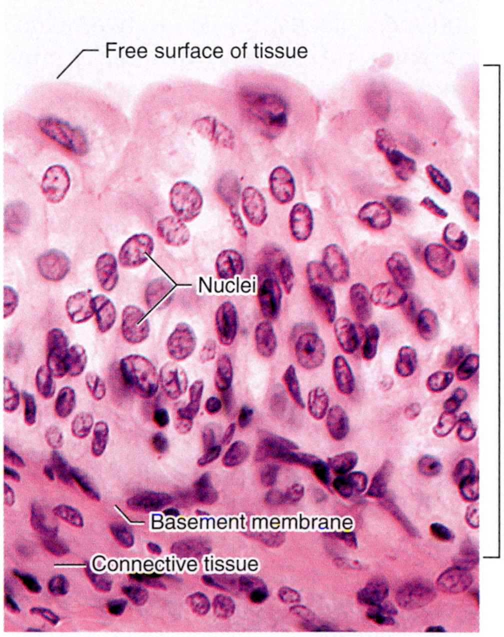

(f) Transitional epithelium (unstretched) (from urinary bladder)

Figure 8.1 Micrographs of epithelial tissues. *Note:* The brackets to the right of each micrograph indicate the tissue.

Name ______________________________

Date ______________________________

Section ______________________________

The ⚠ corresponds to the indicated Learning Outcome(s) listd at the beginning of the laboratory exercise.

Epithelial Tissues

Laboratory Report 8

Part A Assessments

Match the tissues in column A with the characteristics in column B. Place the letter of your choice in the space provided. (Some answers may be used more than once.) ⚠1

Column A

a. Simple columnar epithelium
b. Simple cuboidal epithelium
c. Simple squamous epithelium
d. Pseudostratified columnar epithelium
e. Stratified squamous epithelium
f. Transitional epithelium

Column B

_______ **1.** Consists of several layers of cube-shaped, elongated, and irregular cells
_______ **2.** Commonly possesses cilia that move dust and mucus out of the airways
_______ **3.** Single layer of flattened cells
_______ **4.** Nuclei located at different levels within cells
_______ **5.** Forms walls of capillaries and air sacs of lungs
_______ **6.** Forms linings of trachea and bronchi
_______ **7.** Younger cells cuboidal, older cells flattened
_______ **8.** Forms inner lining of urinary bladder
_______ **9.** Lines kidney tubules and ducts of salivary glands
_______ **10.** Forms lining of stomach and intestines
_______ **11.** Nuclei located near basement membrane
_______ **12.** Forms lining of oral cavity, anal canal, and vagina

Part B Assessments

In the space that follows, sketch a few cells of each type of epithelium you observed. For each sketch, label the major characteristics, indicate the magnification used, write an example of a location in the body, and provide a function. 1 2 3

Simple squamous epithelium (____×) Location: ____________________ Function: ____________________	Simple cuboidal epithelium (____×) Location: ____________________ Function: ____________________
Simple columnar epithelium (____×) Location: ____________________ Function: ____________________	Pseudostratified columnar epithelium with cilia (____×) Location: ____________________ Function: ____________________
Stratified squamous epithelium (____×) Location: ____________________ Function: ____________________	Transitional epithelium (____×) Location: ____________________ Function: ____________________

Critical Thinking Application

As a result of your observations of epithelial tissues, which one(s) provide(s) the best protection? Explain your answer.

____

____

____

____

Learning Extension

Use colored pencils to differentiate various cellular structures in Part B. Select a different color for a nucleus, cytoplasm, cell membrane, basement membrane, goblet cell, and cilia whenever visible.

Laboratory Exercise 9

Connective Tissues

Materials Needed

Textbook
Compound light microscope
Prepared slides of the following:
- Areolar tissue
- Adipose tissue
- Reticular connective tissue
- Dense connective tissue (regular type)
- Elastic connective tissue
- Hyaline cartilage
- Elastic cartilage
- Fibrocartilage
- Bone (compact, ground, cross section)
- Blood (human smear)

For Learning Extension:
Colored pencils

Connective tissues contain a variety of cell types and occur in all regions of the body. They bind structures together, provide support and protection, fill spaces, store fat, and produce blood cells.

Connective tissue cells are often widely scattered in an abundance of extracellular matrix. The matrix consists of fibers and a ground substance of various densities and consistencies. Many of the prepared slides contain more than the tissue to be studied, so be certain that your view matches the correct tissue. Additional study of bone and blood will be found in Laboratory Exercises 12 and 37.

Purpose of the Exercise

To review the characteristics of connective tissues and to observe examples of the major types.

LEARNING OUTCOMES

After completing this exercise, you should be able to

1. Differentiate the special characteristics of each of the major types of connective tissue.
2. Sketch and label the characteristics of connective tissues that you were able to observe.
3. Indicate a location and function of each type of connective tissue.
4. Identify the major types of connective tissues on microscope slides.

EXPLORE

Procedure—Connective Tissues

1. Review the section entitled "Connective Tissues" in chapter 5 of the textbook.
2. Complete Part A of Laboratory Report 9.
3. Use a microscope to observe the prepared slides of various connective tissues. As you observe each tissue, look for its special distinguishing features as described in the textbook. Compare your prepared slides of connective tissues to the micrographs in figure 9.1. As you observe each type of connective tissue, prepare a labeled sketch of a representative portion of the tissue in Part B of the laboratory report.
4. Complete Part B of the laboratory report.
5. Test your ability to recognize each of these connective tissues by having a laboratory partner select a slide, cover its label, and focus the microscope on this tissue. Then see if you correctly identify the tissue. 4

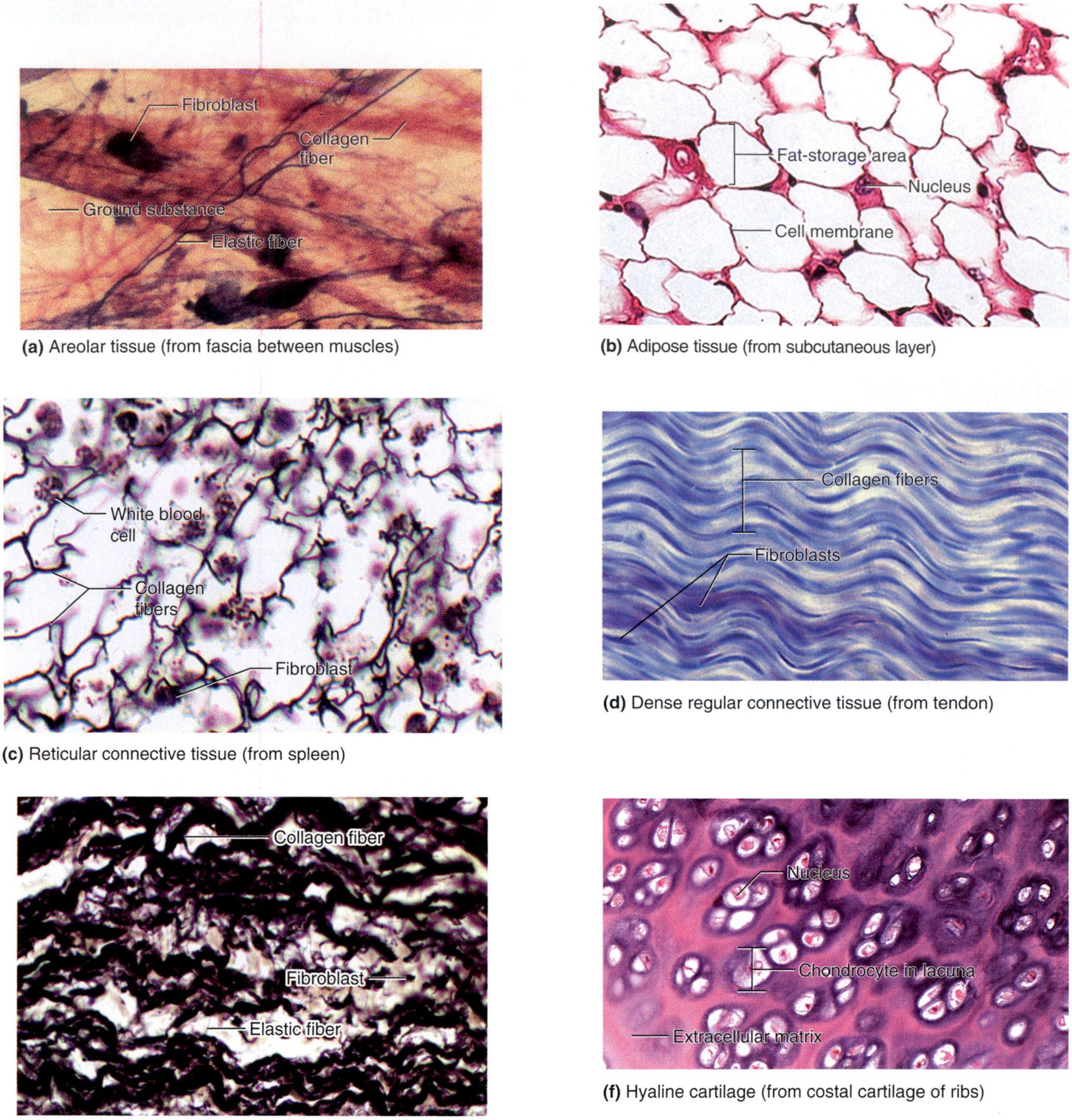

(a) Areolar tissue (from fascia between muscles)

(b) Adipose tissue (from subcutaneous layer)

(c) Reticular connective tissue (from spleen)

(d) Dense regular connective tissue (from tendon)

(e) Elastic connective tissue (from artery wall)

(f) Hyaline cartilage (from costal cartilage of ribs)

Figure 9.1 Micrographs of connective tissues.

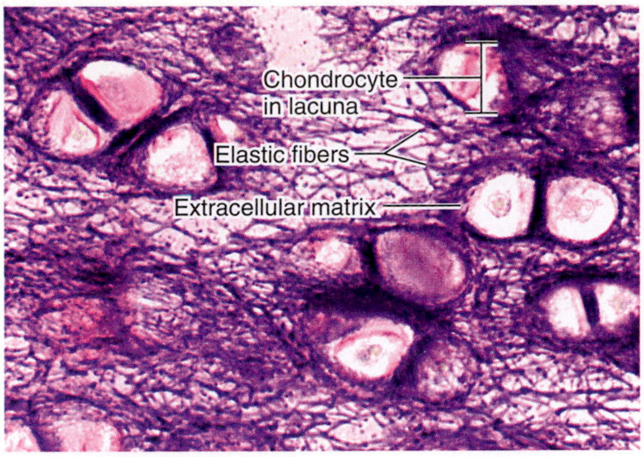

(g) Elastic cartilage (from ear)

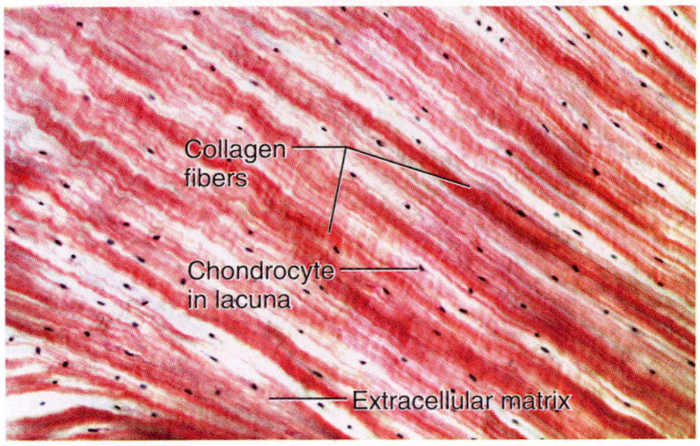

(h) Fibrocartilage (from intervertebral discs)

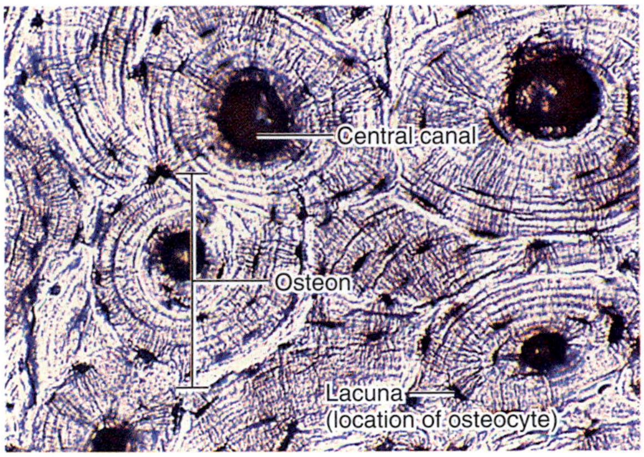

(i) Compact bone (from skeleton)

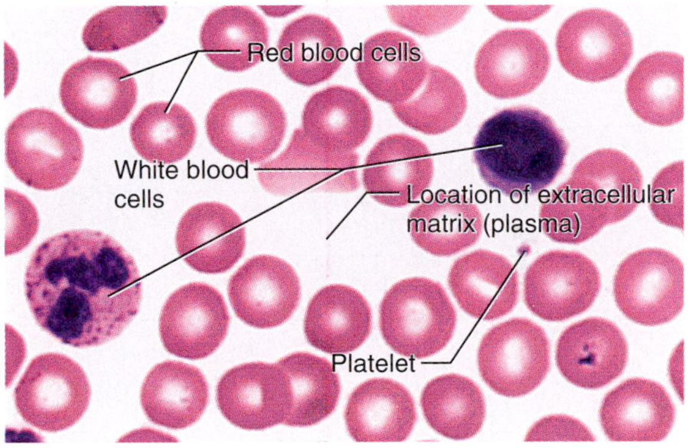

(j) Blood

Figure 9.1 *Continued.*

Notes

Name ______________________________

Date ______________________________

Section ______________________________

The [A] corresponds to the indicated Learning Outcome(s) listed at the beginning of the laboratory exercise.

Laboratory Report 9

Connective Tissues

Part A Assessments

Match the tissues in column A with the characteristics in column B. Place the letter of your choice in the space provided. (Some answers may be used more than once.) [1] [3]

Column A

- **a.** Adipose tissue
- **b.** Areolar tissue
- **c.** Blood
- **d.** Bone (compact)
- **e.** Dense connective tissue (regular)
- **f.** Elastic cartilage
- **g.** Elastic connective tissue
- **h.** Fibrocartilage
- **i.** Hyaline cartilage
- **j.** Reticular connective tissue

Column B

_______ **1.** Forms framework of outer ear

_______ **2.** Functions as heat insulator beneath skin

_______ **3.** Contains large amounts of fluid and lacks fibers

_______ **4.** Cells arranged around central canal

_______ **5.** Binds skin to underlying organs

_______ **6.** Main tissue of tendons and ligaments

_______ **7.** Provides stored energy supply in fat vacuoles

_______ **8.** Forms the flexible part of the nasal septum

_______ **9.** Pads between vertebrae that are shock absorbers

_______ **10.** Matrix contains collagen fibers and mineral salts

_______ **11.** Occurs in ligament attachments between vertebrae and artery walls

_______ **12.** Forms supporting tissue in walls of liver and spleen

Part B Assessments

In the space that follows, sketch a small section of each of the types of connective tissues you observed. For each sketch, label the major characteristics, indicate the magnification used, write an example of a location in the body, and provide a function. 1 2 3

Areolar tissue (____×) Location: ____________ Function: ____________	Adipose tissue (____×) Location: ____________ Function: ____________
Reticular connective tissue (____×) Location: ____________ Function: ____________	Dense connective tissue (regular type) (____×) Location: ____________ Function: ____________
Elastic connective tissue (____×) Location: ____________ Function: ____________	Hyaline cartilage (____×) Location: ____________ Function: ____________

Elastic cartilage (____×)

Location: ______________________________

Function: ______________________________

Fibrocartilage (____×)

Location: ______________________________

Function: ______________________________

Bone (compact) (____×)

Location: ______________________________

Function: ______________________________

Blood (____×)

Location: ______________________________

Function: ______________________________

Critical Thinking Application

Abdominal impact injuries often involve the spleen. Explain the structural tissue characteristics that make the spleen so vulnerable to serious injury.

____

____

____

____

Learning Extension

Use colored pencils to differentiate various cellular structures in Part B. Select a different color for the cells, fibers, and ground substance whenever visible.

Notes

Laboratory Exercise 10

Muscle and Nervous Tissues

Materials Needed

Textbook
Compound light microscope
Prepared slides of the following:
- Skeletal muscle tissue
- Smooth muscle tissue
- Cardiac muscle tissue
- Nervous tissue (spinal cord smear and/or cerebellum)

For Learning Extension:
Colored pencils

Muscle tissues are characterized by the presence of elongated cells or muscle fibers that can contract in response to specific stimuli. As they shorten, these fibers pull at their attached ends and cause body parts to move. The three types of muscle tissues are skeletal, smooth, and cardiac.

Nervous tissues occur in the brain, spinal cord, and peripheral nerves. They consist of neurons (nerve cells), the impulse-conducting cells of the nervous system, and neuroglia, which perform supportive and protective functions.

Purpose of the Exercise

To review the characteristics of muscle and nervous tissues and to observe examples of these tissues.

LEARNING OUTCOMES

After completing this exercise, you should be able to

1. Differentiate the special characteristics of each type of muscle tissue and nervous tissue.
2. Sketch and label the characteristics of muscle tissues and nervous tissues that you were able to observe.
3. Indicate a location and function of each type of muscle tissue and nervous tissue.
4. Identify three types of muscle tissues and nervous tissues on microscope slides.

EXPLORE

Procedure—Muscle and Nervous Tissues

1. Review the sections entitled "Muscle Tissues" and "Nervous Tissues" in chapter 5 of the textbook.
2. Complete Part A of Laboratory Report 10.
3. Using the microscope, observe each of the types of muscle tissues on the prepared slides. Look for the special features of each type, as described in the textbook. Compare your prepared slides of muscle tissues to the micrographs in figure 10.1 and the muscle tissue characteristics in table 10.1. As you observe each type of muscle tissue, prepare a labeled sketch of a representative portion of the tissue in Part B of the laboratory report.

Table 10.1 Muscle Tissue Characteristics

Characteristic	Skeletal Muscle	Smooth Muscle	Cardiac Muscle
Appearance of cells	Unbranched and relatively parallel	Spindle-shaped	Branched and connected in complex networks
Striations	Present and obvious	Absent	Present but faint
Nucleus	Multinucleated	Uninucleated	Uninucleated (usually)
Intercalated discs	Absent	Absent	Present
Control	Voluntary	Involuntary	Involuntary

4. Observe the prepared slide of nervous tissue and identify neurons (nerve cells), neuron cellular processes, and neuroglia. Compare your prepared slide of nervous tissue to the micrograph in figure 10.1. Prepare a labeled sketch of a representative portion of the tissue in Part B of the laboratory report.
5. Complete Part B of the laboratory report.
6. Test your ability to recognize each of these muscle and nervous tissues by having your laboratory partner select a slide, cover its label, and focus the microscope on this tissue. Then see if you correctly identify the tissue.

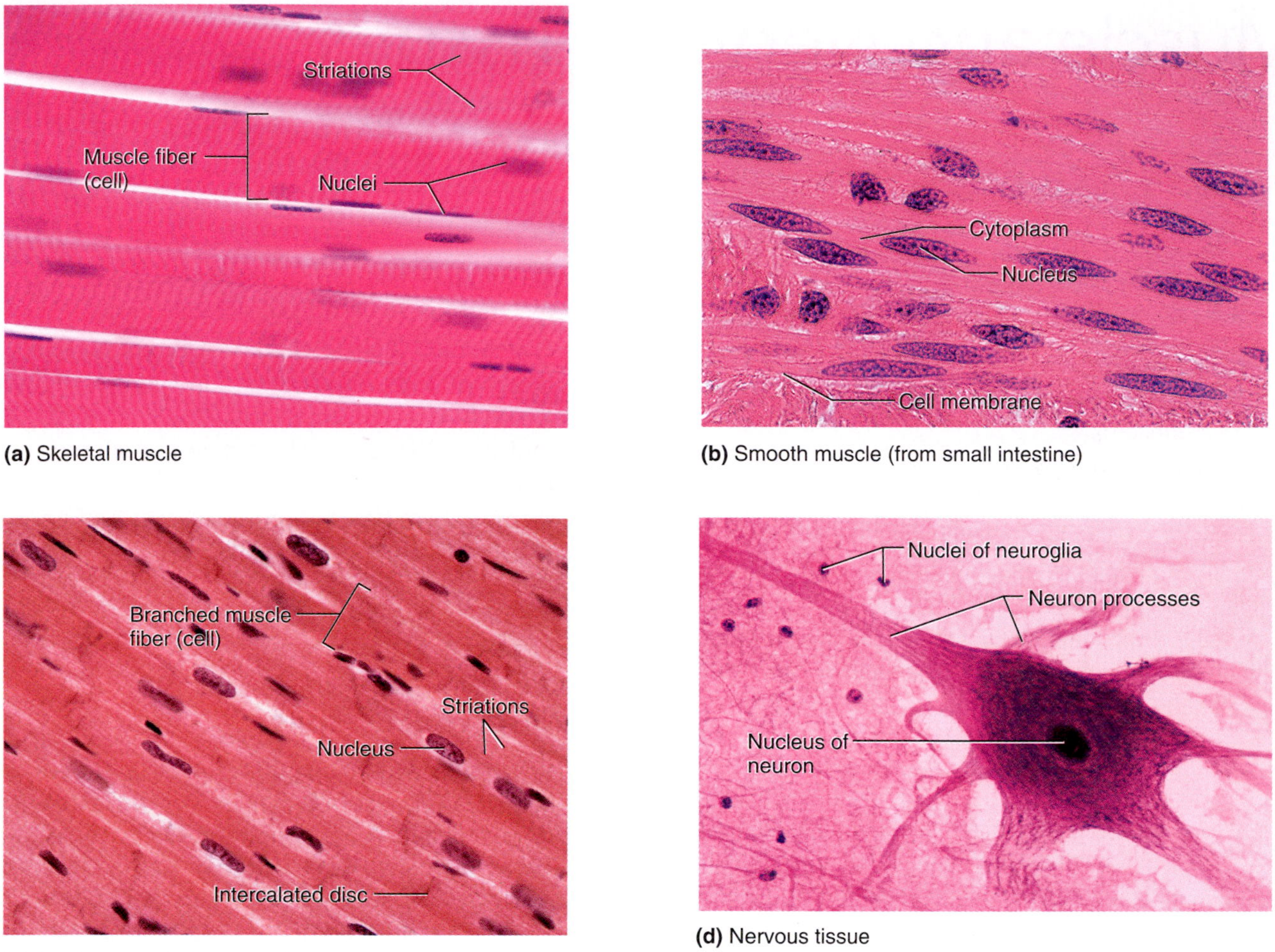

(a) Skeletal muscle

(b) Smooth muscle (from small intestine)

(c) Cardiac muscle (from heart)

(d) Nervous tissue

Figure 10.1 Micrographs of muscle and nervous tissues.

Name ______________________________

Date ______________________________

Section ______________________________

The [A] corresponds to the indicated Learning Outcome(s) listed at the beginning of the laboratory exercise.

Muscle and Nervous Tissues

10 Laboratory Report

Part A Assessments

Match the tissues in column A with the characteristics in column B. Place the letter of your choice in the space provided. (Some answers may be used more than once.) [1] [3]

Column A	Column B
a. Cardiac muscle	_______ **1.** Coordinates, regulates, and integrates body functions
b. Nervous tissue	_______ **2.** Contains intercalated discs
c. Skeletal muscle	_______ **3.** Muscle that lacks striations
d. Smooth muscle	_______ **4.** Striated and involuntary
	_______ **5.** Striated and voluntary
	_______ **6.** Contains neurons and neuroglia
	_______ **7.** Muscle attached to bones
	_______ **8.** Muscle that composes heart
	_______ **9.** Moves food through the digestive tract
	_______ **10.** Transmits impulses along cellular processes

Part B Assessments

In the space that follows, sketch a few cells or fibers of each of the three types of muscle tissues and of nervous tissue as they appear through the microscope. For each sketch, label the major structures of the cells or fibers, indicate the magnification used, write an example of a location in the body, and provide a function. 1 2 3

Skeletal muscle tissue (____×) Location: ______________________ Function: ______________________	Smooth muscle tissue (____×) Location: ______________________ Function: ______________________
Cardiac muscle tissue (____×) Location: ______________________ Function: ______________________	Nervous tissue (____×) Location: ______________________ Function: ______________________

Learning Extension

Use colored pencils to differentiate various cellular structures in Part B.

Laboratory Exercise 11

Integumentary System

Materials Needed

Textbook
Skin model
Hand magnifier or dissecting microscope
Forceps
Microscope slide and coverslip
Compound light microscope
Prepared microscope slide of human scalp or axilla
Prepared slide of dark (heavily pigmented) human skin
Prepared slide of thick skin (plantar or palmar)

For Learning Extension:
Tattoo slide

The integumentary system includes the skin, hair, nails, sebaceous glands, and sweat glands. These structures provide a protective covering for deeper tissues, aid in regulating body temperature, retard water loss, house sensory receptors, synthesize various chemicals, and excrete small quantities of wastes.

The skin consists of two distinct layers. The outer layer, the epidermis, consists of stratified squamous epithelium. The inner layer, the dermis, consists of a thicker layer of dense connective tissue. Beneath the dermis is the subcutaneous layer (not considered a true layer of the skin), composed of adipose and areolar connective tissues.

Purpose of the Exercise

To observe the structures and tissues of the integumentary system and to review the functions of these parts.

LEARNING OUTCOMES

After completing this exercise, you should be able to

1. Locate the structures of the integumentary system.
2. Describe the major functions of these structures.
3. Distinguish the locations and tissues among epidermis, dermis, and the subcutaneous layer.
4. Sketch the layers of the skin and associated structures observed on the prepared slide.

EXPLORE

Procedure—Integumentary System

1. Review the sections entitled "Skin and Its Tissues" and "Accessory Structures of the Skin" in chapter 6 of the textbook.
2. As a review activity, label figures 11.1, 11.2, and 11.3. Locate as many of these structures as possible on a skin model.
3. Complete Part A of Laboratory Report 11.
4. Use the hand magnifier or dissecting microscope and proceed as follows:
   a. Observe the skin, hair, and nails on your hand.
   b. Compare the type and distribution of hairs on the front and back of your forearm.
5. Use low-power magnification of the compound light microscope and proceed as follows:
   a. Pull out a single hair with forceps and mount it on a microscope slide under a coverslip.
   b. Observe the root and shaft of the hair and note the scalelike parts that make up the shaft.
6. Complete Part B of the laboratory report.
7. As vertical sections of human skin are observed, remember that the lenses of the microscope invert and reverse images. It is important to orient the position of the epidermis, dermis, and subcutaneous (hypodermis) layers using scan magnification before continuing with additional observations. Compare all of your skin observations to figure 11.4. Use low-power magnification of the compound light microscope and proceed as follows:
   a. Observe the prepared slide of human scalp or axilla.
   b. Locate the epidermis, dermis, and subcutaneous layer; a hair follicle; an arrector pili muscle; a sebaceous gland; and a sweat gland.
   c. Focus on the epidermis with high power and locate the stratum corneum, stratum granulosum, stratum spinosum, and stratum basale. Note how the shapes of the cells in these layers differ.
   d. Observe the dense connective tissue (irregular type) that makes up the bulk of the dermis.
   e. Observe the adipose tissue that composes most of the subcutaneous layer.

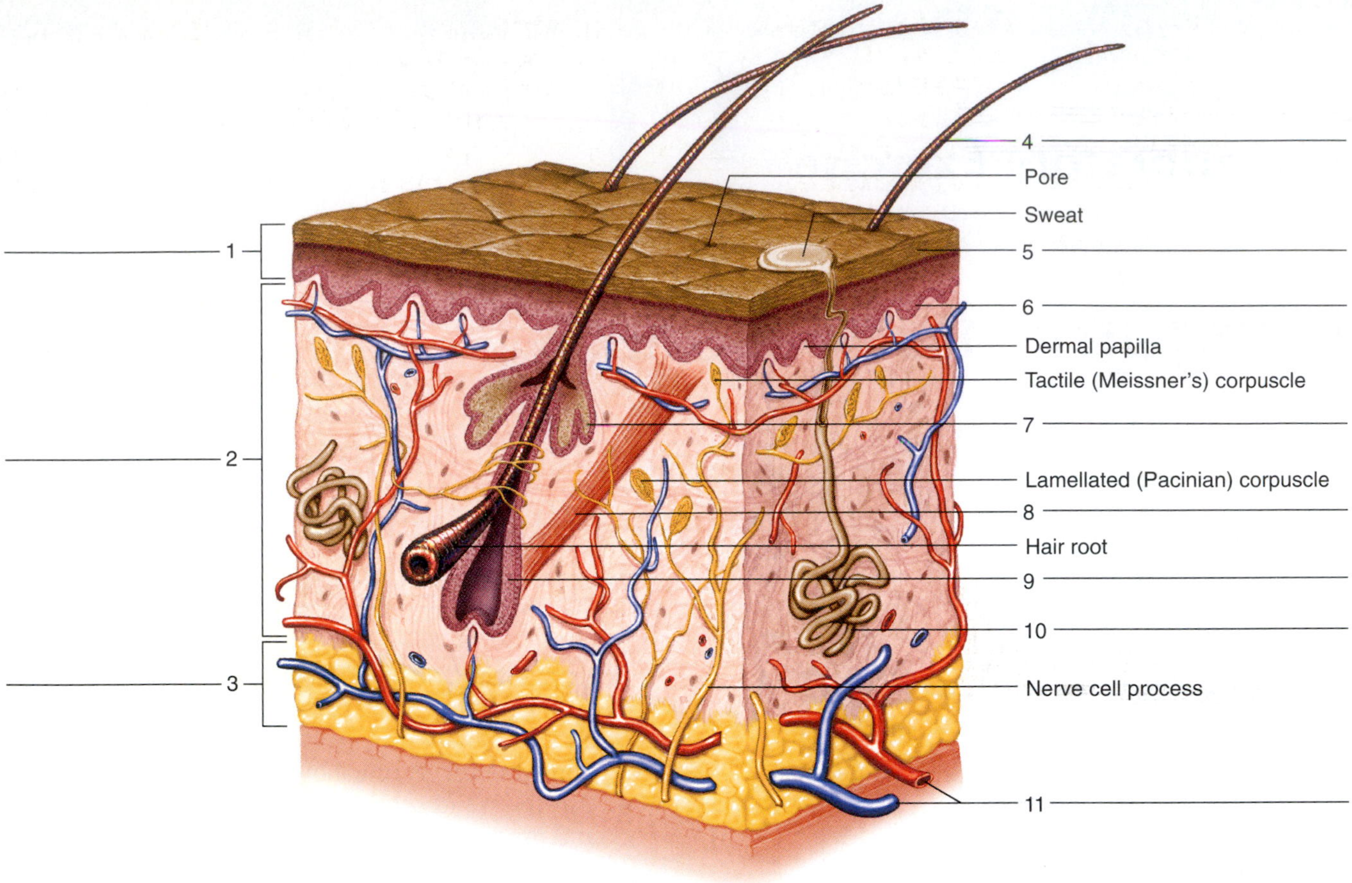

Figure 11.1 Label this vertical section of the skin and subcutaneous layer. 1

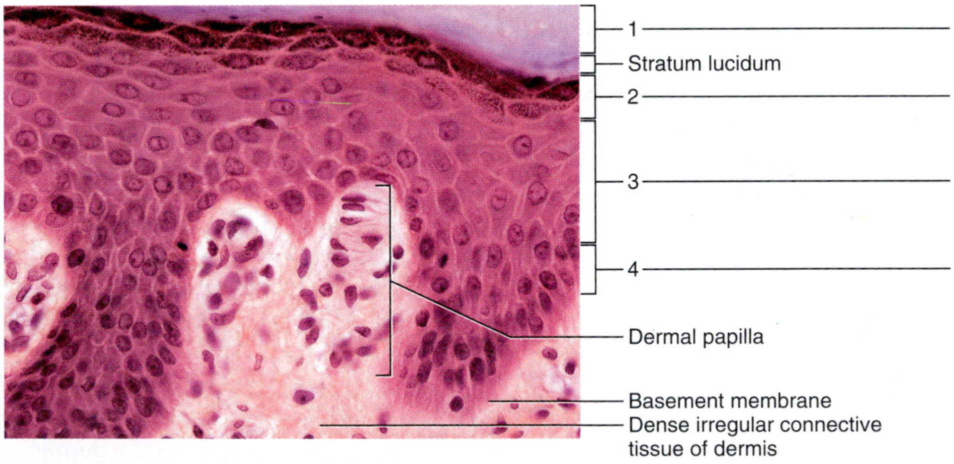

Figure 11.2 Label the epidermal layers in this section of thick skin from a fingertip (50×). 3

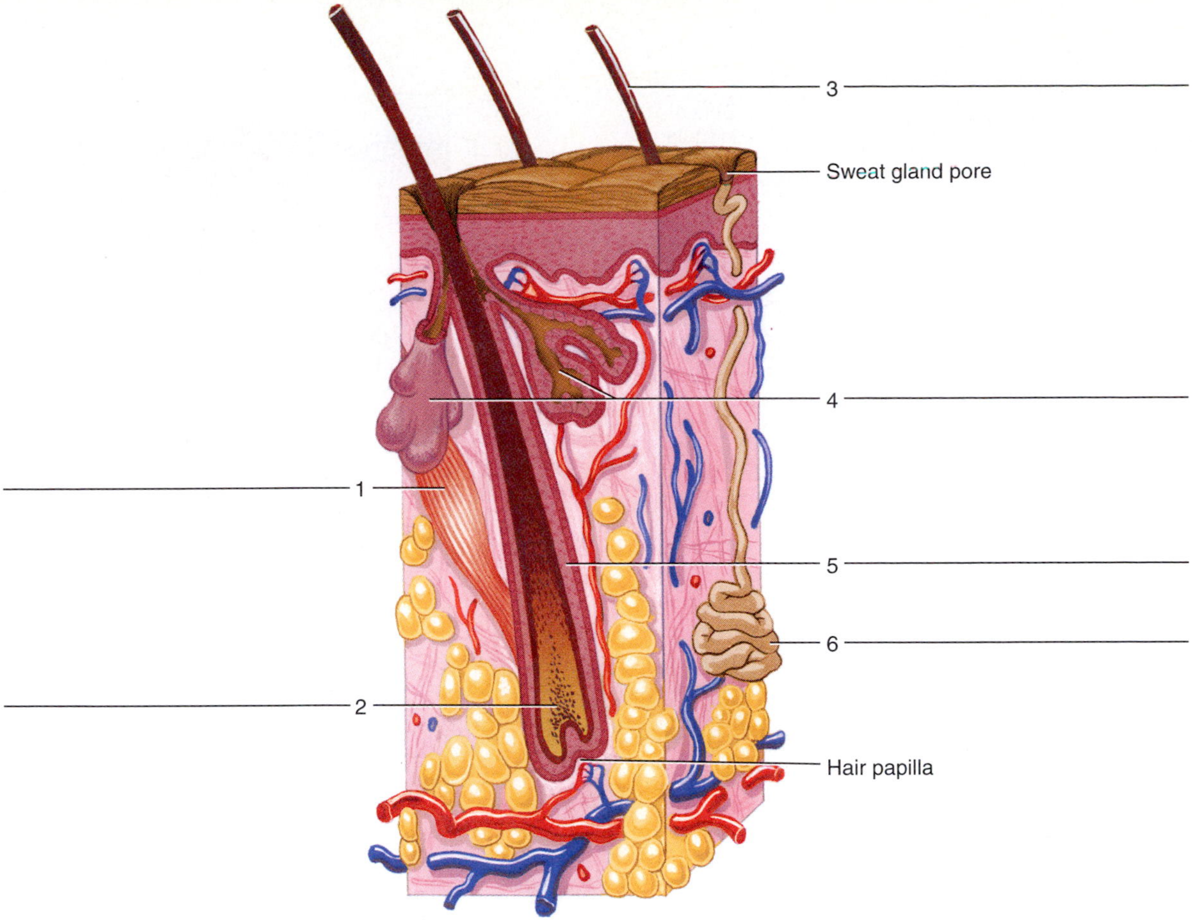

Figure 11.3 Label the features associated with this hair follicle. 1

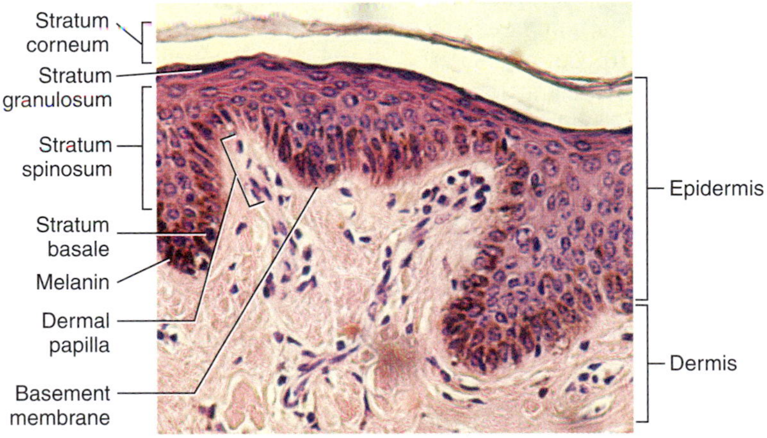

(a)

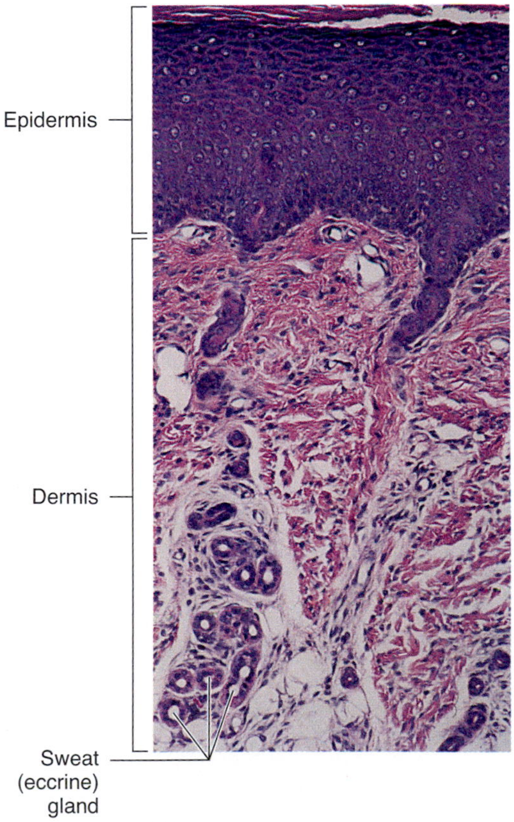

(b)

Figure 11.4 Features of human skin are indicated in these micrographs. Magnifications: (*a*) 290×; (*b*) 30× micrograph enlarged to 280×; (*c*) 45×; (*d*) 110×.

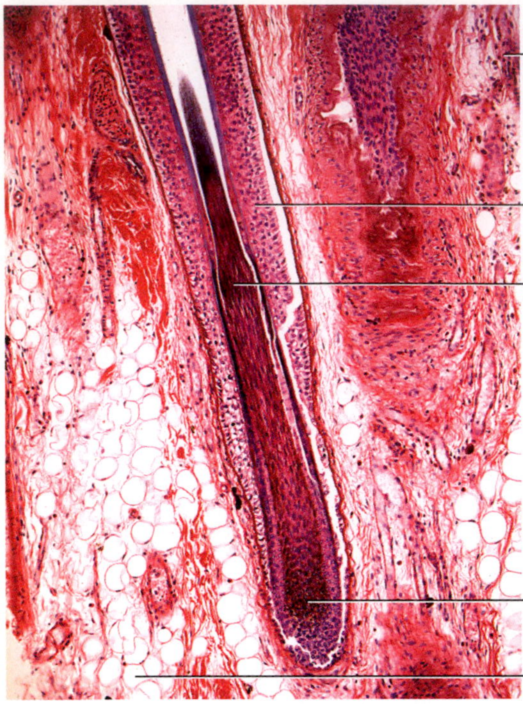

(c)

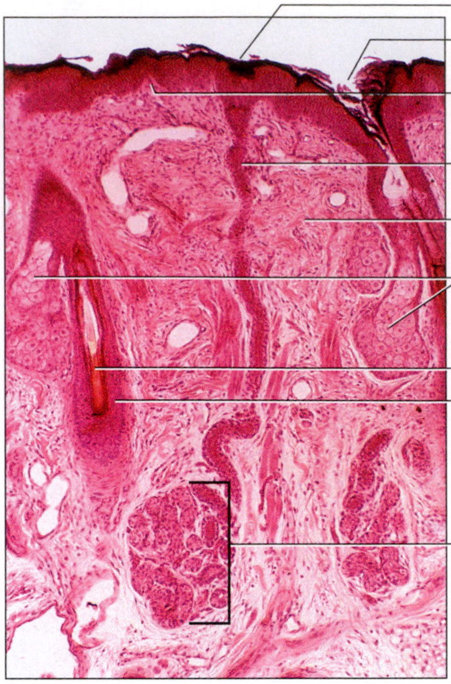

(d)

Figure 11.4 *Continued.*

8. Observe the prepared slide of dark (heavily pigmented) human skin with low-power magnification. The pigment is most abundant in the epidermis. Focus on this region with the high-power objective. The pigment-producing cells, or melanocytes, are located among the deeper layers of epidermal cells. Differences in skin color are primarily due to the amount of pigment (melanin) produced by these cells.

Critical Thinking Application

Explain the advantage for melanin granules being located in the deep layer of the epidermis.

____

____

____

____

____

____

____

9. Observe the prepared slide of thick skin from the palm of a hand or the sole of a foot (fig. 11.2). Locate the stratum lucidum. Note how the stratum corneum compares to your observation of the human scalp.
10. Complete Part C of the laboratory report.
11. Using low-power magnification, locate a hair follicle sectioned longitudinally through its bulblike base. Also locate a sebaceous gland close to the follicle and find a sweat gland (fig. 11.4). Observe the detailed structure of these parts with high-power magnification.
12. Complete Parts D and E of the laboratory report.

Learning Extension

Observe a vertical section of human skin through a tattoo, using low-power magnification. Note the location of the dispersed ink granules within the upper portion of the dermis. From a thin vertical section of a tattoo, it is not possible to determine the figure or word of the entire tattoo as seen on the surface of the skin. Compare this to the location of melanin granules found in dark (heavily pigmented) skin. Suggest reasons why a tattoo is permanent and a suntan is not.

____

____

____

Name ______________________________

Date ______________________________

Section ______________________________

The [A] corresponds to the indicated Learning Outcome(s) listed at the beginning of the laboratory exercise.

Laboratory Report 11

Integumentary System

Part A Assessments

Match the structures in column A with the description and functions in column B. Place the letter of your choice in the space provided. [1] [2]

Column A		Column B
a. Apocrine sweat gland	_______	1. An oily secretion that helps to waterproof body surface
b. Arrector pili muscle	_______	2. Outermost layer of epidermis
c. Dermis	_______	3. Become active at puberty
d. Eccrine sweat gland	_______	4. Epidermal pigment
e. Epidermis	_______	5. Inner layer of skin
f. Hair follicle	_______	6. Responds to elevated body temperature
g. Keratin	_______	7. General name of entire superficial layer of the skin
h. Melanin	_______	8. Gland that secretes an oily substance
i. Sebaceous gland	_______	9. Hard protein of nails and hair
j. Sebum	_______	10. Cell division and deepest layer of epidermis
k. Stratum basale	_______	11. Tubelike part that contains the root of the hair
l. Stratum corneum	_______	12. Causes hair to stand on end and goose bumps to appear

Part B Assessments

Complete the following:

1. How does the skin of your palm differ from that on the back (posterior) of your hand? ______________________________

____

____

2. Describe the differences you observed in the type and distribution of hair on the front (anterior) and back (posterior) of your forearm. ______________________________

____

3. Explain how a hair is formed. [2] ______________________________

____

4. What cells produce the pigment in hair? [2] ______________________________

Part C Assessments

Complete the following:

1. Distinguish the locations and tissues among epidermis, dermis, and subcutaneous layer. 3 ______________________________
2. How do the cells of stratum corneum and stratum basale differ? 3 ______________________________
3. State the specific location of melanin observed in dark (heavily pigmented) skin. 3 ______________________________
4. What special qualities does the connective tissue of the dermis have? 3 ______________________________

Part D Assessments

Complete the following:

1. What part of the hair extends from the hair papilla to the body surface? 1 ______________________________
2. In which layer of skin are sebaceous glands found? 1 ______________________________
3. How are sebaceous glands associated with hair follicles? 1 ______________________________
4. In which layer of skin are sweat glands usually located? 1 ______________________________

Part E Assessments

Using the scanning objective, sketch a vertical section of human skin. Label the skin layers and a hair follicle, a sebaceous gland, and a sweat gland. 4

Laboratory Exercise 12

Bone Structure and Classification

Materials Needed

Textbook
Prepared microscope slide of ground compact bone
Human bone specimens including long, short, flat, irregular, and sesamoid types
Human long bone, sectioned longitudinally
Fresh animal bones, sectioned longitudinally and transversely
Compound light microscope
Dissecting microscope

For Demonstration:

Fresh chicken bones (radius and ulna from wings)
Vinegar or dilute hydrochloric acid

Safety

- Wear disposable gloves for handling fresh bones and for the demonstration of a bone soaked in vinegar or dilute hydrochloric acid.
- Wash your hands before leaving the laboratory.

A bone represents an organ of the skeletal system. As such, it is composed of a variety of tissues including bone tissue, cartilage, dense connective tissue, blood, and nervous tissue. Bones are not only alive, but also multifunctional. They support and protect softer tissues, provide points of attachment for muscles, house blood-producing cells, and store inorganic salts.

Bones are classified according to their shapes as long, short, flat, irregular, or sesamoid. Although bones of the skeleton vary greatly in size and shape, they have much in common structurally and functionally.

Purpose of the Exercise

To review the way bones are classified and to examine the structure of a long bone.

LEARNING OUTCOMES

After completing this exercise, you should be able to

1. Locate the major structures of a long bone.
2. Distinguish between compact and spongy bone.
3. Differentiate the special characteristics of compact bone tissue.
4. Arrange five groups of bones based on their shapes and identify an example for each group.
5. Describe the functions of various structures of a bone.

EXPLORE

Procedure—Bone Structure and Classification

1. Review the section entitled "Connective Tissues" in chapter 5 of the textbook and the section entitled "Bone Structure" in chapter 7 of the textbook.
2. As a review activity, label figures 12.1 and 12.2.
3. Reexamine the microscopic structure of bone tissue by observing a prepared microscope slide of ground compact bone. Use figure 12.3 of bone tissue to locate the following features:

osteon (Haversian system)—cylinder-shaped unit

central canal (Haversian canal)—contains blood vessels and nerves

lamella—concentric ring of matrix around central canal

lacuna—small chamber for an osteocyte

bone extracellular matrix—collagen and calcium phosphate

canaliculus—minute tube containing cellular process

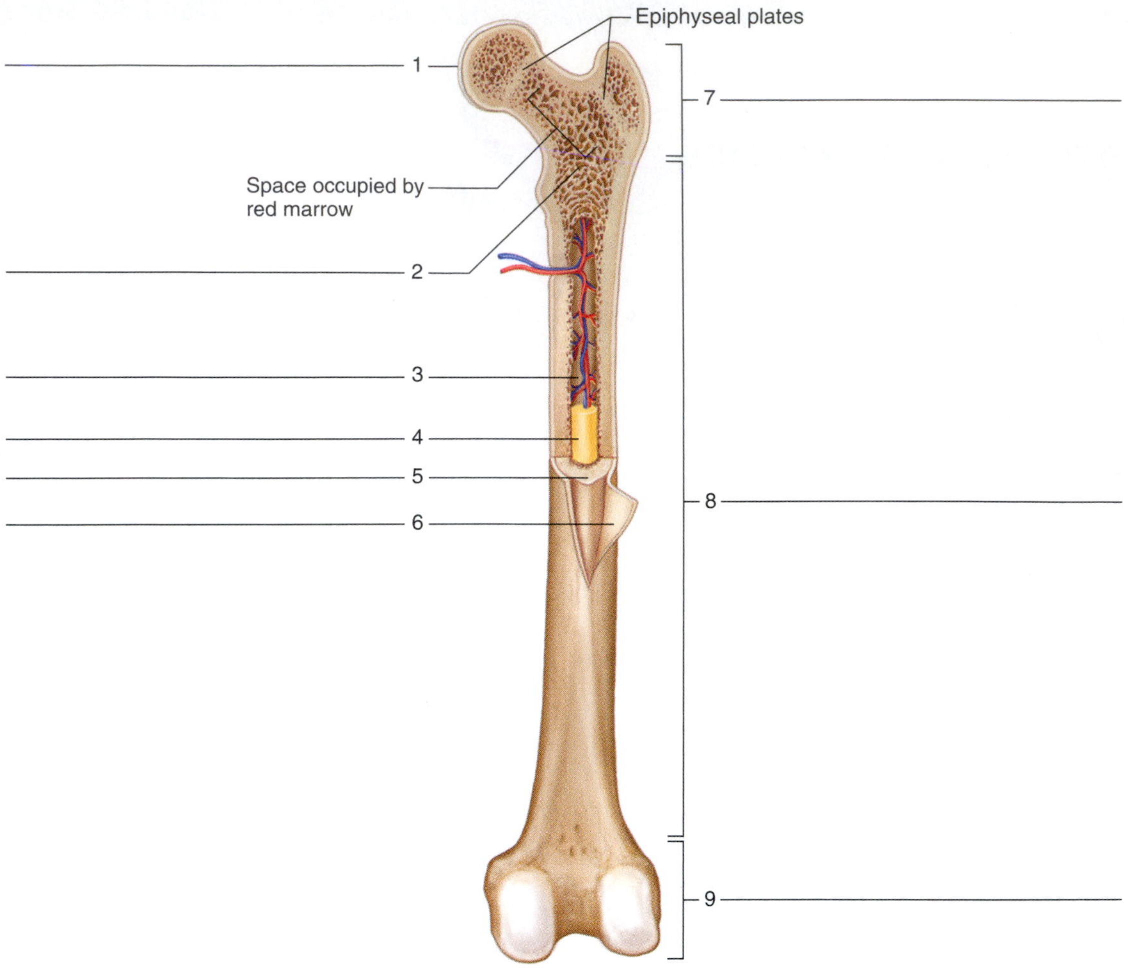

Figure 12.1 Label the major structures of this long bone (femur). 1 2

Critical Thinking Application

Explain how bone cells embedded in a solid ground substance obtain nutrients and eliminate wastes.

____

____

____

____

____

4. Observe the individual bone specimens and arrange them into groups, according to the following shapes and examples: 4

   long—femur; humerus; phalanges
   irregular—vertebrae
   short— carpals; tarsals
   sesamoid (round)—patella
   flat—ribs; most cranial bones

5. Complete Part A of Laboratory Report 12.

6. Examine the sectioned long bones and locate the following:

   epiphysis
   - proximal—nearest torso
   - distal—farthest from torso

   epiphyseal plate—growth zone of hyaline cartilage
   articular cartilage—on ends of epiphyses
   diaphysis—shaft between epiphyses
   periosteum—membrane around bone (except articular cartilage) of dense irregular connective tissue
   compact bone—forms diaphysis and epiphyseal surfaces
   spongy bone—within epiphyses
   - trabeculae—a structural lattice in spongy bone

   medullary cavity—hollow chamber
   endosteum—thin membrane lining medullary cavity of reticular connective tissue
   yellow marrow—occupies medullary cavity
   red marrow—occupies spongy bone in some epiphyses and flat bones

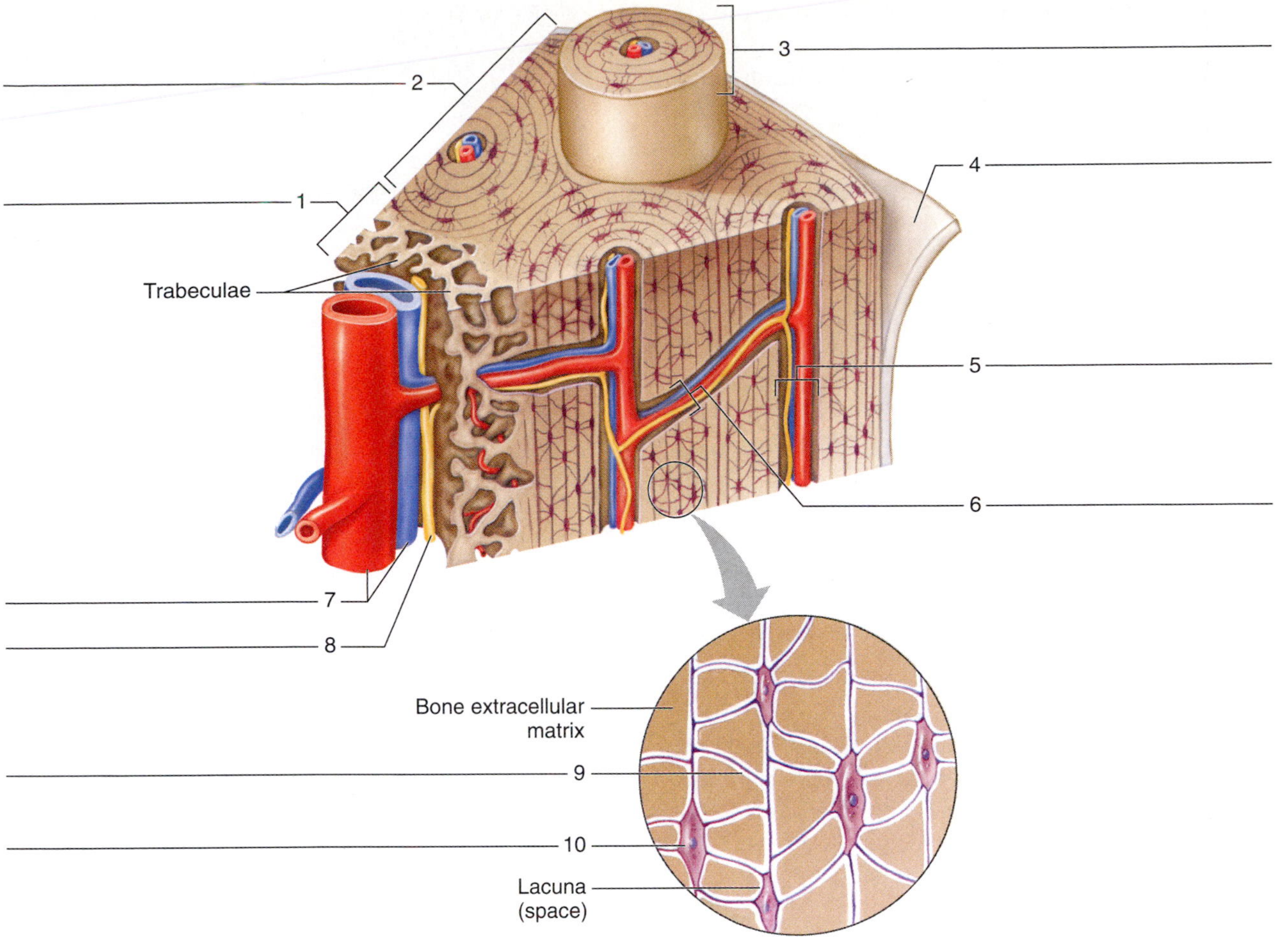

Figure 12.2 Label the features associated with the microscopic structure of bone. 2 3

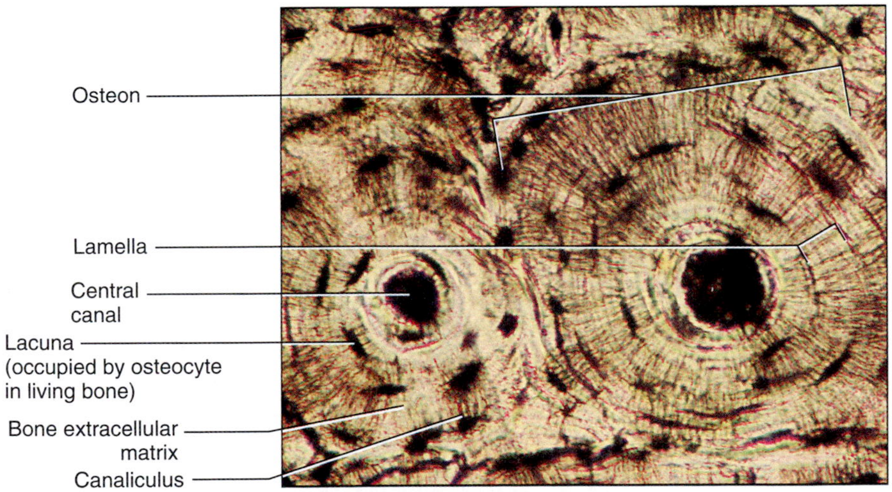

Figure 12.3 Micrograph of ground compact bone tissue (160×).

7. Use the dissecting microscope to observe the compact bone and spongy bone of the sectioned specimens. Also examine the marrow in the medullary cavity and the spaces within the spongy bone of the fresh specimen.
8. Complete Parts B and C of the laboratory report.

Demonstration

Examine a fresh chicken bone and a chicken bone that has been soaked for several days in vinegar or overnight in dilute hydrochloric acid. Wear disposable gloves for handling these bones. This acid treatment removes the inorganic salts from the bone extracellular matrix. Rinse the bones in water and note the texture and flexibility of each (fig. 12.4*a*). Based on your observations, what quality of the fresh bone seems to be due to the inorganic salts removed by the acid treatment? 3

____

____

Examine the specimen of chicken bone that has been exposed to high temperature (baked at 121°C [250°F] for 2 hours). This treatment removes the protein and other organic substances from the bone extracellular matrix (fig.12.4*b*). What quality of the fresh bone seems to be due to these organic materials? 3

____

____

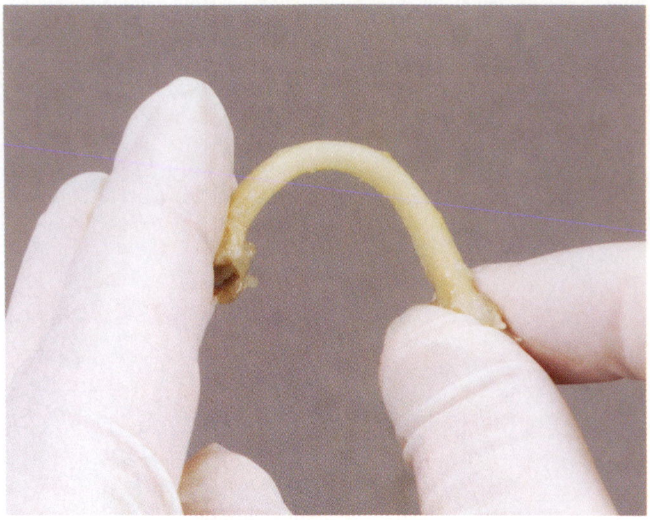

(a)

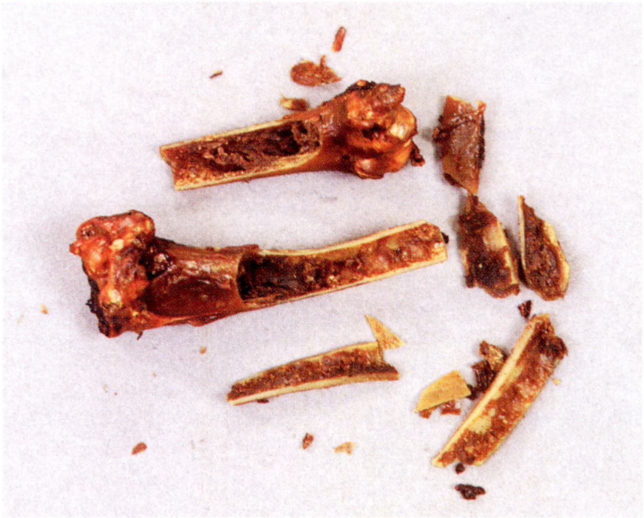

(b)

Figure 12.4 Results of fresh chicken bone demonstration: (*a*) soaked in vinegar; (*b*) baked in oven.

Name ______________________________

Date ______________________________

Section ______________________________

The [A] corresponds to the indicated Learning Outcome(s) listed at the beginning of the laboratory exercise.

Bone Structure and Classification

12 Laboratory Report

Part A Assessments

Complete the following statements. (*Note:* Questions 1-6 pertain to bone classification by shape.)

1. A bone that is platelike is classified as a(an) ______________ bone. [4]
2. The bones of the wrist are examples of ______________ bones. [4]
3. The bone of the thigh is an example of a(an) ______________ bone. [4]
4. Vertebrae are examples of ______________ bones. [4]
5. The patella (kneecap) is an example of a large ______________ bone. [4]
6. The bones of the skull that form a protective covering of the brain are examples of ______________ bones. [4]
7. Distinguish between the epiphysis and the diaphysis of a long bone. [1] ______________
8. Describe where cartilage is found on the surface of a long bone. [1] ______________
9. Describe where dense irregular connective tissue is found on the surface of a long bone. [1] ______________
10. Distinguish the locations and tissues between the periosteum and the endosteum. [1] ______________

Part B Assessments

Complete the following:

1. What structural differences did you note between compact bone and spongy bone? [2] ______________
2. How are these structural differences related to the locations and functions of these two types of bone? [5] ______________
3. From your observations, how does the marrow in the medullary cavity compare with the marrow in the spaces of the spongy bone? [5] ______________

Part C Assessments

Identify the structures indicated in figure 12.5.

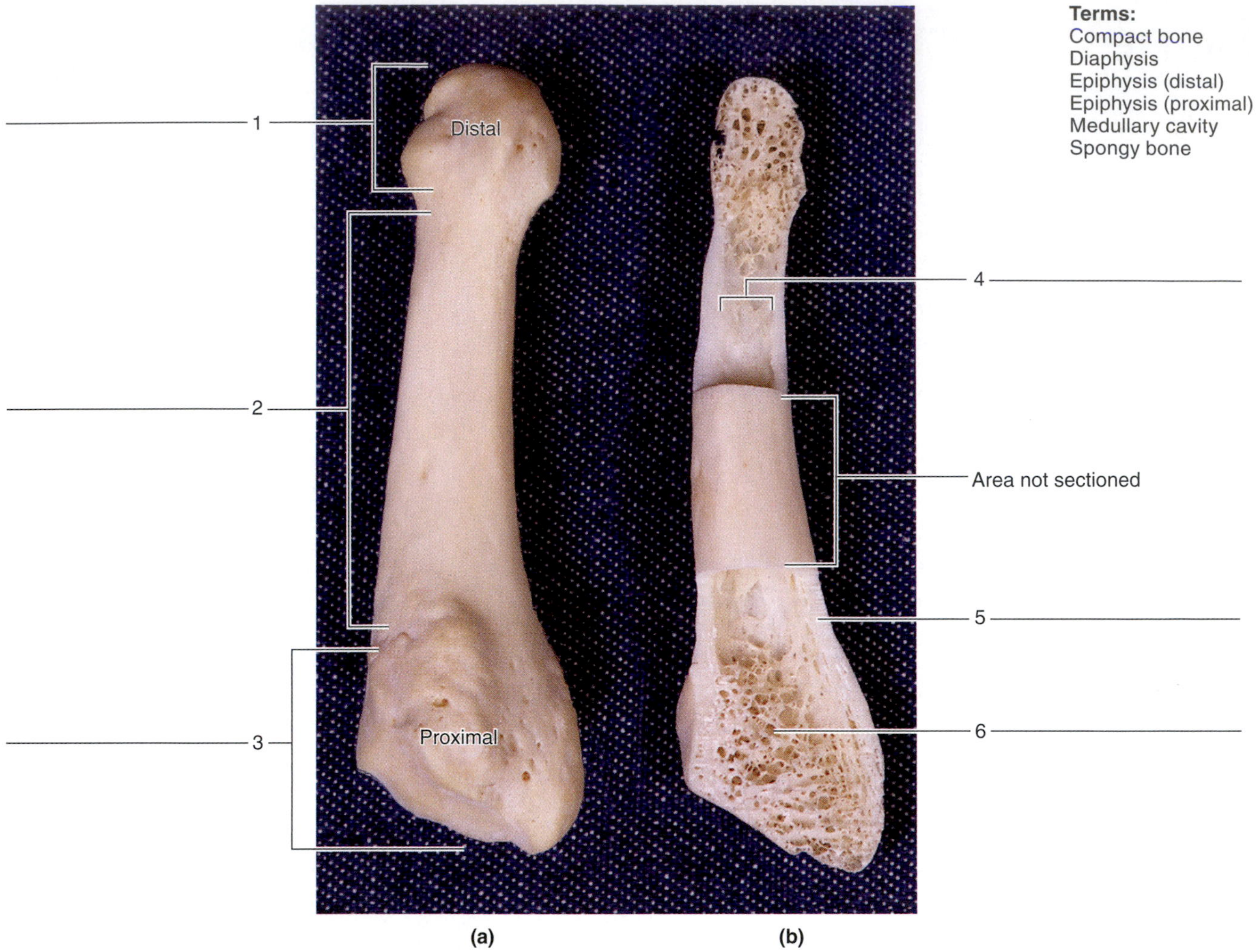

Figure 12.5 Identify the structures indicated in (*a*) the unsectioned long bone (fifth metatarsal) and (*b*) the partially sectioned long bone, using the terms provided. 1 2

Laboratory Exercise **13**

Organization of the Skeleton

Materials Needed

Textbook
Articulated human skeleton

For Learning Extension:
Colored pencils

For Demonstration:
Radiographs (X rays) of skeletal structures

The skeleton can be divided into two major portions: (1) the axial skeleton, which consists of the bones and cartilages of the head, neck, and trunk, and (2) the appendicular skeleton, which consists of the bones of the limbs and those that anchor the limbs to the axial skeleton. The bones that anchor the limbs include the pectoral and pelvic girdles.

Purpose of the Exercise

To review the organization of the skeleton, the major bones of the skeleton, and the terms used to describe skeletal structures.

LEARNING OUTCOMES

After completing this exercise, you should be able to

1. Distinguish between the axial skeleton and the appendicular skeleton.
2. Locate and label the major bones of the human skeleton.
3. Associate the terms used to describe skeletal structures and locate examples of such structures on the human skeleton.

EXPLORE

Procedure—Organization of the Skeleton

1. Review the section entitled "Skeletal Organization" in chapter 7 of the textbook. (Pronunciations of the names for major skeletal structures are included within the narrative of chapter 7.)
2. As a review activity, label figure 13.1.
3. Examine the human skeleton and locate the following parts. As you locate the following bones, note the number of each in the skeleton. Palpate as many of the corresponding bones in your skeleton as possible.

axial skeleton	
skull	
cranium	(8)
face	(14)
middle ear bone	(6)
hyoid bone	(1)
vertebral column	
vertebra	(24)
sacrum	(1)
coccyx	(1)
thoracic cage	
rib	(24)
sternum	(1)
appendicular skeleton	
pectoral (shoulder) girdle	
scapula	(2)
clavicle	(2)
upper limbs	
humerus	(2)
radius	(2)
ulna	(2)
carpal	(16)
metacarpal	(10)
phalanx	(28)
pelvic girdle	
hip bone (coxal bone; pelvic bone; innominate)	(2)
lower limbs	
femur	(2)
tibia	(2)
fibula	(2)
patella	(2)
tarsal	(14)
metatarsal	(10)
phalanx	(28)
Total	**206 bones**

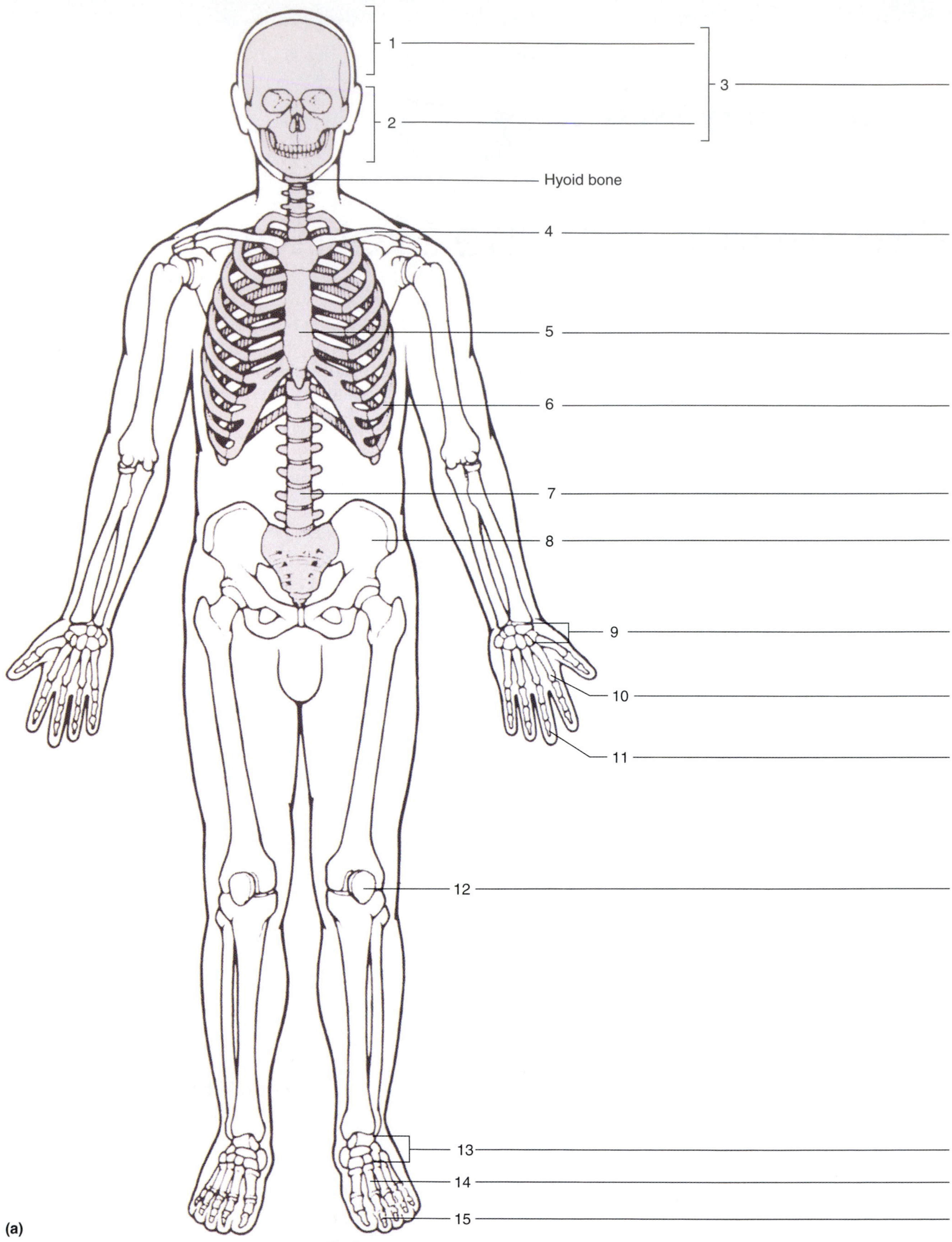

Figure 13.1 Label the major bones of the skeleton: (*a*) anterior view; (*b*) posterior view. 1 2

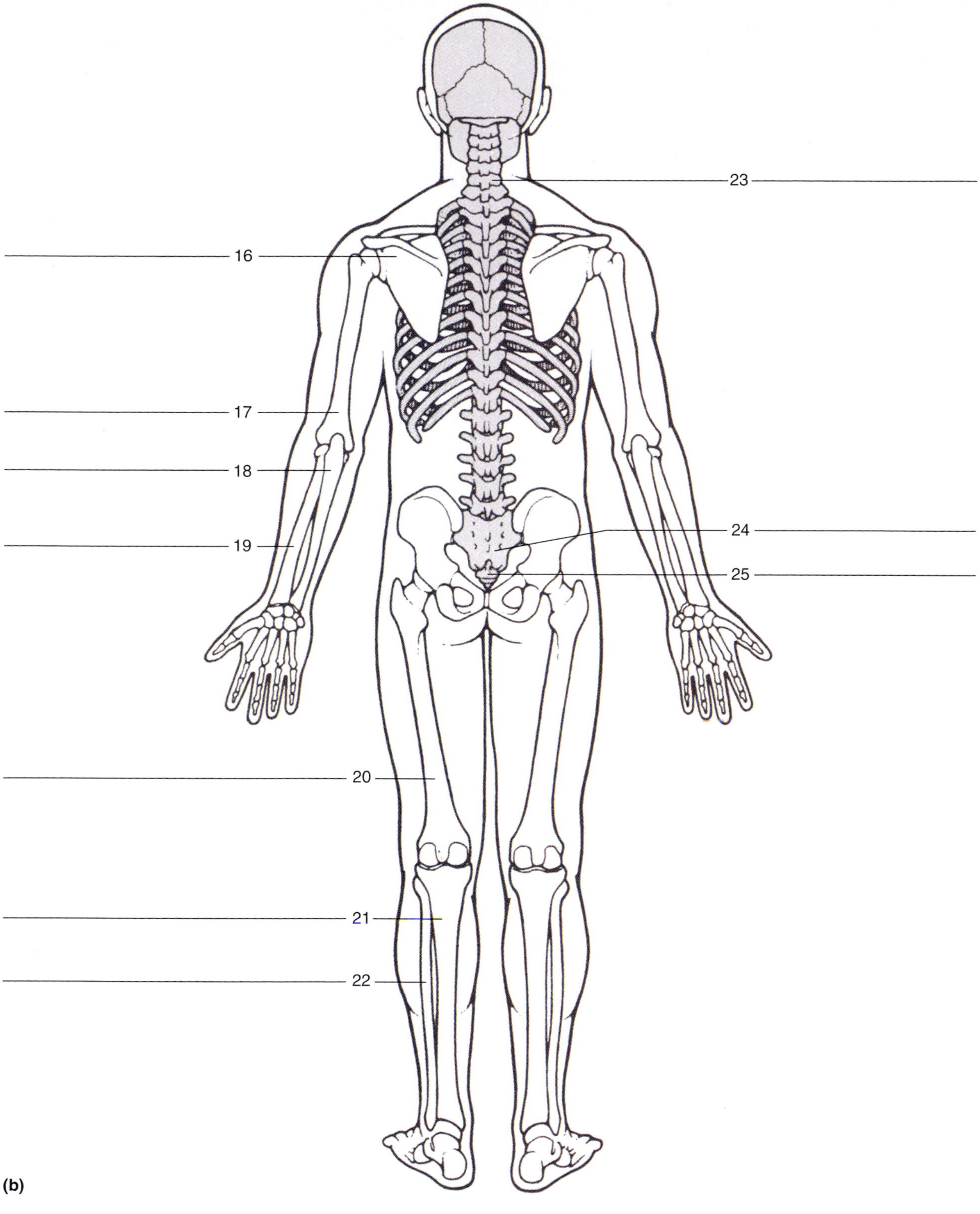

Figure 13.1 *Continued.*

Learning Extension

Use colored pencils to distinguish the individual bones in figure 13.1.

4. Study table 7.4 in chapter 7 of the textbook. Bone features (bone markings) can be grouped together in a category of projections, articulations, depressions, or openings. Within each category more specific examples occur. The bones listed in this section only represent an example of a location in the human body. Locate an example of each of the following features on the example bone listed, noting the size, shape, and location in the human skeleton:

Projections:

- **crest** (ridgelike)—hip bone
- **epicondyle** (superior to condyle)—femur
- **line (linea)** (slightly raised ridge)—femur
- **process** (prominent)—vertebra
- **protuberance** (outgrowth)—skull (occipital)
- **ramus** (extension)—hip bone
- **spine** (thornlike)—scapula
- **trochanter** (large)—femur
- **tubercle** (knoblike)—humerus
- **tuberosity** (rough elevation)—tibia

Articulations:

- **condyle** (rounded process)—skull (occipital)
- **facet** (nearly flat)—vertebra
- **head** (expanded end)—femur

Depressions:

- **fossa** (shallow basin)—humerus
- **fovea** (tiny pit)—femur

Openings:

- **canal** (tubular passage)—skull (occipital)
- **fissure** (slit)—skull (orbit)
- **foramen** (hole)—vertebra
- **meatus** (tubelike)—skull (temporal)
- **sinus** (cavity)—skull (maxilla)

Critical Thinking Application

Locate and name the largest foramen in the skull.

____

____

____

Locate and name the largest foramen in the skeleton.

____

____

____

5. Complete Parts A, B, C, and D of Laboratory Report 13.

Demonstration

Images on radiographs (X rays) are produced by allowing X rays from an X-ray tube to pass through a body part and to expose photographic film positioned on the opposite side of the part. The image that appears on the film after it is developed reveals the presence of parts with different densities. Bone, for example, is very dense tissue and is a good absorber of X rays. Thus, bone generally appears light on the film. Air-filled spaces, on the other hand, absorb almost no X rays and appear as dark areas on the film. Liquids and soft tissues absorb intermediate quantities of X rays, so they usually appear in various shades of gray.

Examine the available radiographs of skeletal structures by holding each film in front of a light source. Identify as many of the bones and features as you can.

Name ______________________________

Date ______________________________

Section ______________________________

The [A] corresponds to the indicated Learning Outcome(s) listed at the beginning of the laboratory exercise.

Organization of the Skeleton

Laboratory Report 13

Part A Assessments

Complete the following statements:

1. The extra bones that sometimes develop between the flat bones of the skull are called [2] ______________.
2. The two divisions of the skeleton are the ______________ skeleton and the appendicular skeleton. [1]
3. The cranium and facial bones compose the ______________. [1]
4. The ______________ bone supports the tongue. [2]
5. The ______________ at the inferior end of the sacrum is composed of several fused vertebrae. [2]
6. Most ribs are attached anteriorly to the ______________. [2]
7. The thoracic cage is composed of ______________ pairs of ribs. [2]
8. The scapulae and clavicles together form the ______________. [1]
9. The humerus, radius, and ______________ articulate to form the elbow joint. [2]
10. The wrist is composed of eight bones called ______________. [2]
11. The hip bones are attached posteriorly to the ______________. [2]
12. The pelvic girdle (hip bones), sacrum, and coccyx together form the ______________. [2]
13. The ______________ covers the anterior surface of the knee. [2]
14. The bones that articulate with the distal ends of the tibia and fibula are called ______________. [2]
15. All finger and toe bones are called ______________. [2]

Part B Assessments

Match the terms in column A with the definitions in column B. Place the letter of your choice in the space provided. [3]

Column A		Column B
a. Condyle	_______	1. Small, nearly flat articular surface
b. Crest	_______	2. Deep depression or shallow basin
c. Facet	_______	3. Rounded process
d. Fontanel	_______	4. Opening or passageway
e. Foramen	_______	5. Interlocking line of union
f. Fossa	_______	6. Narrow, ridgelike projection
g. Suture	_______	7. Soft region between bones of skull

Part C Assessments

Match the terms in column A with the definitions in column B. Place the letter of your choice in the space provided. 3

Column A		Column B
a. Fovea	______	1. Tubelike passageway
b. Head	______	2. Tiny pit or depression
c. Meatus	______	3. Small, knoblike process
d. Sinus	______	4. Thornlike projection
e. Spine	______	5. Rounded enlargement at end of bone
f. Trochanter	______	6. Air-filled cavity within bone
g. Tubercle	______	7. Relatively large process

Part D Assessments

Identify the bones indicated in figure 13.2.

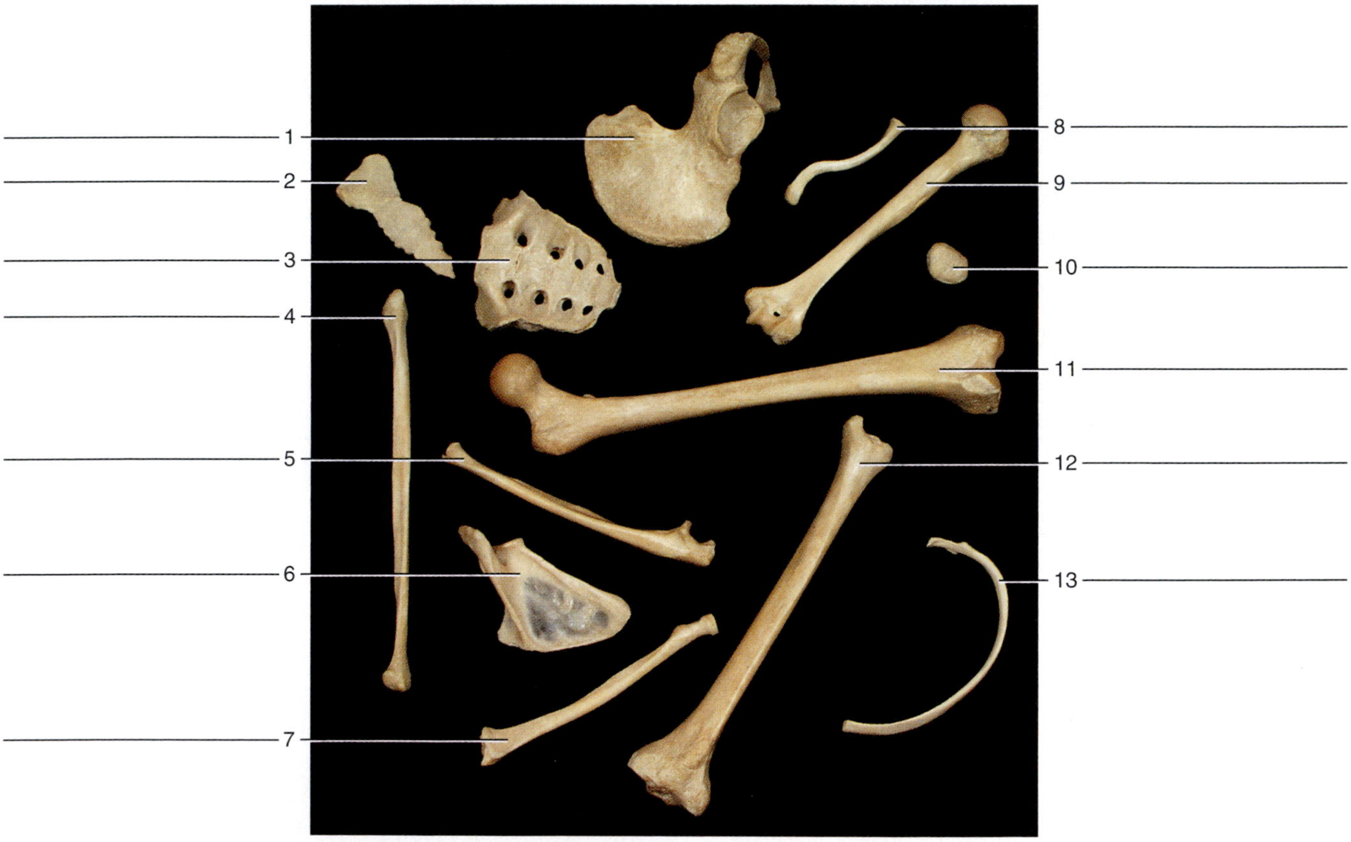

Terms:

Clavicle	Patella	Sternum
Femur	Radius	Tibia
Fibula	Rib	Ulna
Hip bone	Sacrum	
Humerus	Scapula	

Figure 13.2 Identify the bones in this random arrangement, using the terms provided. 2

Laboratory Exercise 14

Skull

Materials Needed

Textbook
Human skull, articulated
Human skull, disarticulated (Beauchene)
Human skull, sagittal section

For Learning Extension:
Colored pencils

For Demonstration:
Fetal skull

A human skull consists of twenty-two bones that, except for the lower jaw, are firmly interlocked along sutures. Eight of these immovable bones make up the braincase, or cranium, and thirteen more immovable bones and the mandible form the facial skeleton.

Purpose of the Exercise

To examine the structure of the human skull and to identify the bones and major features of the skull.

LEARNING OUTCOMES

After completing this exercise, you should be able to

1. Distinguish between the cranium and the facial skeleton.
2. Locate and label the bones of the skull and their major features.
3. Locate and label the major sutures of the cranium.
4. Locate and label the sinuses of the skull.

EXPLORE

Procedure—Skull

1. Review the section entitled "Skull" in chapter 7 of the textbook.
2. As a review activity, label figures 14.1, 14.2, 14.3, 14.4, and 14.5.
3. Examine the **cranial bones** of the articulated human skull and the sectioned skull. Also observe the corresponding disarticulated bones. Locate the following bones and features in the laboratory specimens and, at the same time, palpate as many of these bones and features in your skull as possible.

frontal bone (1)
- supraorbital foramen
- frontal sinus

parietal bone (2)
- sagittal suture
- coronal suture

occipital bone (1)
- lambdoid suture
- foramen magnum
- occipital condyle

temporal bone (2)
- squamous suture
- external acoustic meatus
- mandibular fossa
- mastoid process
- styloid process
- carotid canal
- jugular foramen
- internal acoustic meatus
- zygomatic process

sphenoid bone (1)
- sella turcica
- greater and lesser wings
- sphenoidal sinus

ethmoid bone (1)
- cribriform plate
- perpendicular plate
- superior nasal concha
- middle nasal concha
- ethmoidal sinus
- crista galli

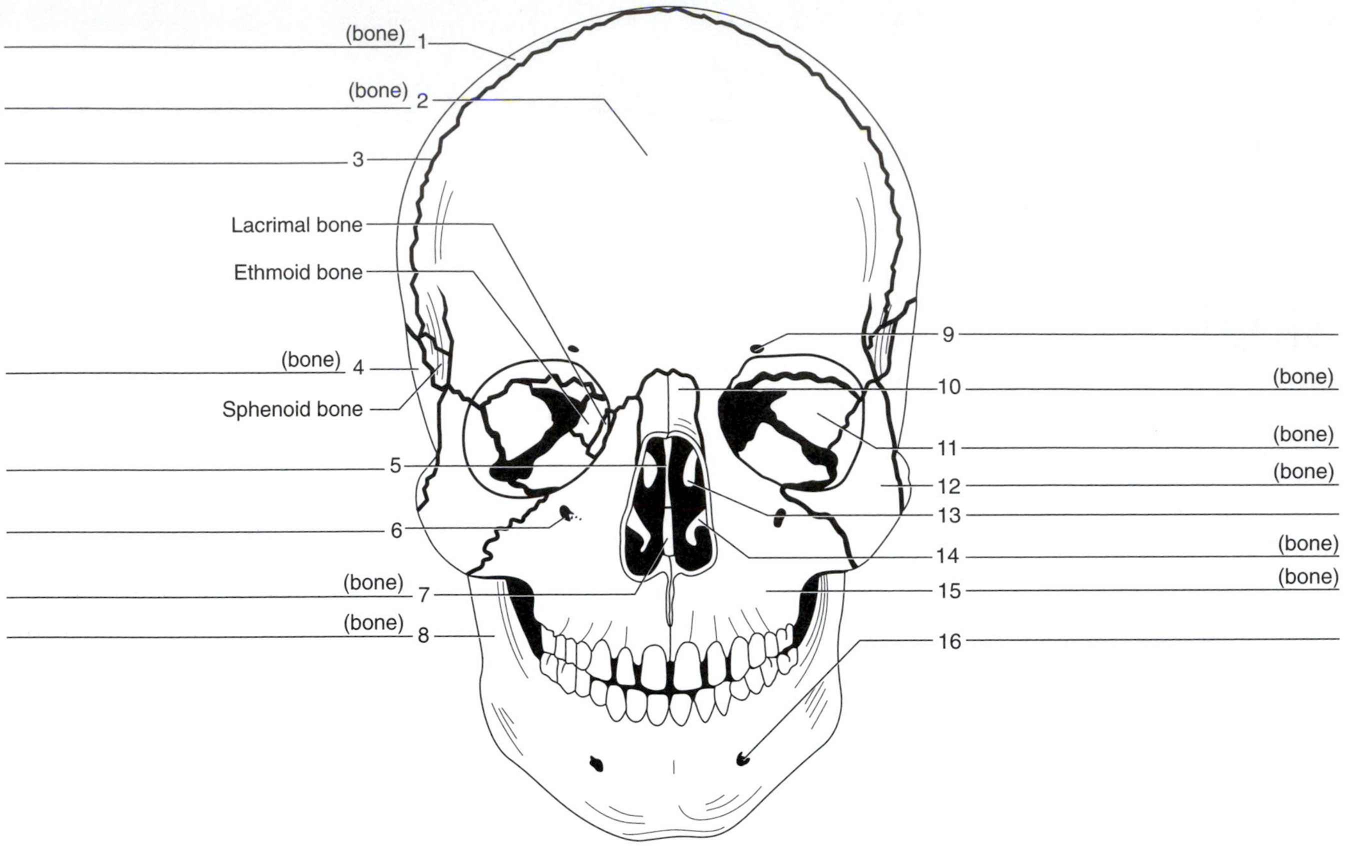

Figure 14.1 Label the anterior bones and features of the skull. (If the line lacks the word *bone,* label the particular feature of that bone.) 2

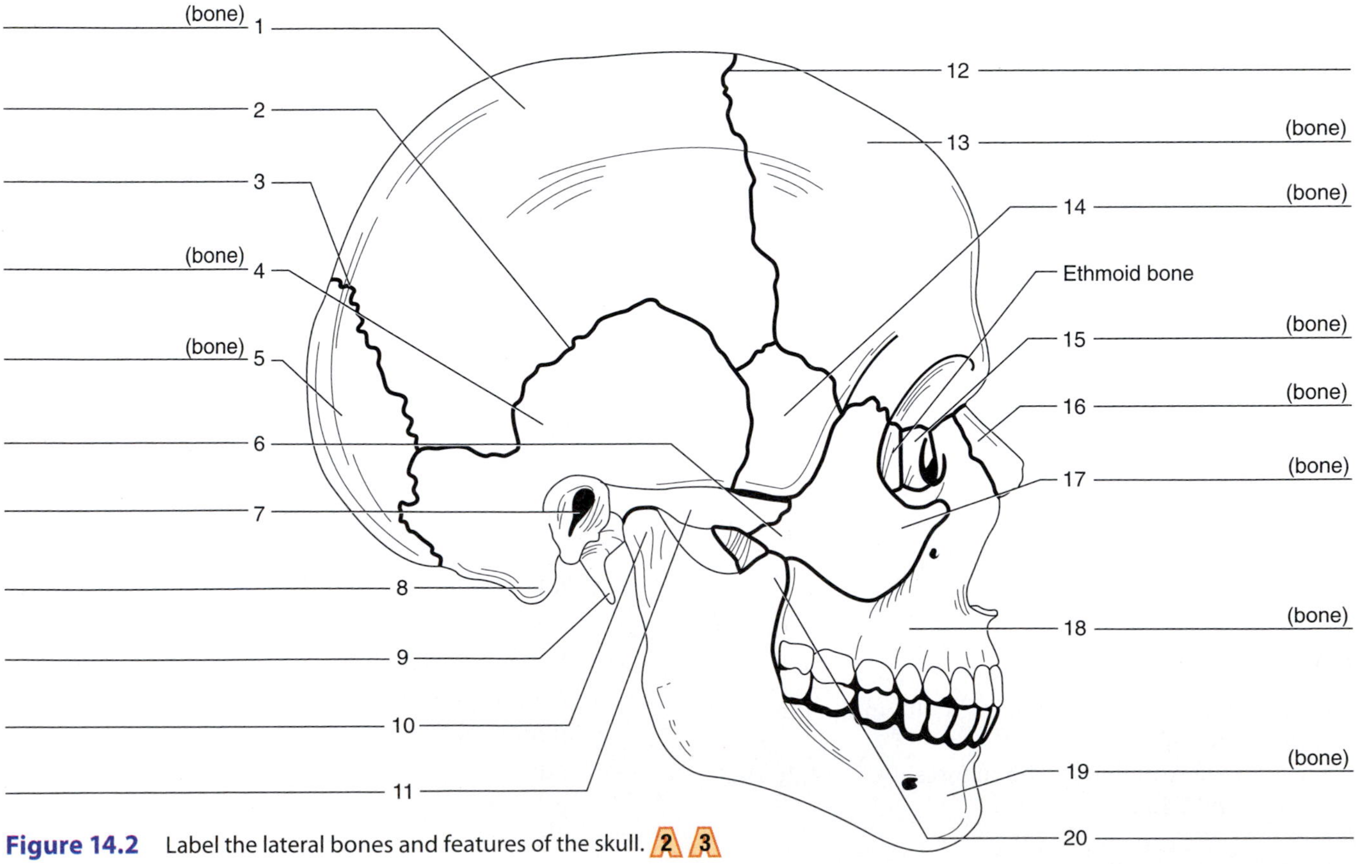

Figure 14.2 Label the lateral bones and features of the skull. 2 3

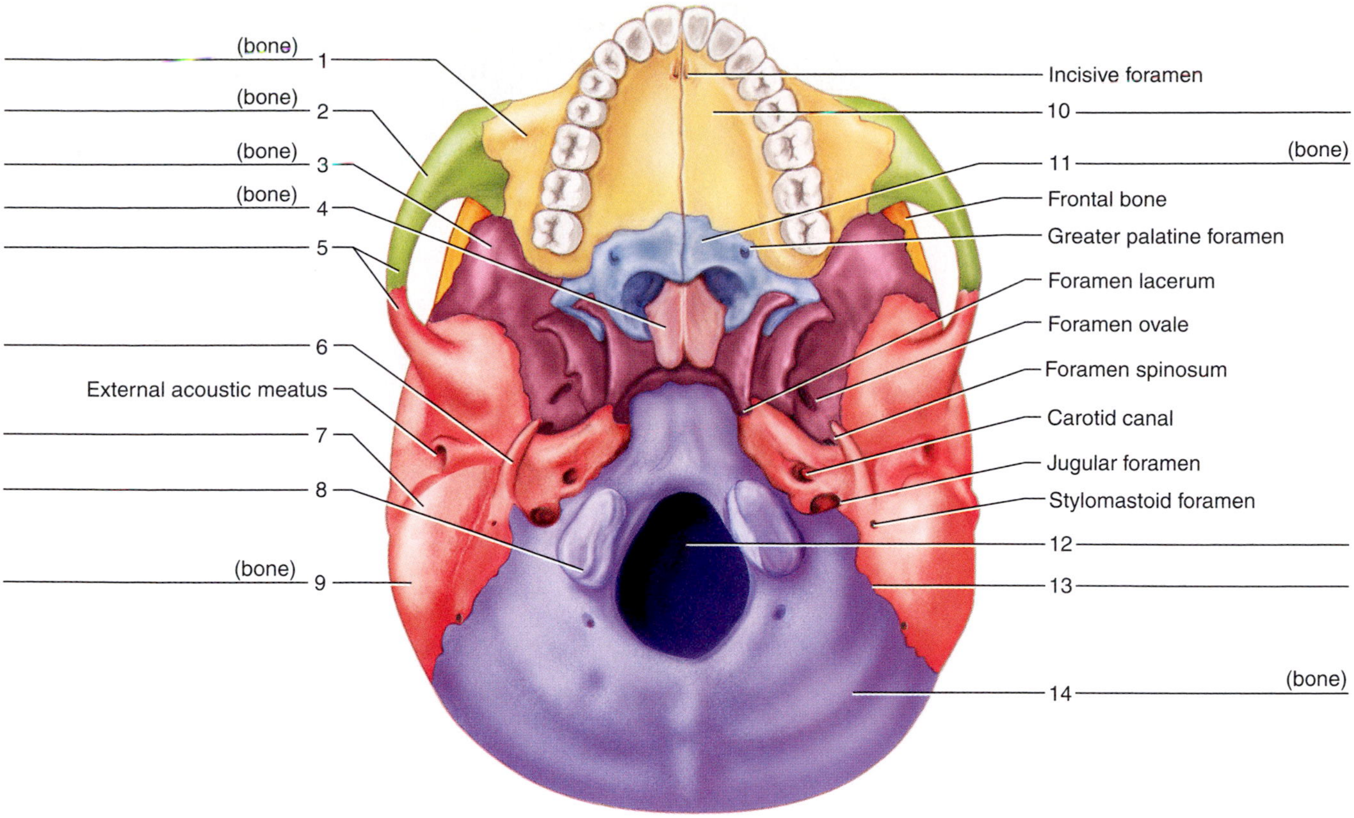

Figure 14.3 Label the inferior bones and features of the skull. 2

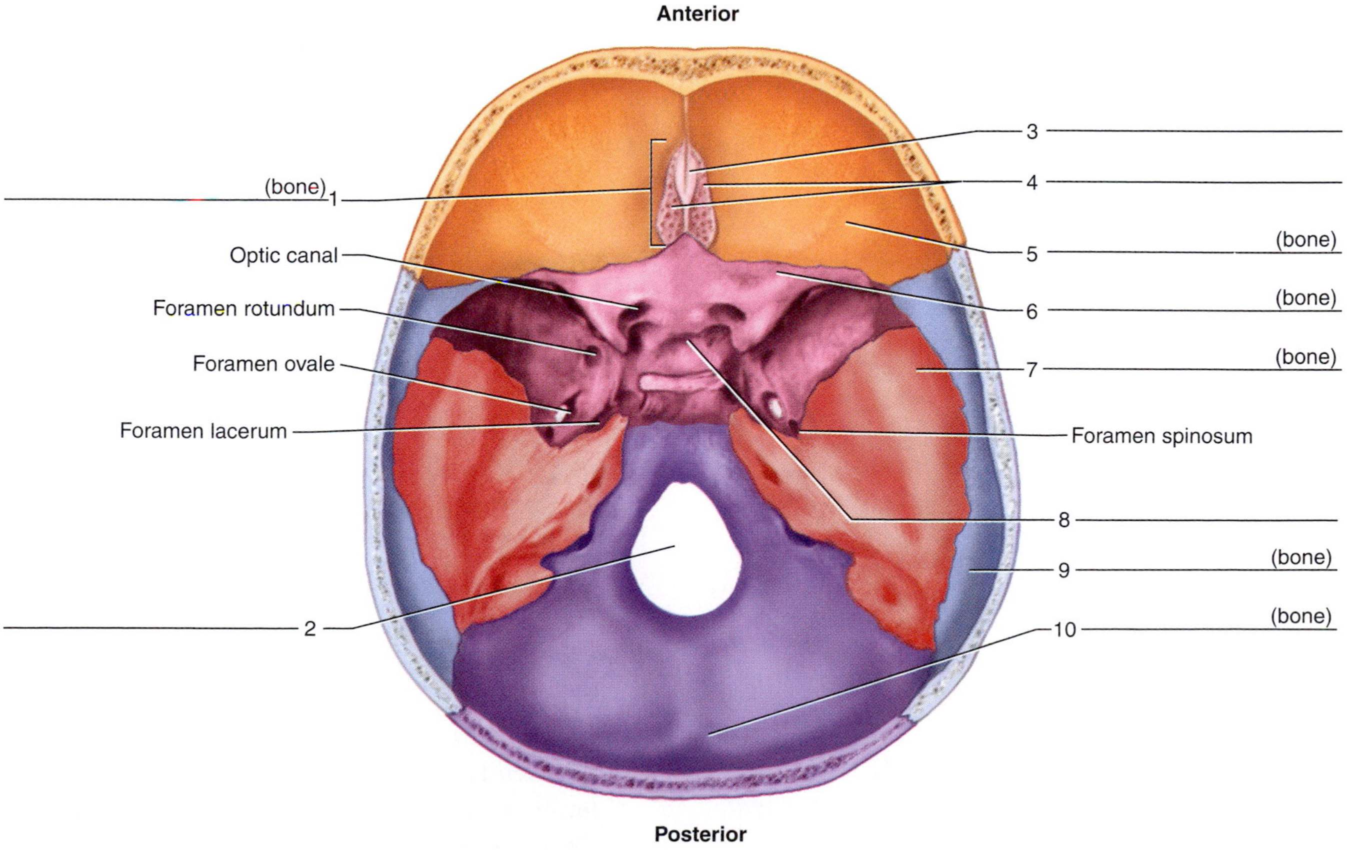

Figure 14.4 Label the bones and features of the floor of the cranial cavity as viewed from above. 2

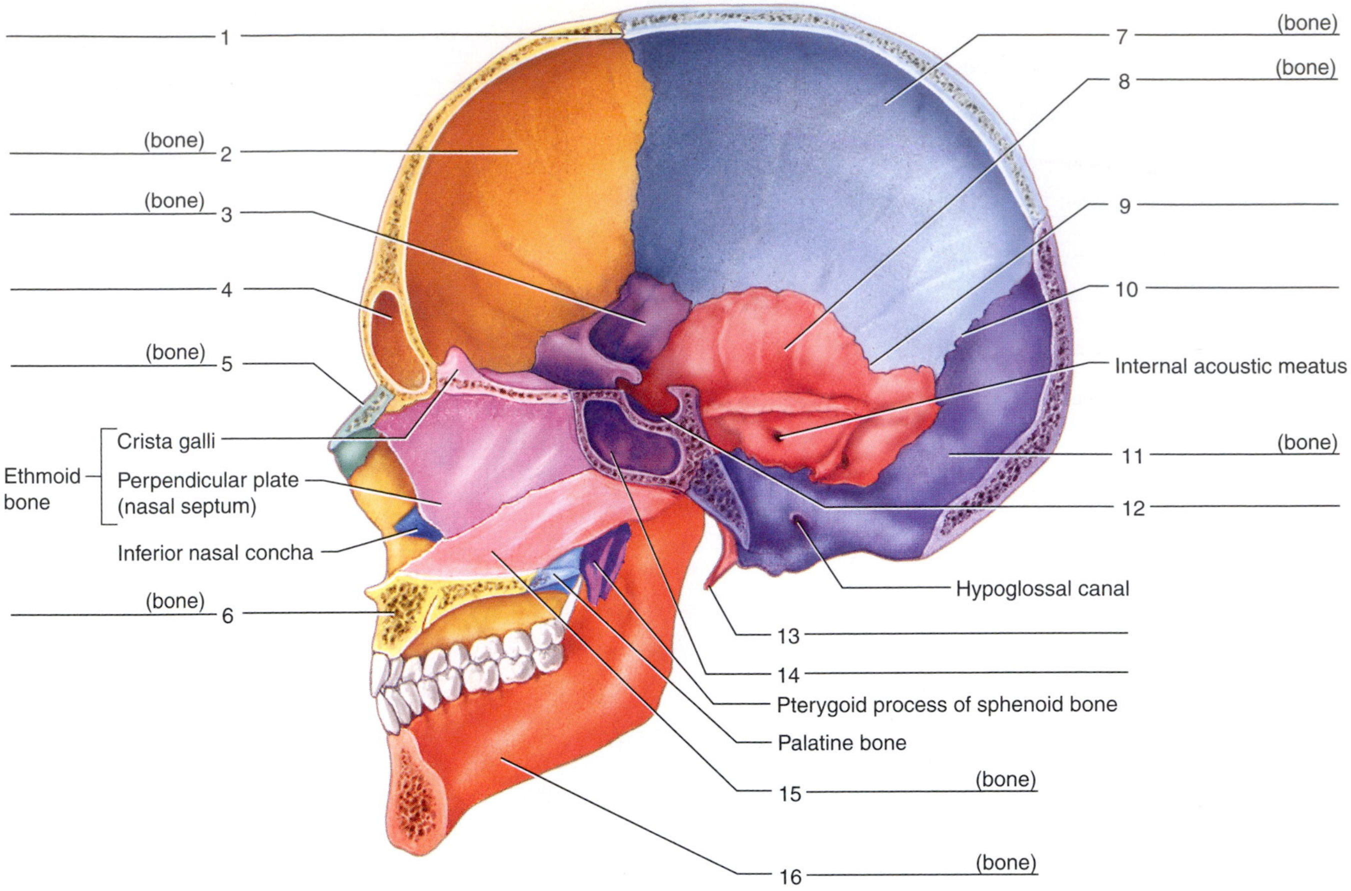

Figure 14.5 Label the bones and features of the sagittal section of the skull. 2 3 4

4. Complete Parts A and B of Laboratory Report 14.
5. Examine the **facial bones** of the articulated and sectioned skulls and the corresponding disarticulated bones. Locate the following:

maxilla (2)
- maxillary sinus
- palatine process
- alveolar process
- alveolar arch

palatine bone (2)

zygomatic bone (2)
- temporal process
- zygomatic arch

lacrimal bone (2)

nasal bone (2)

vomer bone (1)

inferior nasal concha (2)

mandible (1)
- ramus
- mandibular condyle
- coronoid process
- alveolar border
- mandibular foramen
- mental foramen

6. Complete Part C of the laboratory report.

Learning Extension

Use colored pencils to differentiate the bones illustrated in figures 14.1 and 14.2. Select a different color for each bone in the series. This activity should help you locate various bones shown in different views in the figures. You can check your work by referring to the corresponding figures in the textbook, presented in full color.

7. Study table 7.8 of chapter 7 in the textbook. Locate the following features of the human skull:

carotid canal
foramen lacerum
foramen magnum
foramen ovale
foramen rotundum

foramen spinosum
greater palatine foramen
hypoglossal canal
incisive foramen
inferior orbital fissure
infraorbital foramen
internal acoustic meatus
jugular foramen
mandibular foramen
mental foramen
optic canal
stylomastoid foramen
superior orbital fissure
supraorbital foramen

8. Complete Parts D and E of the laboratory report.

Critical Thinking Application

Examine the inside of the cranium on a sectioned skull. What area appears to be the weakest area? Explain your answer.

____

____

____

____

Demonstration

Examine the fetal skull and skeleton (figs. 14.6 and 14.7). The skull is incompletely developed and the cranial bones are separated by fibrous membranes. These membranous areas are called fontanels, or "soft spots." The fontanels close as the cranial bones grow together. The posterior and lateral fontanels usually close during the first year after birth, whereas the anterior fontanel may not close until the middle or end of the second year. What other features characterize the fetal skull?

____

____

____

____

____

____

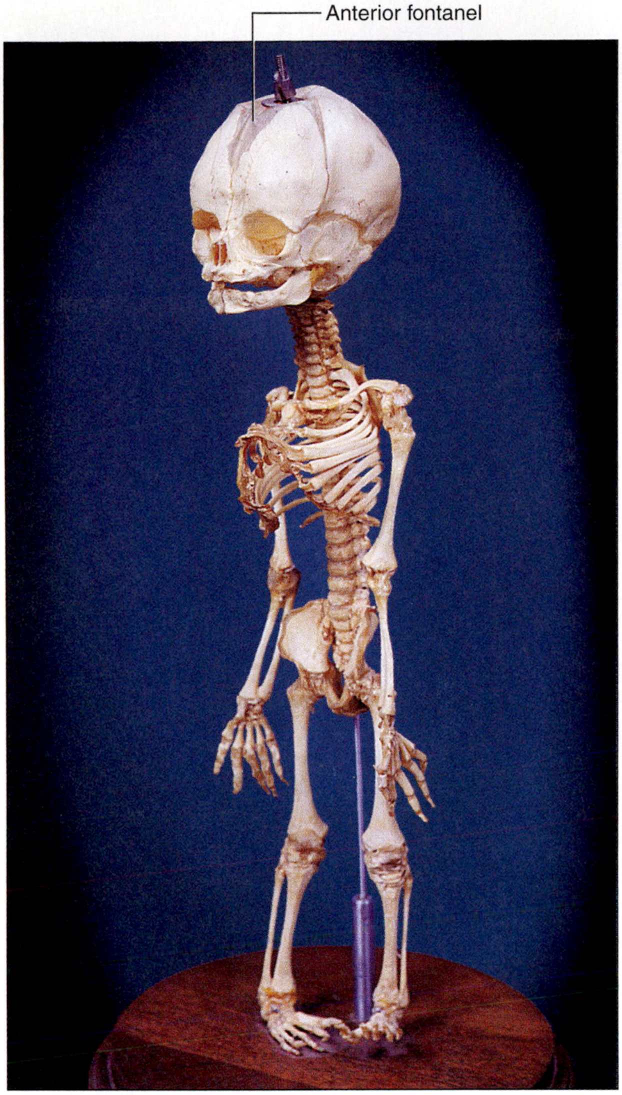

Figure 14.6 Human fetal skeleton. The gestational age of this skeleton is 7–8 months.

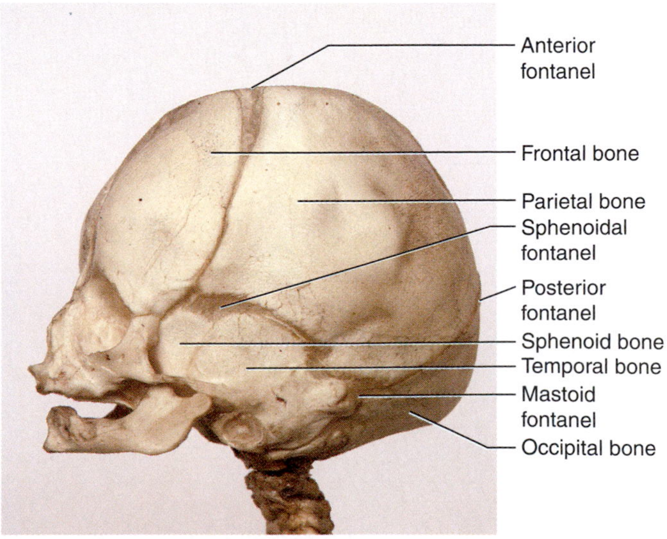

Figure 14.7 Lateral view of fetal skull.

Notes

Name ______________________________

Date ______________________________

Section ______________________________

The [A] corresponds to the indicated Learning Outcome(s) listed at the beginning of the laboratory exercise.

Skull

Laboratory Report 14

Part A Assessments

Match the bones in column A with the features in column B. Place the letter of your choice in the space provided. (Some answers are used more than once.) [2]

Column A		Column B
a. Ethmoid bone	_______	**1.** Forms sagittal, coronal, squamous, and lambdoid sutures
b. Frontal bone	_______	**2.** Cribriform plate
c. Occipital bone	_______	**3.** Crista galli
d. Parietal bone	_______	**4.** External acoustic meatus
e. Sphenoid bone	_______	**5.** Foramen magnum
f. Temporal bone	_______	**6.** Mandibular fossa
	_______	**7.** Mastoid process
	_______	**8.** Middle nasal concha
	_______	**9.** Occipital condyle
	_______	**10.** Sella turcica
	_______	**11.** Styloid process
	_______	**12.** Supraorbital foramen

Part B Assessments

Complete the following statements:

1. The ______________________ suture joins the frontal bone to the parietal bones. [3]
2. The parietal bones are firmly interlocked along the midline by the ______________________ suture. [3]
3. The ______________________ suture joins the parietal bones to the occipital bone. [3]
4. The temporal bones are joined to the parietal bones along the ______________________ sutures. [3]
5. Name the three cranial bones that contain sinuses. [1] [4] ____

   ____
6. Name a facial bone that contains a sinus. [1] [4] ____

Part C Assessments

Match the bones in column A with the characteristics in column B. Place the letter of your choice in the space provided. 2

Column A		Column B
a. Inferior nasal concha	_______	1. Forms bridge of nose
b. Lacrimal bone	_______	2. Only movable bone in the facial skeleton
c. Mandible	_______	3. Contains coronoid process
d. Maxilla	_______	4. Creates prominence of cheek inferior and lateral to the eye
e. Nasal bone	_______	5. Contains sockets of upper teeth
f. Palatine bone	_______	6. Forms inferior portion of nasal septum
g. Vomer bone	_______	7. Forms anterior portion of zygomatic arch
h. Zygomatic bone	_______	8. Scroll-shaped bone
	_______	9. Forms anterior roof of mouth
	_______	10. Contains mental foramen
	_______	11. Forms posterior roof of mouth
	_______	12. Scalelike part in medial wall of orbit

Part D Assessments

Match the passageways in column A with the structures transmitted through them in column B. Place the letter of your choice in the space provided. 2

Column A		Column B
a. Foramen magnum	_______	1. Maxillary division of trigeminal nerve
b. Foramen ovale	_______	2. Nerve fibers of spinal cord
c. Foramen rotundum	_______	3. Optic nerve
d. Incisive foramen	_______	4. Vagus and accessory nerves
e. Internal acoustic meatus	_______	5. Nasopalatine nerves
f. Jugular foramen	_______	6. Mandibular division of trigeminal nerve
g. Optic canal	_______	7. Vestibular and cochlear nerves

Part E Assessments

Identify the numbered bones and features of the skulls indicated in figures 14.8, 14.9, 14.10, 14.11, and 14.12. 2 3

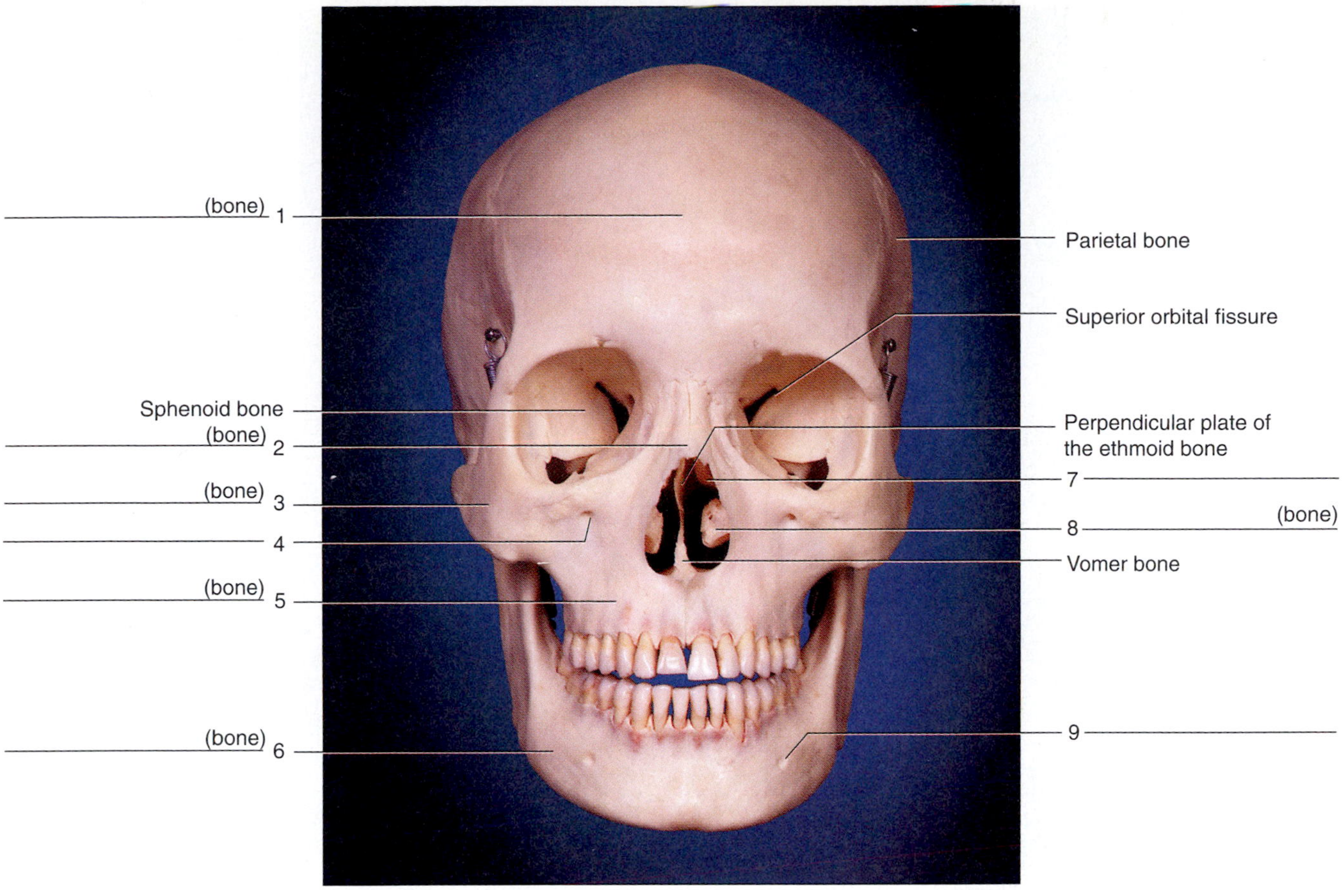

Terms:

Frontal bone
Inferior nasal concha
Infraorbital foramen
Mandible
Maxilla
Mental foramen
Middle nasal concha (of ethmoid bone)
Nasal bone
Zygomatic bone

Figure 14.8 Identify the bones and features indicated on this anterior view of the skull, using the terms provided.

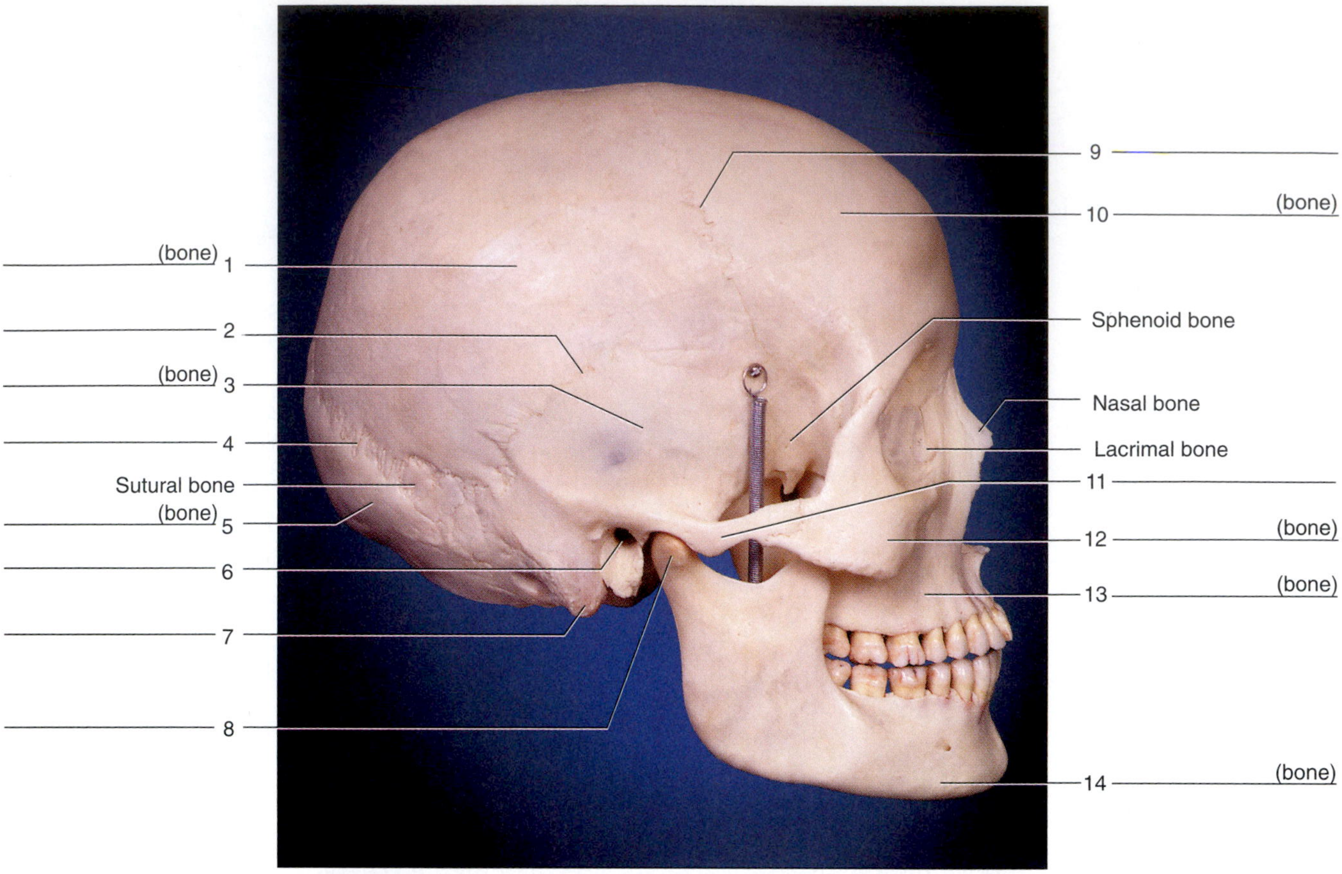

Terms:

Coronal suture
External acoustic meatus
Frontal bone
Lambdoid suture
Mandible
Mandibular condyle
Mastoid process
Maxilla
Occipital bone
Parietal bone
Squamous suture
Temporal bone
Zygomatic bone
Zygomatic process (of temporal bone)

Figure 14.9 Identify the bones and features indicated on this lateral view of the skull, using the terms provided.

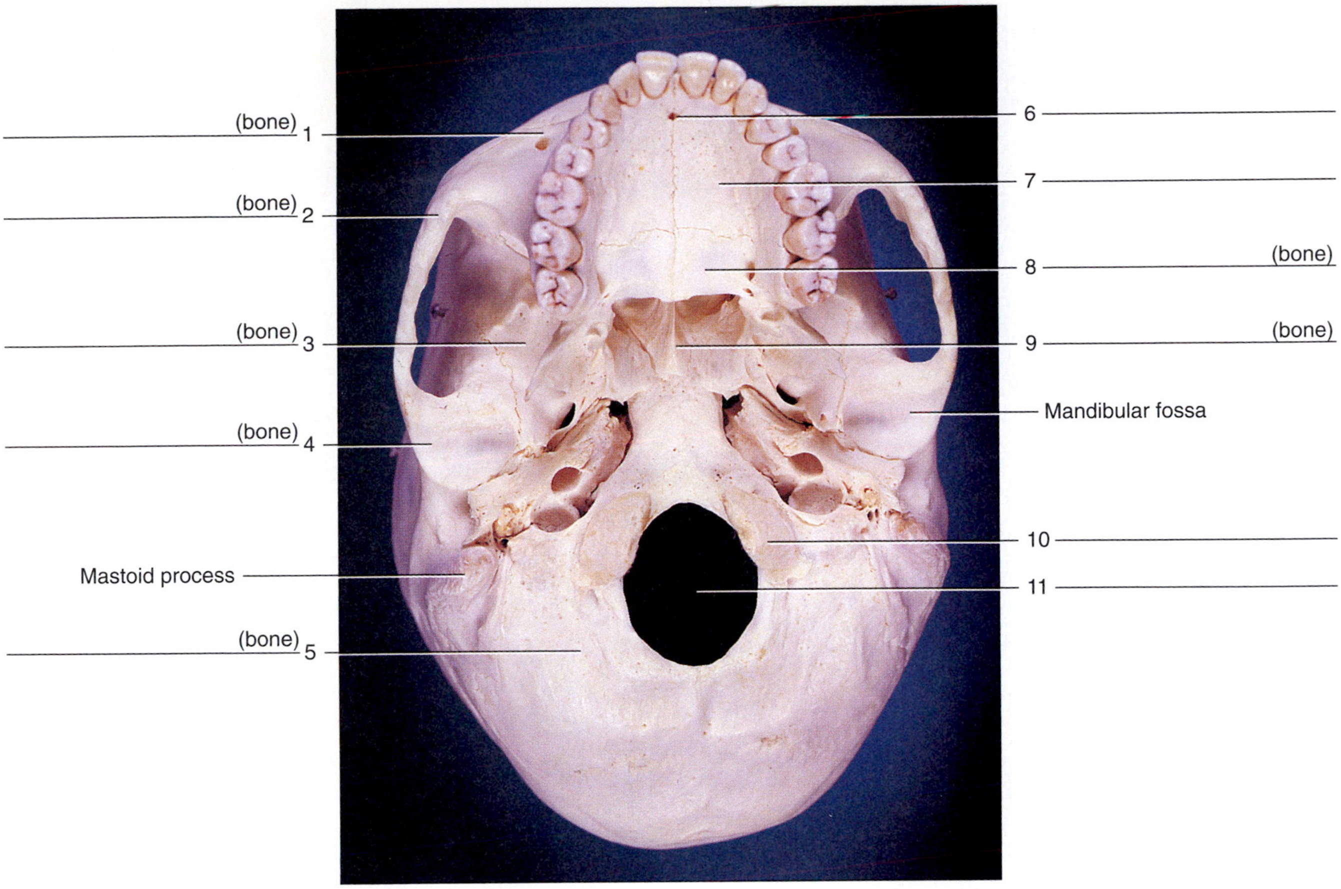

Terms:

Foramen magnum	Occipital condyle	Temporal bone
Incisive foramen	Palatine bone	Vomer bone
Maxilla	Palatine process of maxilla	Zygomatic bone
Occipital bone	Sphenoid bone	

Figure 14.10 Identify the bones and features indicated on this inferior view of the skull, using the terms provided.

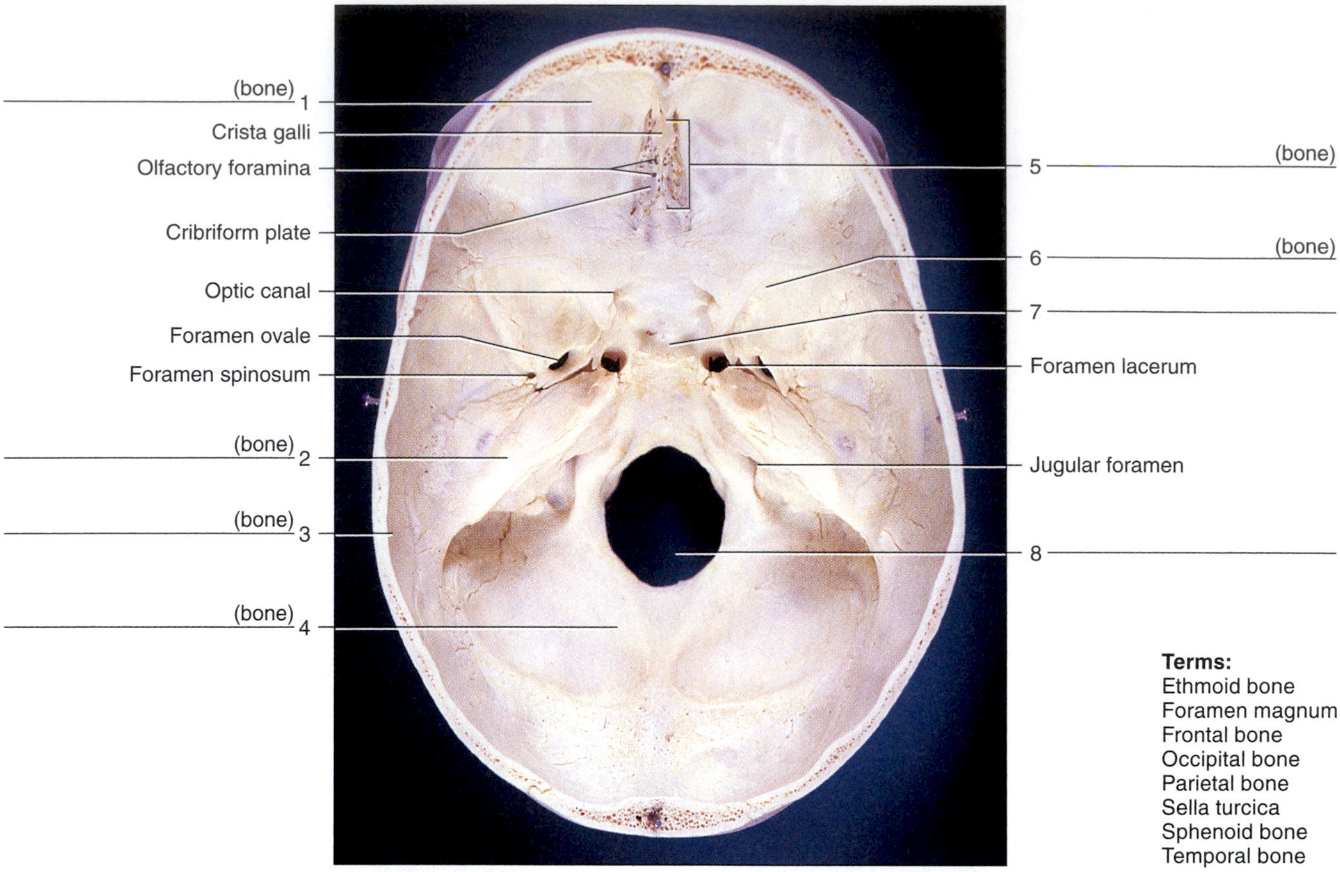

Terms:
Ethmoid bone
Foramen magnum
Frontal bone
Occipital bone
Parietal bone
Sella turcica
Sphenoid bone
Temporal bone

Figure 14.11 Identify the bones and features on this floor of the cranial cavity of a skull, using the terms provided.

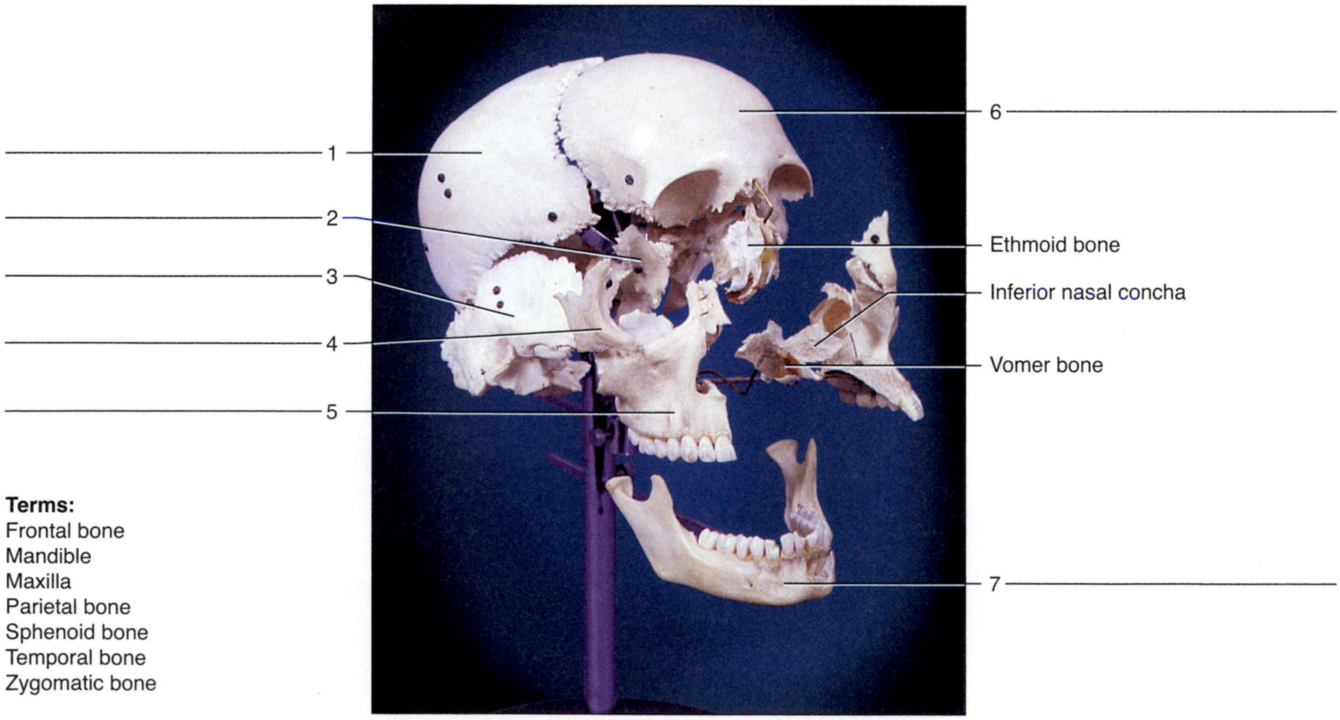

Terms:
Frontal bone
Mandible
Maxilla
Parietal bone
Sphenoid bone
Temporal bone
Zygomatic bone

Figure 14.12 Identify the bones on this disarticulated skull, using the terms provided.

Laboratory Exercise 15

Vertebral Column and Thoracic Cage

Materials Needed

Textbook
Human skeleton, articulated
Samples of cervical, thoracic, and lumbar vertebrae
Human skeleton, disarticulated

The vertebral column, consisting of twenty-six bones, extends from the skull to the pelvis and forms the vertical axis of the human skeleton. The vertebral column includes seven cervical vertebrae, twelve thoracic vertebrae, five lumbar vertebrae, one sacrum of five fused vertebrae, and one coccyx of usually four fused vertebrae. To help you to remember the number of cervical, thoracic, and lumbar vertebrae from superior to inferior, consider this saying: breakfast at 7, lunch at 12, and dinner at 5. These vertebrae are separated from one another by cartilaginous intervertebral discs and are held together by ligaments.

The thoracic cage surrounds the thoracic and upper abdominal cavities. It includes the ribs, the thoracic vertebrae, the sternum, and the costal cartilages. Men and women, although variation can exist, possess the same total bone number of 206.

Purpose of the Exercise

To examine the vertebral column and the thoracic cage of the human skeleton and to identify the bones and major features of these parts.

LEARNING OUTCOMES

After completing this exercise, you should be able to

1. Identify the major features of the vertebral column.
2. Locate the features of a vertebra.
3. Distinguish between cervical, thoracic, and lumbar vertebrae, and locate the sacrum and coccyx.
4. Identify the structures and functions of the thoracic cage.
5. Distinguish between true and false ribs.

EXPLORE

Procedure A—The Vertebral Column

1. Review the section entitled "Vertebral Column" in chapter 7 of the textbook.
2. As a review activity, label figures 15.1, 15.2, 15.3, and 15.4.
3. Examine the vertebral column of the human skeleton and locate the following bones and features. At the same time, locate as many of the corresponding bones and features in your skeleton as possible.

atlas (C1)	(1)
axis (C2)	(1)
vertebra prominens (C7)	(1)
cervical vertebrae (includes atlas, axis, and vertebra prominens)	(7)
thoracic vertebrae	(12)
lumbar vertebrae	(5)
intervertebral discs (fibrocartilage)	
vertebral canal (contains spinal cord)	
sacrum	(1)
coccyx	(1)
cervical curvature	
thoracic curvature	
lumbar curvature	
sacral curvature	
intervertebral foramina (passageway for spinal nerves)	

Critical Thinking Application

Note the four curvatures of the vertebral column. What functional advantages exist with curvatures for skeletal structure instead of a straight vertebral column?

____

____

____

____

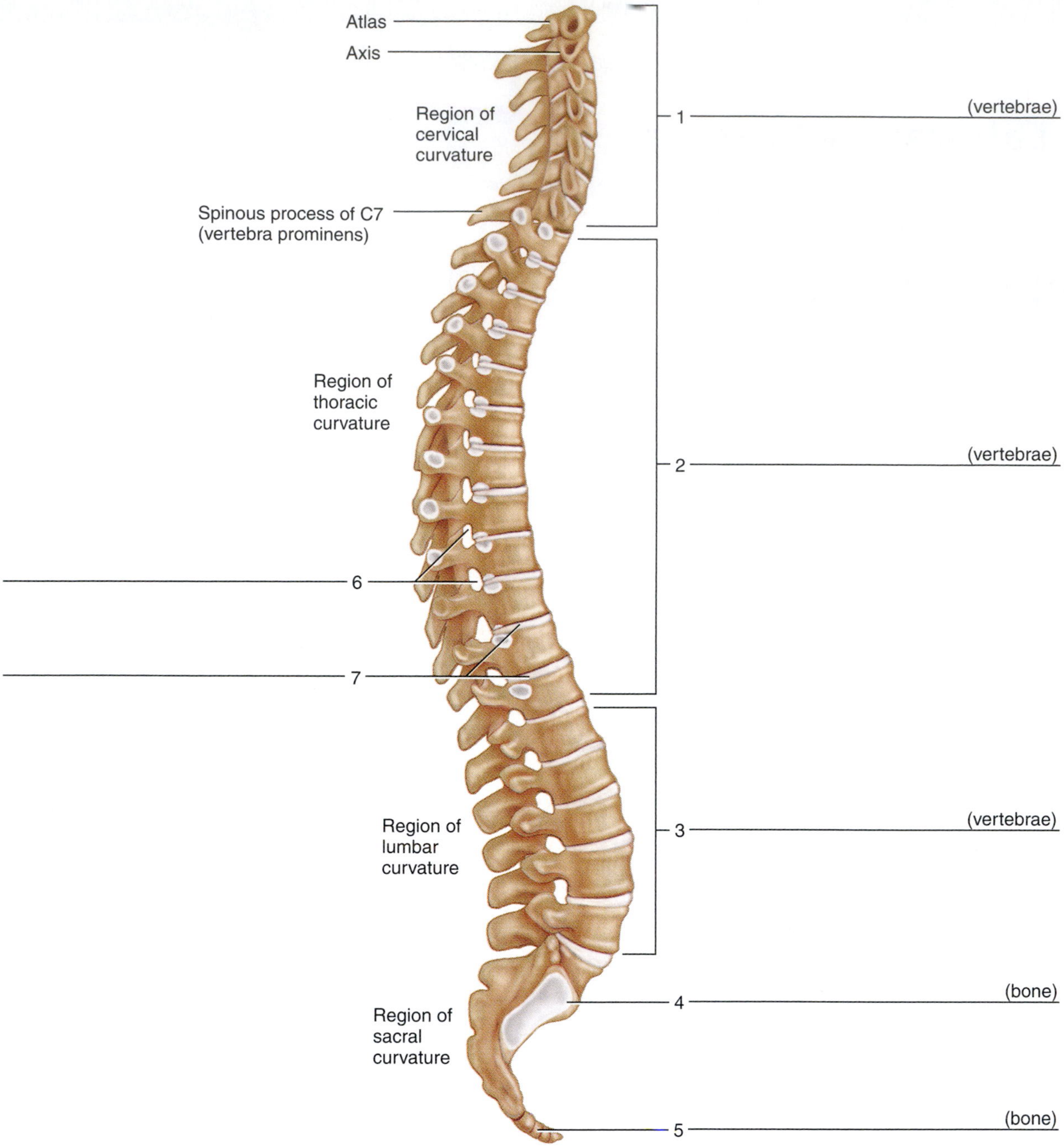

Figure 15.1 Label the bones and features of the vertebral column (right lateral view). 1

____ Body
____ Dens
____ Facet that articulates with occipital condyle
____ Spinous process
____ Superior articular facet
____ Transverse foramen
____ Transverse process
____ Vertebral foramen

(a) Atlas (C1)

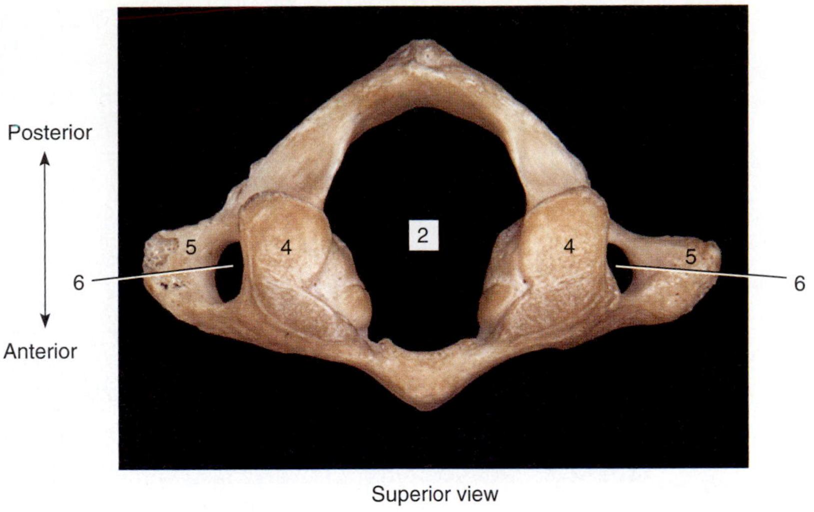

(b) Axis (C2)

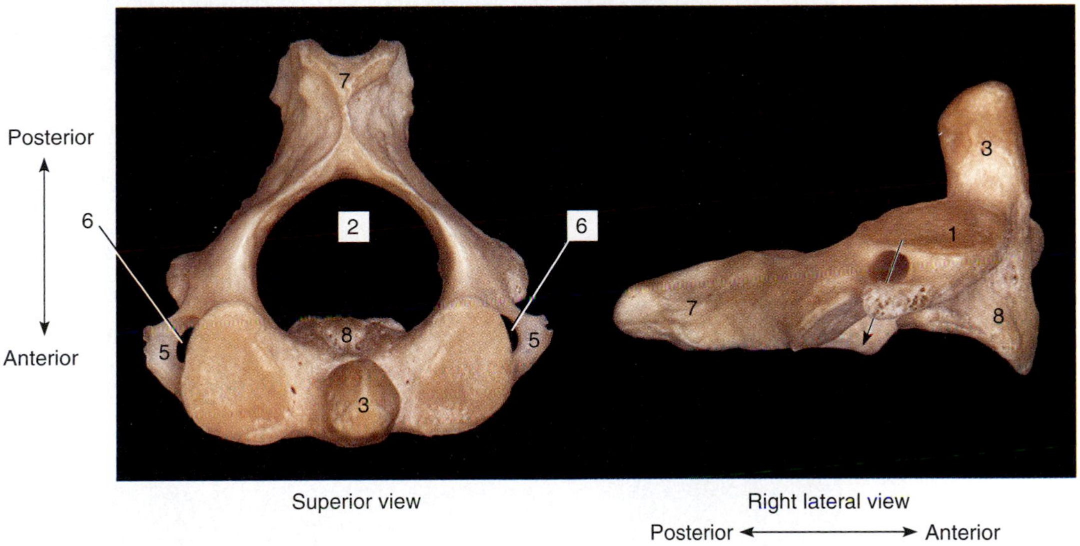

Figure 15.2 Label the superior features of (*a*) the atlas and the superior and right lateral features of (*b*) the axis by placing the correct numbers in the spaces provided. (The broken arrow indicates a transverse foramen.) 2 3

_____ Body
_____ Inferior notch
_____ Lamina
_____ Pedicle
_____ Spinous process
_____ Superior articular process
_____ Transverse foramen
_____ Transverse process
_____ Vertebral foramen

Superior views　　Right lateral views

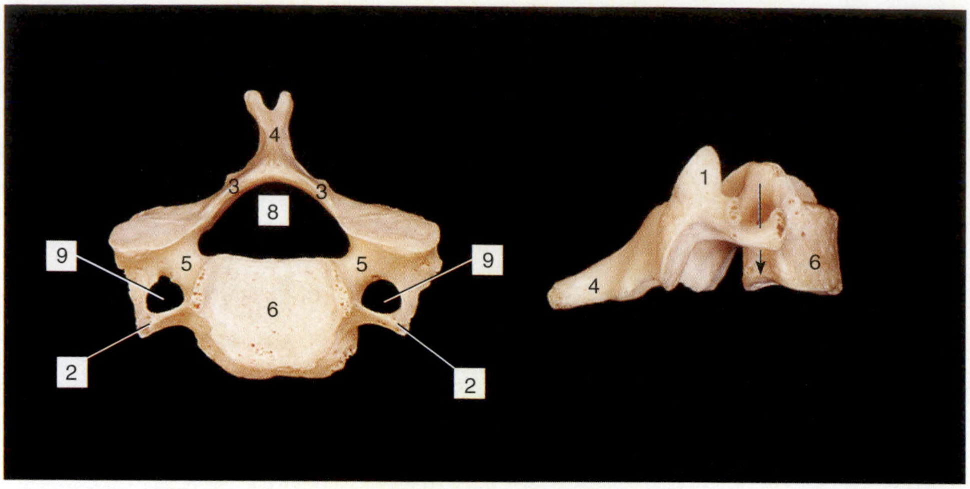

(a) Cervical vertebra

Posterior ↕ Anterior

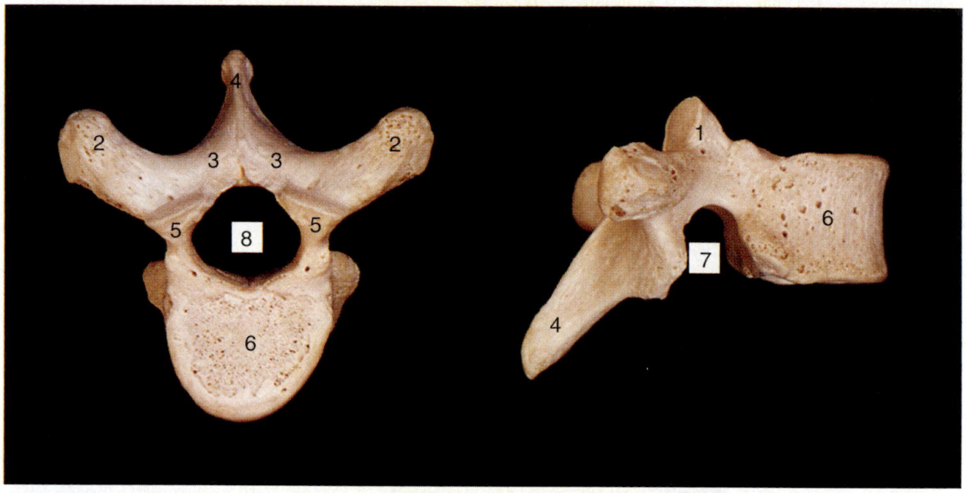

(b) Thoracic vertebra

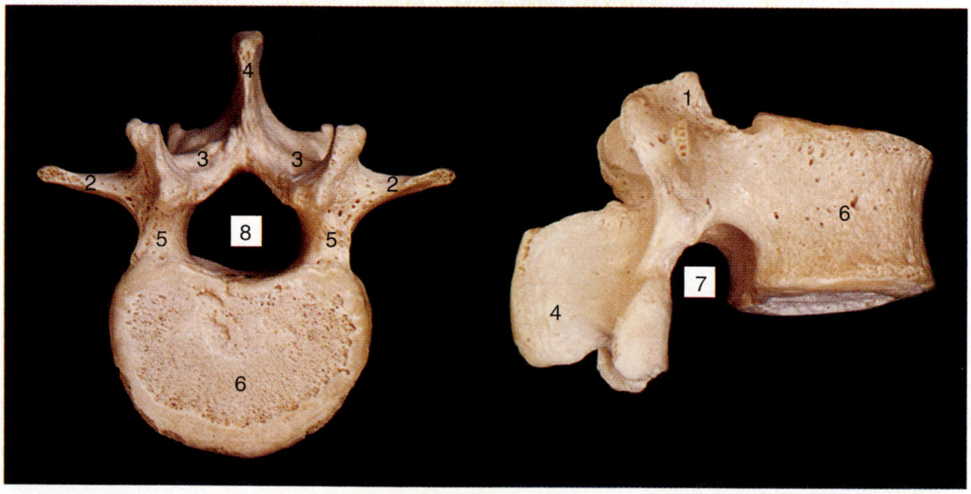

(c) Lumbar vertebra

Posterior ⟷ Anterior

Figure 15.3 Label the superior and right lateral features of the (*a*) cervical, (*b*) thoracic, and (*c*) lumbar vertebrae by placing the correct numbers in the spaces provided. (The broken arrow indicates a transverse foramen.) 2 3

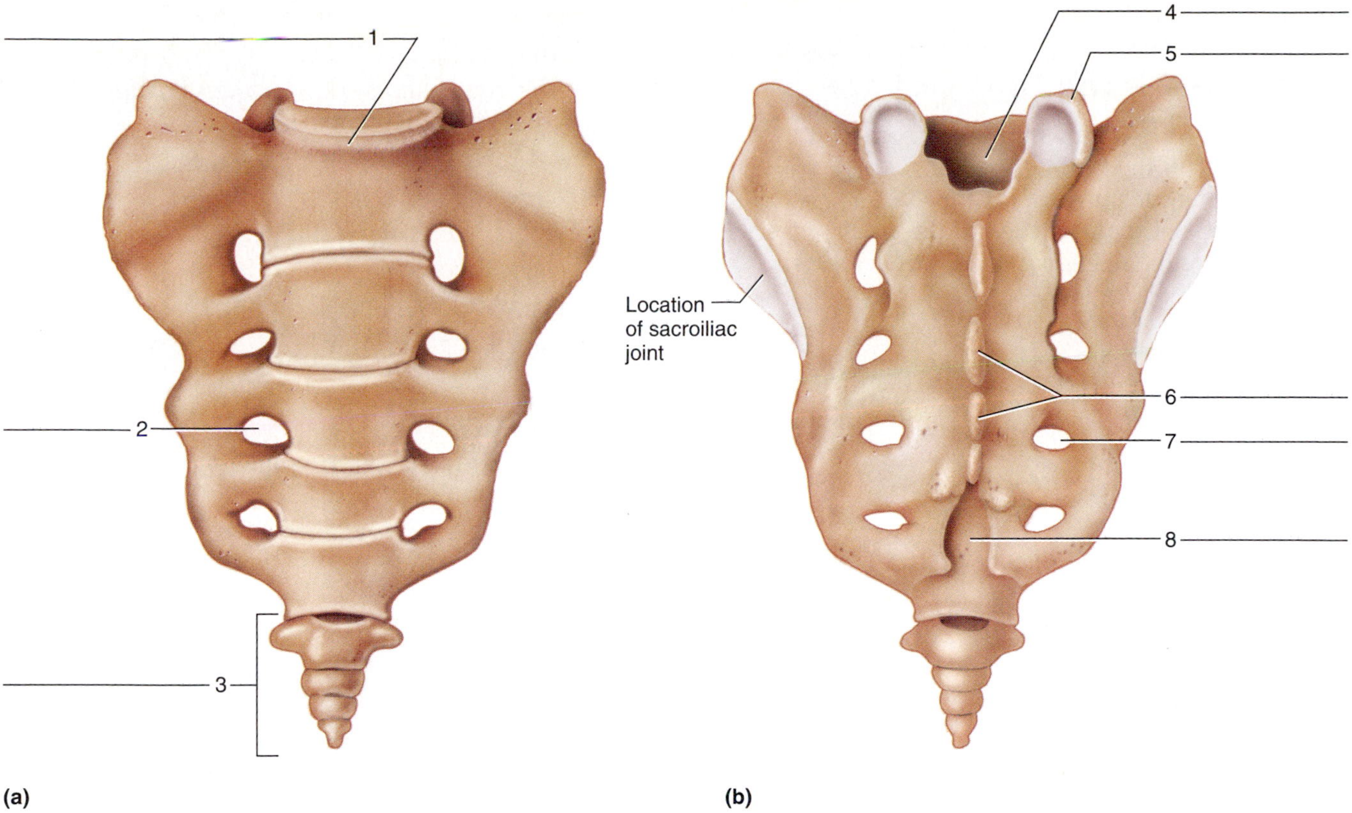

Figure 15.4 Label the coccyx and the features of the sacrum: (*a*) anterior view; (*b*) posterior view. 2 3

4. Compare the available samples of cervical, thoracic, and lumbar vertebrae by noting differences in size and shape and by locating the following features:

   vertebral foramen
   body
   pedicles
   laminae
   spinous process
   vertebral arch
   transverse processes
   superior articular processes
   inferior articular processes
   inferior notch
   transverse foramina
   facets
   dens of axis

5. Examine the sacrum and coccyx. Locate the following features:

   sacrum
   - superior articular process
   - posterior sacral foramen
   - anterior sacral foramen
   - sacral promontory
   - sacral canal
   - tubercles
   - median sacral crest
   - sacral hiatus

   coccyx

6. Complete Parts A and B of Laboratory Report 15.

EXPLORE

Procedure B—The Thoracic Cage

1. Review the section entitled "Thoracic Cage" in chapter 7 of the textbook.
2. As a review activity, label figure 15.5.
3. Examine the thoracic cage of the human skeleton and locate the following bones and features:

   rib
   - head
   - tubercle
   - neck
   - shaft
   - anterior (sternal) end
   - facets
   - true ribs (pairs 1–7)
   - false ribs (pairs 8–12) (includes floating ribs)
     - floating ribs (pairs 11–12)

   costal cartilages (hyaline cartilage)

   sternum
   - jugular (suprasternal) notch
   - clavicular notch
   - manubrium
   - sternal angle
   - body
   - xiphoid process
4. Complete Parts C and D of the laboratory report.

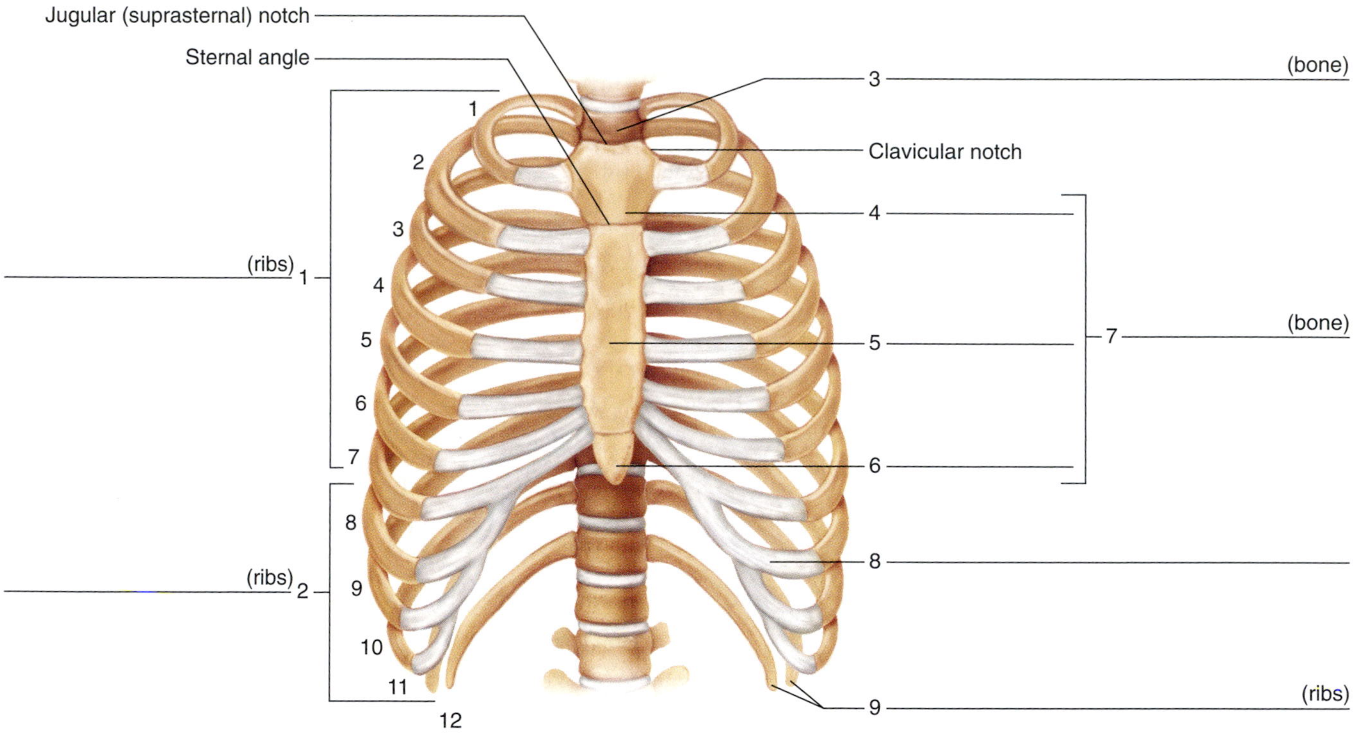

Figure 15.5 Label the bones and features of the thoracic cage (anterior view). 4 5

Name ______________________________

Date ______________________________

Section ______________________________

The [A] corresponds to the indicated Learning Outcome(s) listed at the beginning of the laboratory exercise.

Vertebral Column and Thoracic Cage

Laboratory Report 15

Part A Assessments

Complete the following statements:

1. The vertebral column encloses and protects the ______________________. 1
2. The number of separate bones in the vertebral column of an adult is ______________________. 3
3. The ______________________ of the vertebrae support the weight of the head and trunk. 1
4. The ______________________ separate adjacent vertebrae, and they soften the forces created by walking. 1
5. The pedicles, laminae, and ______________________ of a vertebra form the vertebral arch. 2
6. The intervertebral foramina provide passageways for ______________________. 1
7. Transverse foramina of cervical vertebrae serve as passageways for ______________________ leading to the brain. 2
8. The first vertebra is also called the ______________________. 3
9. The second vertebra is also called the ______________________. 1
10. When the head is moved from side to side, the first vertebra pivots around the ______________________ of the second vertebra. 2
11. The ______________________ vertebrae have the largest and strongest bodies. 3
12. The number of vertebrae that fuse to form the sacrum is ______________________. 1
13. The upper, anterior margin of the sacrum that projects forward is called the ______________________. 1
14. An opening called the ______________________ exists at the tip of the sacral canal. 1

Part B Assessments

Based on your observations, compare typical cervical, thoracic, and lumbar vertebrae in relation to the characteristics indicated in the table. For your responses, consider characteristics such as size, shape, presence or absence, and unique features. 2 3

Vertebra	Number	Size	Body	Spinous Process	Transverse Foramina
Cervical					
Thoracic					
Lumbar					

Part C Assessments

Complete the following statements:

1. The adult skeleton of most men and women contains a total of ________________ (number) bones.
2. The last two pairs of ribs that have no cartilaginous attachments to the sternum are sometimes called ________________________ ribs. 5
3. There are ______________________ pairs of true ribs. 5
4. Costal cartilages are composed of ______________________ tissue. 4
5. The manubrium articulates with the ______________________ on its superior border. 4
6. List three general functions of the thoracic cage. 4 ____

____

Part D Assessments

Identify the bones and features indicated in the radiograph of the neck in figure 15.6. 1 2

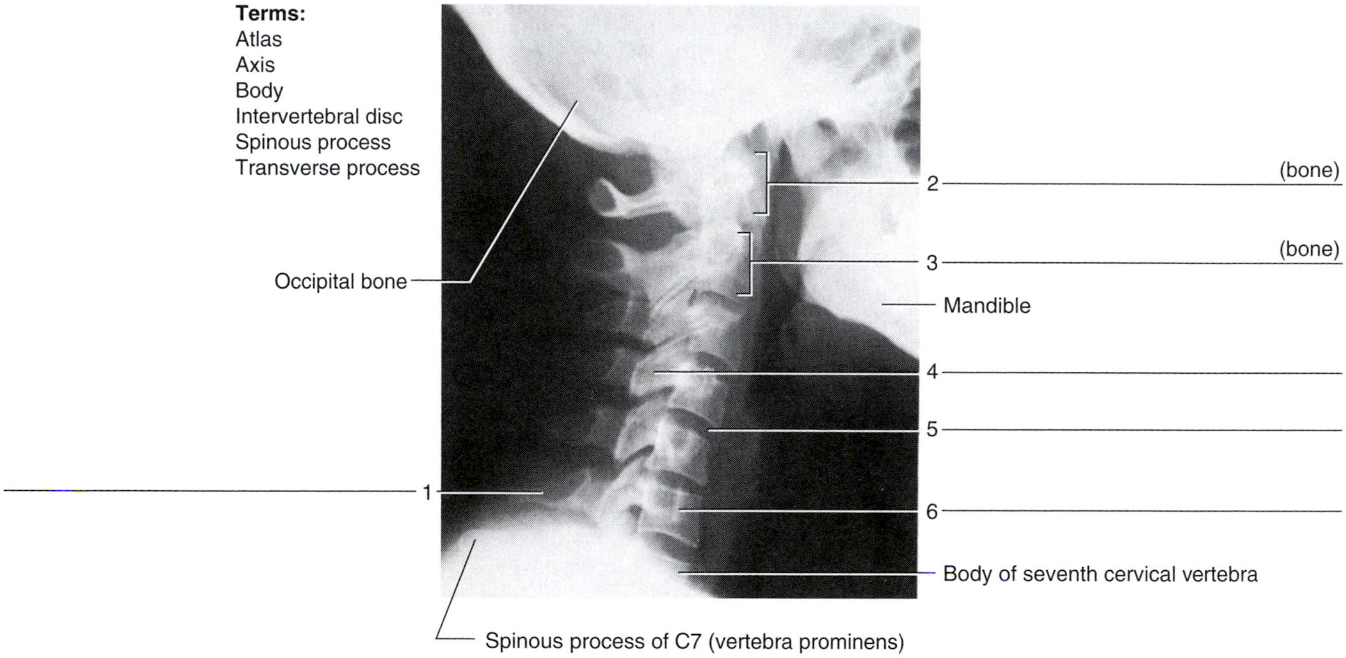

Figure 15.6 Identify the bones and features indicated in this radiograph of the neck (lateral view), using the terms provided.

Laboratory Exercise 16

Pectoral Girdle and Upper Limb

Materials Needed

Textbook
Human skeleton, articulated
Human skeleton, disarticulated

For Learning Extension:
Colored pencils

The pectoral girdle (shoulder girdle) consists of two clavicles and two scapulae. These parts support the upper limbs and serve as attachments for various muscles that move these limbs.

Each upper limb includes a humerus, a radius, an ulna, eight carpals, five metacarpals, and fourteen phalanges. These bones form the framework of the arm, forearm, and hand. They also function as parts of levers when the limbs are moved.

Purpose of the Exercise

To examine the bones of the pectoral girdle and upper limb and to identify the major features of these bones.

LEARNING OUTCOMES

After completing this exercise, you should be able to

1. Locate and identify the bones of the pectoral girdle and their major features.
2. Locate and identify the bones of the upper limb and their major features.

EXPLORE

Procedure A—The Pectoral Girdle

1. Review the section entitled "Pectoral Girdle" in chapter 7 of the textbook.
2. As a review activity, label figures 16.1 and 16.2.
3. Examine the bones of the pectoral girdle and locate the following features. At the same time, locate as many of the corresponding surface bones and features of your own skeleton as possible.

clavicle
- medial (sternal) end
- lateral (acromial) end

scapula
- spine
- acromion process
- coracoid process
- glenoid cavity
- borders
  - superior border
  - medial (vertebral) border
  - lateral (axillary) border
- fossae
  - supraspinous fossa
  - infraspinous fossa

Critical Thinking Application

Why is the clavicle a bone that can easily fracture?

____

____

____

____

____

4. Complete Part A of Laboratory Report 16.

EXPLORE

Procedure B—The Upper Limb

1. Review the section entitled "Upper Limb" in chapter 7 of the textbook.
2. As a review activity, label figures 16.3, 16.4, 16.5, and 16.6.

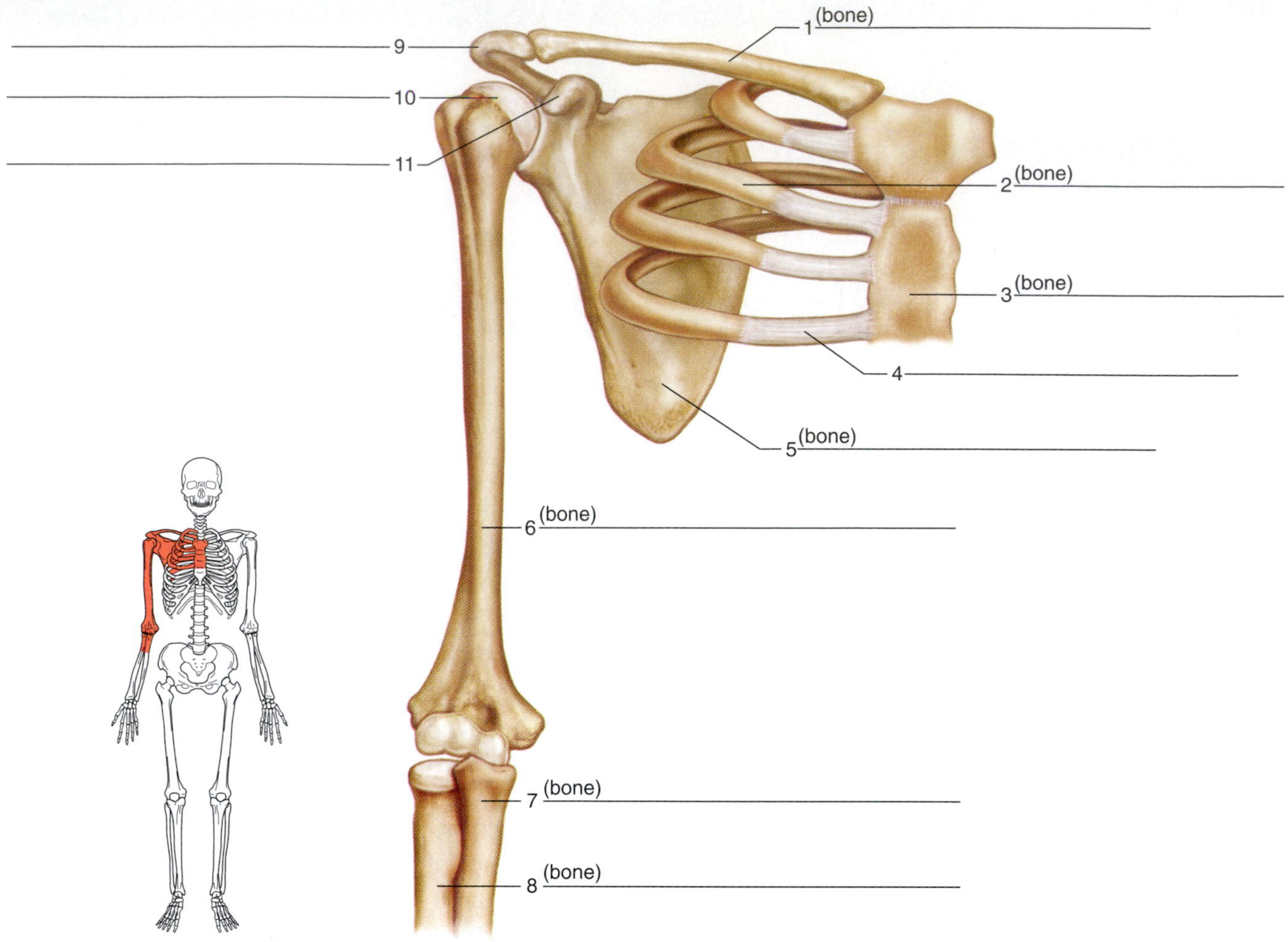

Figure 16.1 Label the bones and features of the right shoulder and upper limb (anterior view). 1 2

3. Examine the following bones and features of the upper limb:

humerus

- proximal features
  - head
  - greater tubercle
  - lesser tubercle
  - anatomical neck
  - surgical neck
  - intertubercular groove (sulcus)
- shaft
  - deltoid tuberosity
- distal features
  - capitulum
  - trochlea
  - medial epicondyle
  - lateral epicondyle
  - coronoid fossa
  - olecranon fossa

radius

- head
- radial tuberosity
- styloid process
- ulnar notch

ulna

- trochlear notch (semilunar notch)
- radial notch
- olecranon process
- coronoid process
- styloid process
- head

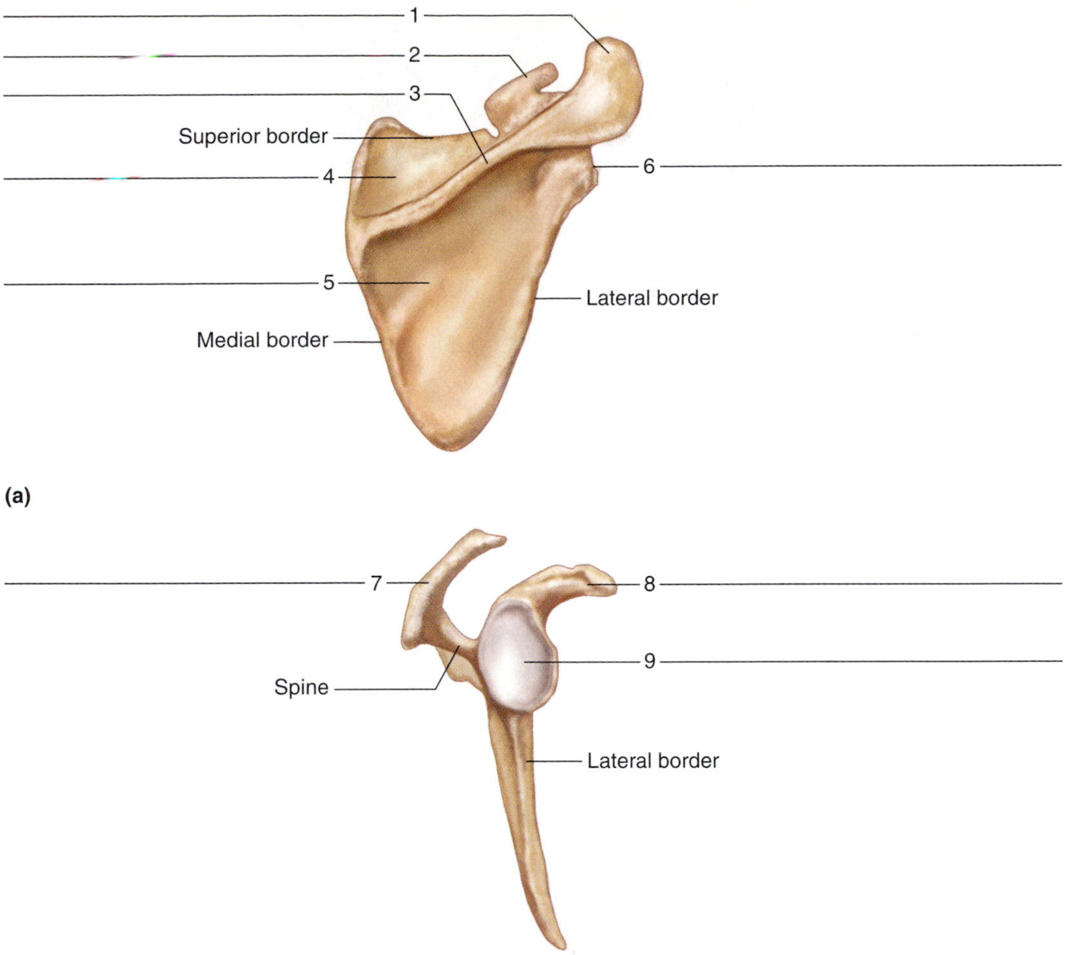

Figure 16.2 Label (*a*) the posterior surface of the right scapula and (*b*) the lateral aspect of the right scapula. 1

carpal bones
- proximal row (listed lateral to medial)
  - scaphoid
  - lunate
  - triquetrum
  - pisiform
- distal row (listed medial to lateral)
  - hamate
  - capitate
  - trapezoid
  - trapezium

metacarpal bones

phalanges
- proximal phalanx
- middle phalanx
- distal phalanx

4. Complete Parts B, C, and D of the laboratory report.

The following mnemonic device will help you learn the eight carpals:

So **L**ong **T**op **P**art
Here **C**omes **T**he **T**humb

The first letter of each word corresponds to the first letter of a carpal. This device arranges the carpals in order for the proximal, transverse row of four bones from lateral to medial, followed by the distal, transverse row from medial to lateral, which ends nearest the thumb. This arrangement assumes the hand is in the anatomical position.

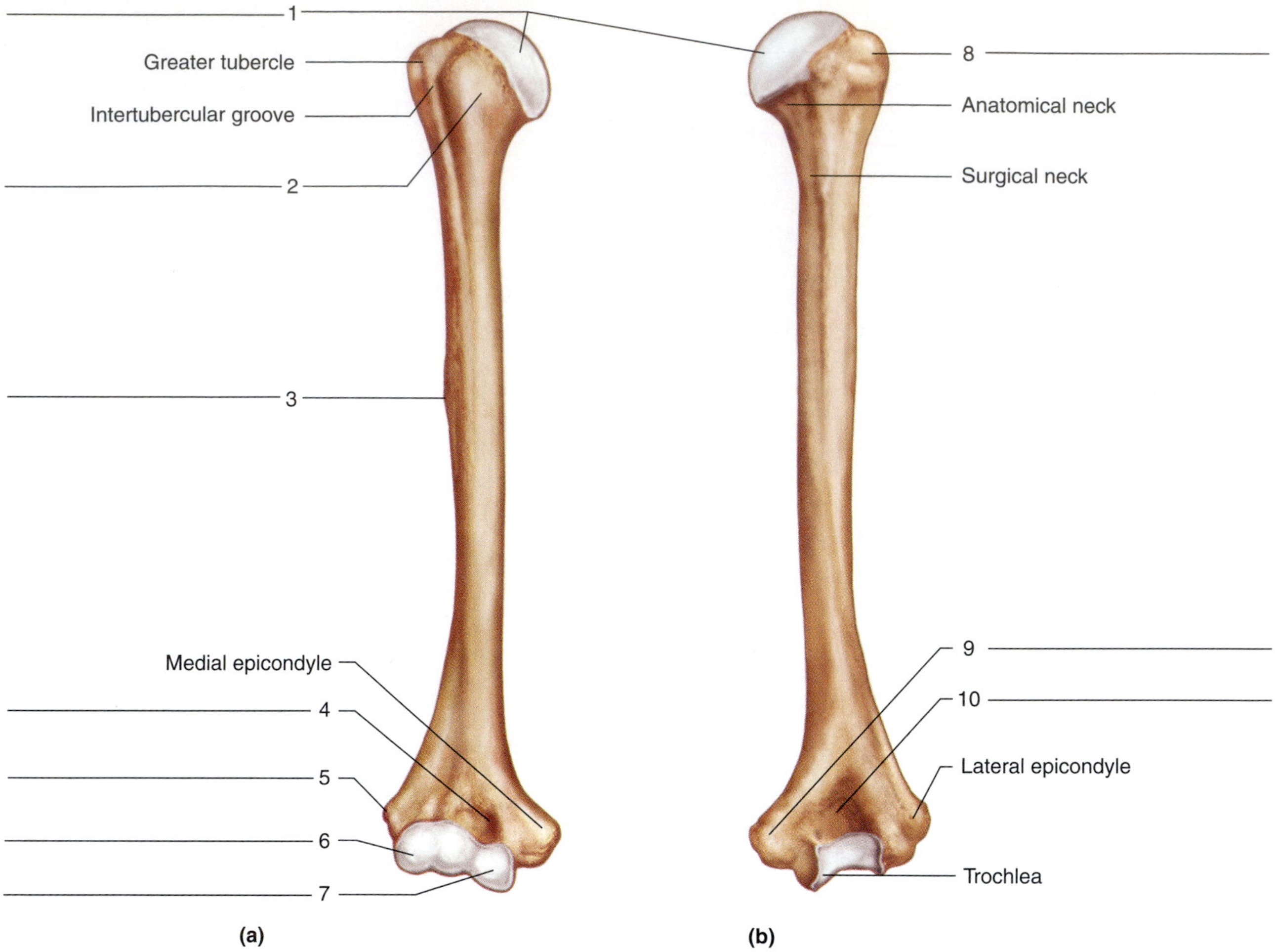

Figure 16.3 Label the (*a*) anterior surface and (*b*) posterior surface of the right humerus. 2

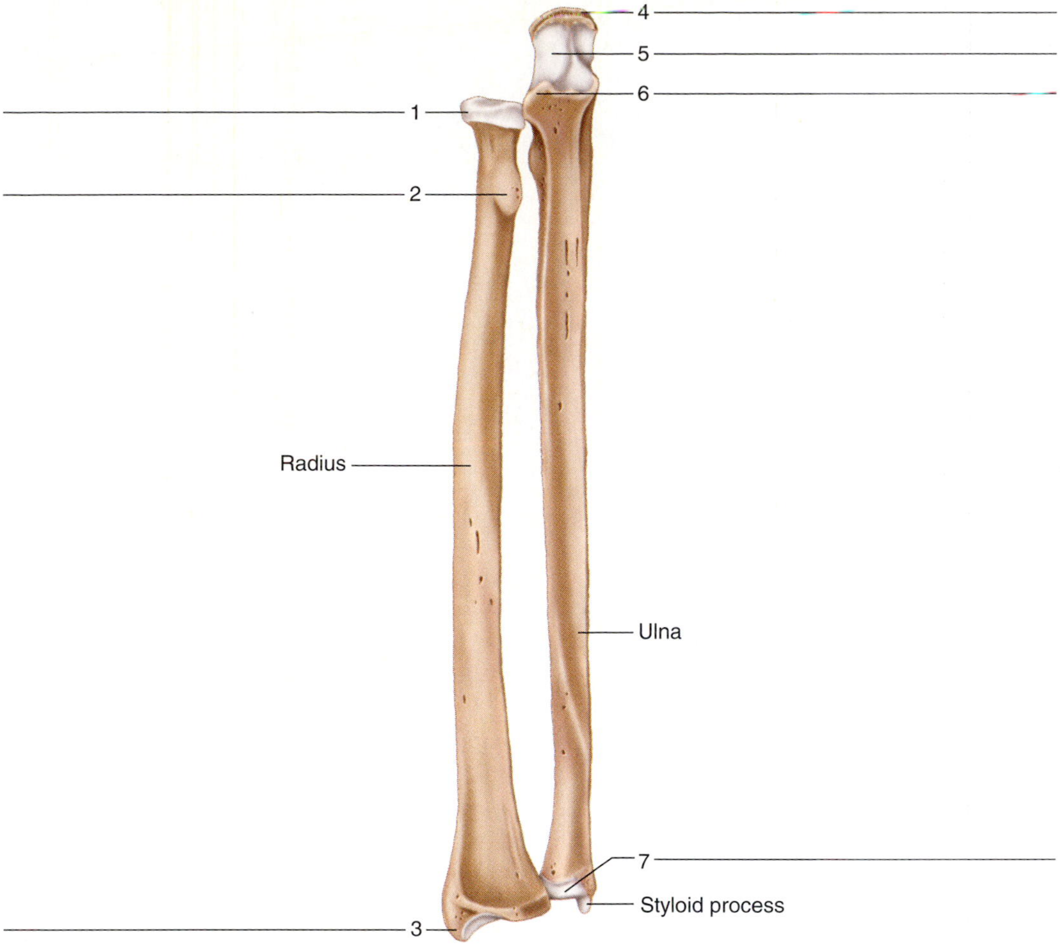

Figure 16.4 Label the major anterior features of the right radius and ulna (anterior view). 2

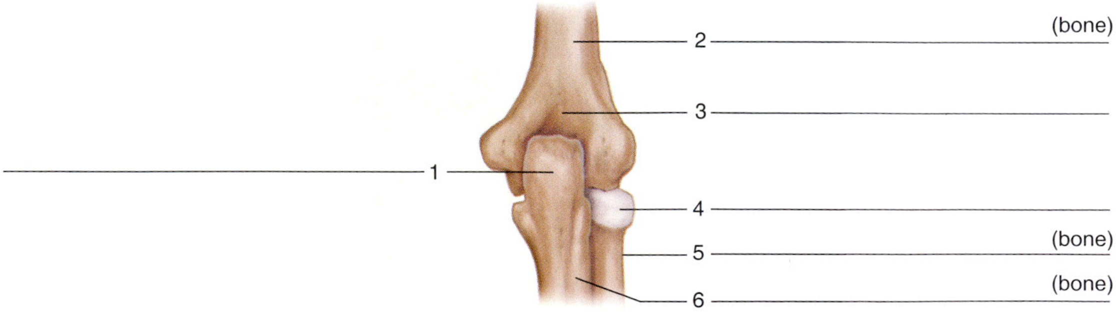

Figure 16.5 Label the bones and features of the right elbow, posterior view. 2

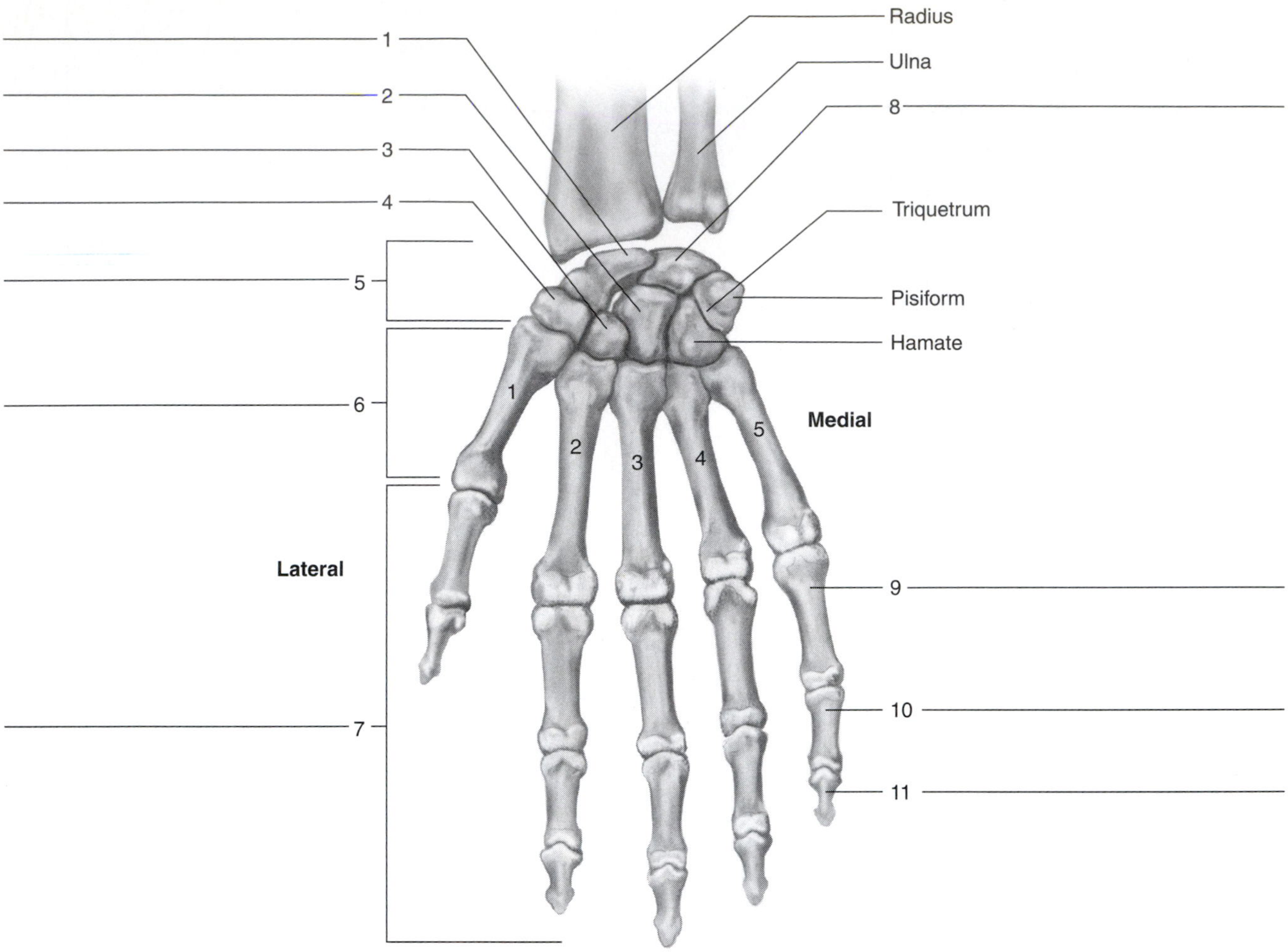

Figure 16.6 Label the bones and groups of bones in this anterior view of the right hand. 2

Learning Extension

Use different colored pencils to distinguish the individual bones in figure 16.6.

Name ________________________________

Date ________________________________

Section ________________________________

The [A] corresponds to the indicated Learning Outcome(s) listed at the beginning of the laboratory exercise.

Laboratory Report 16

Pectoral Girdle and Upper Limb

Part A Assessments

Complete the following statements:

1. The pectoral girdle is an incomplete ring because it is open in the back between the ________________ . [1]
2. The medial ends of the clavicles articulate with the ________________ of the sternum. [1]
3. The lateral ends of the clavicles articulate with the ________________ of the scapulae. [1]
4. The ________________ is a bone that serves as a brace between the sternum and the scapula. [1]
5. The ________________ divides the scapula into unequal portions. [1]
6. The tip of the shoulder is the ________________ of the scapula. [1]
7. Near the lateral end of the scapula, the ________________ curves anteriorly and inferiorly from the clavicle. [1]
8. The glenoid cavity of the scapula articulates with the ________________ of the humerus. [1]

Part B Assessments

Match the bones in column A with the bones and features in column B. Place the letter of your choice in the space provided. [2]

Column A

a. Carpals
b. Humerus
c. Metacarpals
d. Phalanges
e. Radius
f. Ulna

Column B

_______ 1. Capitate
_______ 2. Coronoid fossa
_______ 3. Deltoid tuberosity
_______ 4. Greater tubercle
_______ 5. Five palmar bones
_______ 6. Fourteen bones in digits
_______ 7. Intertubercular groove (sulcus)
_______ 8. Lunate
_______ 9. Olecranon fossa
_______ 10. Radial tuberosity
_______ 11. Trapezium
_______ 12. Trochlear notch

Part C Assessments

Identify the bones and features indicated in the radiographs of figures 16.7, 16.8, and 16.9. 1 2

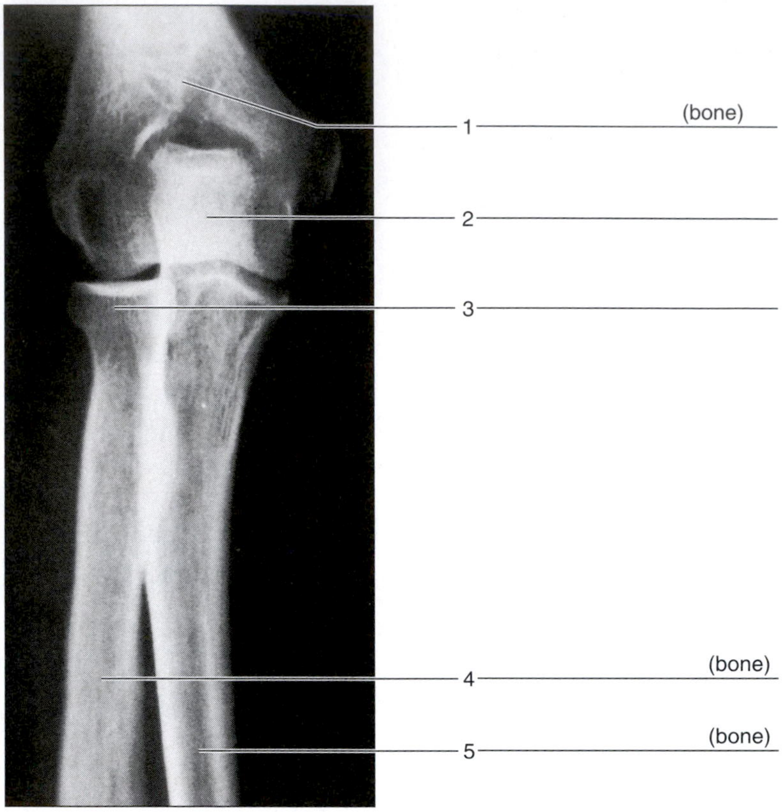

Terms:
Head of radius
Humerus
Olecranon process
Radius
Ulna

Figure 16.7 Identify the bones and features indicated on this radiograph of the right elbow (anterior view), using the terms provided.

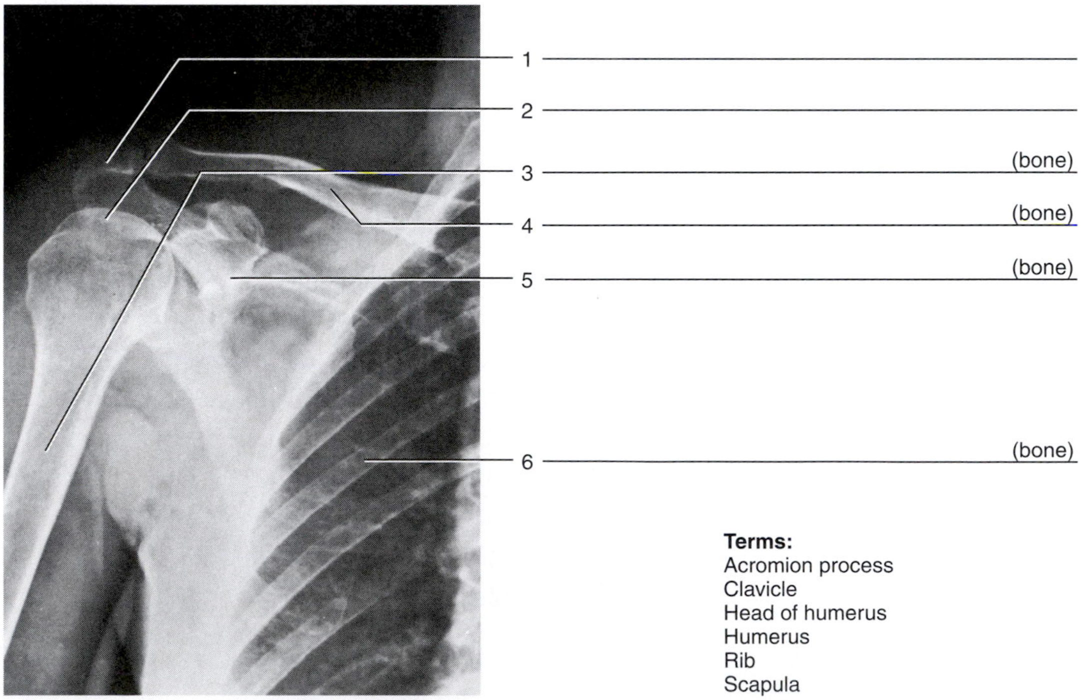

Terms:
Acromion process
Clavicle
Head of humerus
Humerus
Rib
Scapula

Figure 16.8 Identify the bones and features indicated on this radiograph of the shoulder (anterior view), using the terms provided.

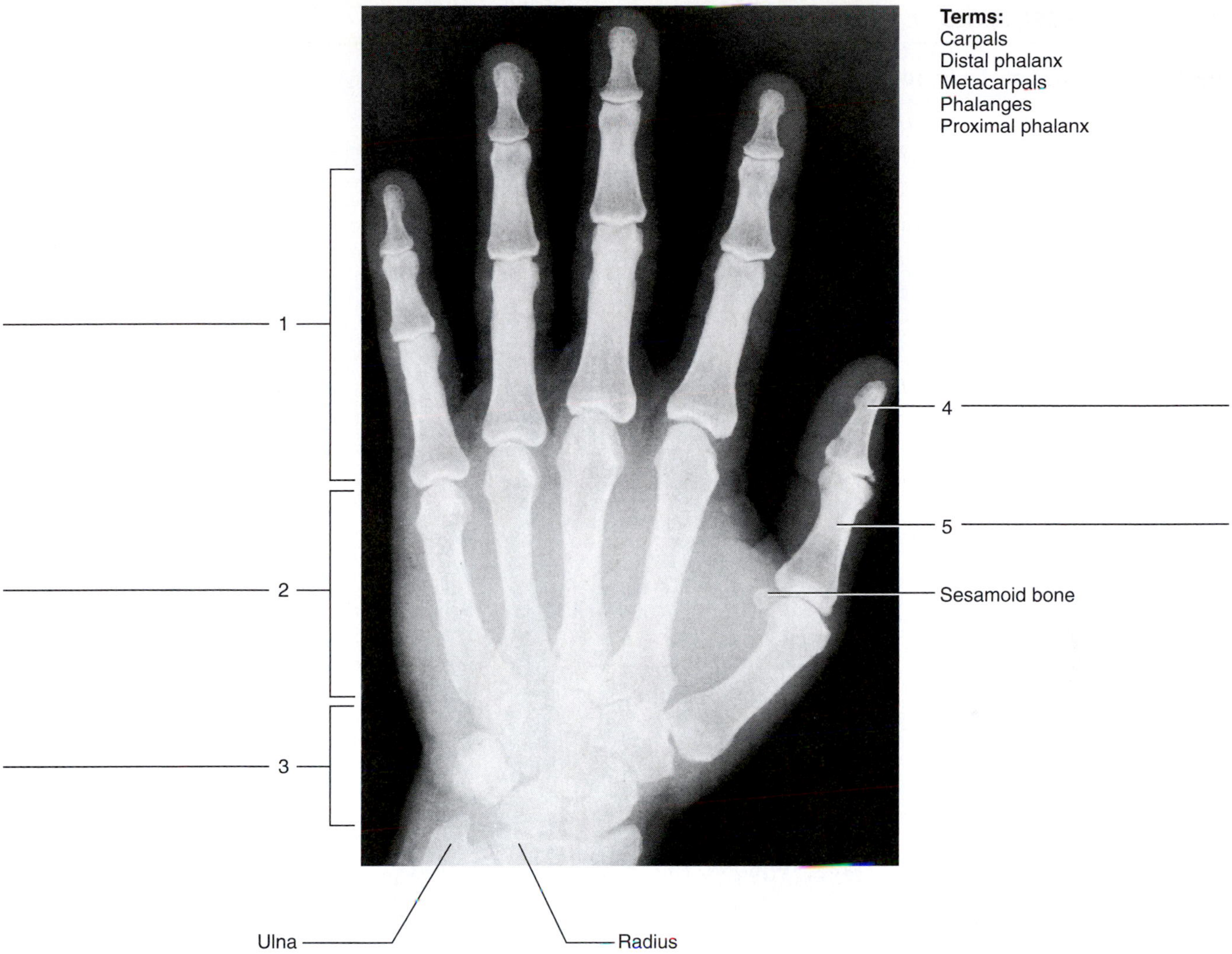

Figure 16.9 Identify the bones indicated on this radiograph of the right hand (anterior view), using the terms provided.

Part D Assessments

Identify the bones of the hand in figure 16.10. 2

________ Capitate
________ Distal phalanges
________ Hamate
________ Lunate
________ Metacarpals
________ Middle phalanges
________ Pisiform
________ Proximal phalanges
________ Scaphoid
________ Trapezium
________ Trapezoid
________ Triquetrum

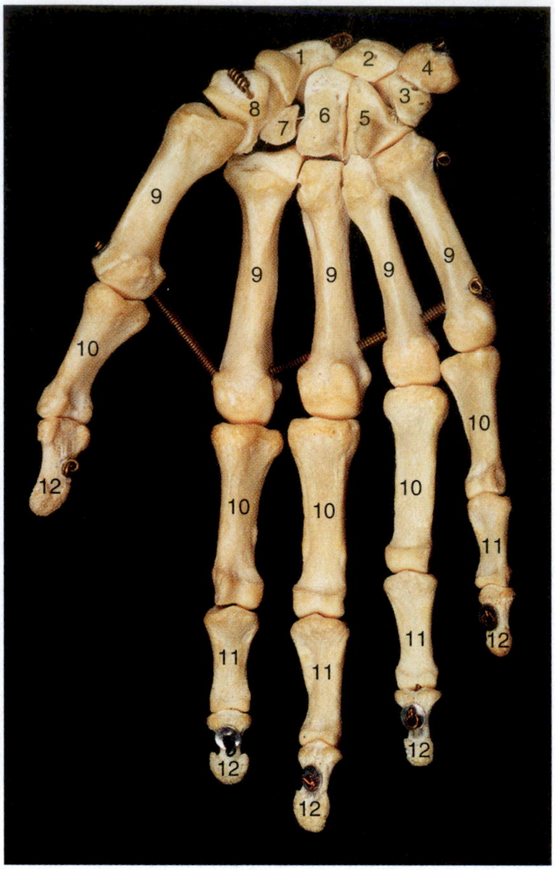

Figure 16.10 Label the bones numbered on this anterior view of the right hand by placing the correct numbers in the spaces provided.

Laboratory Exercise 17

Pelvic Girdle and Lower Limb

Materials Needed

Textbook
Human skeleton, articulated
Human skeleton, disarticulated
Male and female pelves

For Learning Extension:

Colored pencils

The pelvic girdle includes two hip bones that articulate with each other anteriorly at the symphysis pubis and posteriorly with the sacrum. Together, the pelvic girdle, sacrum, and coccyx comprise the pelvis. The pelvis, in turn, provides support for the trunk of the body and provides attachments for the lower limbs.

The bones of the lower limb form the framework of the thigh, leg, and foot. Each limb includes a femur, a patella, a tibia, a fibula, seven tarsals, five metatarsals, and fourteen phalanges.

Purpose of the Exercise

To examine the bones of the pelvic girdle and lower limb and to identify the major features of these bones.

LEARNING OUTCOMES

After completing this exercise, you should be able to

1. Locate and identify the bones of the pelvic girdle and their major features.
2. Locate and identify the bones of the lower limb and their major features.

EXPLORE

Procedure A—The Pelvic Girdle

1. Review the section entitled "Pelvic Girdle" in chapter 7 of the textbook.
2. As a review activity, label figures 17.1 and 17.2.
3. Examine the bones of the pelvic girdle and locate the following:

   hip bone (coxal bone; pelvic bone; innominate)
   - acetabulum
   - ilium
   - iliac crest
   - iliac fossa
   - sacroiliac joint
   - anterior superior iliac spine
   - posterior superior iliac spine
   - greater sciatic notch
   - lesser sciatic notch
   - ischium
   - ischial tuberosity
   - ischial spine
   - pubis
   - symphysis pubis
   - pubic arch
   - obturator foramen
4. Complete Part A of Laboratory Report 17.

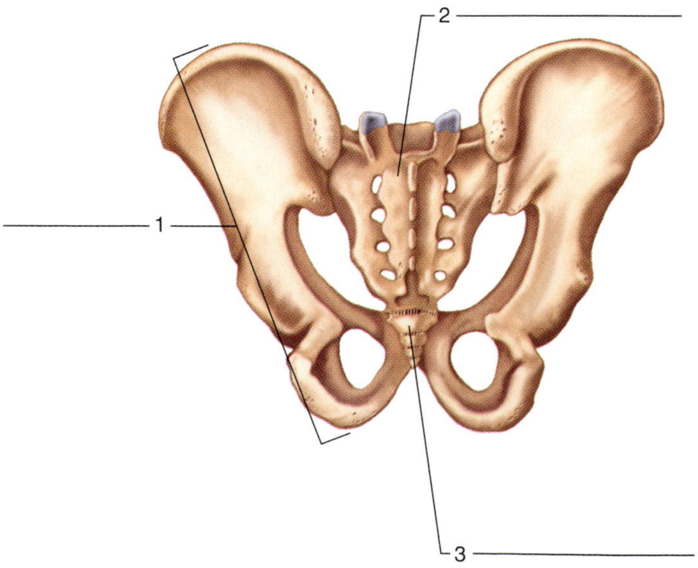

Figure 17.1 Label the bones of the pelvis (posterior view). 1

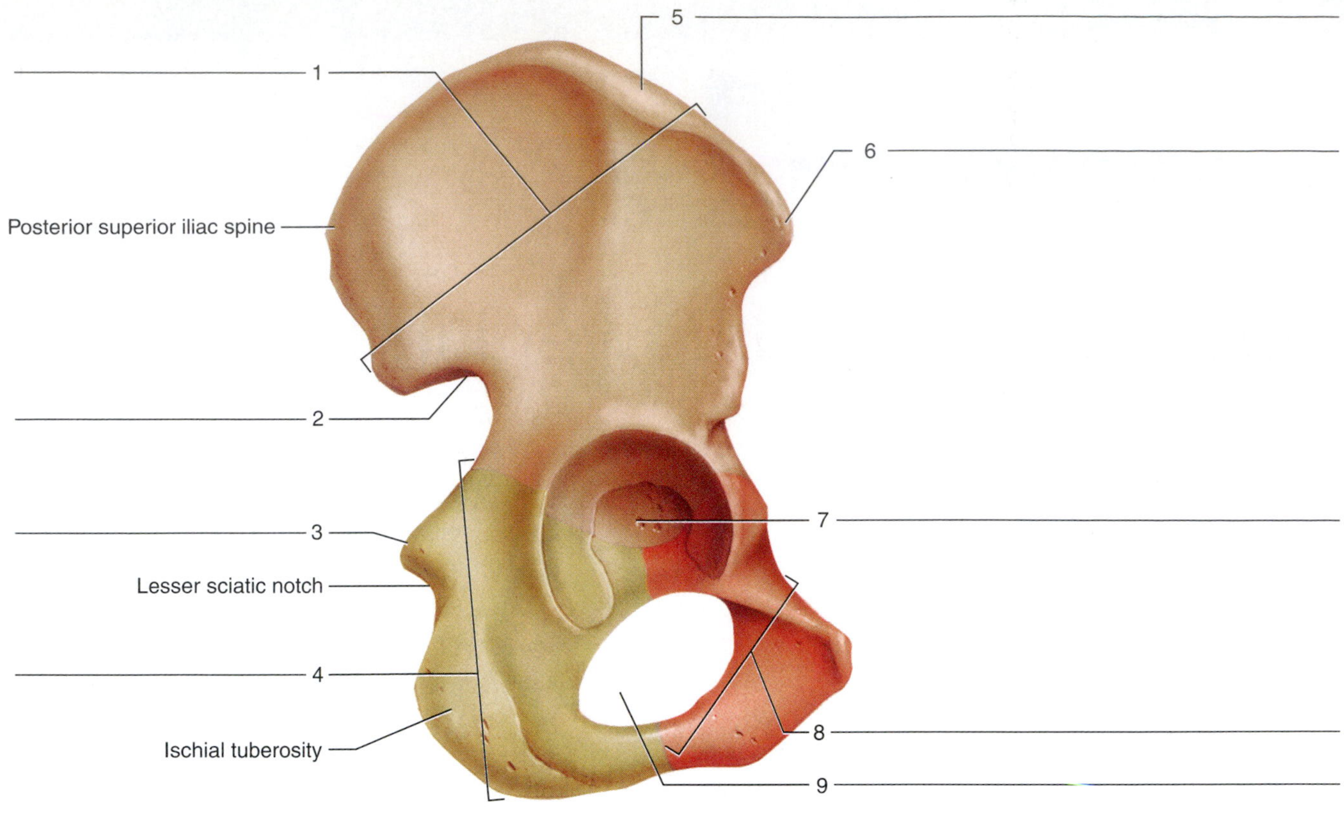

(a)

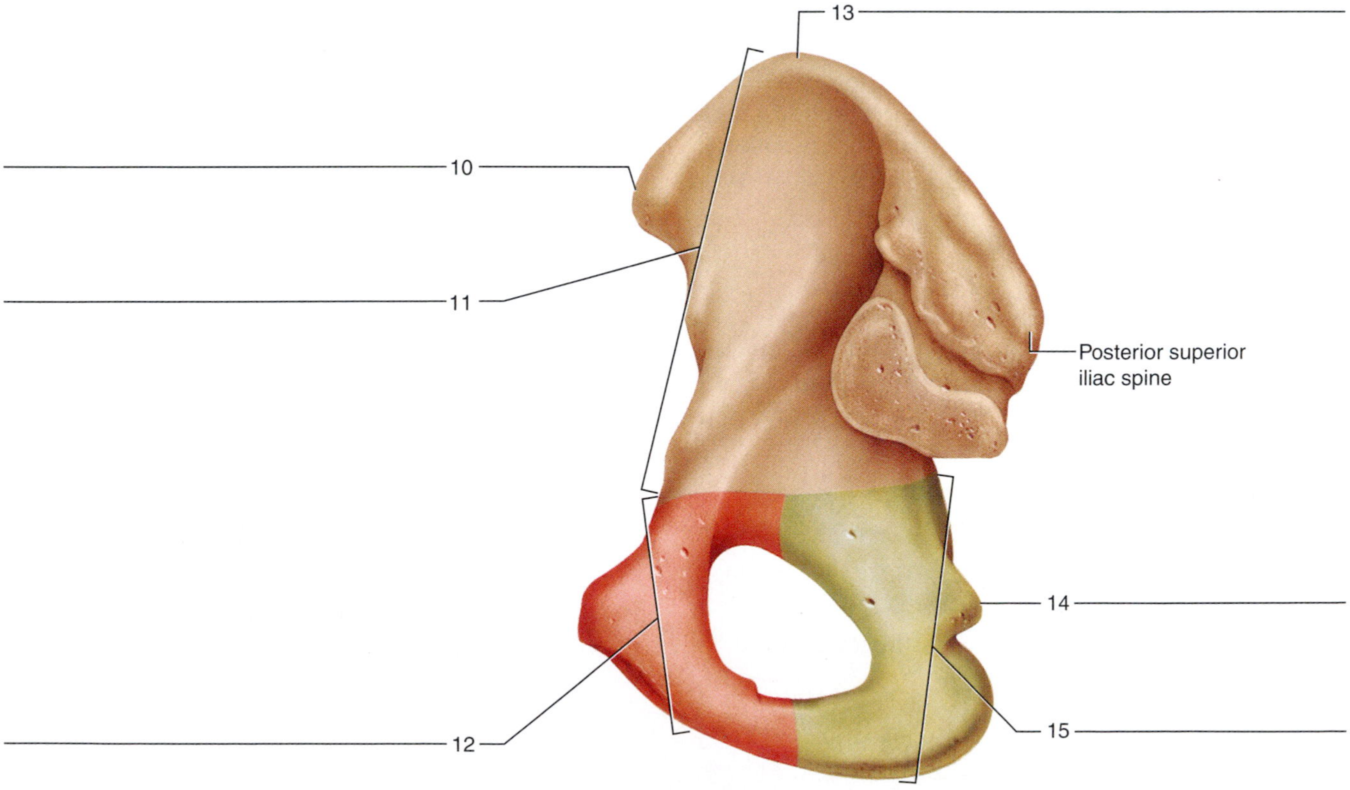

(b)

Figure 17.2 Label (*a*) the lateral and (*b*) the medial features of the right hip bone. 1

Critical Thinking Application

Examine the male and female pelves. Look for major differences between them. Note especially the flare of the iliac bones, the angle of the pubic arch, the distance between the ischial spines and ischial tuberosities, and the curve and width of the sacrum. In what ways are the differences you observed related to the function of the female pelvis as a birth canal?

____

____

____

____

EXPLORE

Procedure B—The Lower Limb

1. Review the section entitled "Lower Limb" in chapter 7 of the textbook.
2. As a review activity, label figures 17.3, 17.4, 17.5, and 17.6.

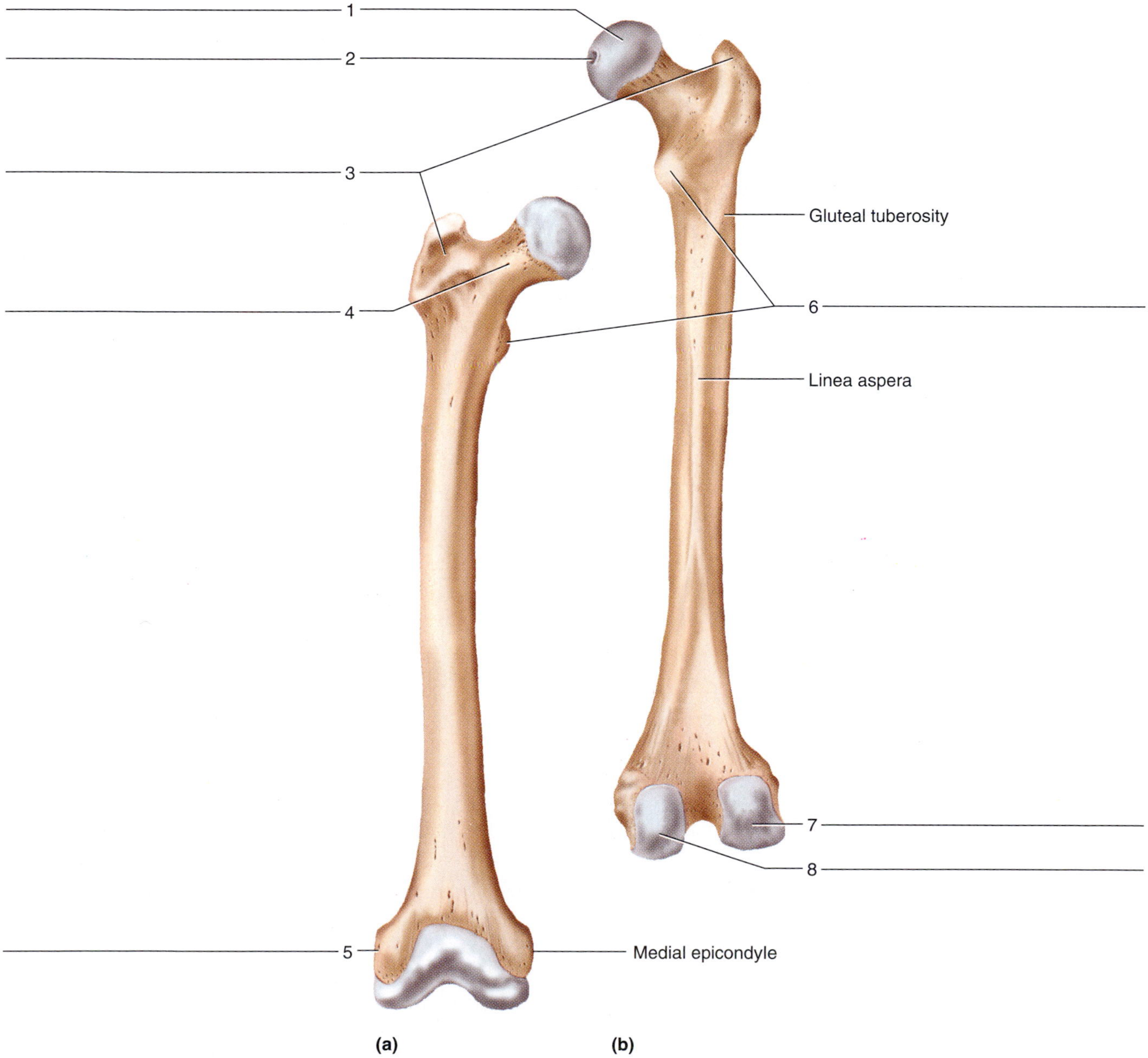

Figure 17.3 Label the features of (*a*) the anterior surface and (*b*) the posterior surface of the right femur. 2

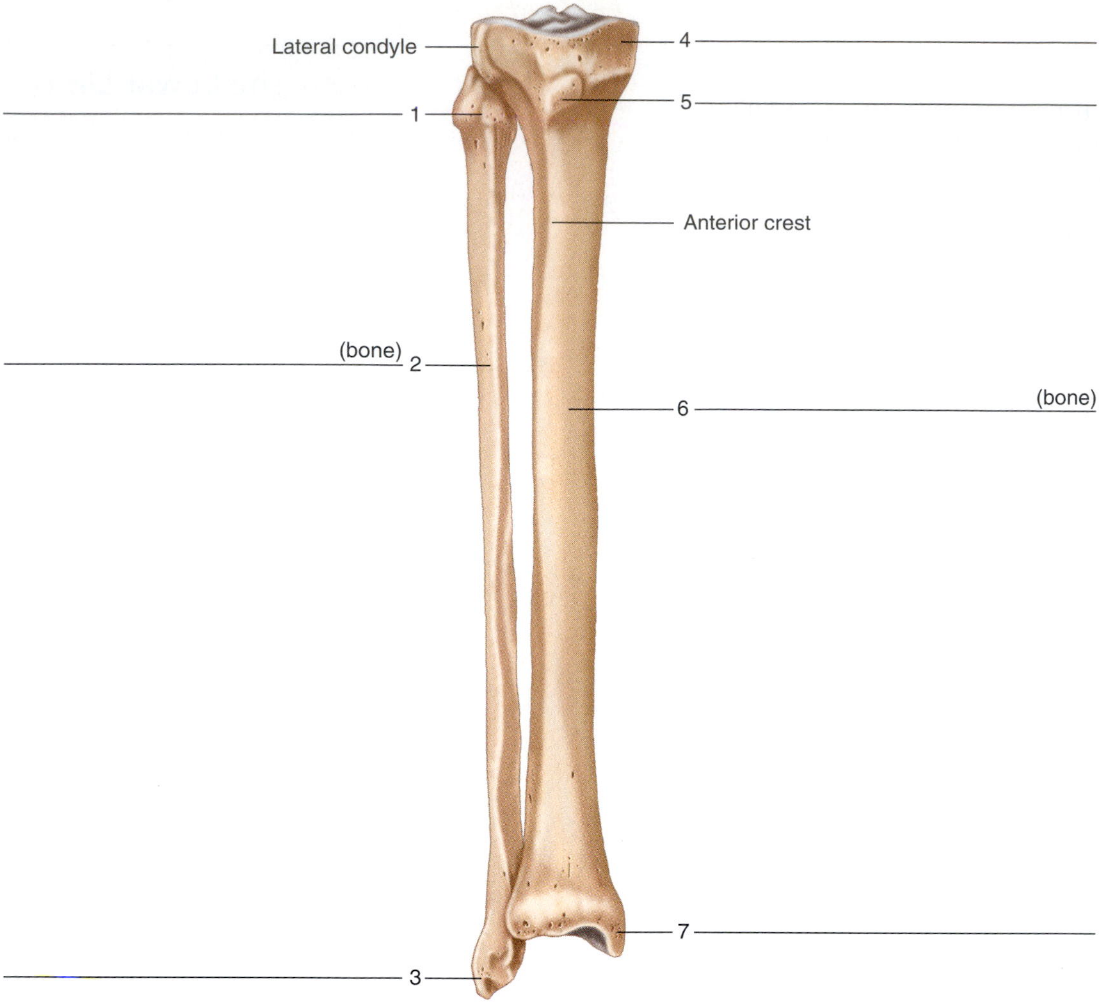

Figure 17.4 Label the bones and features of the right tibia and fibula in this anterior view. 2

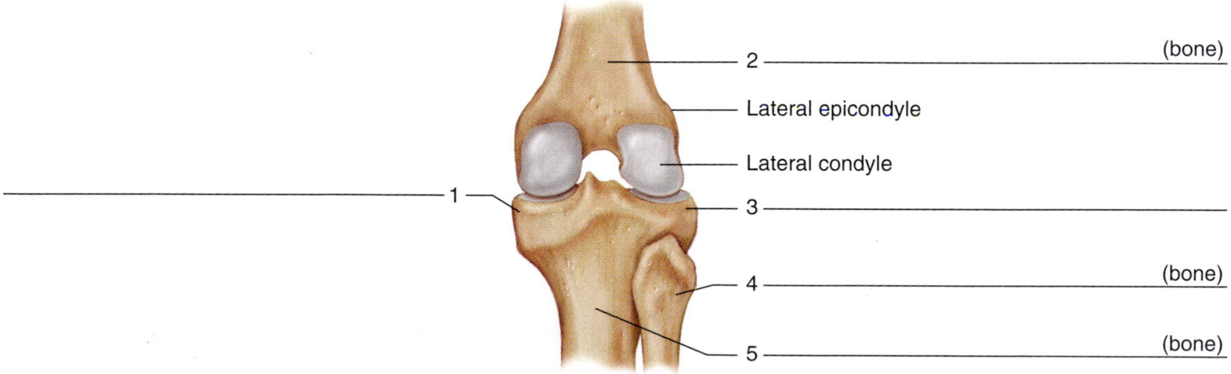

Figure 17.5 Label the bones and features of the right knee, posterior view. 2

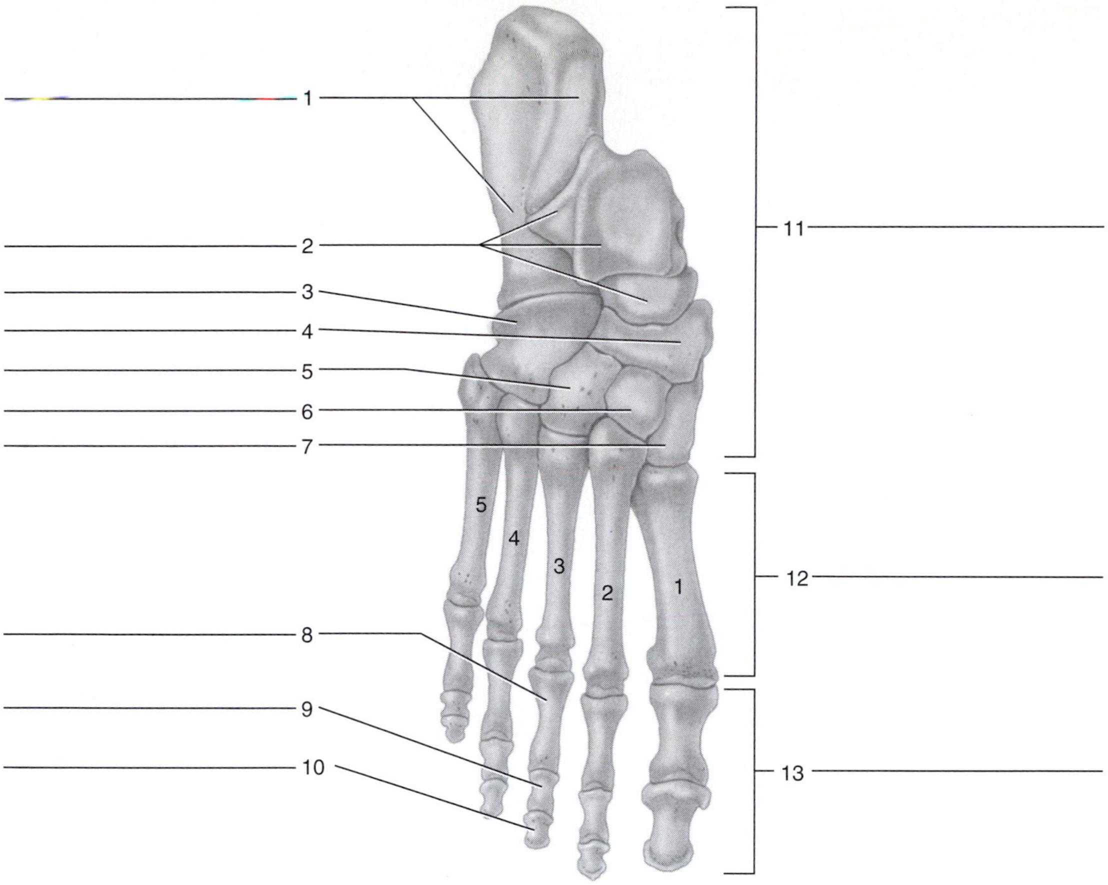

Figure 17.6 Label the bones of the superior surface of the right foot. 2

3. Examine the bones of the lower limb and locate each of the following:

femur
- proximal features
  - head
  - fovea capitis
  - neck
  - greater trochanter
  - lesser trochanter
- shaft
  - gluteal tuberosity
  - linea aspera
- distal features
  - lateral epicondyle
  - medial epicondyle
  - lateral condyle
  - medial condyle

patella

tibia
- medial condyle
- lateral condyle
- tibial tuberosity
- anterior crest (border)
- medial malleolus

fibula
- head
- lateral malleolus

tarsal bones
- talus
- calcaneus
- navicular
- cuboid
- lateral cuneiform
- intermediate cuneiform
- medial cuneiform

metatarsal bones

phalanges
- proximal phalanx
- middle phalanx
- distal phalanx

4. Complete Parts B, C, and D of the laboratory report.

Learning Extension

Use different colored pencils to distinguish the individual bones in figure 17.6.

Notes

Name ______________________________

Date ______________________________

Section ______________________________

The [A] corresponds to the indicated Learning Outcome(s) listed at the beginning of the laboratory exercise.

Laboratory Report 17

Pelvic Girdle and Lower Limb

Part A Assessments

Complete the following statements:

1. The pelvic girdle consists of two ______________________. [1]
2. The head of the femur articulates with the ______________________ of the hip bone. [1]
3. The ______________________ is the largest portion of the hip bone. [1]
4. The distance between the ______________________ represents the shortest diameter of the pelvic outlet. [1]
5. The pubic bones come together anteriorly to form the joint called the ______________________. [1]
6. The ______________________ is the portion of the ilium that causes the prominence of the hip. [1]
7. When a person sits, the ______________________ of the ischium supports the weight of the body. [1]
8. The angle formed by the pubic bones below the symphysis pubis is called the ______________________. [1]
9. The ______________________ is the largest foramen in the skeleton. [1]
10. The ilium joins the sacrum at the ______________________ joint. [1]

Part B Assessments

Match the bones in column A with the features in column B. Place the letter of your choice in the space provided. [2]

Column A		Column B
a. Femur	_______	1. Middle phalanx
b. Fibula	_______	2. Lesser trochanter
c. Metatarsals	_______	3. Medial malleolus
d. Patella	_______	4. Fovea capitis
e. Phalanges	_______	5. Calcaneus
f. Tarsals	_______	6. Lateral cuneiform
g. Tibia	_______	7. Tibial tuberosity
	_______	8. Talus
	_______	9. Linea aspera
	_______	10. Lateral malleolus
	_______	11. Sesamoid bone
	_______	12. Five bones that form the instep

Part C Assessments

Identify the bones and features indicated in the radiographs of figures 17.7, 17.8, and 17.9. 1 2

Terms:
Head of femur
Ilium
Obturator foramen
Pubis
Sacrum
Symphysis pubis

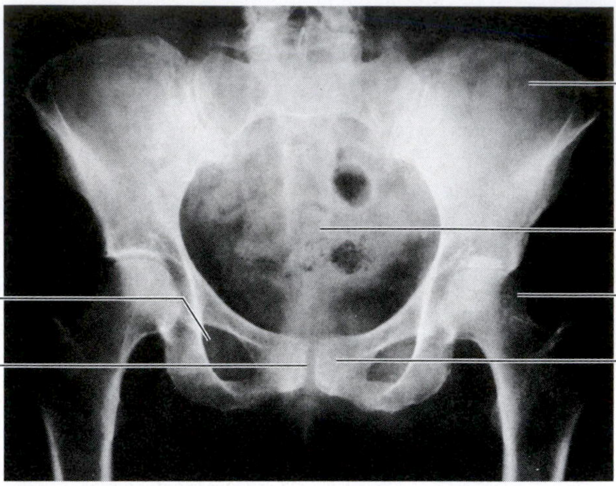

Figure 17.7 Identify the bones and features indicated on this radiograph of the anterior view of the pelvic region, using the terms provided.

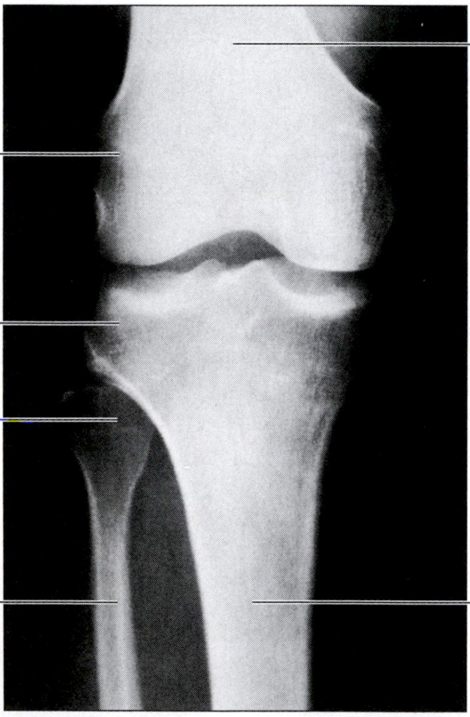

Terms:
Femur
Fibula
Head of fibula
Lateral condyle
Lateral epicondyle
Tibia

Figure 17.8 Identify the bones and features indicated in this radiograph of the right knee (anterior view), using the terms provided.

Terms:
Calcaneus
Distal phalanx
Metatarsal
Proximal phalanx
Talus
Tibia

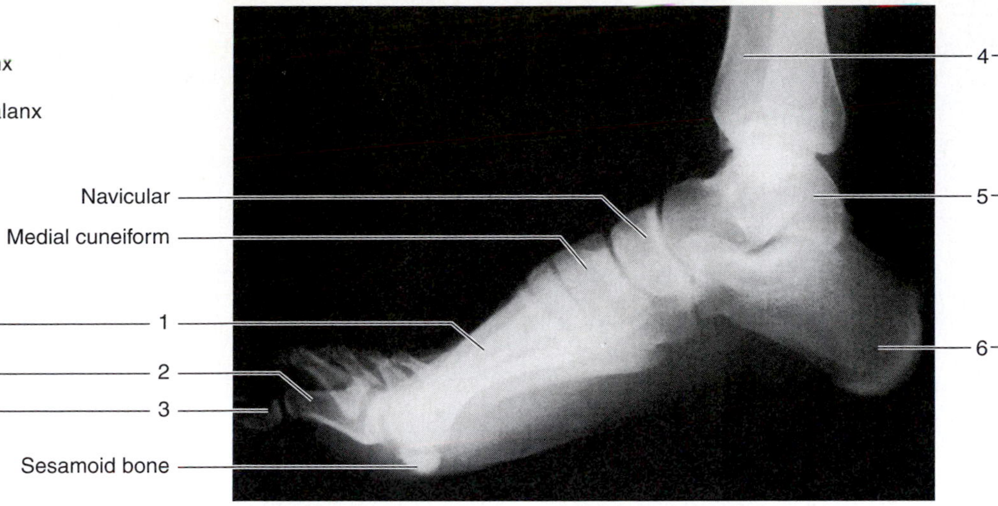

Figure 17.9 Identify the bones indicated on this radiograph of the right foot (medial side), using the terms provided.

Part D Assessments

Identify the bones of the foot in figure 17.10. 2

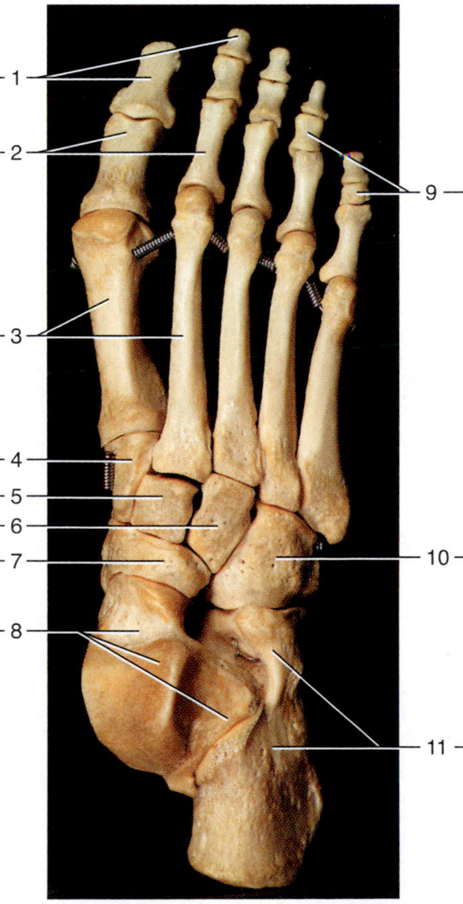

Terms:
Calcaneus
Cuboid
Distal phalanges
Intermediate cuneiform
Lateral cuneiform
Medial cuneiform
Metatarsals
Middle phalanges
Navicular
Proximal phalanges
Talus

Figure 17.10 Identify the bones indicated on this superior view of the right foot, using the terms provided.

Notes

Laboratory Exercise 18

Joint Structure and Movements

Materials Needed

Textbook
Human skull
Human skeleton, articulated
Models of synovial joints (shoulder, elbow, hip, and knee)

For Demonstrations:

Fresh animal joint (knee joint preferred)
Radiographs of major joints

Safety

- Wear disposable gloves when handling the fresh animal joint.
- Wash your hands before leaving the laboratory.

Joints are junctions between bones. Although they vary considerably in structure, they can be classified according to the type of tissue that binds the bones together. Thus, the three groups of structural joints can be identified as fibrous joints, cartilaginous joints, and synovial joints. Fibrous joints are filled with dense fibrous connective tissue, cartilaginous joints are filled with a type of cartilage, and synovial joints contain synovial fluid inside a joint cavity.

Joints can also be classified by the degree of functional movement allowed: synarthroses are immovable, amphiarthroses allow slight movement, and diarthroses allow free movement. Movements occurring at freely movable synovial joints are due to the contractions of skeletal muscles. In each case, the type of movement depends on the type of joint involved and the way in which the muscles are attached to the bones on either side of the joint.

Purpose of the Exercise

To examine examples of the three types of joints, to identify the major features of these joints, and to review the types of movements produced at synovial joints.

LEARNING OUTCOMES

After completing this exercise, you should be able to

1. Identify the structural and functional characteristics of fibrous, cartilaginous, and synovial joints.
2. Locate examples of each type of joint.
3. Locate examples of each of the six types of synovial joints.
4. Demonstrate the types of movements that occur at synovial joints.
5. Examine the structure of the shoulder, elbow, hip, and knee joints.

EXPLORE

Procedure A—Types of Joints

1. Review the section entitled "Classification of Joints" in chapter 8 of the textbook.
2. Examine the human skull and articulated skeleton to locate examples of the following types of joints:

   fibrous joints

   suture (synarthrotic joint)—skull

   gomphosis (synarthrotic joint)—tooth sockets

   syndesmosis (amphiarthrotic joint)—tibiofibular joint

   cartilaginous joints

   synchondrosis (synarthrotic joint)—epiphyseal plates

   symphysis (amphiarthrotic joint)—intervertebral discs

3. Complete Part A of Laboratory Report 18.
4. Locate examples of the following types of synovial joints (diarthrotic joints) in the skeleton. Experiment with each joint to experience its range of movements. 3 4

synovial joints

ball-and-socket joint—hip; shoulder

condylar (ellipsoidal) joint—radiocarpal; metacarpophalangeal

plane (gliding) joint—intercarpal; intertarsal

hinge joint—elbow; knee; interphalangeal

pivot joint—radioulnar at elbow; dens at atlas

saddle joint—base of thumb with trapezium

5. Examine models of the shoulder, elbow, hip, and knee joints. Locate the major knee joint structures of figure 18.1 that are visible on the knee joint model.
6. Complete Parts B and C of the laboratory report.

Demonstration

Examine a longitudinal section of a fresh synovial animal joint. Locate the dense connective tissue that forms the joint capsule and the hyaline cartilage that forms the articular cartilage on the ends of the bones. Locate the synovial membrane on the inside of the joint capsule. Does the joint have any semilunar cartilages (menisci)? 5

____

What is the function of such cartilages?

____

____

____

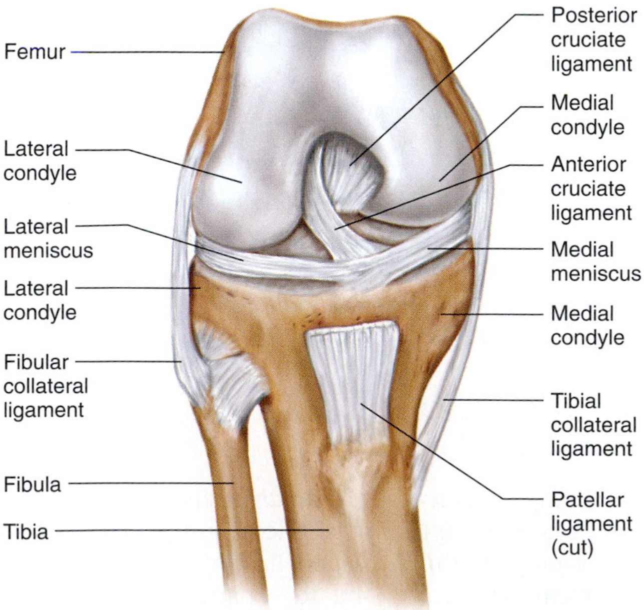

Figure 18.1 Anterior view of right knee (patella removed).

EXPLORE

Procedure B—Joint Movements

1. Review the section entitled "Types of Joint Movements" in chapter 8 of the textbook and study figures 18.2 and 18.3.
2. When the body is in anatomical position, most joints are extended and/or adducted. Skeletal muscle action involves the movable end (*insertion*) being pulled toward the stationary end (*origin*). In the limbs, the origin is usually proximal to the insertion; in the trunk, the origin is usually medial to the insertion. Use these concepts as reference points as you move joints. Move various parts of your body to demonstrate the joint movements described in tables 18.1 and 18.2. 4
3. Have your laboratory partner do some of the preceding movements and see if you can correctly identify the movements made. 4

Critical Thinking Application

Describe a body position that can exist when all major body parts are flexed.

____

____

4. Complete Parts D and E of the laboratory report.

Demonstration

Study the available radiographs of joints by holding the films in front of a light source. Identify the type of joint and the bones incorporated in the joint. Also identify other major visible features.

Table 18.1 Angular and Rotational Movements

Movement	Description
Flexion	Decrease of an angle (usually in the sagittal plane)
Extension Hyperextension	Increase of an angle (usually in the sagittal plane Extension beyond anatomical position
Abduction	Movement away from the midline (usually in the frontal plane)
Adduction	Movement toward the midline (usually in the frontal plane)
Circumduction	Circular movement (combines flexion, abduction extension, and adduction)
Rotation Medial (internal) Lateral (external)	Movement of part around its long axis Inward rotation Outward rotation

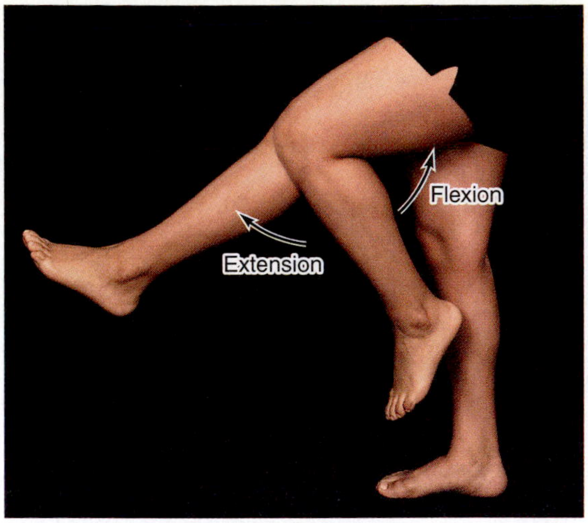

(a) Flexion and extension of the knee joint

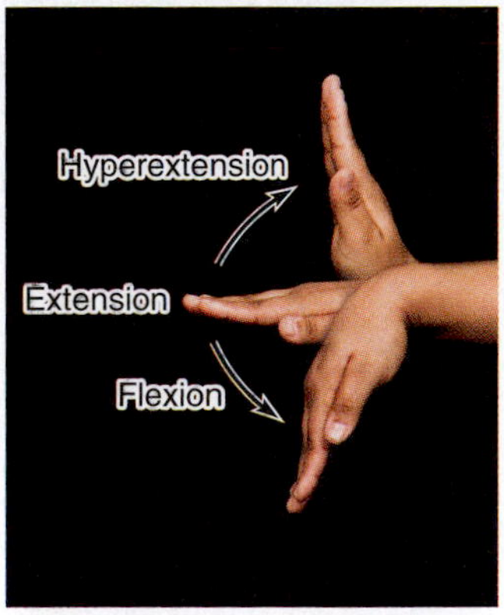

(b) Flexion, extension, and hyper-extension of the hand at the wrist joint

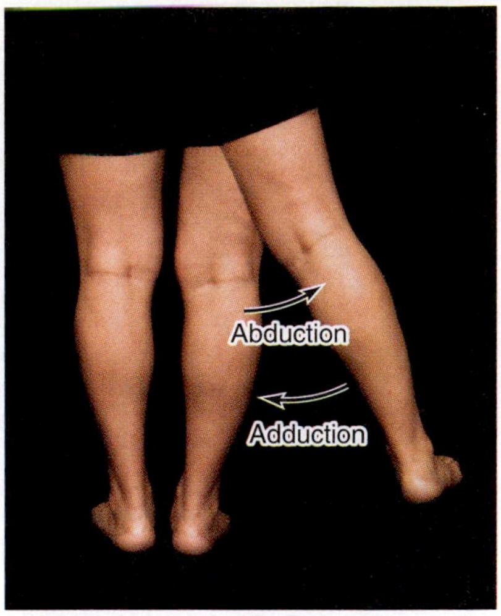

(c) Abduction and adduction of the hip joint

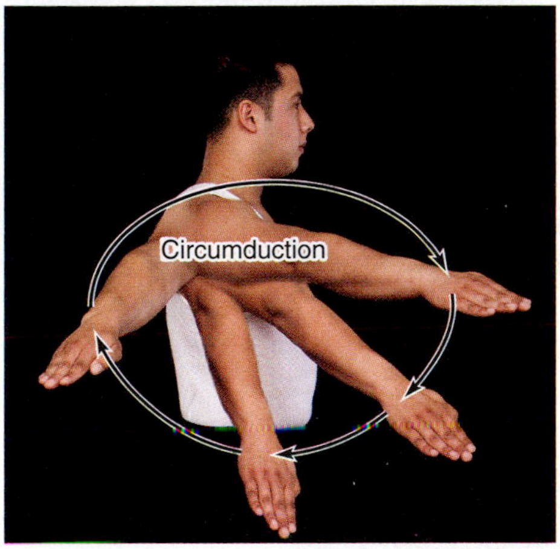

(d) Circumduction of the shoulder joint

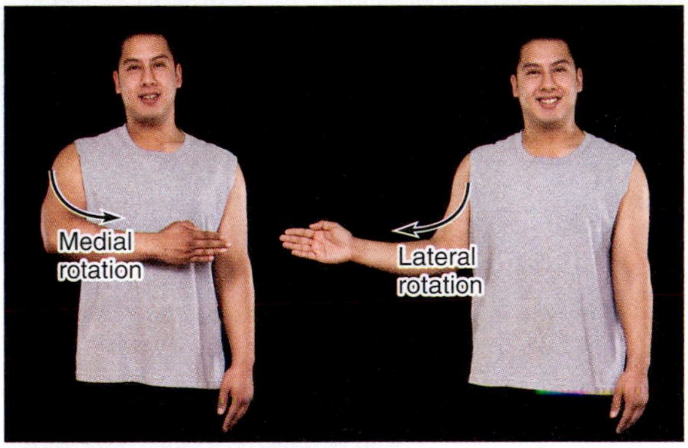

(e) Medial and lateral rotation of the arm at the shoulder joint

Figure 18.2 Examples of angular and rotational movements of synovial joints.

Table 18.2 **Special Movements (pertain to specific joints)**

Movement	Description
Supination	Movement of palm of hand anteriorly or upward
Pronation	Movement of palm of hand posteriorly or downward
Elevation	Movement of body part upward
Depression	Movement of body part downward
Dorsiflexion	Movement of ankle joint so dorsum (superior) of foot becomes closer to anterior surface of leg (as standing on heels)
Plantar flexion	Movement of ankle joint so the plantar surface of foot becomes closer to the posterior surface of leg (as standing on toes)
Inversion (called supination in some health professions)	Medial movement of sole of foot at ankle joint
Eversion (called pronation in some health professions)	Lateral movement of sole of foot at ankle joint
Protraction	Anterior movement in the transverse plane
Retraction	Posterior movement in the transverse plane

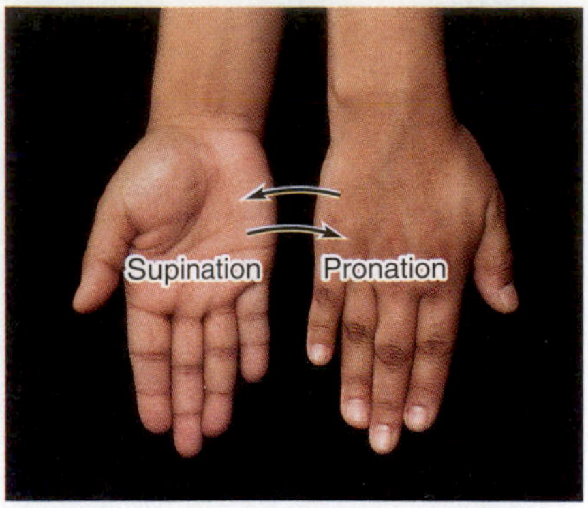

(a) Supination and pronation of the hand involving movement at the radioulnar joint

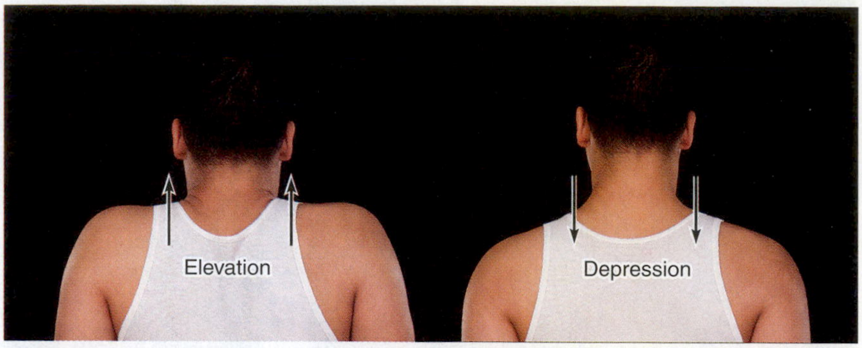

(b) Elevation and depression of the shoulder (scapula)

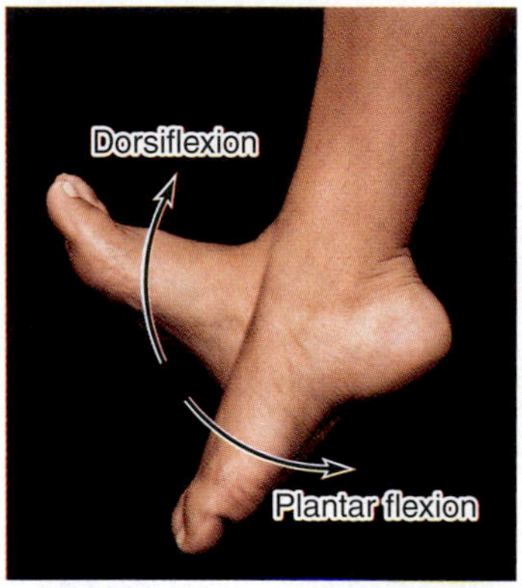

(c) Dorsiflexion and plantar flexion of the foot at the ankle joint

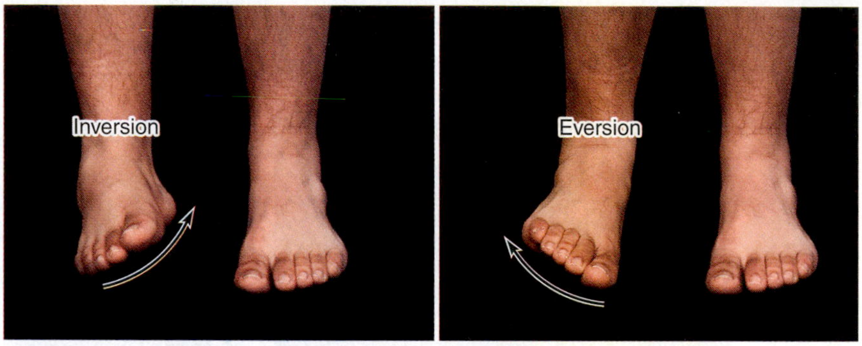

(d) Inversion and eversion of the right foot at the ankle joint (The left foot is unchanged for comparison.)

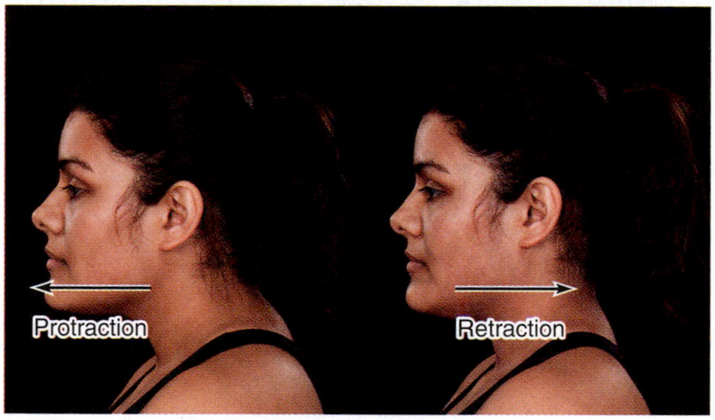

(e) Protraction and retraction of the head

Figure 18.3 Special movements of synovial joints.

Name ______________________________

Date ______________________________

Section ______________________________

The [A] corresponds to the indicated Learning Outcome(s) listed at the beginning of the laboratory exercise.

Laboratory Report 18

Joint Structure and Movements

Part A Assessments

Match the terms in column A with the descriptions in column B. Place the letter of your choice in the space provided. [1] [2]

Column A		Column B
a. Gomphosis	_______	1. Immovable joint between flat bones of the skull united by a thin layer of dense connective tissue
b. Suture	_______	2. Fibrocartilage pad fills the slightly movable joint
c. Symphysis	_______	3. Temporary joint in which bones are united by bands of hyaline cartilage
d. Synchondrosis	_______	4. Slightly movable joint in which bones are united by interosseous membrane
e. Syndesmosis	_______	5. Joint formed by union of tooth root in bony socket

Part B Assessments

Identify the types of structural and functional joints numbered in figure 18.4. [1] [2]

1. ______________________________
2. ______________________________
3. ______________________________
4. ______________________________
5. ______________________________
6. ______________________________
7. ______________________________
8. ______________________________
9. ______________________________

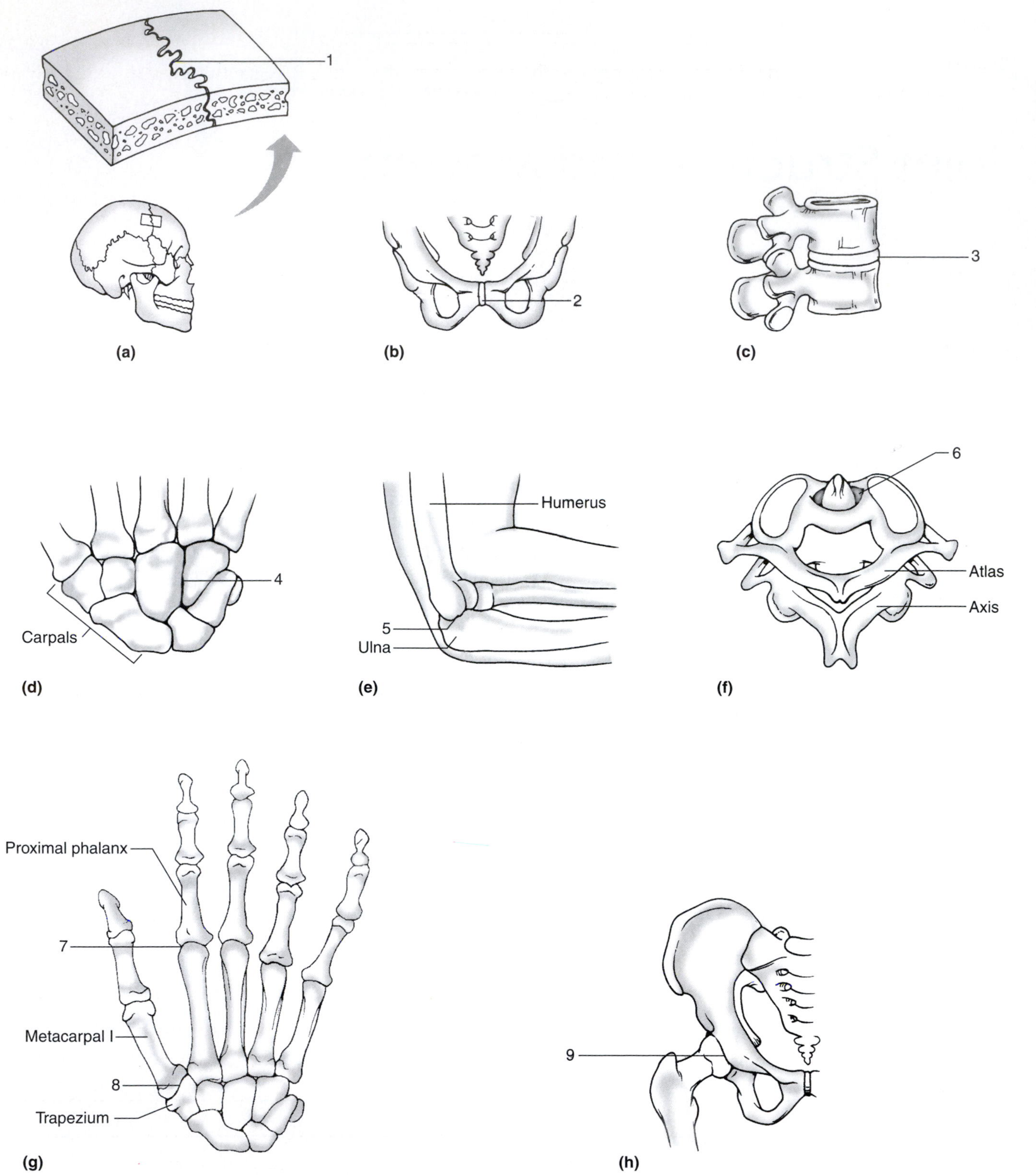

Figure 18.4 Identify the types of structural and functional joints numbered in these illustrations.

Part C Assessments

Match the types of synovial joints in column A with the examples in column B. Place the letter of your choice in the space provided. 3

Column A

a. Ball-and-socket
b. Condylar (ellipsoidal)
c. Hinge
d. Pivot
e. Plane (gliding)
f. Saddle

Column B

_______ **1.** Hip joint
_______ **2.** Metacarpal-phalanx
_______ **3.** Proximal radius-ulna
_______ **4.** Humerus-ulna of the elbow joint
_______ **5.** Phalanx-phalanx
_______ **6.** Shoulder joint
_______ **7.** Knee joint
_______ **8.** Carpal-metacarpal of the thumb
_______ **9.** Carpal-carpal
_______ **10.** Tarsal-tarsal

Part D Assessments

Complete the following table: 4 5

Name of Joint	Type of Joint	Bones Included	Types of Movement Possible
Shoulder joint			
Elbow joint			
Hip joint			
Knee joint			

Part E Assessments

Identify the types of joint movements numbered in figure 18.5. 4

1. ______________ (of head)
2. ______________ (of shoulder)
3. ______________ (of shoulder)
4. ______________ (of hand)
5. ______________ (of hand)
6. ______________ (of arm at shoulder)
7. ______________ (of arm at shoulder)
8. ______________ (of hand at wrist)
9. ______________ (of hand at wrist)
10. ______________ (of thigh at hip)
11. ______________ (of thigh at hip)
12. ______________ (of lower limb at hip)
13. ______________ (of chin/mandible)
14. ______________ (of chin/mandible)
15. ______________ (of vertebral column/trunk)
16. ______________ (of vertebral column/trunk)
17. ______________ (of head and neck)
18. ______________ (of head and neck)
19. ______________ (of arm at shoulder)
20. ______________ (of arm at shoulder)
21. ______________ (of forearm at elbow)
22. ______________ (of forearm at elbow)
23. ______________ (of thigh at hip)
24. ______________ (of thigh at hip)
25. ______________ (of leg at knee)
26. ______________ (of leg at knee)
27. ______________ (of foot at ankle)
28. ______________ (of foot at ankle)

Figure 18.5 Identify each of the types of movements numbered and illustrated: (*a*) anterior view; (*b*) lateral view of head; (*c*) lateral view. (Original illustration drawn by Ross Martin)

Laboratory Exercise 19

Skeletal Muscle Structure

Materials Needed

Textbook
Compound light microscope
Prepared microscope slide of skeletal muscle tissue
Human torso model with musculature
Model of skeletal muscle fiber

For Demonstration:

Fresh round beefsteak

Safety

- Wear disposable gloves when handling the fresh beefsteak.
- Wash your hands before leaving the laboratory.

A skeletal muscle represents an organ of the muscular system and is composed of several types of tissues. These tissues include skeletal muscle tissue, nervous tissue, blood, and various connective tissues.

Each skeletal muscle is encased and permeated with connective tissue sheaths. The connective tissue often extends beyond the end of a muscle, providing an attachment to other muscles or to bones. The connective tissues also extend into the structure of a muscle and separate it into compartments. The outer layer, called *epimysium,* extends deeper into the muscle as *perimysium,* containing bundles of cells (fascicles), and farther inward extends around each muscle cell (fiber) as a thin *endomysium.* The connective tissues provide support and reinforcement during muscular contractions and allow portions of a muscle to contract somewhat independently. Some of the collagen fibers are continuous with the tendon, periosteum, and inward extensions as bone fibers, allowing a strong structural continuity.

Muscles are named according to their location, size, action, shape, attachments, or the direction of the fibers. Examples of how muscles are named include: gluteus maximus (location and size); adductor longus (action and shape); sternocleidomastoid (attachments); and orbicularis oculi (direction of fibers and location).

Purpose of the Exercise

To review the structure of a skeletal muscle.

LEARNING OUTCOMES

After completing this exercise, you should be able to

1. Locate the major structures of a skeletal muscle fiber on a microscope slide of skeletal muscle tissue and on a model.
2. Describe how connective tissue is associated with muscle tissue within a skeletal muscle.
3. Distinguish between the origin and insertion of a muscle.
4. Describe and demonstrate the general actions of prime movers (agonists), synergists, and antagonists.

EXPLORE

Procedure—Skeletal Muscle Structure

1. Study the section entitled "Skeletal Muscle Tissue" in chapter 5 of the textbook.
2. Reexamine the microscopic structure of skeletal muscle by observing a prepared microscope slide of this tissue. Use figure 19.1 of skeletal muscle tissue to locate the following features:

   skeletal muscle fiber (cell)
   nuclei
   striations (alternating light and dark)

3. Review the section entitled "Structure of a Skeletal Muscle" in chapter 9 of the textbook.
4. As a review activity, label figure 19.2 and study figure 19.3.
5. Examine the human torso model and locate examples of fascia, tendons, and aponeuroses. An origin tendon is attached to a fixed location, while the insertion tendon is attached to a more movable location. Sheets of connective tissue, called aponeuroses, also serve for some muscle attachments. Locate examples of tendons in your body.
6. Complete Part A of Laboratory Report 19.

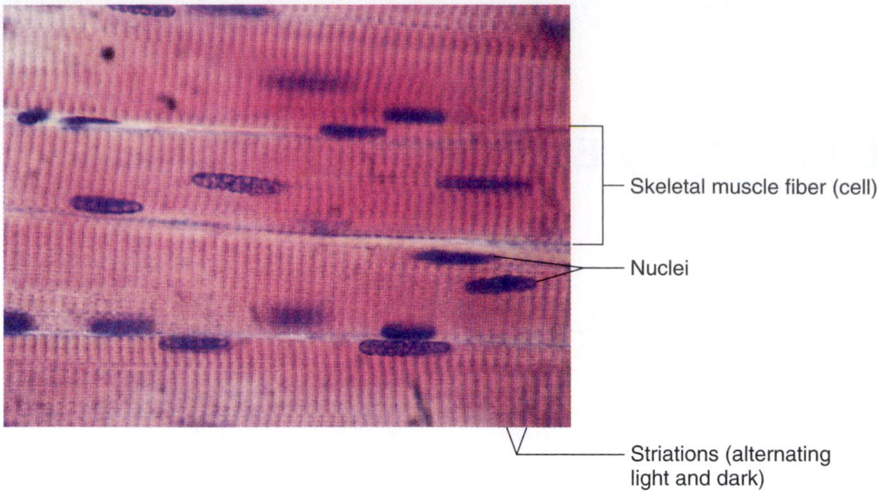

Figure 19.1 Structures found in skeletal muscle fibers (cells) (250× micrograph enlarged to 700×).

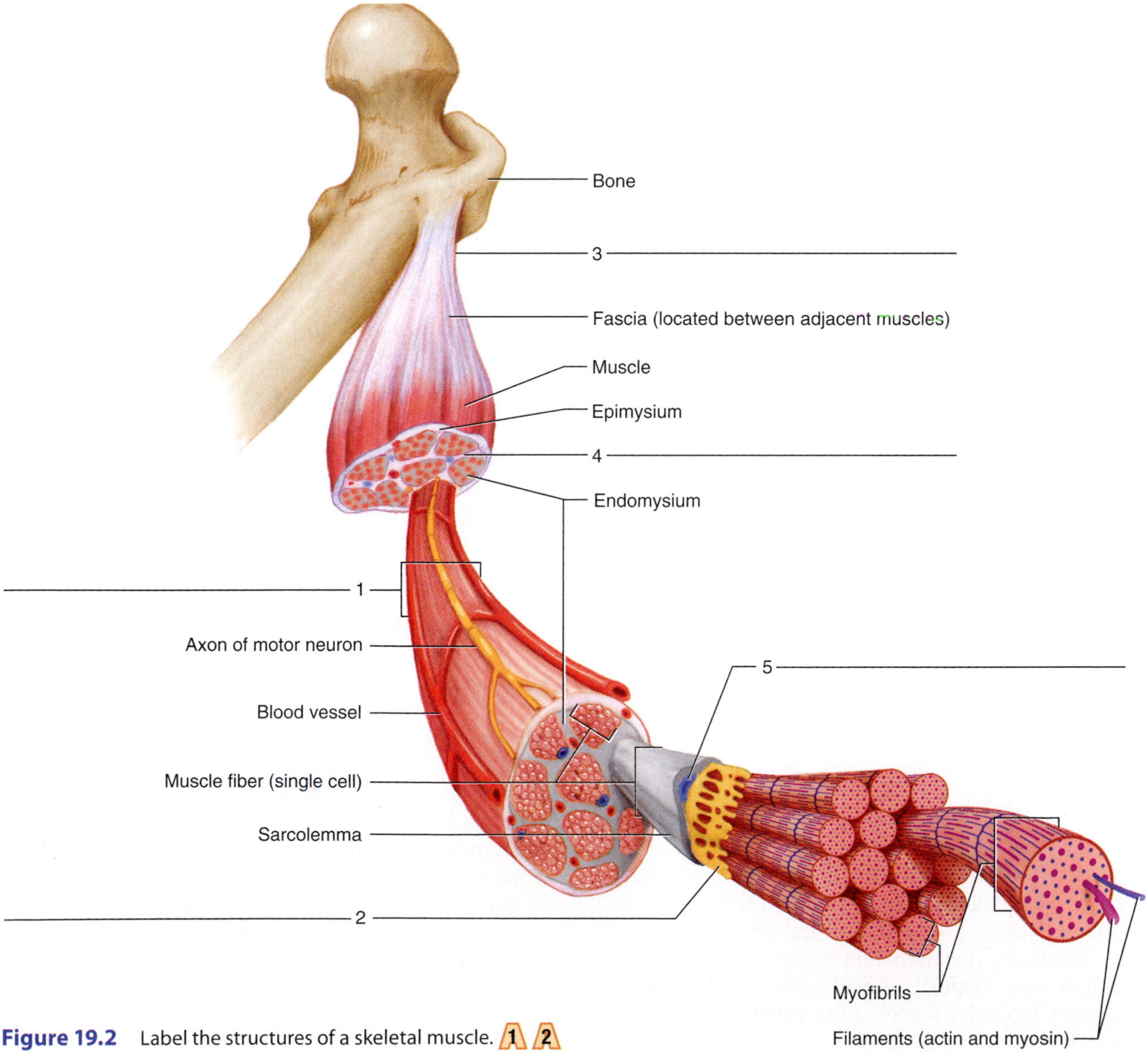

Figure 19.2 Label the structures of a skeletal muscle. 1 2

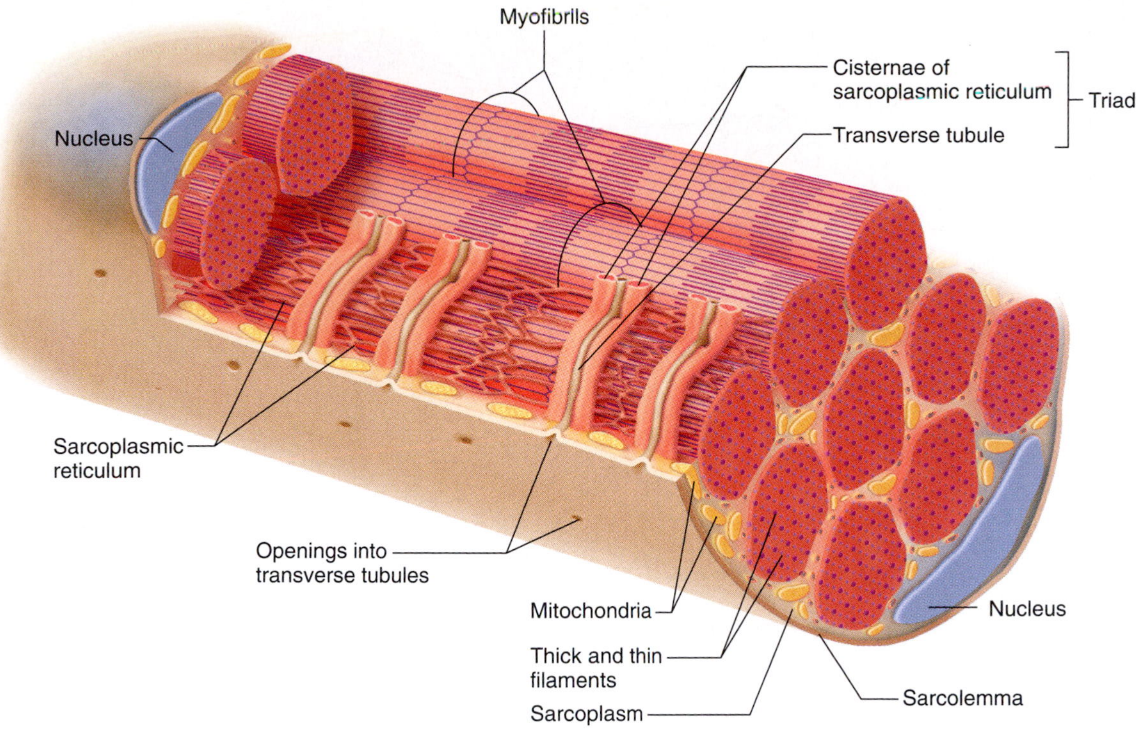

Figure 19.3 Structures of a skeletal muscle fiber.

Demonstration

Examine the fresh round beefsteak. It represents a cross section through the beef thigh muscles. Note the white lines of connective tissue that separate the individual skeletal muscles. Also note how the connective tissue extends into the structure of a muscle and separates it into small compartments of muscle tissue. Locate the epimysium and the perimysium of an individual muscle.

7. Examine the model of the skeletal muscle fiber and locate the following:

   sarcolemma
   sarcoplasm
   myofibril
   - thick (myosin) filament
   - thin (actin, tropomyosin, and troponin) filament

   sarcomere (functional contractile unit)
   - A band (dark)
   - I band (light)
   - H zone
   - M line
   - Z line (disc)

   sarcoplasmic reticulum
   - cisternae

   transverse (T) tubules

8. Complete Part B of the laboratory report.
9. Review the section entitled "Skeletal Muscle Actions" in chapter 9 of the textbook.
10. Provide labels for figure 19.4.
11. Locate the biceps brachii and its origins and insertions in the human torso model and in your body.
12. Make various movements with your upper limb at the shoulder and elbow. For each movement, determine the location of the muscles that function as prime movers (agonists) and as antagonists. **Remember, when a prime mover contracts (shortens) and a joint moves, its antagonist relaxes (lengthens).** Antagonistic muscle pairs pull from opposite sides of the same body region. Synergistic muscles often supplement the contraction force of a prime mover, or by acting as fixators, they might also stabilize nearby joints. The role of a muscle as a prime mover, antagonist, or synergist depends upon the movement under consideration, as their roles can change.
13. Complete Part C of the laboratory report.

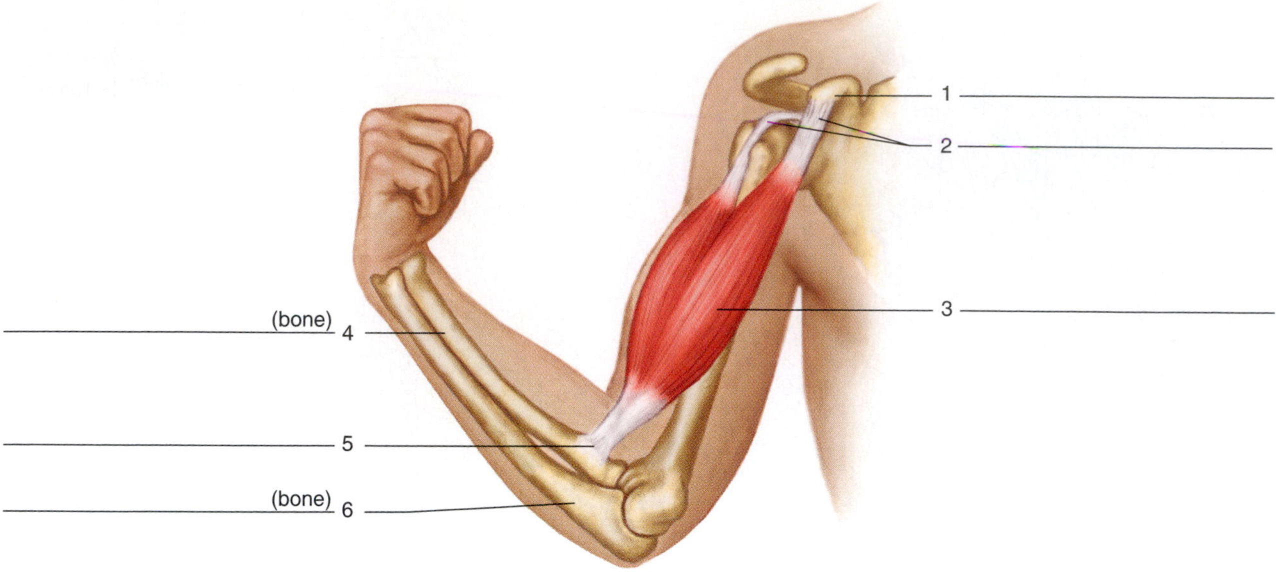

Figure 19.4 Label the major features of the upper limb.

Name ______________________________

Date ______________________________

Section ______________________________

The A corresponds to the indicated Learning Outcome(s) listed at the beginning of the laboratory exercise.

Laboratory Report 19

Skeletal Muscle Structure

Part A Assessments

Match the terms in column A with the definitions in column B. Place the letter of your choice in the space provided. 1 2

Column A		Column B
a. Endomysium	_______	1. Membranous channel extending inward from muscle fiber membrane
b. Epimysium	_______	2. Cytoplasm of a muscle fiber
c. Fascia	_______	3. Connective tissue located between adjacent muscles
d. Fascicle	_______	4. Layer of connective tissue that separates a muscle into small bundles called fascicles
e. Myosin	_______	5. Cell membrane of a muscle fiber
f. Perimysium	_______	6. Layer of connective tissue that surrounds a skeletal muscle
g. Sarcolemma	_______	7. Unit of alternating light and dark striations between Z lines
h. Sarcomere	_______	8. Layer of connective tissue that surrounds an individual muscle fiber
i. Sarcoplasm	_______	9. Cellular organelle in muscle fiber corresponding to the endoplasmic reticulum
j. Sarcoplasmic reticulum	_______	10. Cordlike part that attaches a muscle to a bone
k. Tendon	_______	11. Protein found within thick filament
l. Transverse (T) tubule	_______	12. A small bundle of muscle fibers

Part B Assessments

Provide the labels for the electron micrograph in figure 19.5. 1

1. ______________________________
2. ______________________________
3. ______________________________
4. ______________________________

Terms:
A band (dark)
I band (light)
Sarcomere
Z line

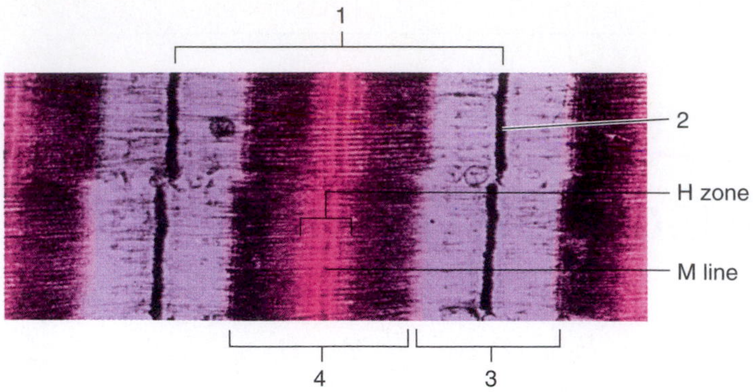

Figure 19.5 Identify the bands and lines of the striations of a transmission electron micrograph of relaxed sarcomeres (16,000×), using the terms provided.

Part C Assessments

Complete the following statements:

1. The ________________________ of a muscle is usually attached to a fixed location. 3
2. The ________________________ of a muscle is usually attached to a movable location. 3
3. When the forearm is extended at the elbow joint, the ________________________ muscle acts as the prime mover (agonist). 4
4. The forearm is flexed at the elbow when the ________________________ muscle contracts. 4
5. A muscle responsible for most of a movement is called a(n) ________________________ . 4
6. Assisting muscles are called ________________________ . 4
7. Antagonists are muscles that resist the actions of ________________________ and cause movement in the opposite direction. 4

Laboratory Exercise 20

Muscles of the Face, Head, and Neck

Materials Needed

Textbook
Human torso model with musculature
Human skull
Human skeleton, articulated

For Learning Extension:
Long rubber bands

The skeletal muscles of the face and head include the muscles of facial expression, which lie just beneath the skin; the muscles of mastication, attached to the mandible; and the muscles that move the head, located in the neck.

Purpose of the Exercise

To review the locations, actions, origins, and insertions of the muscles of the face, head, and neck.

LEARNING OUTCOMES

After completing this exercise, you should be able to

1. Locate and identify the muscles of facial expression, the muscles of mastication, and the muscles that move the head.
2. Describe and demonstrate the action of each of these muscles.
3. Locate the origin and insertion of each of these muscles in a human skeleton and the musculature of the human torso model.

EXPLORE

Procedure—Muscles of the Face, Head, and Neck

1. Review the sections entitled "Muscles of Facial Expression," "Muscles of Mastication," and "Muscles That Move the Head and Vertebral Column" in chapter 9 of the textbook.
2. As a review activity, label figures 20.1, 20.2, and 20.3.
3. Locate the following muscles in the human torso model and in your body whenever possible:

   muscles of facial expression
   - epicranius
     - frontalis
     - occipitalis
   - orbicularis oculi
   - orbicularis oris
   - buccinator
   - zygomaticus major
   - zygomaticus minor
   - platysma

   muscles of mastication
   - masseter
   - temporalis
   - medial pterygoid
   - lateral pterygoid

   muscles that move the head and neck
   - sternocleidomastoid
   - splenius capitis
   - semispinalis capitis
   - erector spinae (cervical components)
4. Demonstrate the action of these muscles in your body.
5. Locate the origins and insertions of these muscles in the human skull and skeleton.
6. Complete Parts A, B, and C of Laboratory Report 20.

Learning Extension

A long rubber band can be used to simulate muscle locations, origins, insertions, and actions on the human torso model, the skeleton, or a laboratory partner. Hold one end of the rubber band firmly on the origin location of a muscle, then slightly stretch the rubber band and hold the other end on the insertion site. Allow the insertion end to slowly move toward the origin end to simulate the contraction and action of the muscle. 1 2 3

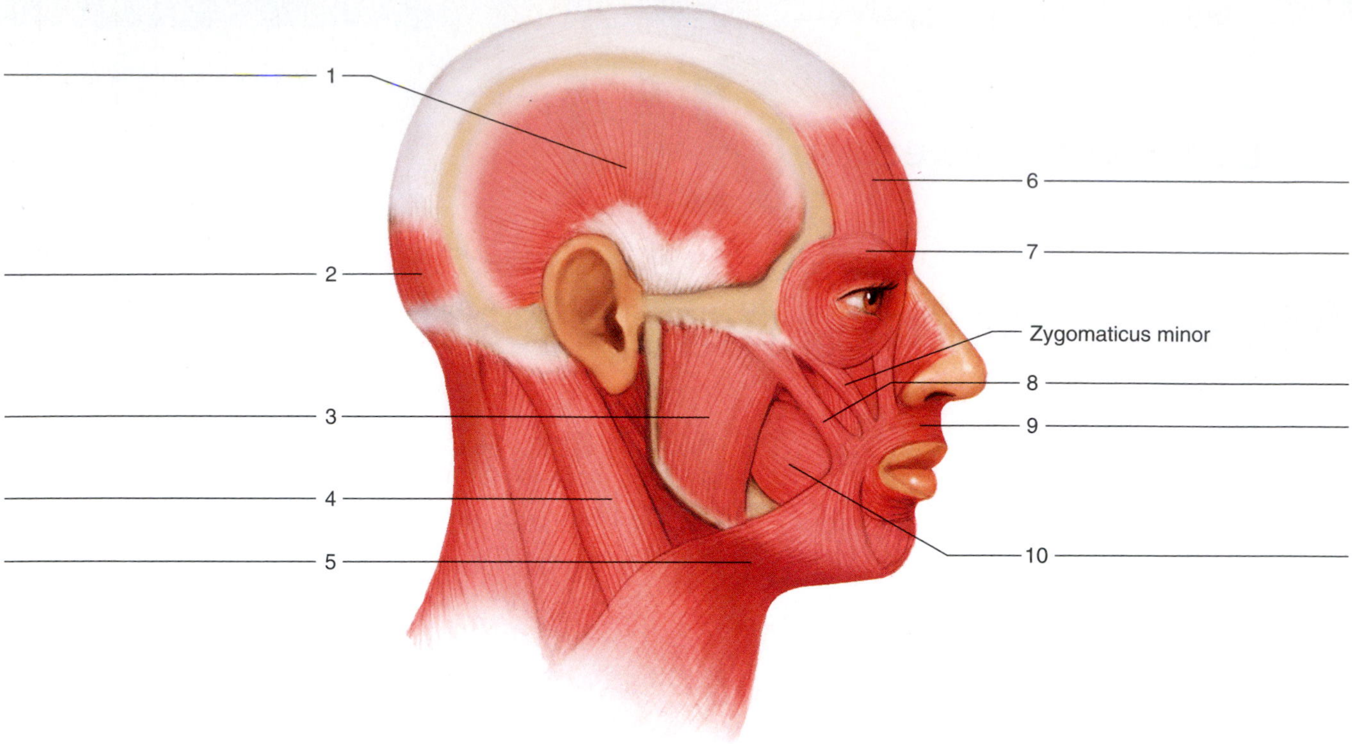

Figure 20.1 Label the muscles of expression and mastication. 1

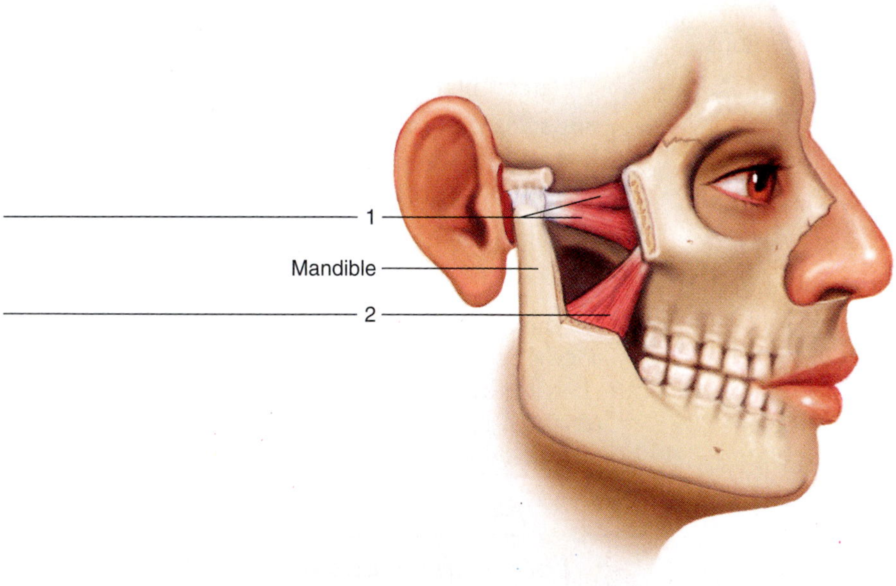

Figure 20.2 Label these deep muscles of mastication. 1

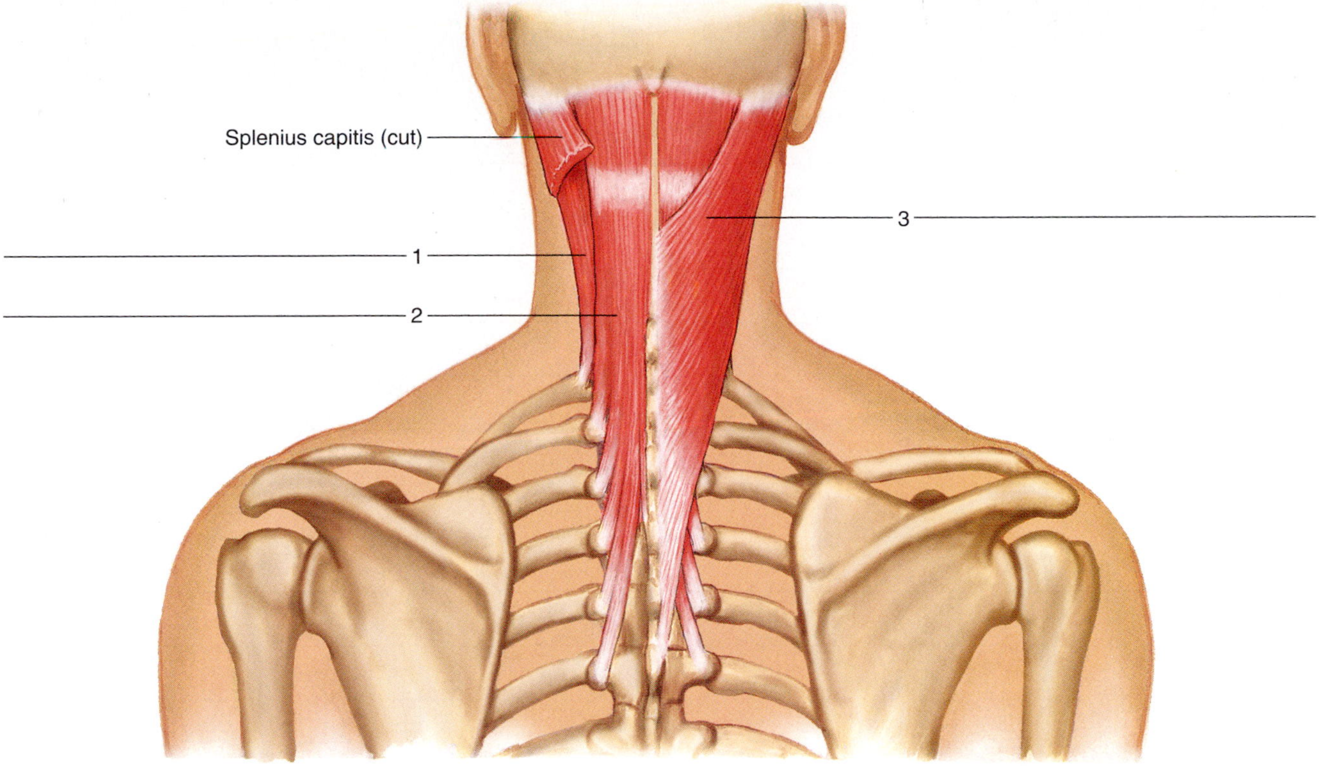

Terms:
Longissimus capitis (erector spinae)
Semispinalis capitis
Splenius capitis

Figure 20.3 Label these deep muscles of the posterior neck (trapezius removed), using the terms provided.

Notes

Name ____

Date ____

Section ____

The A corresponds to the indicated Learning Outcome(s) listed at the beginning of the laboratory exercise.

Laboratory Report 20

Muscles of the Face, Head, and Neck

Part A Assessments

Complete the following statements:

1. When the ____________________ contracts, the corner of the mouth is drawn upward. 2
2. The ____________________ acts to compress the wall of the cheeks when air is blown out of the mouth. 2
3. The ____________________ causes the lips to close and pucker during kissing, whistling, and speaking. 2
4. The ____________________ and platysma help to lower the mandible. 2
5. The temporalis acts to ____________________ . 2
6. The ____________________ pterygoid can close the jaw and pull it sideways. 2
7. The ____________________ pterygoid can protrude the jaw, pull the jaw sideways, and open the mouth. 2
8. The ____________________ can close the eye, as in blinking. 2
9. The ____________________ can pull the head toward the chest. 2
10. The ____________________ can pull the head to one side, rotate it, or bring it into an upright position (extend head). 2
11. The muscle used for pouting and to express horror is the ____________________ . 2
12. The muscle used to smile and laugh is the ____________________ . 2

Part B Assessments

Name the muscle indicated by the following combinations of origin and insertion. 3

Origin	Insertion	Muscle
1. Occipital bone	Skin and muscle around eye	____________
2. Zygomatic bone	Orbicularis oris	____________
3. Zygomatic arch	Lateral surface of mandible	____________
4. Sphenoid bone	Anterior surface of mandibular condyle	____________
5. Anterior surface of sternum and upper clavicle	Mastoid process of temporal bone	____________
6. Outer surfaces of mandible and maxilla	Orbicularis oris	____________
7. Fascia in upper chest	Lower border of mandible and skin around corner of mouth	____________
8. Temporal bone	Coronoid process and anterior ramus of mandible	____________
9. Spinous processes of cervical and thoracic vertebrae	Mastoid process of temporal bone and occipital bone	____________
10. Processes of cervical and thoracic vertebrae	Occipital bone	____________

Critical Thinking Application

Identify the muscles of various facial expressions in the photographs of figure 20.4. 1 2

1. ______
2. ______
3. ______
4. ______
5. ______

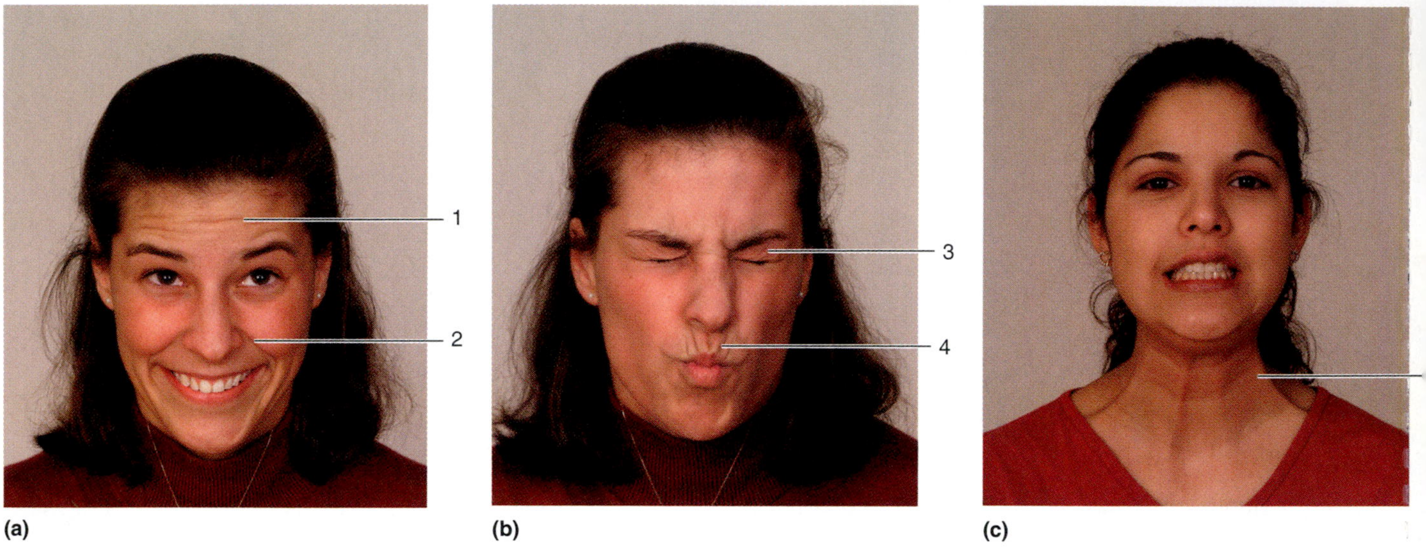

Terms:
Epicranius/frontalis
Orbicularis oculi
Orbicularis oris
Platysma
Zygomaticus major

Figure 20.4 Identify the muscles of facial expression being contracted in each of these photographs (*a–c*), using the terms provided.

Laboratory Exercise 21

Muscles of the Chest, Shoulder, and Upper Limb

Materials Needed

Textbook
Human torso model
Human skeleton, articulated
Muscular models of the upper limb

For Learning Extension:

Long rubber bands

The muscles of the chest and shoulder are responsible for moving the scapula and arm, whereas those within the arm and forearm move joints in the elbow and hand.

Purpose of the Exercise

To review the locations, actions, origins, and insertions of the muscles in the chest, shoulder, and upper limb.

LEARNING OUTCOMES

After completing this exercise, you should be able to

1. Locate and identify the muscles of the chest, shoulder, and upper limb.
2. Describe and demonstrate the action of each of these muscles.
3. Locate the origin and insertion of each of these muscles in a human skeleton and on muscular models.

EXPLORE

Procedure—Muscles of the Chest, Shoulder, and Upper Limb

1. Review the sections entitled "Muscles That Move the Pectoral Girdle," "Muscles That Move the Arm," "Muscles That Move the Forearm," and "Muscles That Move the Hand" in chapter 9 of the textbook.
2. As a review activity, label figures 21.1, 21.2, 21.3, and 21.4.
3. Locate the following muscles in the human torso model and models of the upper limb. Also locate in your body as many of the muscles as you can.

muscles that move the pectoral girdle

- anterior muscles
  - pectoralis minor
  - serratus anterior
- posterior muscles
  - trapezius
  - rhomboid major
  - rhomboid minor
  - levator scapulae

muscles that move the arm

- origins on axial skeleton
  - pectoralis major
  - latissimus dorsi
- origins on scapula
  - deltoid
  - teres major
  - coracobrachialis
- rotator cuff (SITS) muscles (origins also on scapula)
  - supraspinatus
  - infraspinatus
  - teres minor
  - subscapularis

muscles that move the forearm

- muscle bellies in arm
  - biceps brachii
  - brachialis
  - triceps brachii
- muscle bellies in forearm
  - brachioradialis
  - supinator
  - pronator teres
  - pronator quadratus

muscles that move the hand

- anterior flexor muscles
  - flexor carpi radialis
  - flexor carpi ulnaris
  - palmaris longus
  - flexor digitorum profundus
  - flexor digitorum superficialis
- posterior extensor muscles
  - extensor carpi radialis longus
  - extensor carpi radialis brevis
  - extensor carpi ulnaris
  - extensor digitorum

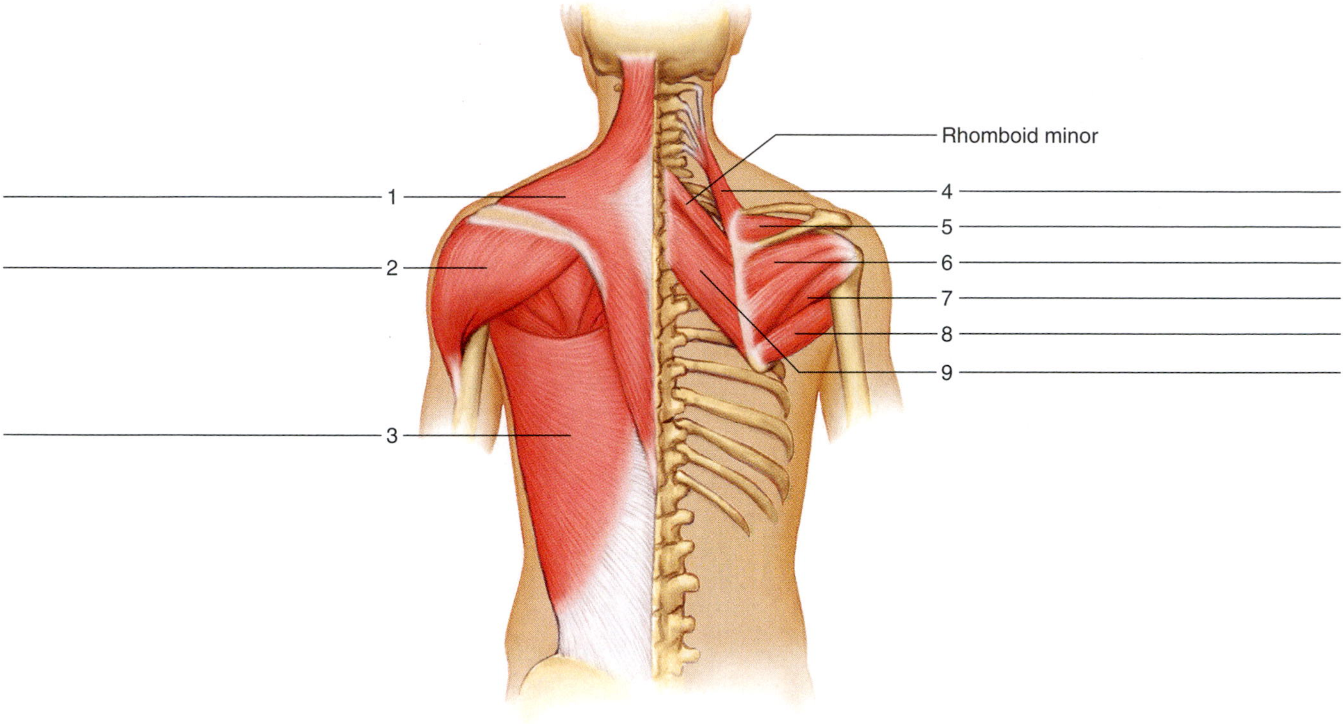

Figure 21.1 Label the muscles of the posterior shoulder. Superficial muscles are illustrated on the left side and deep muscles on the right side. 1

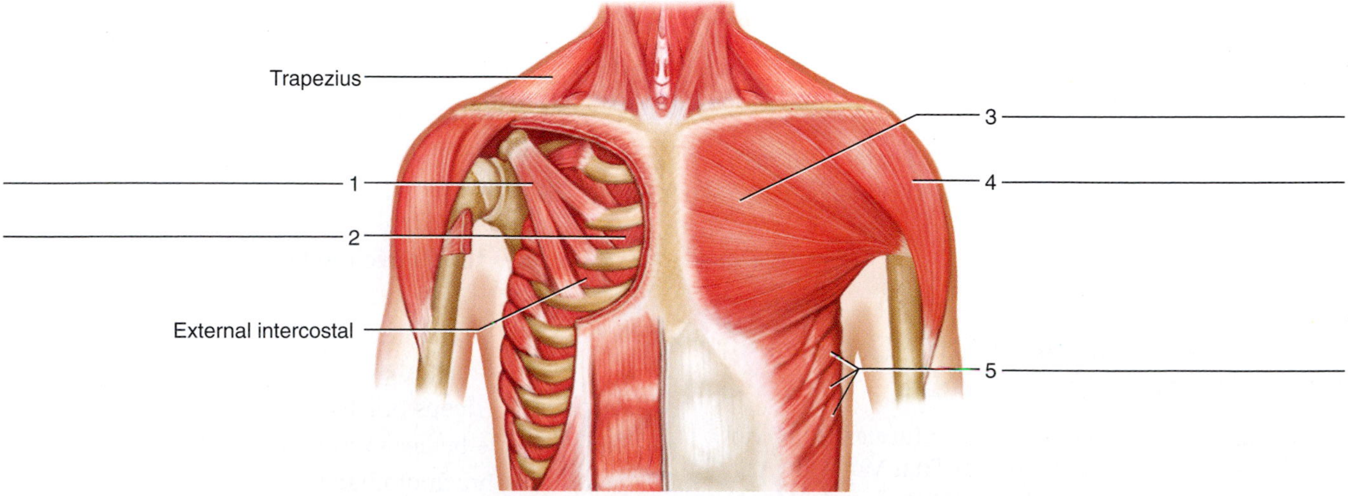

Figure 21.2 Label the muscles of the anterior chest. Superficial muscles are illustrated on the left side and deep muscles on the right side. 1

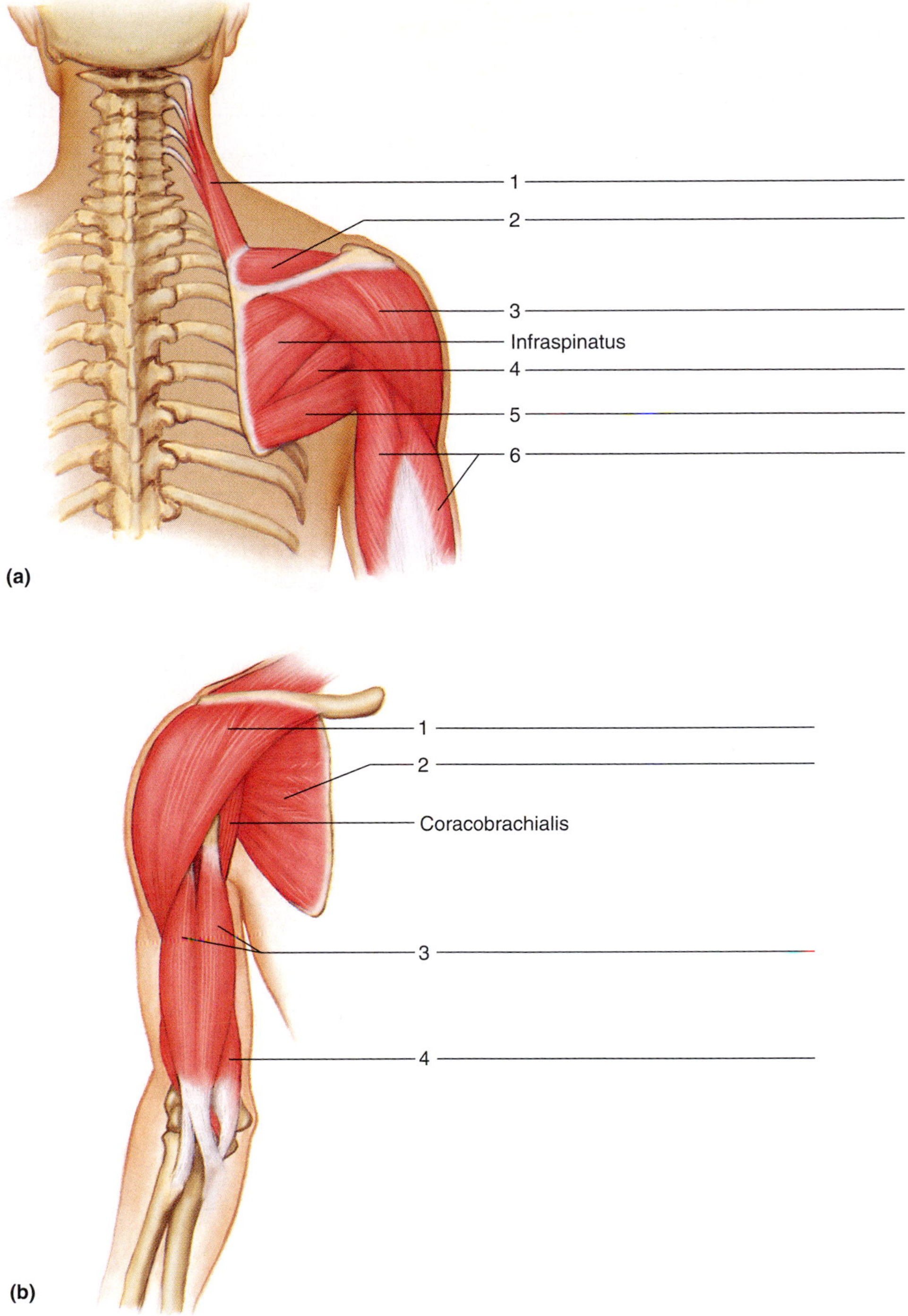

Figure 21.3 Label (*a*) the muscles of the posterior shoulder and arm and (*b*) the muscles of the anterior shoulder and arm, with the rib cage removed. 1

4. Demonstrate the action of these muscles in your body.
5. Locate the origins and insertions of these muscles in the human skeleton.
6. Complete Parts A, B, and C of Laboratory Report 21.

Learning Extension

A long rubber band can be used to simulate muscle locations, origins, insertions, and actions on muscular models, the skeleton, or laboratory partner. Hold one end of the rubber band firmly on the origin location of a muscle, then slightly stretch the rubber band and hold the other end on the insertion site. Allow the insertion end to slowly move toward the origin end to simulate the contraction and action of the muscle. 1 2 3

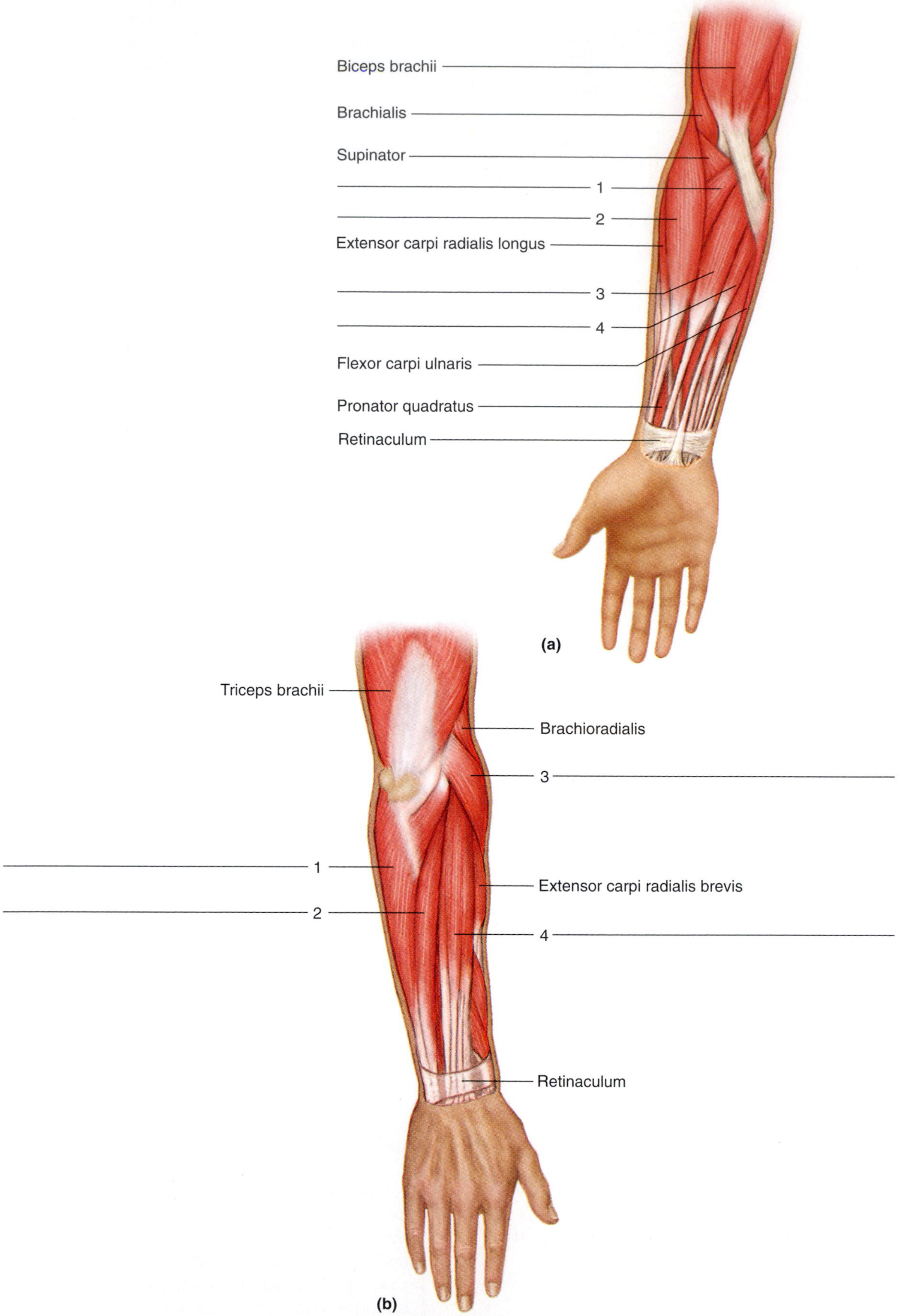

Figure 21.4 Label (*a*) the muscles of the anterior forearm and (*b*) the muscles of the posterior forearm. 1

Name ____________________

Date ____________________

Section ____________________

The [A] corresponds to the indicated Learning Outcome(s) listed at the beginning of the laboratory exercise.

Laboratory Report 21

Muscles of the Chest, Shoulder, and Upper Limb

Part A Assessments

Match the muscles in column A with the actions in column B. Place the letter of your choice in the space provided. [2]

Column A	Column B
a. Brachialis	_______ 1. Abducts arm
b. Coracobrachialis	_______ 2. Pulls arm forward (flexion) and across chest (adduction) and rotates arm medially
c. Deltoid	_______ 3. Flexes and adducts hand at the wrist
d. Extensor carpi ulnaris	_______ 4. Raises and adducts scapula
e. Flexor carpi ulnaris	_______ 5. Raises ribs in forceful inhalation or pulls scapula forward and downward
f. Infraspinatus	_______ 6. Used to thrust shoulder anteriorly (protraction), as when pushing something
g. Pectoralis major	_______ 7. Flexes the forearm at the elbow
h. Pectoralis minor	_______ 8. Flexes and adducts arm at the shoulder along with pectoralis major
i. Rhomboid major	_______ 9. Extends the forearm at the elbow
j. Serratus anterior	_______ 10. Extends, adducts, and rotates arm medially
k. Teres major	_______ 11. Extends and adducts hand at the wrist
l. Triceps brachii	_______ 12. Rotates arm laterally

Part B Assessments

Name the muscle indicated by the following combinations of origin and insertion. 3

	Origin	Insertion	Muscle
1.	Spines of upper thoracic vertebrae	Medial border of scapula	______
2.	Outer surfaces of upper ribs	Ventral surface of scapula	______
3.	Sternal ends of upper ribs	Coracoid process of scapula	______
4.	Coracoid process of scapula	Shaft of humerus	______
5.	Lateral border of scapula	Intertubercular groove of humerus	______
6.	Anterior surface of scapula	Lesser tubercle of humerus	______
7.	Lateral border of scapula	Greater tubercle of humerus	______
8.	Anterior shaft of humerus	Coronoid process of ulna	______
9.	Medial epicondyle of humerus and coronoid process of ulna	Lateral surface of radius	______
10.	Distal lateral end of humerus	Lateral surface of radius above styloid process	______
11.	Medial epicondyle of humerus	Base of second and third metacarpals	______
12.	Medial epicondyle of humerus	Fascia of palm	______

Part C Assessments

Critical Thinking Application

Identify the muscles indicated in figure 21.5. 1

1. ______
2. ______
3. ______
4. ______
5. ______
6. ______
7. ______
8. ______
9. ______
10. ______
11. ______
12. ______
13. ______
14. ______
15. ______
16. ______
17. ______
18. ______
19. ______
20. ______
21. ______

Terms:
Biceps brachii
Deltoid
External oblique
Pectoralis major
Rectus abdominis
Serratus anterior
Sternocleidomastoid
Trapezius

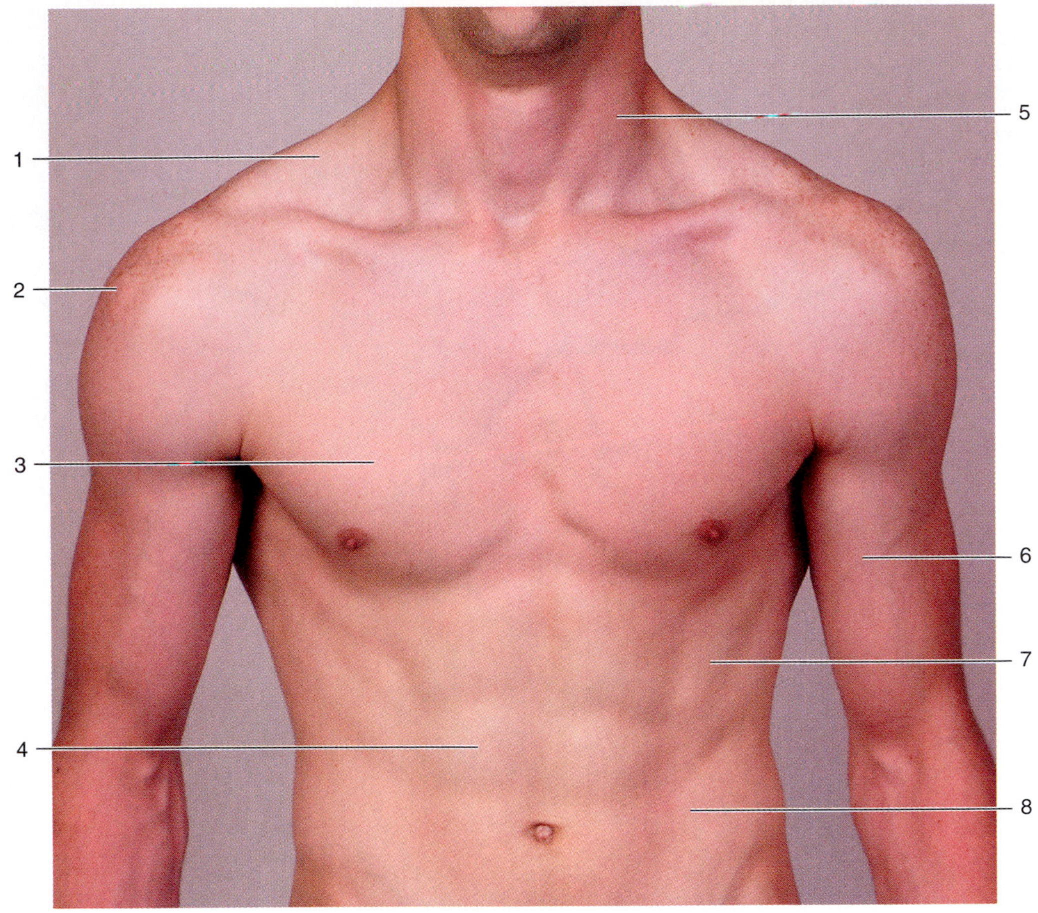

(a)

Terms:
Biceps brachii
Deltoid
Infraspinatus
Latissimus dorsi
Trapezius
Triceps brachii

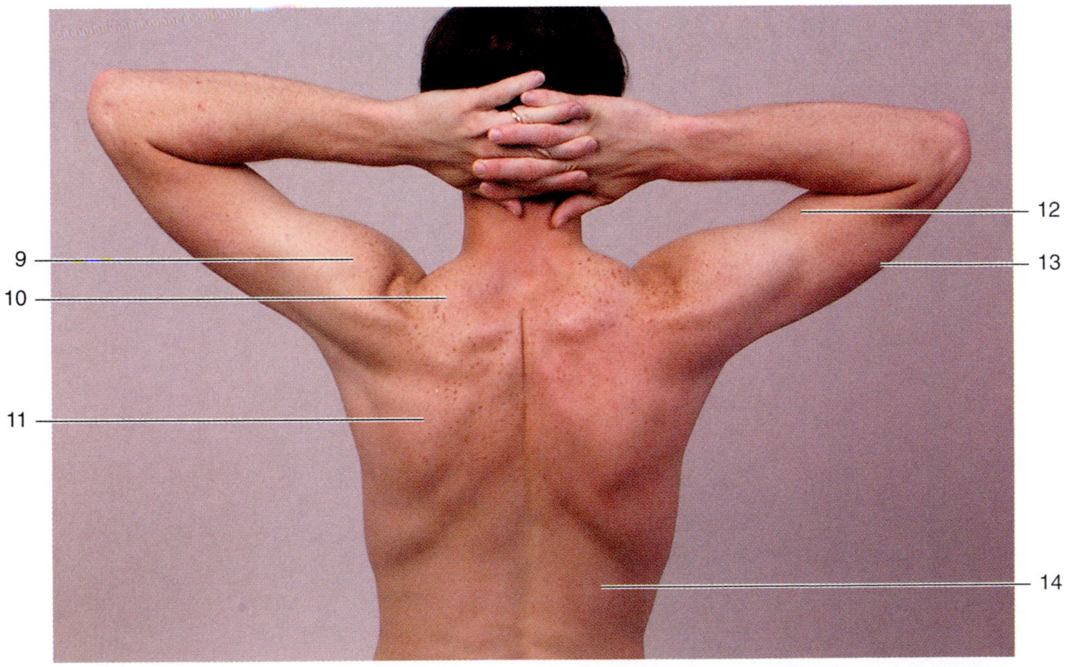

(b)

Figure 21.5 Identify the muscles that appear as body surface features in these photographs (*a*, *b*, and *c*), using the terms provided.

Terms:
Biceps brachii
Brachioradialis
Deltoid
Pectoralis major
Serratus anterior
Trapezius
Triceps brachii

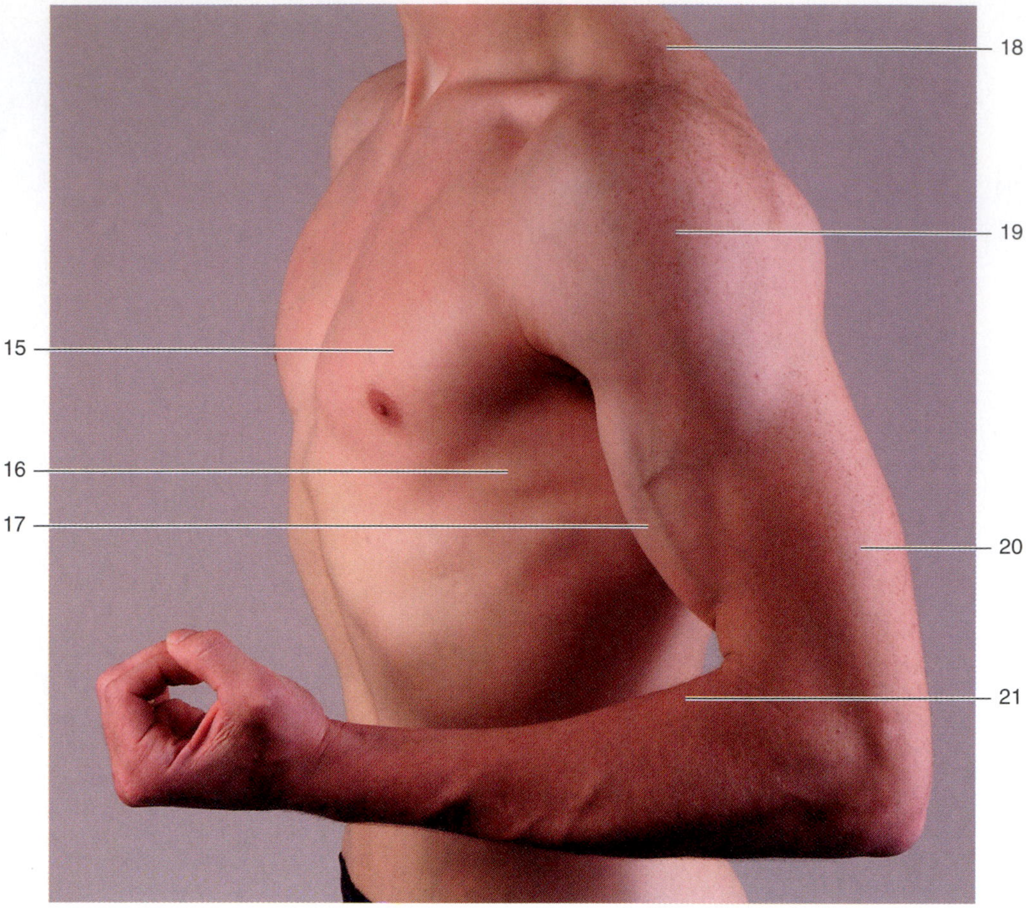

(c)

Figure 21.5 *Continued.*

Laboratory Exercise 22

Muscles of the Deep Back, Abdominal Wall, and Pelvic Outlet

Materials Needed

Textbook
Human torso model with musculature
Human skeleton, articulated
Muscular models of male and female pelves

The deep muscles of the back extend the vertebral column. Because the muscles have numerous origins, insertions, and subgroups, the muscles overlap each other. The deep back muscles can extend the spine when contracting as a group but also help to maintain posture and normal spine curvatures.

The anterior and lateral walls of the abdomen contain broad, flattened muscles arranged in layers. These muscles connect the rib cage and vertebral column to the pelvic girdle. The abdominal wall muscles compress the abdominal visceral organs, help maintain posture, assist in forceful exhalation, and contribute to trunk flexion and waist rotation.

The muscles of the pelvic outlet are arranged in two muscular sheets: (1) a deeper pelvic diaphragm that forms the floor of the pelvic cavity and (2) a urogenital diaphragm that fills the space within the pubic arch. The pelvic floor is penetrated by the urethra, vagina, and anus in a female, and thus are important in obstetrics.

Purpose of the Exercise

To review the actions, origins, and insertions of the muscles of the deep back, abdominal wall, and pelvic outlet.

LEARNING OUTCOMES

After completing this exercise, you should be able to

(1) Locate and identify the muscles of the deep back, abdominal wall, and pelvic outlet.

(2) Describe the action of each of these muscles.

(3) Locate the origin and insertion of each of these muscles in a human skeleton or on muscular models.

EXPLORE

Procedure A—Muscles of the Deep Back and Abdominal Wall

1. Review the section entitled "Muscles of the Abdominal Wall" in chapter 9 of the textbook.
2. As a review activity, label figures 22.1 and 22.2.
3. Locate the following muscles in the human torso model:

   erector spinae group
   - iliocostalis (lateral group)
   - longissimus (intermediate group)
   - spinalis (medial group)

   external oblique
   internal oblique
   transversus abdominis
   rectus abdominis
4. Demonstrate the actions of these muscles in your body. (2)
5. Locate the origin and insertion of each of these muscles in the human skeleton. (3)
6. Complete Part A of Laboratory Report 22.

Critical Thinking Application

List the muscles from superficial to deep for an appendectomy incision. (1)

____

____

____

____

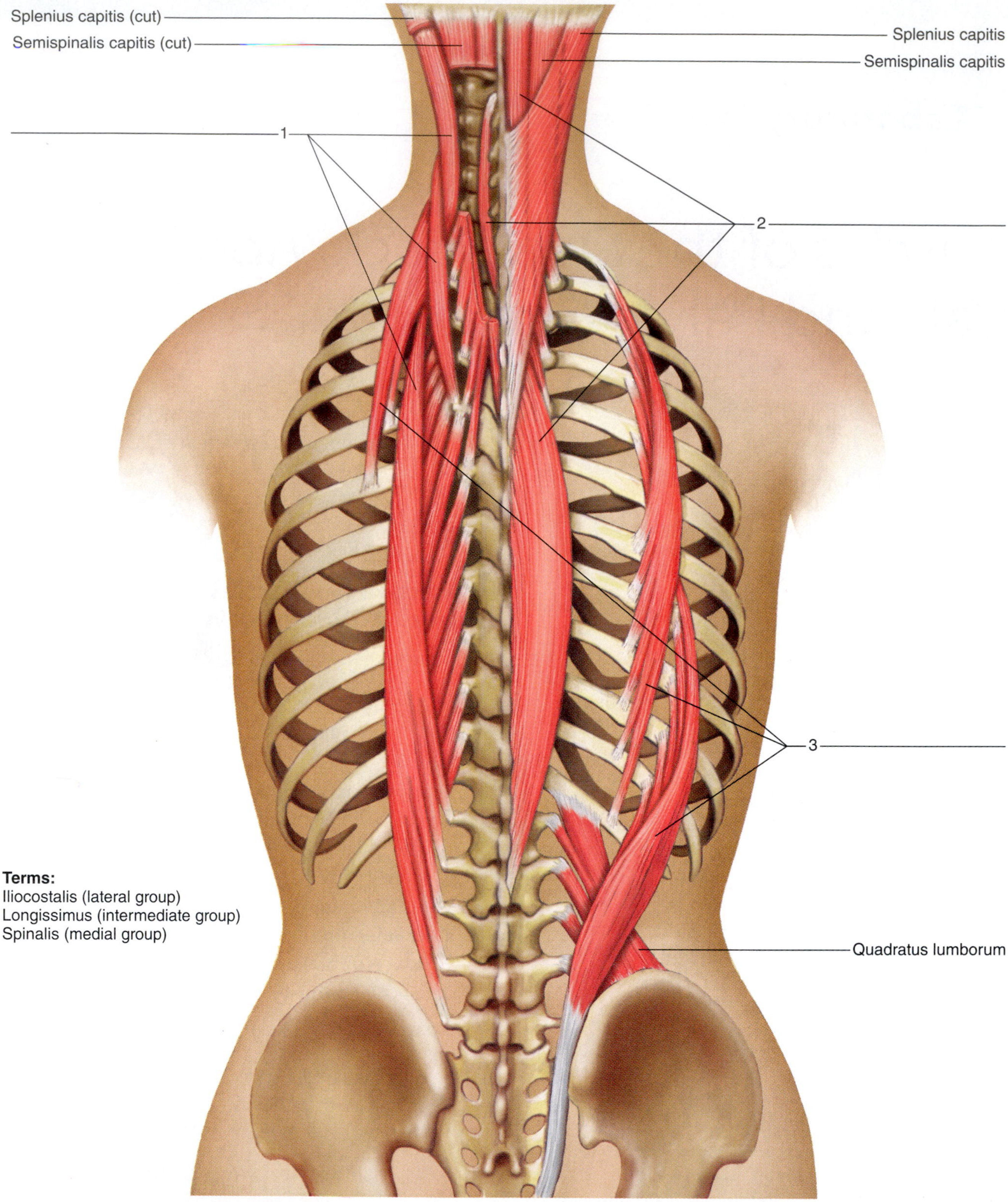

Terms:
Iliocostalis (lateral group)
Longissimus (intermediate group)
Spinalis (medial group)

Figure 22.1 Label the three deep back muscle groups of the erector spinae group, using the terms provided. (*Note:* The three erector spinae muscle groups are not illustrated on both sides of the body.) 1

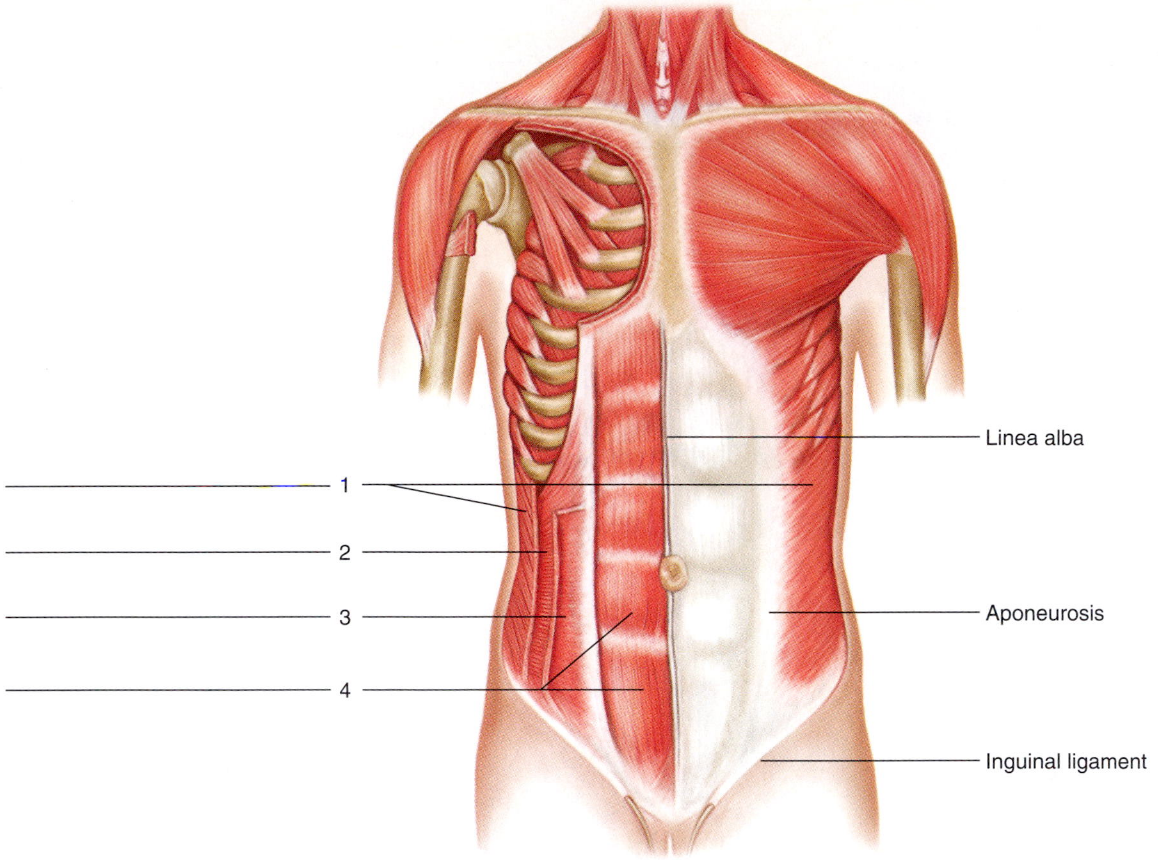

Figure 22.2 Label the muscles of the abdominal wall. 1

EXPLORE

Procedure B—Muscles of the Pelvic Outlet

1. Review the section entitled "Muscles of the Pelvic Outlet" in chapter 9 of the textbook.
2. As a review activity, label figures 22.3 and 22.4.
3. Locate the following muscles in the models of the male and female pelves:

   levator ani
   coccygeus
   superficial transversus perinei
   bulbospongiosus
   ischiocavernosus

4. Locate the origin and insertion of each of these muscles in the human skeleton. 3
5. Complete Part B of the laboratory report.

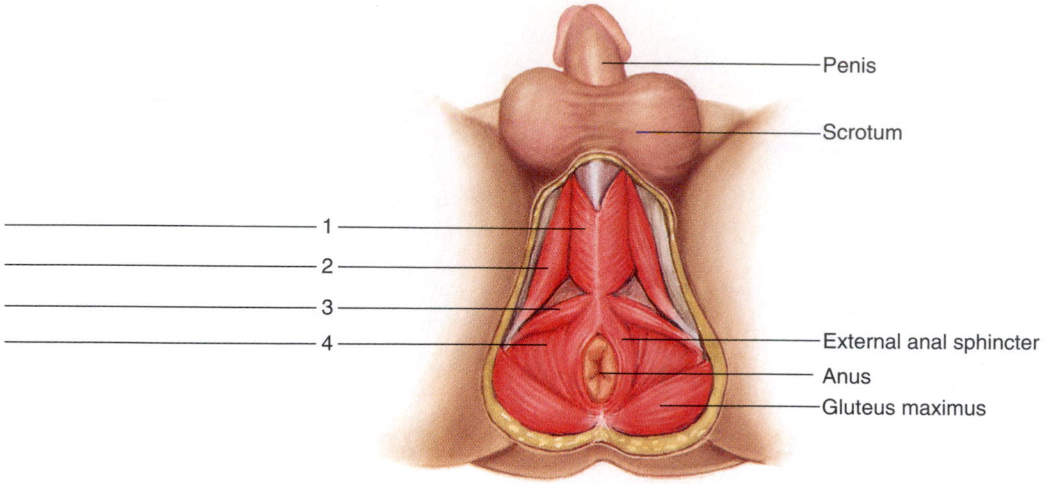

Figure 22.3 Label the muscles of the male pelvic outlet. 1

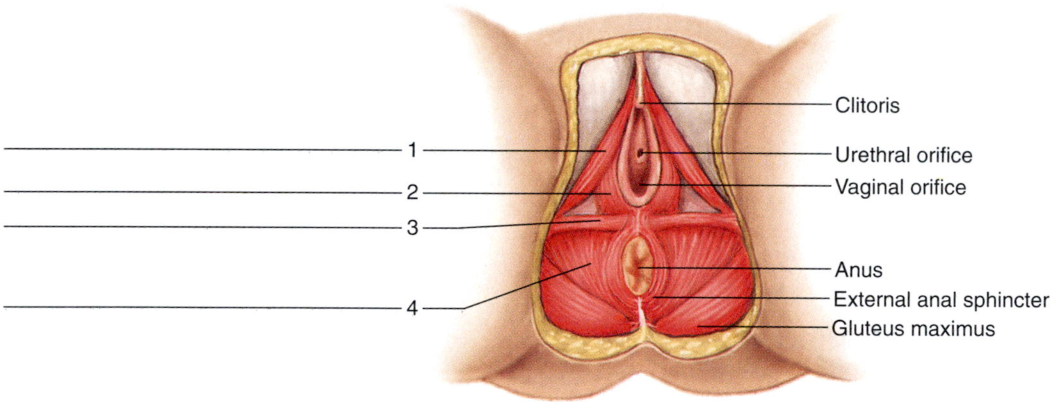

Figure 22.4 Label the muscles of the female pelvic outlet. 1

Name ____________________

Date ____________________

Section ____________________

The [A] corresponds to the indicated Learning Outcome(s) listed at the beginning of the laboratory exercise.

Muscles of the Deep Back, Abdominal Wall, and Pelvic Outlet

Laboratory Report 22

Part A Assessments

Complete the following statements:

1. A band of tough connective tissue in the midline of the anterior abdominal wall called the ____________ serves as a muscle attachment. [1]
2. The ____________ muscle spans from the ribs and sternum to the pubic bones. [3]
3. The ____________ forms the third layer (deepest layer) of the abdominal wall muscles. [1]
4. The action of the external oblique muscle is to ____________. [2]
5. The action of the rectus abdominis is to ____________. [2]
6. The iliocostalis, longissimus, and spinalis muscles together form the ____________. [1]

Part B Assessments

Complete the following statements:

1. The levator ani and coccygeus together form the ____________. [1]
2. The levator ani provides a sphincterlike action in the ____________. [2]
3. The action of the coccygeus is to ____________ ____________. [2]
4. The ____________ surrounds the base of the penis. [1]
5. In females, the bulbospongiosus acts to ____________. [2]
6. The ischiocavernosus extends from the margin of the pubic arch to the ____________. [1]
7. In the female, the ____________ muscles are separated by the vagina, urethra, and anal canal. [1]
8. The action of the superficial transversus perinei is to ____________. [2]
9. The coccygeus extends from the coccyx and sacrum to the ____________. [3]
10. The ____________ assists in closing the urethra. [2]

Notes

Laboratory Exercise 23

Muscles of the Hip and Lower Limb

Materials Needed

Textbook
Human torso model with musculature
Human skeleton, articulated
Muscular models of the lower limb

For Learning Extension:
Long rubber bands

The muscles that move the thigh are attached to the femur and to some part of the pelvic girdle. Those attached anteriorly primarily act to flex the thigh at the hip, whereas those attached posteriorly act to extend, abduct, or rotate the thigh.

The muscles that move the leg connect the tibia or fibula to the femur or to the pelvic girdle. They function to flex or extend the leg at the knee. Other muscles, located in the leg, act to move the foot.

Purpose of the Exercise

To review the actions, origins, and insertions of the muscles that move the thigh, leg, and foot.

LEARNING OUTCOMES

After completing this exercise, you should be able to

1. Locate and identify the muscles that move the thigh, leg, and foot.
2. Describe and demonstrate the actions of each of these muscles.
3. Locate the origin and insertion of each of these muscles in a human skeleton and on muscular models.

EXPLORE

Procedure—Muscles of the Hip and Lower Limb

1. Review the sections entitled "Muscles That Move the Thigh," "Muscles That Move the Leg," and "Muscles That Move the Foot" in chapter 9 of the textbook.
2. As a review activity, label figures 23.1, 23.2, 23.3, 23.4, 23.5, and 23.6.
3. Locate the following muscles in the human torso model and in the lower limb models. Also locate as many of them as possible in your body.

muscles that move the thigh
- anterior hip muscles
  - iliopsoas group
    - psoas major
    - iliacus
- posterior and lateral hip muscles
  - gluteus maximus
  - gluteus medius
  - gluteus minimus
  - tensor fasciae latae
- medial adductor muscles
  - pectineus
  - adductor longus
  - adductor magnus
  - adductor brevis
  - gracilis

muscles that move the leg
- anterior thigh muscles
  - sartorius
  - quadriceps femoris group
    - rectus femoris
    - vastus lateralis
    - vastus medialis
    - vastus intermedius
- posterior thigh muscles
  - hamstring group
    - biceps femoris
    - semitendinosus
    - semimembranosus

muscles that move the foot
- anterior leg muscles
  - tibialis anterior
  - fibularis (peroneus) tertius
  - extensor digitorum longus

posterior leg muscles
- gastrocnemius
- soleus
- flexor digitorum longus
- tibialis posterior

lateral leg muscles
- fibularis (peroneus) longus
- fibularis (peroneus) brevis

4. Demonstrate the action of each of these muscles in your body.
5. Locate the origin and insertion of each of these muscles in the human skeleton.
6. Complete Parts A, B, and C of Laboratory Report 23.

Learning Extension

A long rubber band can be used to simulate muscle locations, origins, insertions, and actions on muscular models, the skeleton, or on a laboratory partner. Hold one end of the rubber band firmly on the origin location of a muscle, then slightly stretch the rubber band and hold the other end on the insertion site. Allow the insertion end to slowly move toward the origin end to simulate the contraction and action of the muscle. 1 2 3

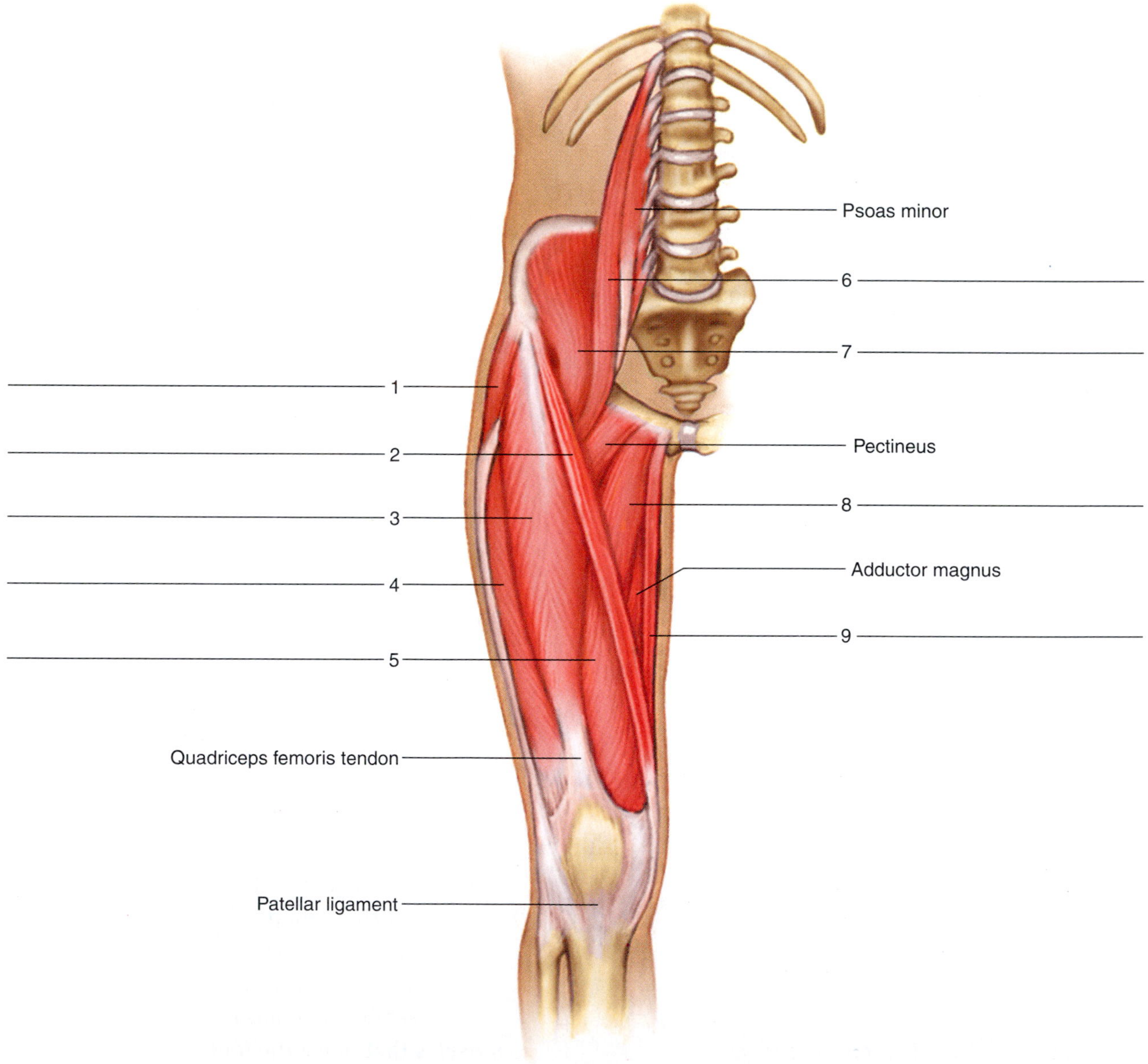

Figure 23.1 Label the muscles of the anterior right hip and thigh. 1

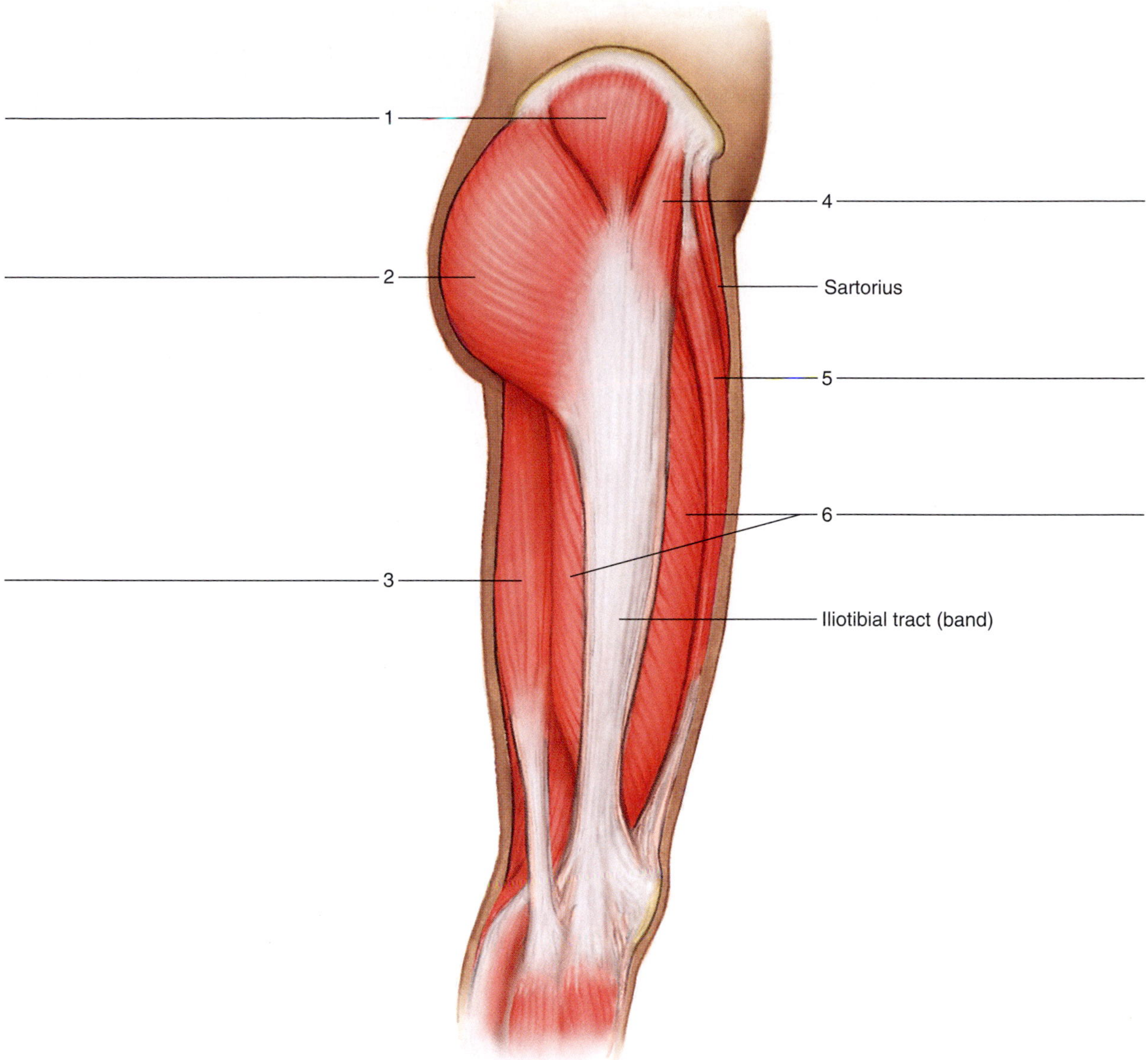

Figure 23.2 Label the muscles of the lateral right hip and thigh.

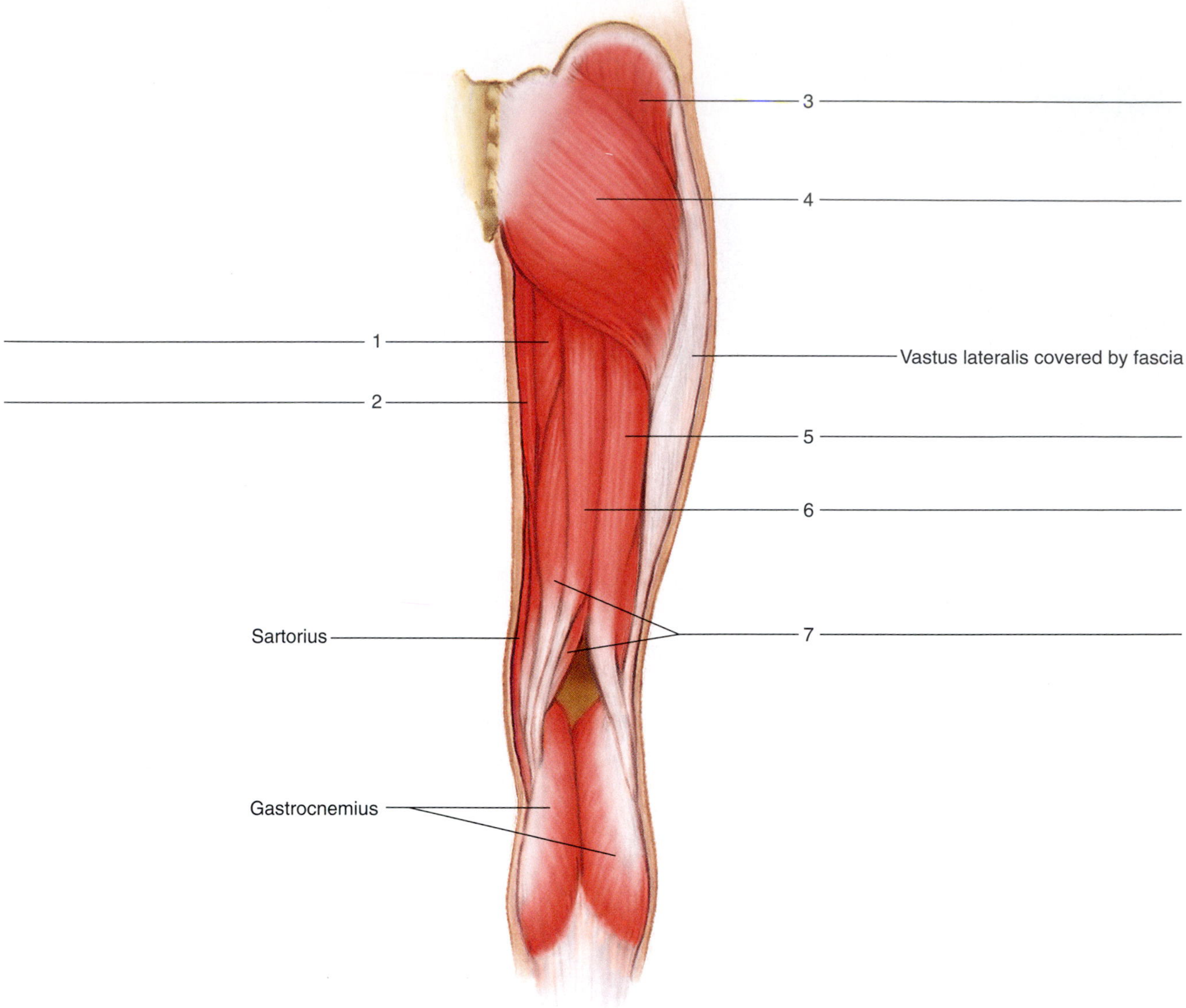

Figure 23.3 Label the muscles of the posterior right hip and thigh.

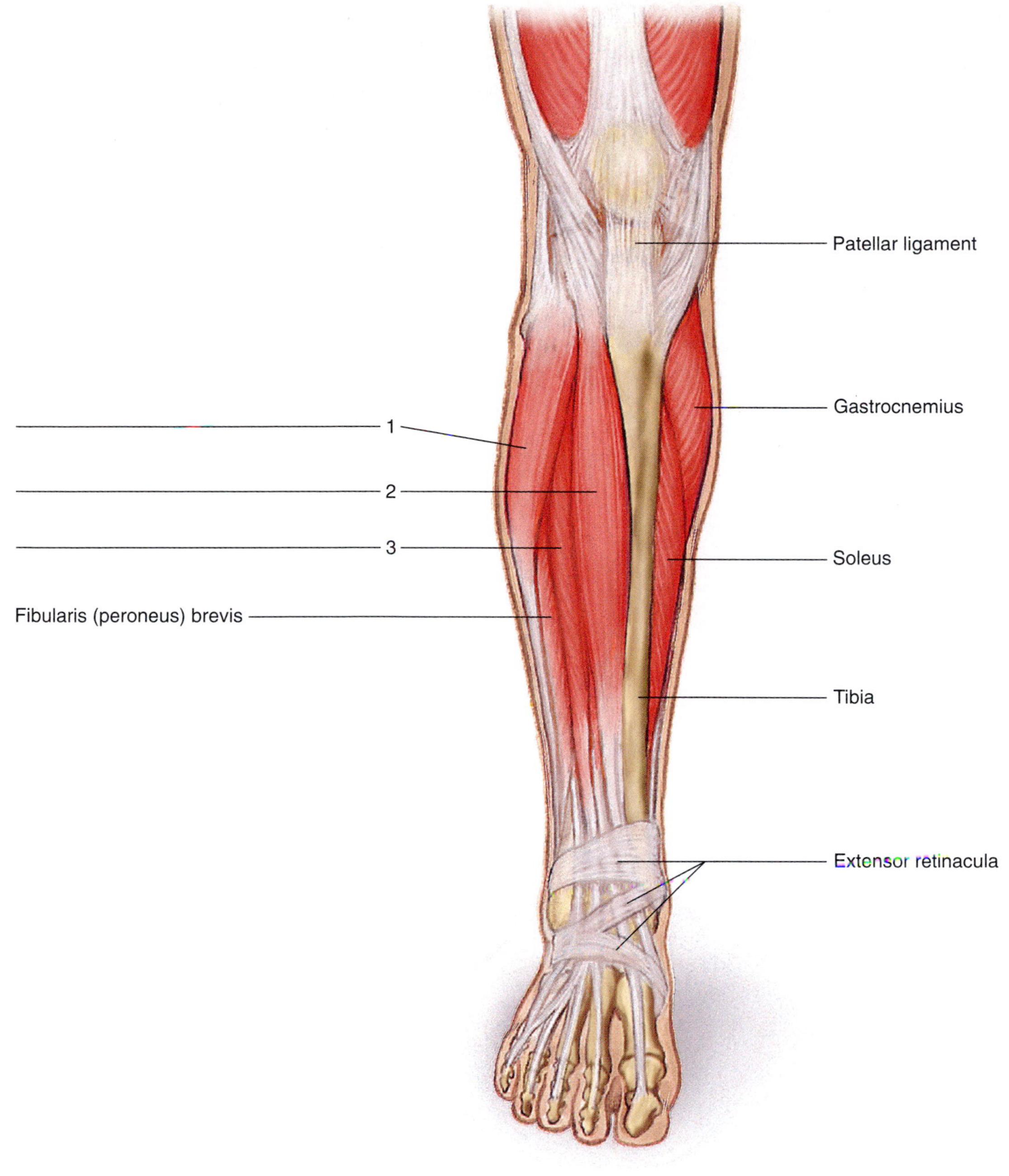

Figure 23.4 Label the muscles of the anterior right leg. 1

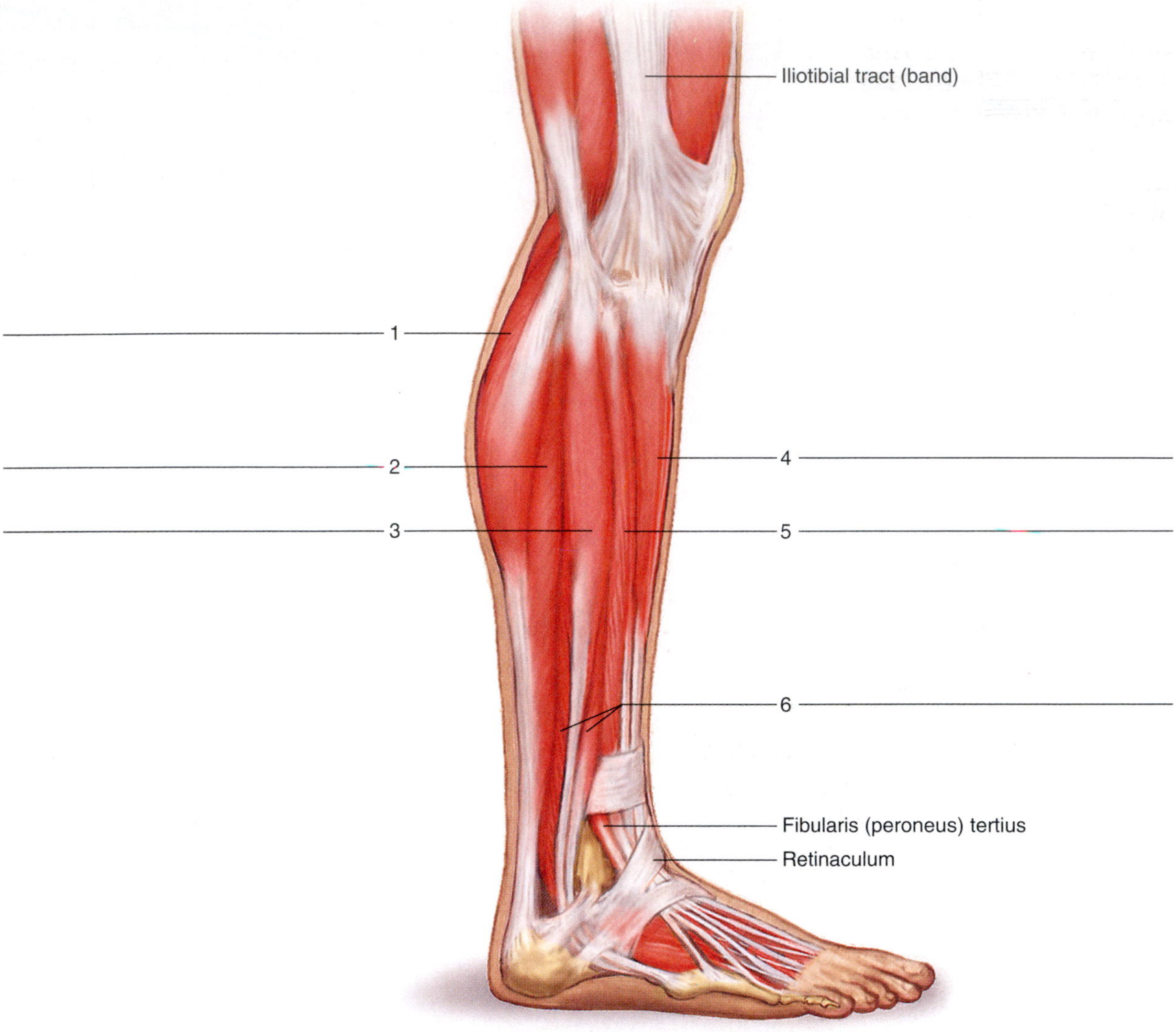

Figure 23.5 Label the muscles of the lateral right leg. 1

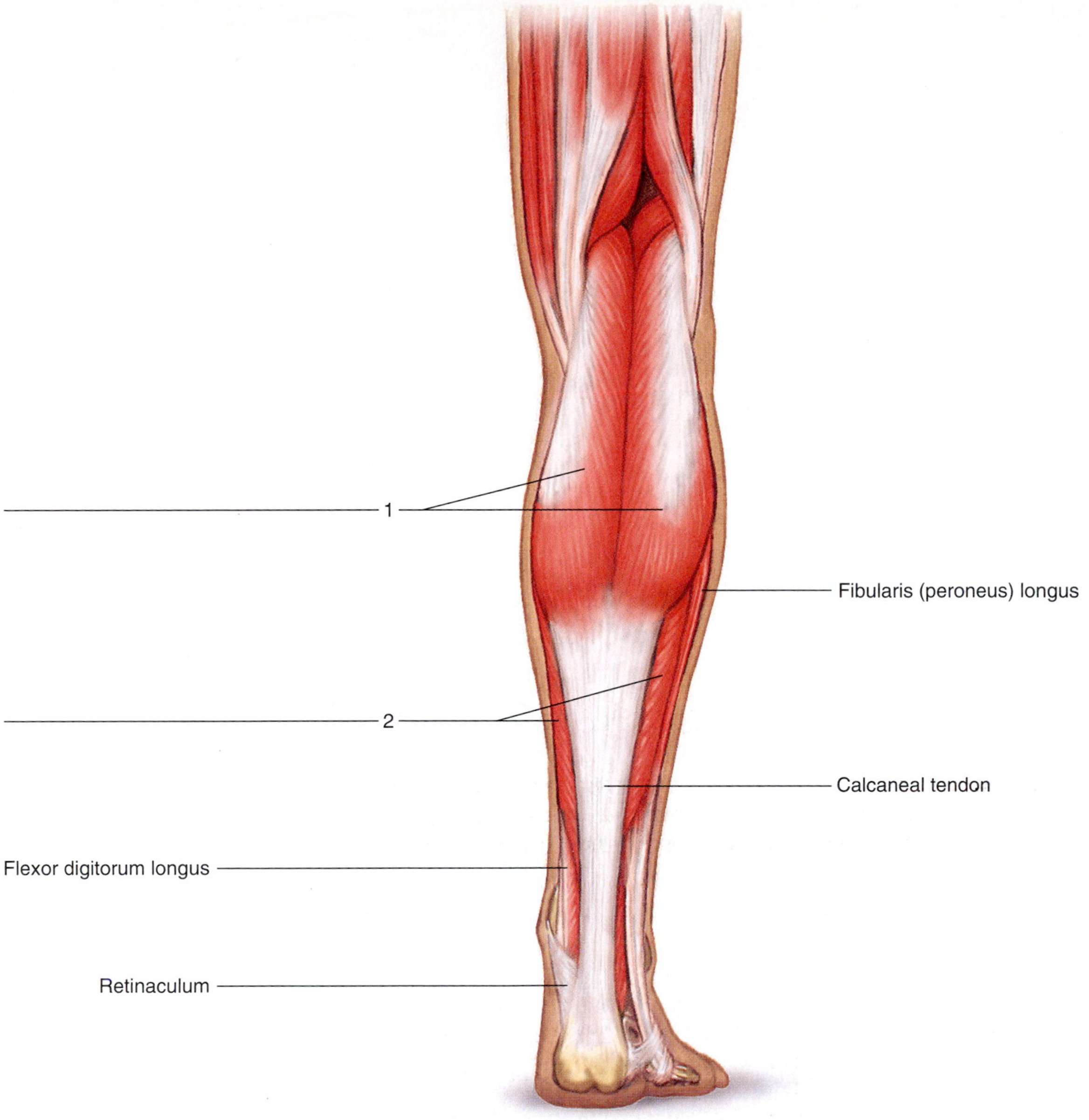

Figure 23.6 Label the muscles of the posterior right leg. 1

Notes

Name ________________

Date ________________

Section ________________

The ⚠ corresponds to the indicated Learning Outcome(s) listed at the beginning of the laboratory exercise.

Laboratory Report 23

Muscles of the Hip and Lower Limb

Part A Assessments

Match the muscles in column A with the actions in column B. Place the letter of your choice in the space provided. 2

Column A		Column B
a. Biceps femoris	_____	1. Adducts thigh
b. Fibularis (peroneus) longus	_____	2. Plantar flexion and eversion of foot
c. Fibularis (peroneus) tertius	_____	3. Flexes thigh at the hip
d. Gluteus medius	_____	4. Abducts thigh and rotates it laterally
e. Gracilis	_____	5. Dorsiflexion and eversion of foot
f. Psoas major and iliacus	_____	6. Abducts thigh and rotates it medially
g. Quadriceps femoris group	_____	7. Plantar flexion and inversion of foot
h. Sartorius	_____	8. Flexes leg at the knee
i. Tibialis anterior	_____	9. Extends leg at the knee
j. Tibialis posterior	_____	10. Dorsiflexion and inversion of foot

Part B Assessments

Name the muscle indicated by the following combinations of origin and insertion. 3

Origin	Insertion	Muscle
1. Lateral surface of ilium	Greater trochanter of femur	_____
2. Ischial tuberosity	Posterior surface of femur	_____
3. Anterior superior iliac spine	Medial surface of proximal tibia	_____
4. Lateral and medial condyles of femur	Posterior surface of calcaneus	_____
5. Anterior iliac crest	Fascia (iliotibial tract) of the thigh	_____
6. Greater trochanter and posterior surface of femur	Patella to tibial tuberosity	_____
7. Ischial tuberosity	Medial surface of proximal tibia	_____
8. Medial surface of femur	Patella to tibial tuberosity	_____
9. Posterior surface of tibia	Distal phalanges of four lateral toes	_____
10. Lateral condyle and lateral surface of tibia	Medial cuneiform and first metatarsal	_____

Critical Thinking Application

Identify the muscles indicated in figure 23.7. 1

1. ______________________________
2. ______________________________
3. ______________________________
4. ______________________________
5. ______________________________
6. ______________________________
7. ______________________________
8. ______________________________

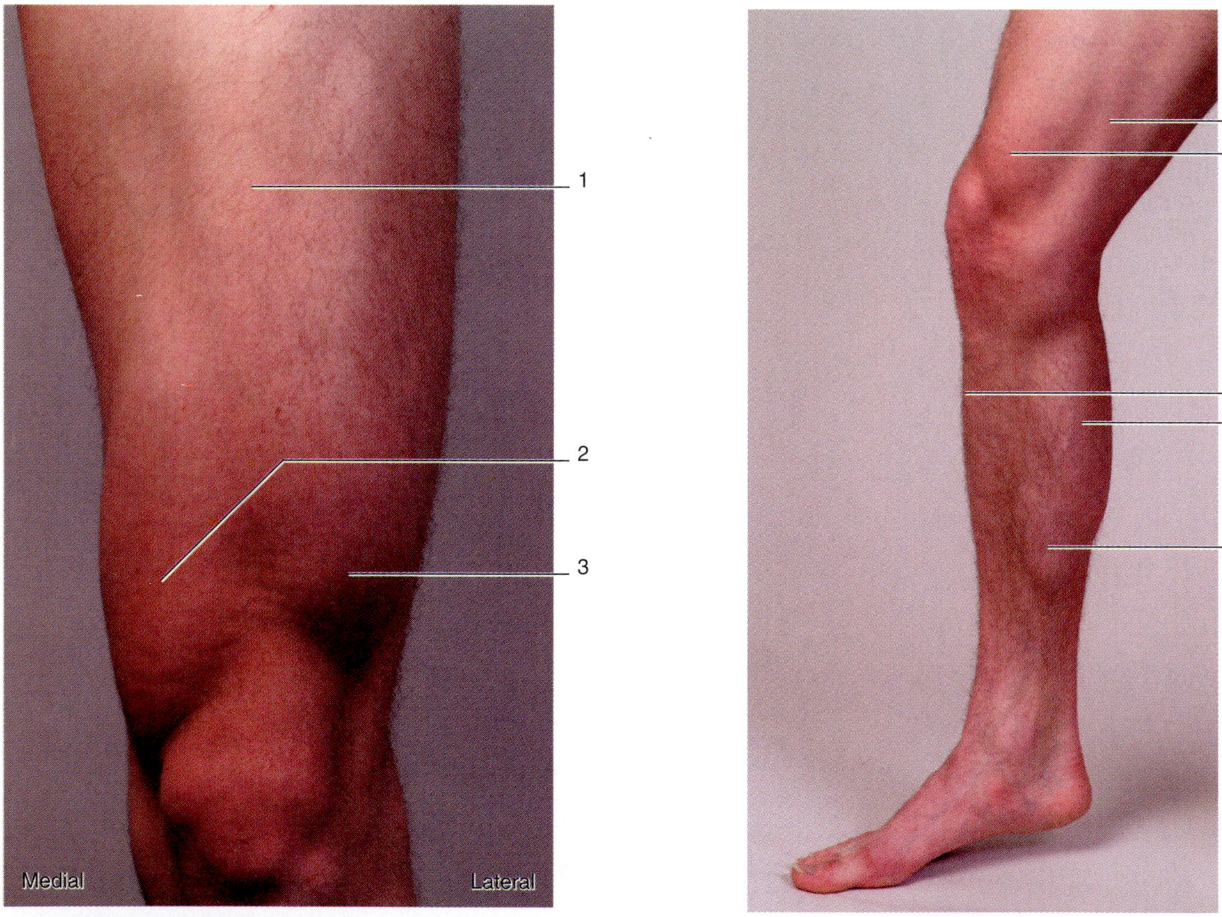

(a) Left thigh, anterior view

(b) Right lower limb, medial view

Terms:

Gastrocnemius	Tibialis anterior
Rectus femoris	Vastus lateralis
Sartorius	Vastus medialis
Soleus	Vastus medialis

Figure 23.7 Identify the muscles that appear as lower limb surface features in these photographs (*a* and *b*), using the terms provided.

Laboratory Exercise 24

Surface Anatomy

Materials Needed

Textbook
Small round stickers
Colored pencils (black and red)

External landmarks, called surface anatomy, located on the human body provide an opportunity to examine surface features and to help locate other internal structures. This exercise will focus on bony landmarks and superficial soft tissue structures, primarily related to the skeletal and muscular systems. The technique used for the examination of surface features is called *palpation,* accomplished by touching with the hands or fingers. This enables us to determine the location, size, and texture of the structure. Some of the surface features can be visible without the need of palpation, especially on body builders and very thin individuals. If the individual involved has considerable subcutaneous adipose tissue, additional pressure might be necessary to palpate some of the structures.

These surface features will help us to locate additional surface features and deeper structures during the study of the additional systems. Some of the respiratory structures in the anterior neck can be palpated. Surface features will help us to determine the pulse locations of superficial arteries and for proper assessments of injection sites; descriptions of pain and injury sites; insertion of examination tubes; and listening to heart, lung, and bowel sounds with a stethoscope. For those who choose careers in emergency medical services, nursing, physicians, physical education, physical therapists, chiropractors, and massage therapists, surface anatomy is especially valuable.

Purpose of the Exercise

To examine the surface features of the human body and the terms used to describe them.

LEARNING OUTCOMES

After completing this exercise you should be able to

1. Locate and identify major surface features of the human body.
2. Arrange surface features by body region.
3. Distinguish between surface features that are bony landmarks and soft tissue.

EXPLORE

Procedure—Surface Anatomy

1. Review textbook figures on body regions, human organism reference plates, skeletal system, and muscular system. These figures are located in chapters 1, 7, 8, and 9.
2. Use the textbook figures and the other figures provided to complete figures 24.1 through 24.8. Many features have already been labeled. Features in boldface are bony features. Other labeled features (not boldface) are soft tissue features. Bony features are a part of a bone and the skeletal system. Soft tissue features are comprised of tissue other than bone such as muscle, connective tissue, or skin. The missing features are listed by each figure.
3. Complete Parts A and B of Laboratory Report 24.
4. Use figures 24.1 through 24.8 to determine which items listed in table 24.1 are bony features and which items are soft tissue features. If the feature listed is a bony surface feature, place a ✓ mark in the second column; if the feature listed is a soft tissue feature, place a ✓ mark in the third column. 3
5. Complete Part C of the laboratory report.

Learning Extension

Select fifteen surface features presented in the laboratory exercise figures. Using small, round stickers, write the name of a surface feature on each sticker. Working with a lab partner, locate the selected surface features on yourself and on your lab partner. Use the stickers to accurately "label" your partner. 1

- Were you able to find all the features on both of you? ____________________
- Was it easier to find the surface features on yourself or your lab partner? ____________________

Terms:
Hyoid bone
Mandible
Sternocleidomastoid
Trapezius (upper)

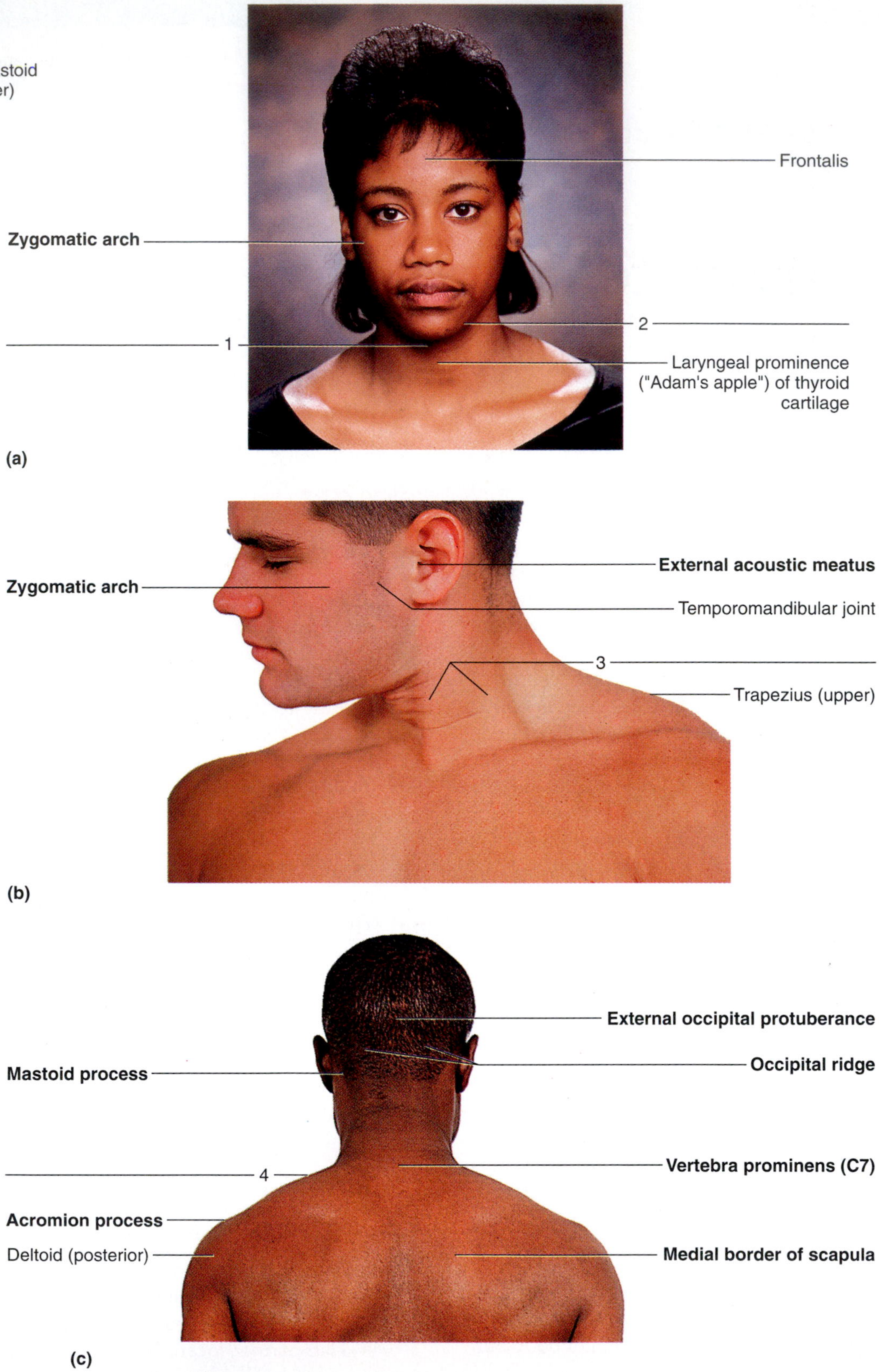

Figure 24.1 Label the surface features of (*a*) the anterior head and neck, (*b*) the lateral head and neck, and (*c*) the posterior head and neck, using the terms provided. (**Boldface** indicates bony features; not boldface indicates soft tissue features.) 3

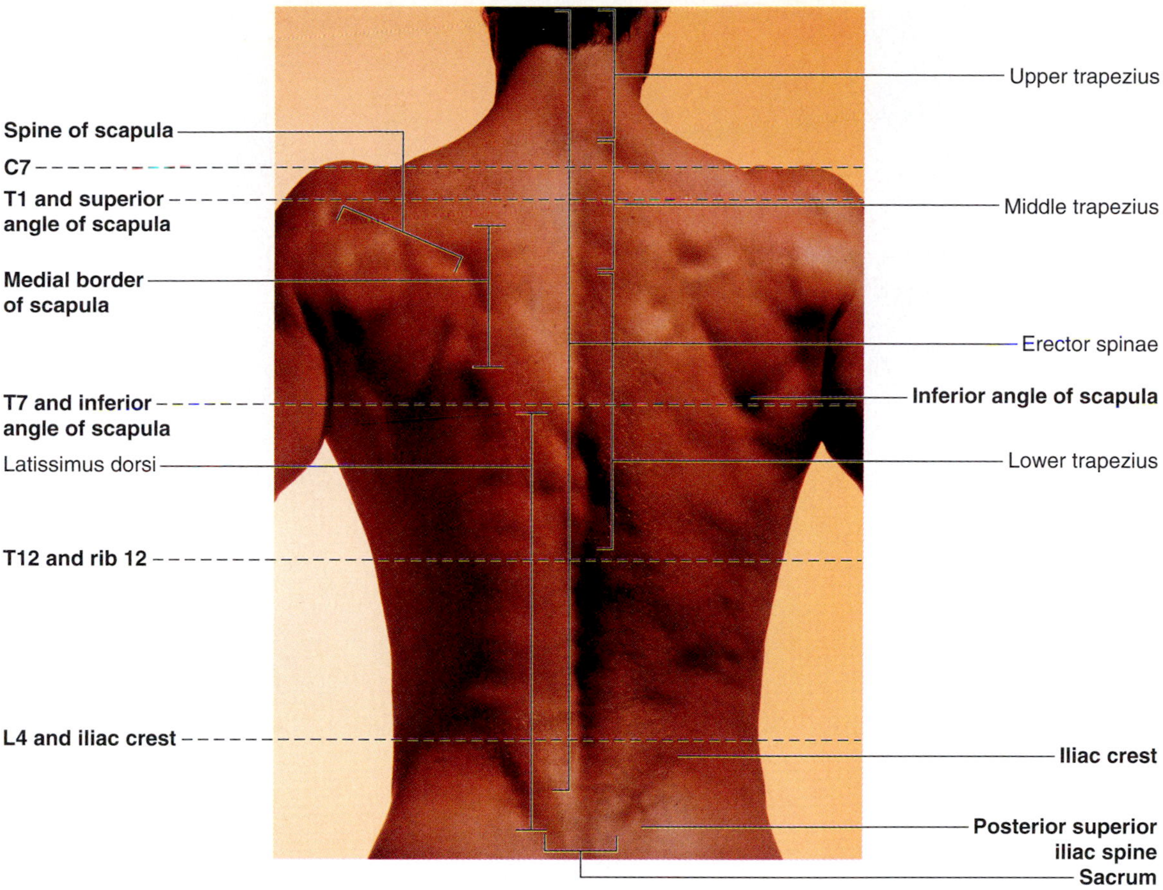

Figure 24.2 Surface features of the posterior shoulder and torso. (**Boldface** indicates bony features; not boldface indicates soft tissue features.)

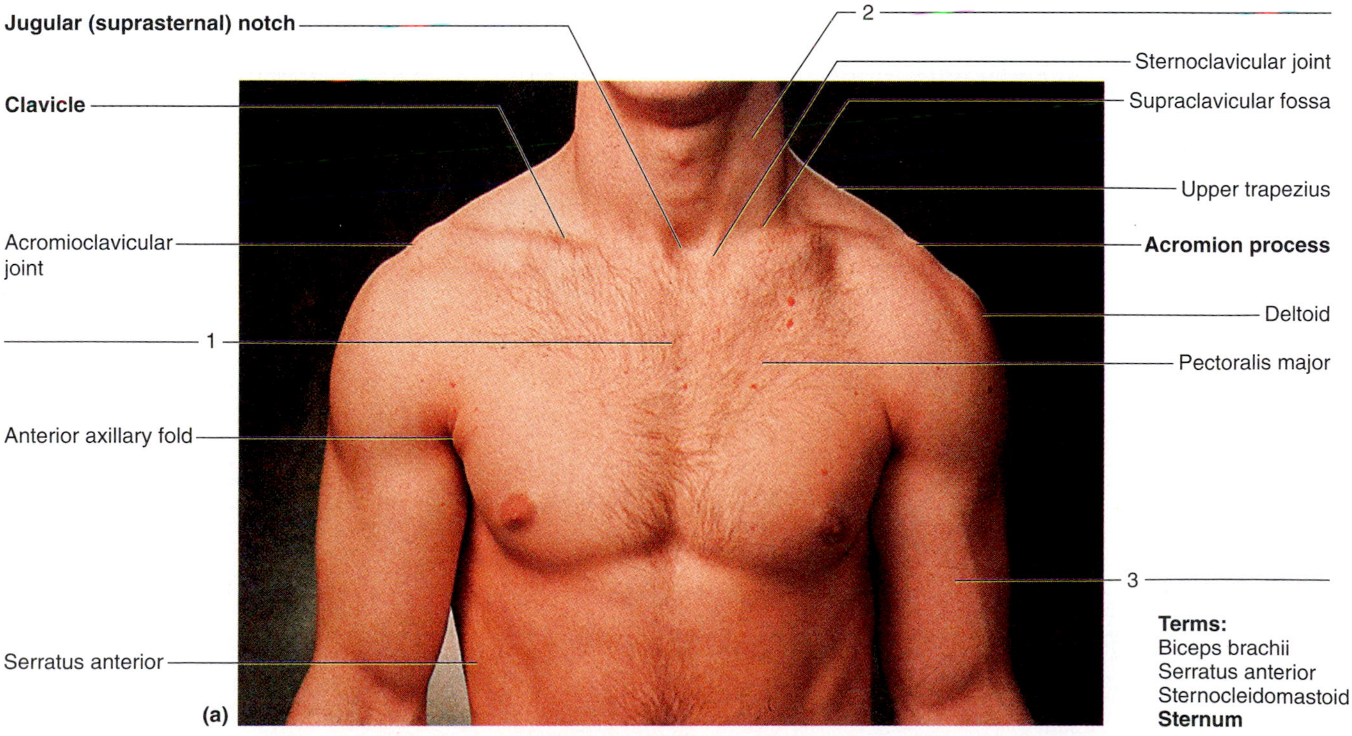

Figure 24.3 Label the anterior surface features of (*a*) upper torso and (*b*) lower torso, using the terms provided. (**Boldface** indicates bony features; not boldface indicates soft tissue features.) 3

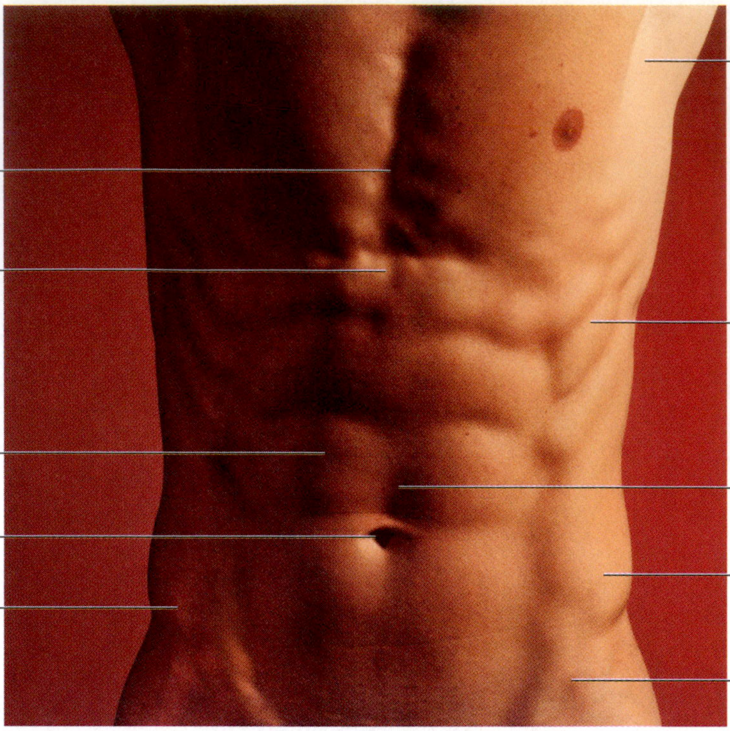

Figure 24.3 *Continued.*

Table 24.1 Representative Bony and Soft Tissue Surface Features

Bony and Soft Tissues	Bony Features	Soft Tissue Features
Biceps brachii		
Calcaneal tendon		
Deltoid		
Gastrocnemius		
Greater trochanter		
Head of fibula		
Iliac crest		
Iliotibial tract		
Lateral malleolus		
Mandible		
Mastoid process		
Medial border of scapula		
Rectus abdominis		
Sacrum		
Serratus anterior		
Sternocleidomastoid		
Sternum		
Thenar eminence		
Tibialis anterior		
Zygomatic arch		

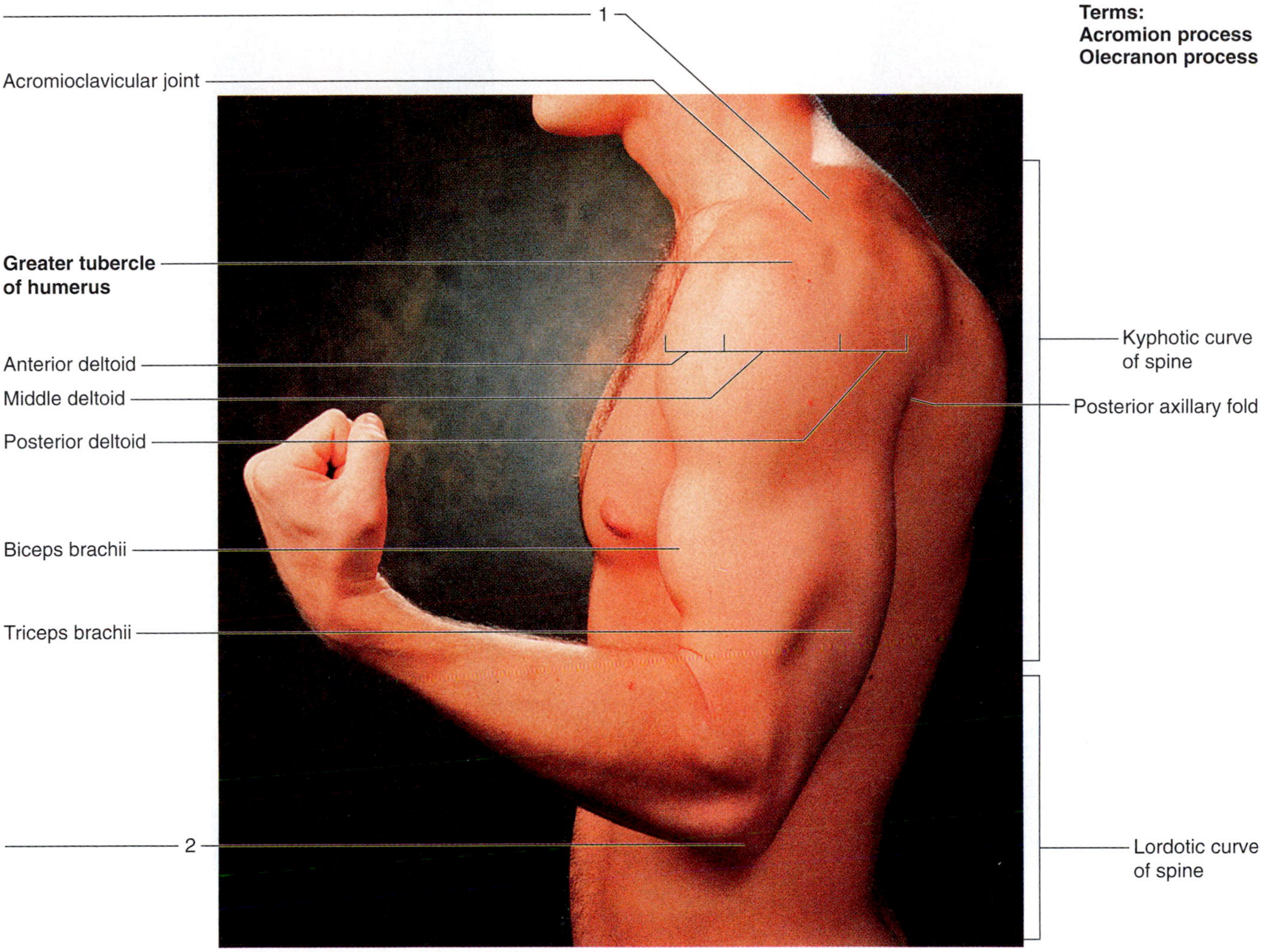

Figure 24.4 Label the surface features of the lateral shoulder and upper limb, using the terms provided. (**Boldface** indicates bony features; not boldface indicates soft tissue features.) 1

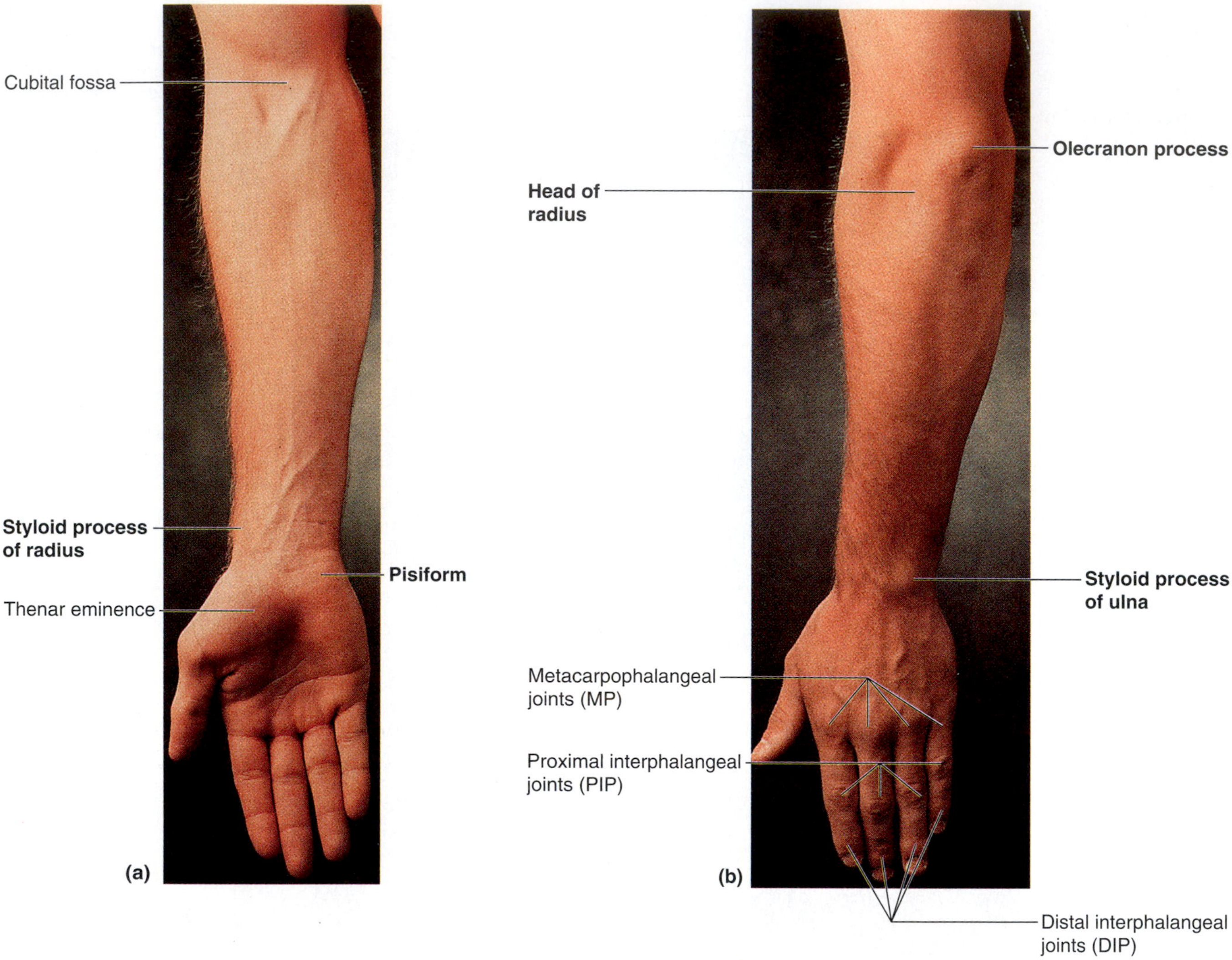

Figure 24.5 Surface features of the upper limb (*a*) anterior view and (*b*) posterior view. (**Boldface** indicates bony features; not boldface indicates soft tissue features.)

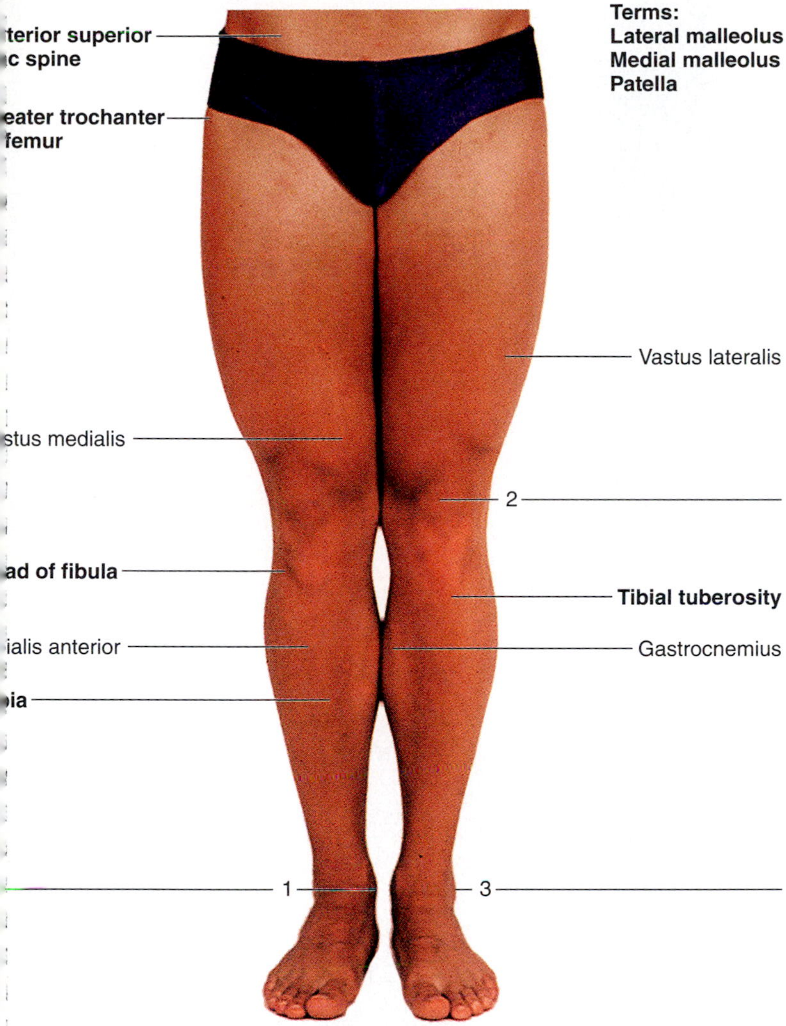

Figure 24.6 Label the surface features of the anterior view of the body, using the terms provided. (**Boldface** indicates bony features; not boldface indicates soft tissue features.) 1

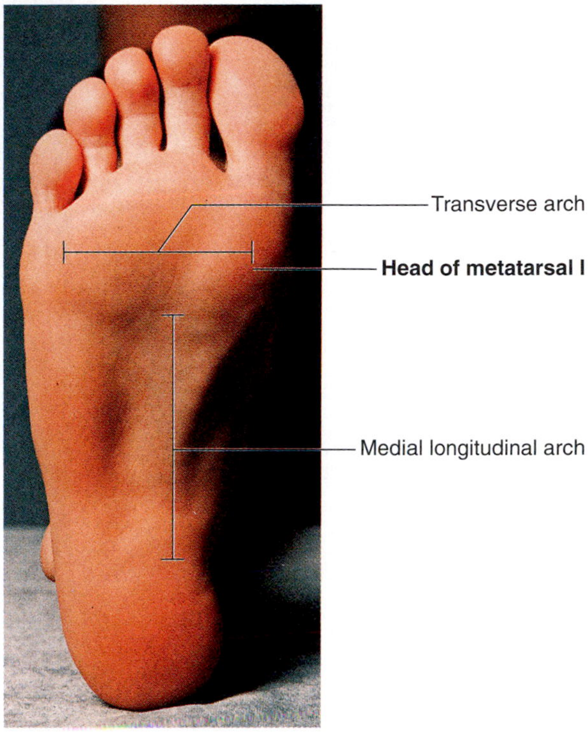

Figure 24.7 Surface features of the plantar surface of the foot. (**Boldface** indicates bony features; not boldface indicates soft tissue features.)

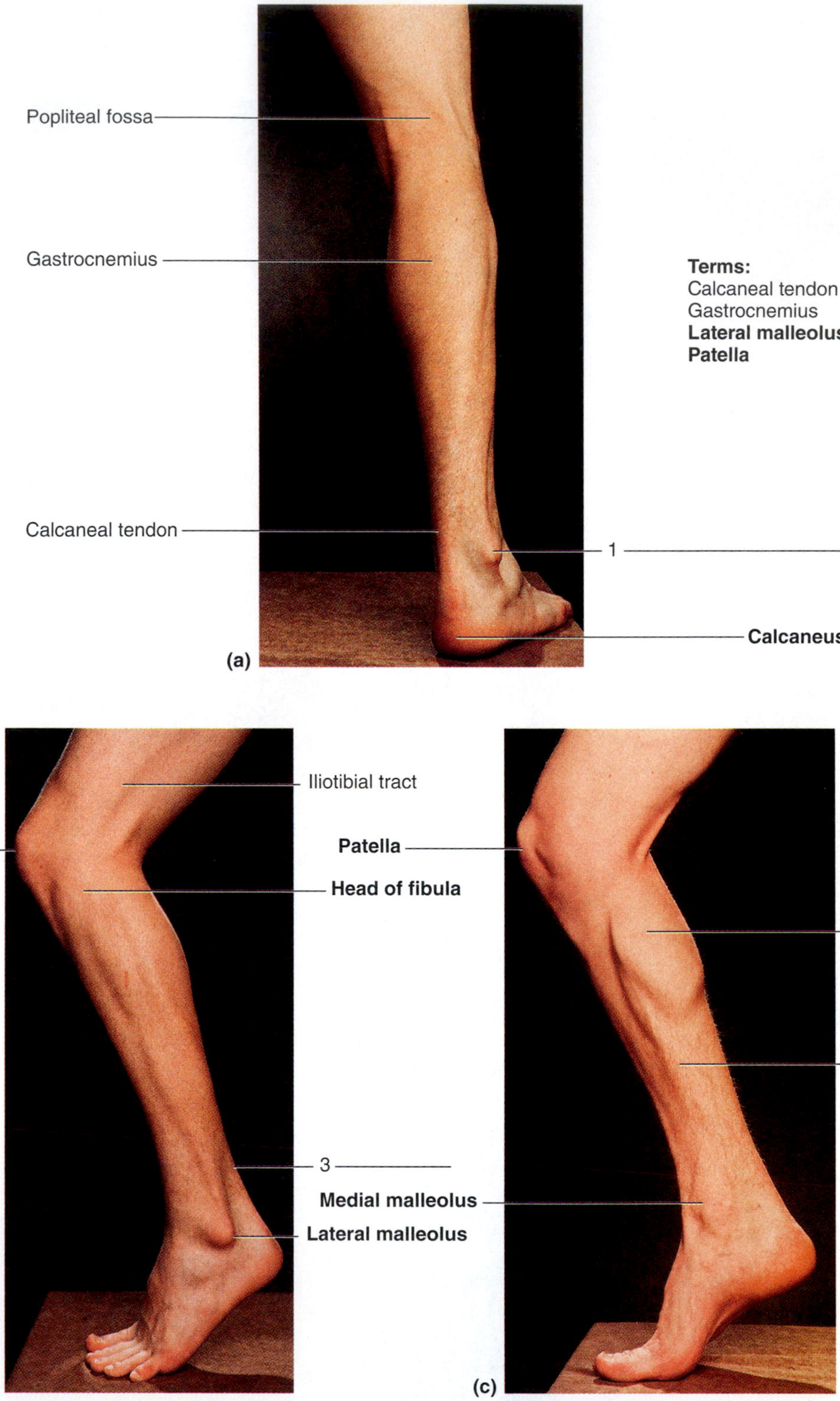

Figure 24.8 Label the surface features of the (*a*) posterior lower limb, (*b*) lateral lower limb, and (*c*) medial lower limb, using the terms provided. (**Boldface** indicates bony features; not boldface indicates soft tissue features.) 3

Name ______________________________

Date ______________________________

Section ______________________________

The [A] corresponds to the indicated Learning Outcome(s) listed at the beginning of the laboratory exercise.

Laboratory Report 24

Surface Anatomy

Part A Assessments

Match the body region in column A with the surface feature in column B. Place the letter of your choice in the space provided. [2]

Column A	Column B
a. Head	_______ **1.** Umbilicus
b. Upper limb	_______ **2.** Medial malleolus
c. Lower limb	_______ **3.** Iliac crest
d. Trunk	_______ **4.** Transverse arch
	_______ **5.** Spine of scapula
	_______ **6.** External occipital protuberance
	_______ **7.** Metacarpophalangeal joints
	_______ **8.** Sternum
	_______ **9.** Olecranon process
	_______ **10.** Zygomatic arch
	_______ **11.** Mastoid process
	_______ **12.** Thenar eminence

Label figure 24.9 with the surface features provided.

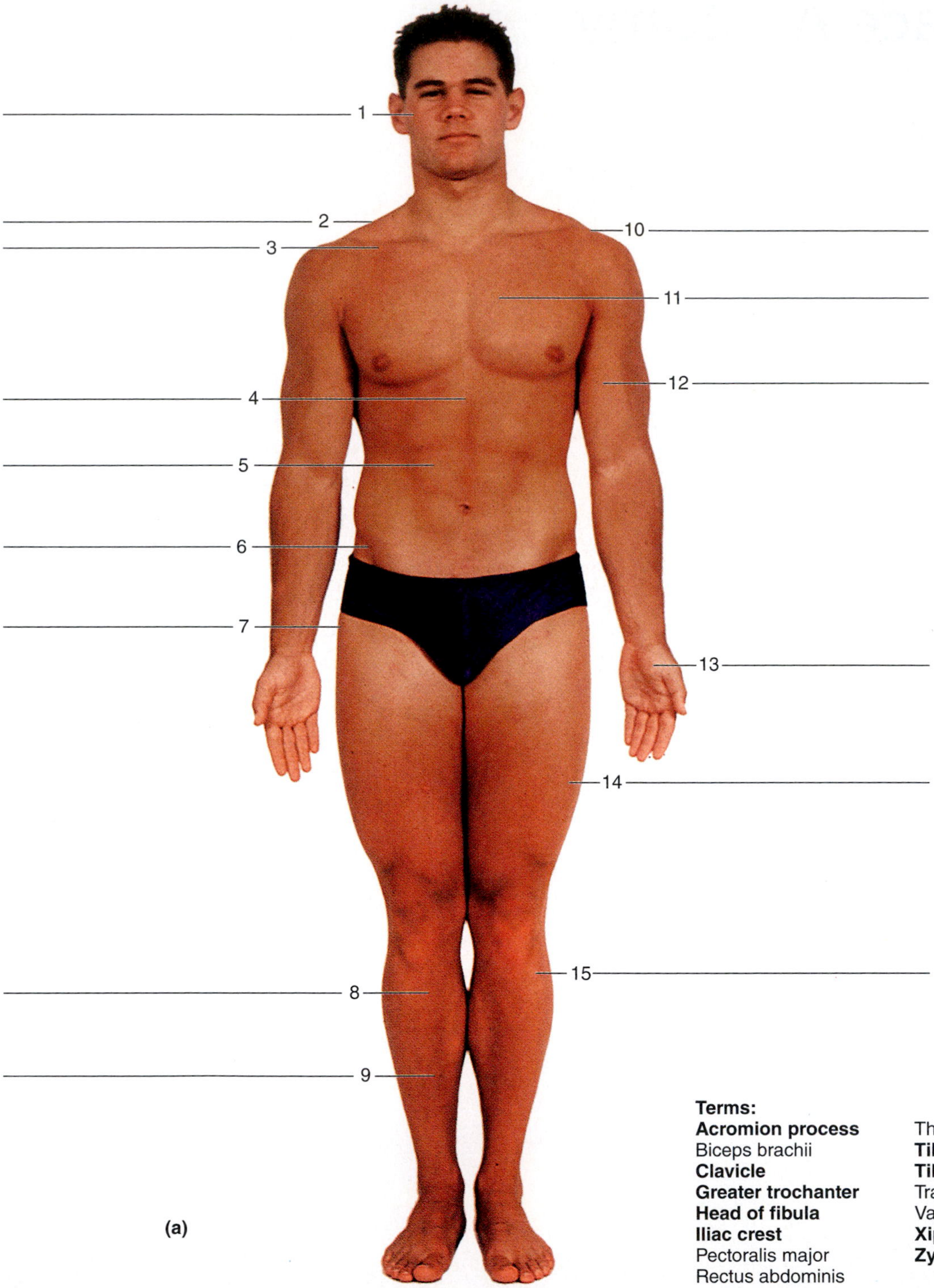

Terms:

Acromion process	Thenar eminence
Biceps brachii	**Tibia**
Clavicle	**Tibial tuberosity**
Greater trochanter	Trapezius (upper)
Head of fibula	Vastus lateralis
Iliac crest	**Xiphoid process**
Pectoralis major	**Zygomatic arch**
Rectus abdominis	

Figure 24.9 Label the surface features of the (*a*) anterior view of the body and (*b*) posterior view of the body, using the terms provided. (**Boldface** indicates bony features; not boldface indicates soft tissue features.) 1

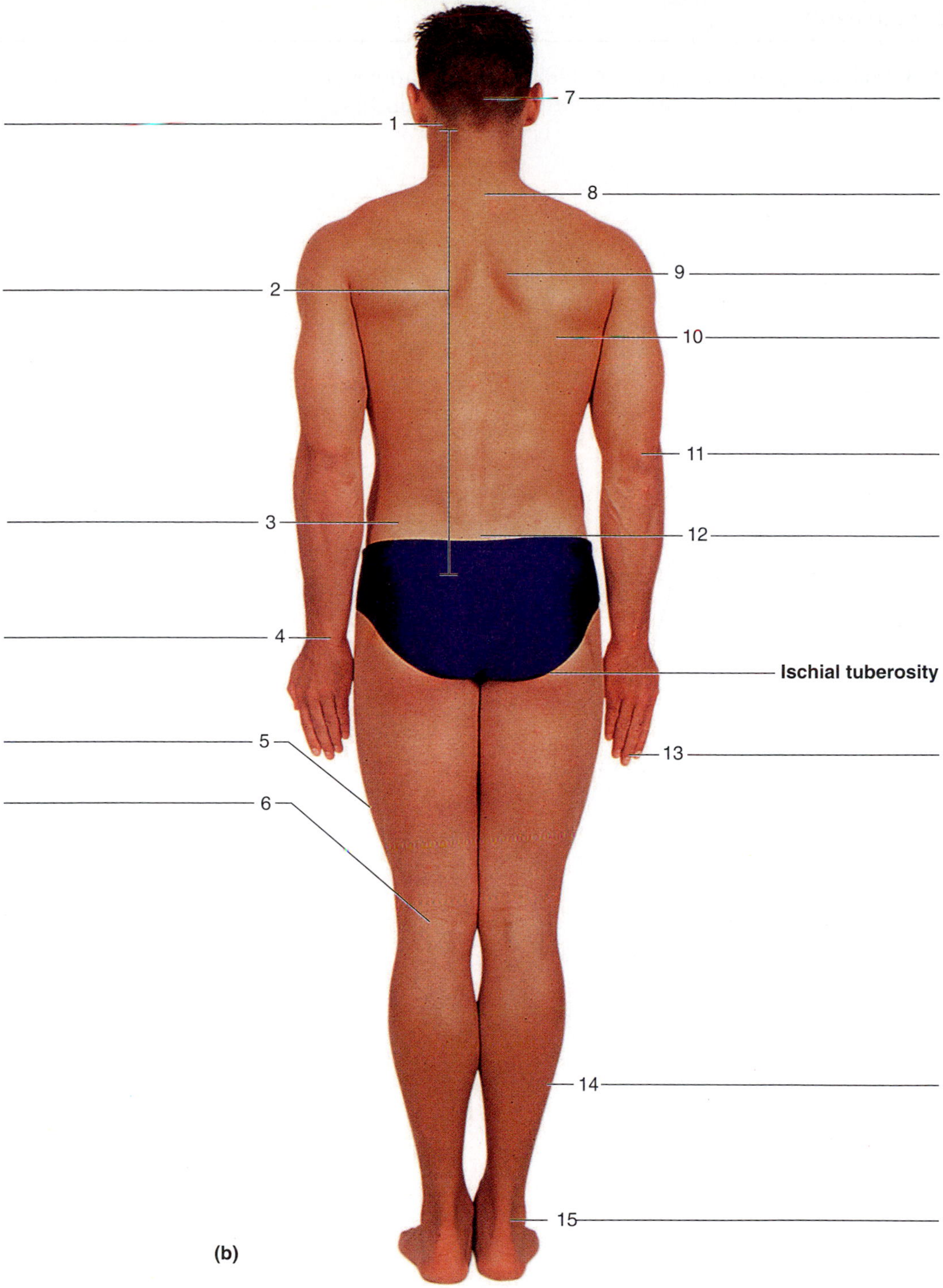

Figure 24.9 *Continued.*

Terms:
C7
Calcaneal tendon
Distal interphalageal joint (DIP)
Erector spinae
External occipital protuberance
Iliac crest
Iliotibial tract
Inferior angle of scapula
Mastoid process
Medial border of scapula
Olecranon process
Popliteal fossa
Sacrum
Soleus
Styloid process of ulna

Part C Assessments

Using table 24.1 from Laboratory Exercise 24, indicate the locations of the surface features with an "X" on figure 24.10. Use a black X for the bony features and a red X for the soft tissue features.

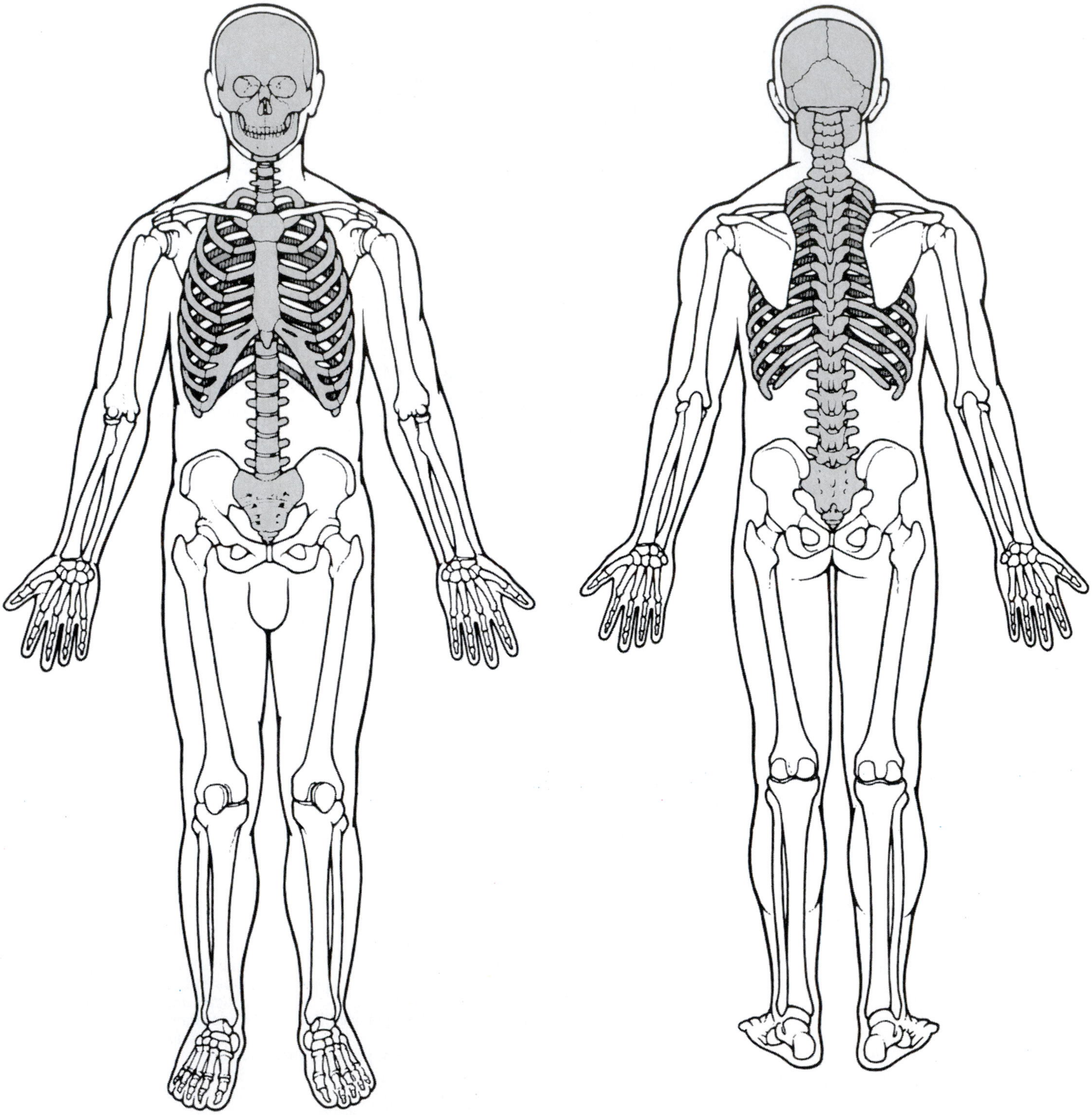

Figure 24.10 Indicate bony surface features with a black "X" and soft tissue surface features with a red "X" using the results from table 24.1, on the (*a*) anterior and (*b*) posterior diagrams of the body. 3

Laboratory Exercise 25

Nervous Tissue and Nerves

Materials Needed

Textbook
Compound light microscope
Prepared microscope slides of the following:
- Spinal cord (smear)
- Dorsal root ganglion (section)
- Neuroglia (astrocytes)
- Peripheral nerve (cross section and longitudinal section)

Neuron model

For Learning Extension:
Prepared microscope slide of Purkinje cells from cerebellum

Nervous tissue, which occurs in the brain, spinal cord, and nerves, contains neurons and neuroglia. The neurons are the basic structural and functional units of the nervous system involved in decision-making processes, detecting stimuli, and conducting messages. The neuroglia perform various supportive and protective functions for neurons.

Purpose of the Exercise

To review the characteristics of nervous tissue and to observe neurons, neuroglia, and various features of the nerves.

LEARNING OUTCOMES

After completing this exercise, you should be able to

1. Describe and locate the general characteristics of nervous tissue.
2. Distinguish structural and functional characteristics of neurons and neuroglia.
3. Identify and sketch the major structures of a neuron and a nerve.

EXPLORE

Procedure—Nervous Tissue and Nerves

1. Review the sections entitled "General Functions of the Nervous System" and "Classification of Cells of the Nervous System" in chapter 10 of the textbook.
2. As a review activity, label figures 25.1 and 25.2.
3. Complete Parts A and B of Laboratory Report 25.
4. Obtain a prepared microscope slide of a spinal cord smear. Using low-power magnification, search the slide and locate the relatively large, deeply stained cell bodies of motor neurons (multipolar neurons).
5. Observe a single motor neuron, using high-power magnification, and note the following features:

   cell body
   - nucleus
   - nucleolus
   - chromatophilic substance (Nissl bodies)
   - neurofibrils (threadlike structures extending into the nerve fibers)

   dendrites

   axon (nerve fiber)

   Compare the slide to the neuron model and to figure 25.3. You also may note small, darkly stained nuclei of neuroglia around the motor neuron.
6. Sketch and label a motor (efferent) neuron in the space provided in Part C of the laboratory report.
7. Obtain a prepared microscope slide of a dorsal root ganglion. Search the slide and locate a cluster of sensory neuron cell bodies. You also may note bundles of nerve fibers passing among groups of neuron cell bodies (fig. 25.4).
8. Sketch and label a sensory (afferent) neuron cell body in the space provided in Part C of the laboratory report.

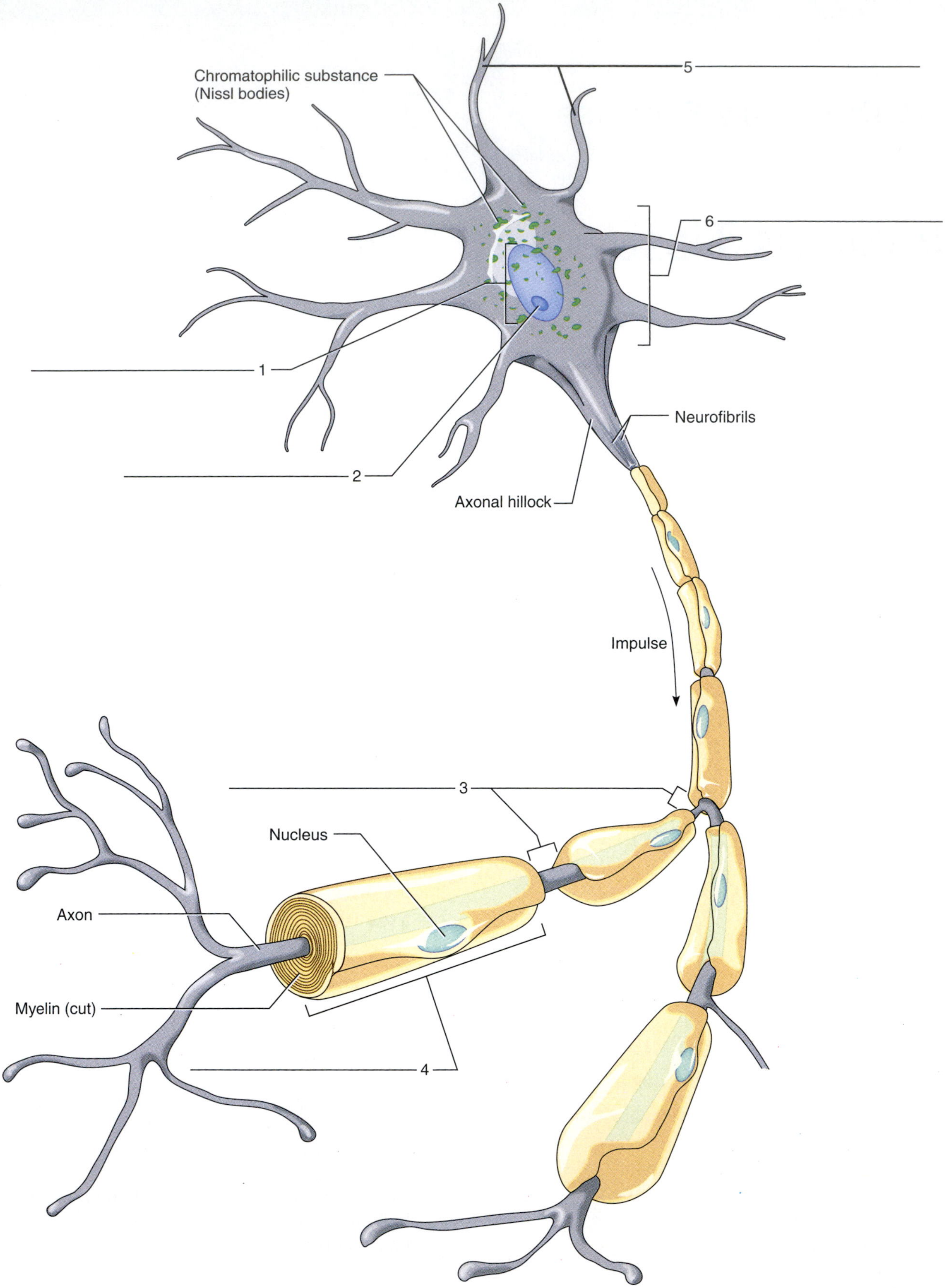

Figure 25.1 Label this diagram of a motor neuron. 3

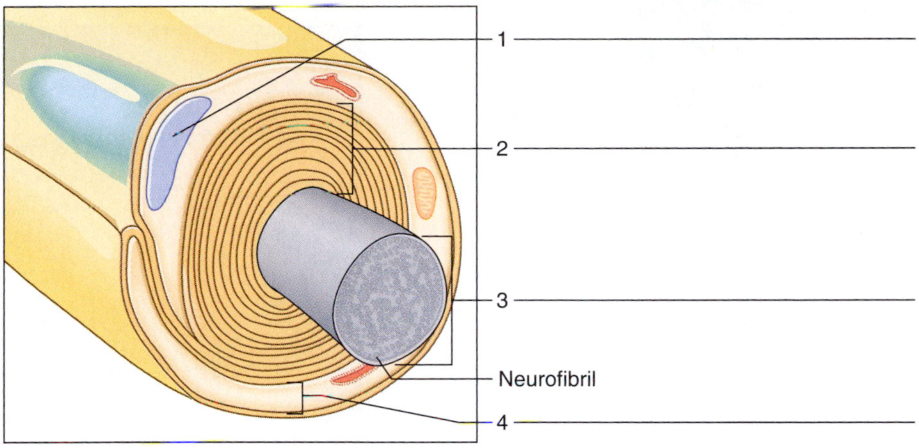

Figure 25.2 Label the features of the myelinated axon (nerve fiber). 3

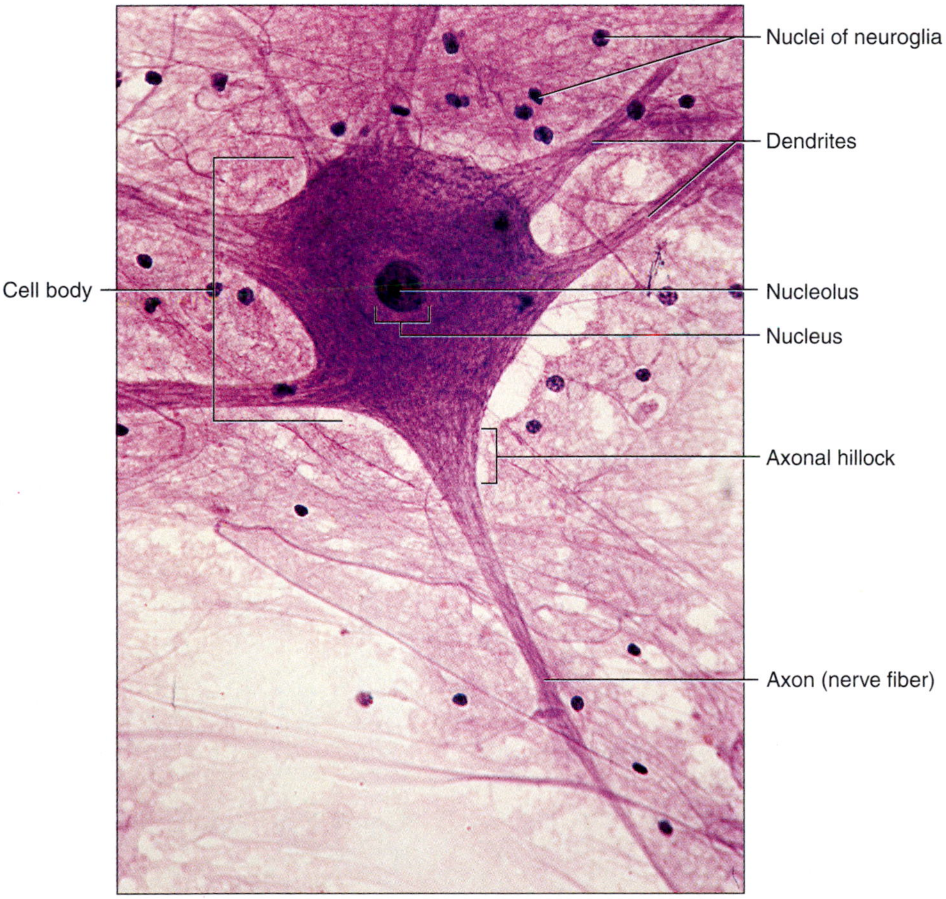

Figure 25.3 Micrograph of a multipolar neuron and neuroglia from a spinal cord smear (100× micrograph enlarged to 600×).

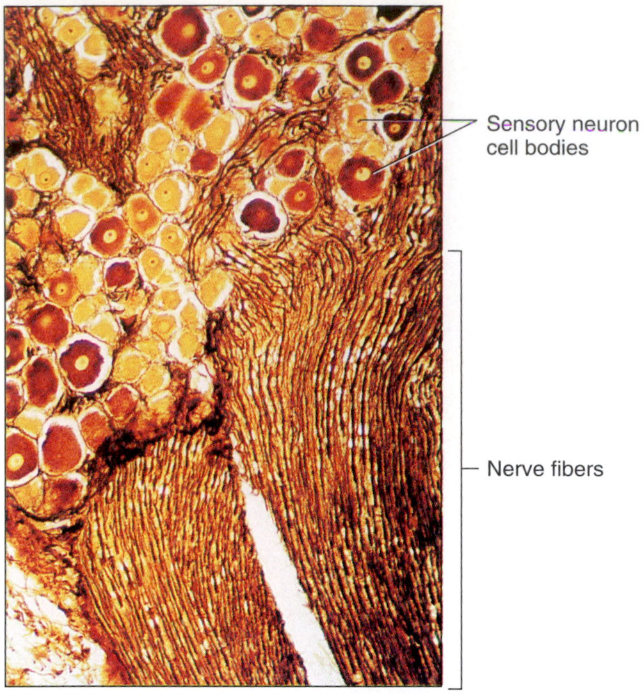

Figure 25.4 Micrograph of a dorsal root ganglion (50× micrograph enlarged to 100×).

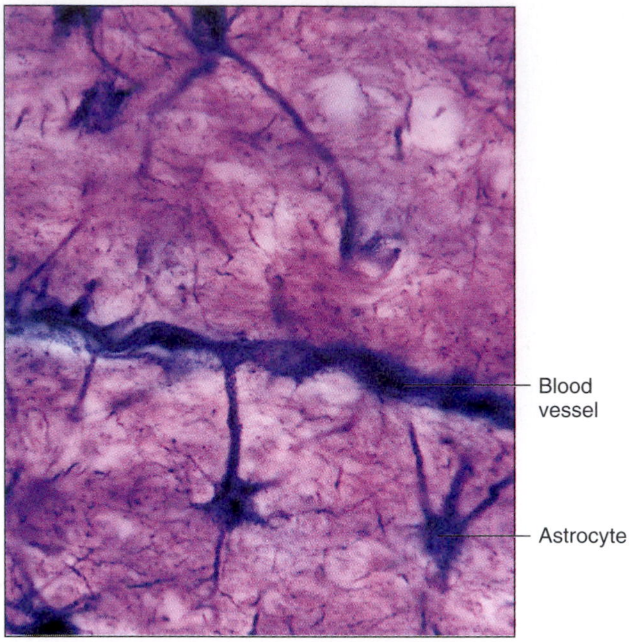

Figure 25.5 Micrograph of astrocytes (250× micrograph enlarged to 1,000×).

9. Obtain a prepared microscope slide of neuroglia. Search the slide and locate some darkly stained astrocytes with numerous long, slender processes (fig. 25.5).
10. Sketch several neuroglia in the space provided in Part C of the laboratory report.
11. Obtain a prepared microscope slide of a nerve. Locate the cross section of the nerve and note the many round nerve fibers inside. Also note the dense layer of connective tissue (perineurium) that encircles a fascicle of nerve fibers and holds them together in a bundle. The individual nerve fibers are surrounded by a layer of more delicate connective tissue (endoneurium) (fig. 25.6).
12. Using high-power magnification, observe a single nerve fiber and note the following features:

    central axon

    myelin sheath around the axon of Schwann cell (most of the myelin may have been dissolved and lost during the slide preparation)

    neurilemma of Schwann cell

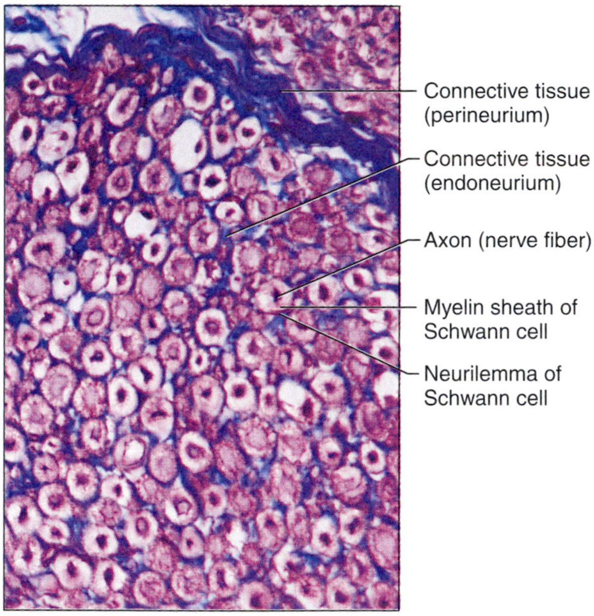

Figure 25.6 Cross section of a bundle of neurons within a nerve (400×).

13. Sketch and label a nerve fiber with Schwann cell (cross section) in the space provided in Part D of the laboratory report.
14. Locate the longitudinal section of the nerve on the slide (fig. 25.7). Note the following:

    central axons
    myelin sheath of Schwann cells
    neurilemma of Schwann cells
    nodes of Ranvier

15. Sketch and label a nerve fiber with Schwann cell (longitudinal section) in the space provided in Part D of the laboratory report.

Learning Extension

Obtain a prepared microscope slide of Purkinje cells. To locate these neurons, search the slide for large, flask-shaped cell bodies. Each cell body has one or two large, thick dendrites that give rise to branching networks of fibers. These cells are found in a particular region of the brain (cerebellar cortex).

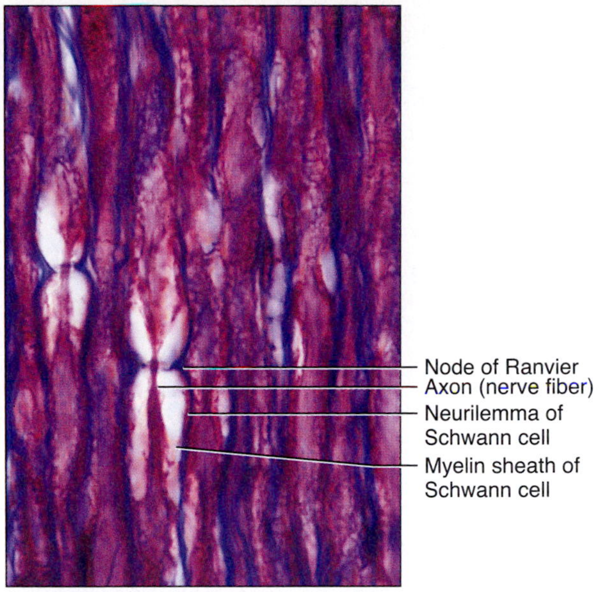

Figure 25.7 Longitudinal section of a nerve (250× micrograph enlarged to 2,000×).

Notes

Name ______________________________

Date ______________________________

Section ______________________________

The [A] corresponds to the indicated Learning Outcome(s) listed at the beginning of the laboratory exercise.

Laboratory Report 25

Nervous Tissue and Nerves

Part A Assessments

Match the terms in column A with the descriptions in column B. Place the letter of your choice in the space provided. [1] [2]

Column A

- **a.** Astrocyte
- **b.** Axon
- **c.** Chromatophilic substance (Nissl bodies)
- **d.** Collateral
- **e.** Dendrite
- **f.** Myelin sheath
- **g.** Neurilemma
- **h.** Neurofibrils

Column B

_______ **1.** Sheath of Schwann cell containing cytoplasm and nucleus that encloses myelin

_______ **2.** Corresponds to rough endoplasmic reticulum in other cells

_______ **3.** Network of fine threads extending into nerve fiber

_______ **4.** Substance of Schwann cell composed of lipoprotein that insulates axons and increases impulse speed

_______ **5.** Neuron process with many branches that conducts an impulse toward the cell body

_______ **6.** Branch of an axon

_______ **7.** Star-shaped neuroglia between neurons and blood vessels

_______ **8.** Nerve fiber arising from a slight elevation of the cell body that conducts an impulse away from the cell body

Part B Assessments

Match the terms in column A with the descriptions in column B. Place the letter of your choice in the space provided. [1] [2]

Column A

- **a.** Effector
- **b.** Ependymal cell
- **c.** Ganglion
- **d.** Interneuron
- **e.** Microglia
- **f.** Motor (efferent) neuron
- **g.** Oligodendrocyte
- **h.** Sensory (afferent) neuron

Column B

_______ **1.** Transmits impulse from sensory to motor neuron within central nervous system

_______ **2.** Transmits impulse out of the brain or spinal cord to effectors (muscles and glands)

_______ **3.** Transmits impulse into brain or spinal cord from receptors

_______ **4.** Myelin-forming neuroglia in brain and spinal cord

_______ **5.** Phagocytic neuroglia

_______ **6.** Structure capable of responding to motor impulse

_______ **7.** Specialized mass of neuron cell bodies outside the brain or spinal cord

_______ **8.** Cells that cover the inside spaces of the brain ventricles and help regulate cerebrospinal fluid

Part C Assessments

1. Sketch a motor neuron. Label the cell body, nucleus, nucleolus, and cell processes (dendrite and axon). 3

2. Sketch and label a sensory (afferent) neuron cell body. 3

3. Sketch several neuroglia. 2

Part D Assessments

1. Sketch and label a nerve fiber with Schwann cell (cross section). 3

2. Sketch and label a nerve fiber with Schwann cell (longitudinal section). 3

Laboratory Exercise 26

Spinal Cord and Meninges

Materials Needed

Textbook
Compound light microscope
Prepared microscope slide of a spinal cord cross section with spinal nerve roots
Spinal cord model with meninges

For Demonstration:

Preserved spinal cord with meninges intact

The spinal cord is a column of nerve fibers that extends down through the vertebral canal. Together with the brain, it makes up the central nervous system.

Neurons within the spinal cord provide a two-way communication system between the brain and body parts outside the central nervous system. The cord also contains the processing centers for spinal reflexes.

The meninges consist of layers of membranes located between the bones of the skull and vertebral column and the soft tissues of the central nervous system. They include the dura mater, the arachnoid mater, and the pia mater.

Purpose of the Exercise

To review the characteristics of the spinal cord and the meninges and to observe the major features of these structures.

LEARNING OUTCOMES

After completing this exercise, you should be able to

1. Identify the major features and functions of the spinal cord.
2. Locate the ascending and descending tracts of the spinal cord.
3. Arrange the layers of the meninges and describe the structure of each.

EXPLORE

Procedure A—Structure of the Spinal Cord

1. Review the section entitled "Spinal Cord" in chapter 11 of the textbook.
2. As a review activity, label figures 26.1 and 26.2.
3. Complete Part A of Laboratory Report 26.
4. Study figure 26.3 and complete Part B of the laboratory report.
5. Obtain a prepared microscope slide of a spinal cord cross section. Use the low power of the microscope to locate the following features:

   posterior median sulcus
   anterior median fissure
   central canal
   gray matter
   - gray commissure
   - posterior (dorsal) horn
   - lateral horn
   - anterior (ventral) horn

   white matter
   - posterior (dorsal) funiculus (column)
   - lateral funiculus (column)
   - anterior (ventral) funiculus (column)

   roots of spinal nerve
   - dorsal roots
   - dorsal root ganglia
   - ventral roots

6. Observe the model of the spinal cord, and locate the features listed in step 5.
7. Complete Part C of the laboratory report.

EXPLORE

Procedure B—Meninges

1. Review the section entitled "Meninges" in chapter 11 of the textbook.
2. Complete Part D of the laboratory report.

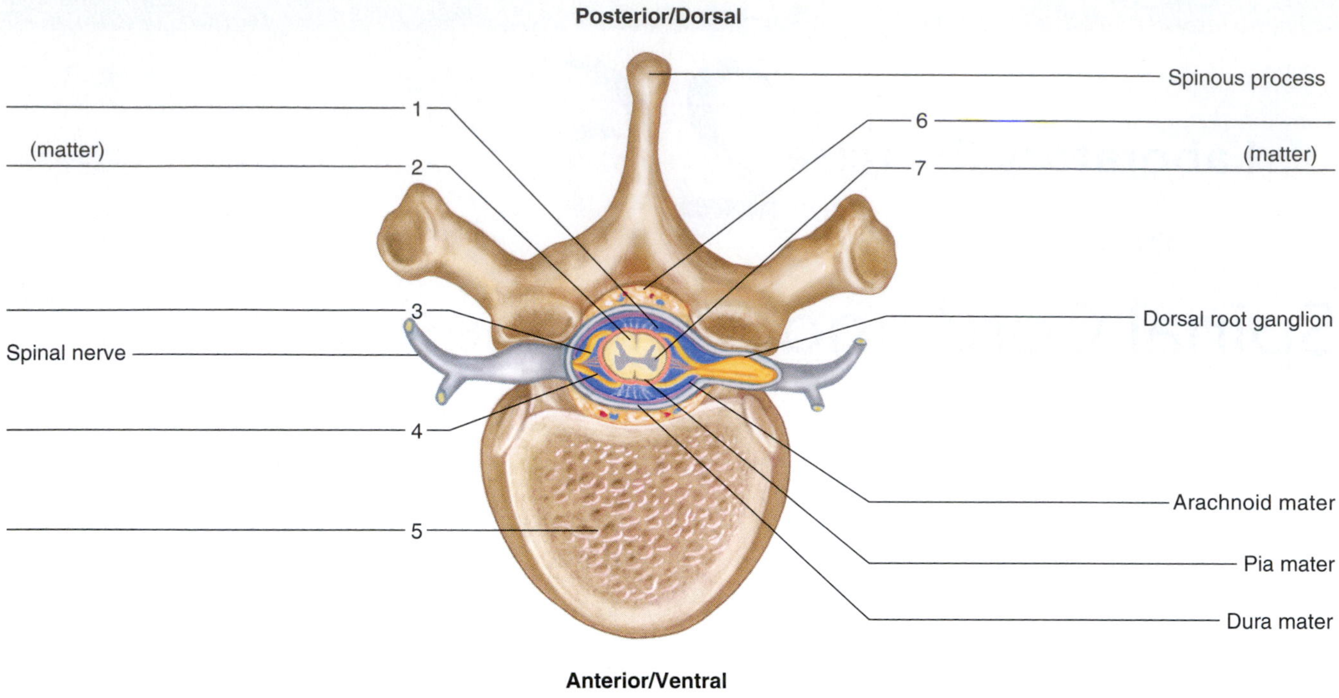

Figure 26.1 Label the features of the spinal cord and the surrounding structures. 1

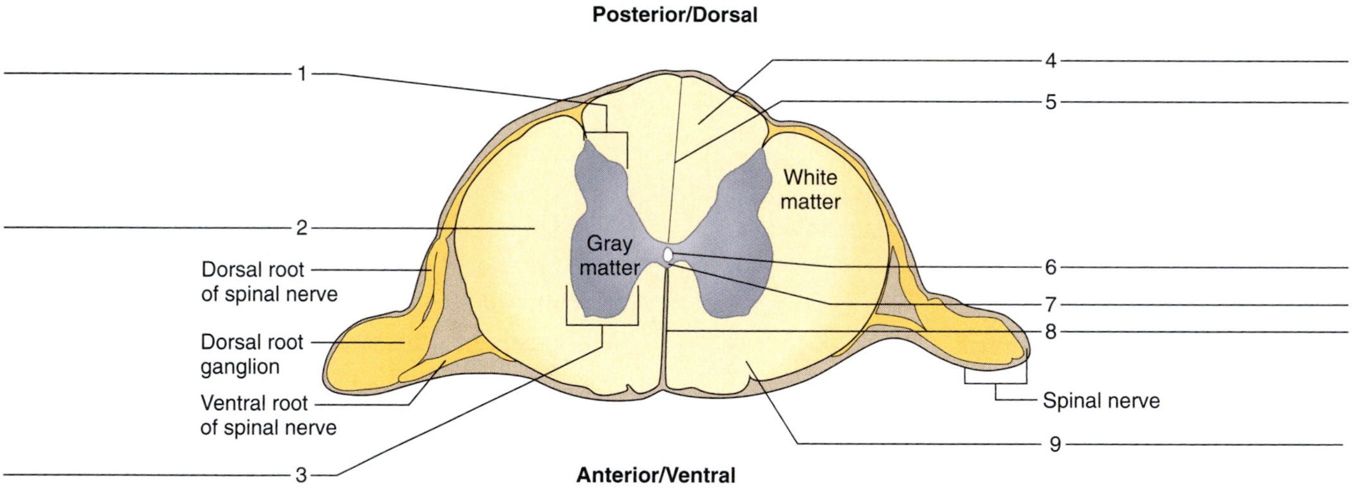

Figure 26.2 Label this cross section of the spinal cord, including the features of the white and gray matter. (*Note:* A lateral horn of the gray matter is not present at this level of the spinal cord illustrated.) 1

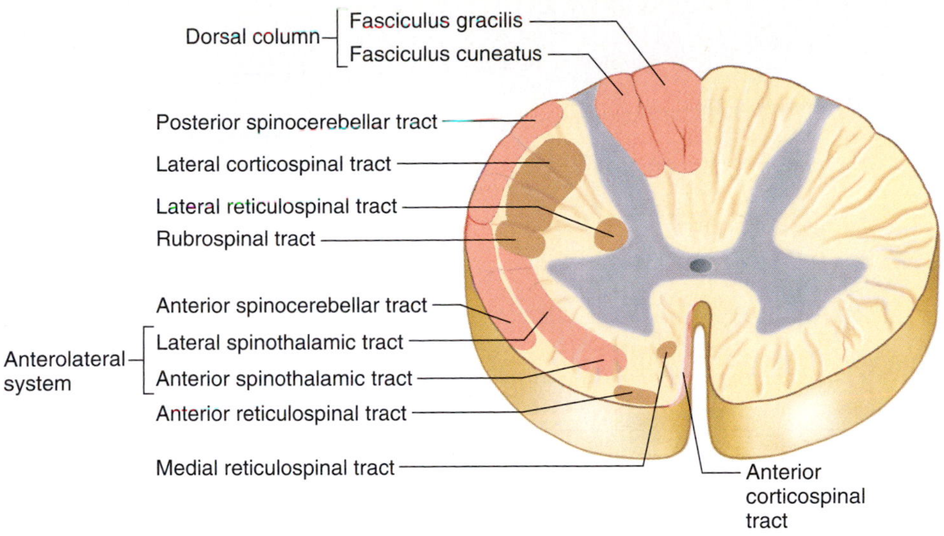

Figure 26.3 Major ascending (sensory) and descending (motor) tracts within a cross section of the spinal cord. Ascending tracts are in orange, descending tracts in tan, and are shown only on one side. This pattern varies with the level of the spinal cord. This pattern is representative of the midcervical region. (*Note:* These tracts are not visible as individually stained structures on microscope slides.)

Demonstration

Observe the preserved section of spinal cord. Note the heavy covering of dura mater, firmly attached to the cord on each side by a set of ligaments (denticulate ligaments) originating in the pia mater. The intermediate layer of meninges, the arachnoid mater, is devoid of blood vessels, but in a live human being, the space beneath this layer contains cerebrospinal fluid. The pia mater, closely attached to the surface of the spinal cord, contains many blood vessels. What are the functions of these layers?

____

____

____

____

Notes

Name ________________________________

Date ________________________________

Section ________________________________

The [A] corresponds to the indicated Learning Outcome(s) listed at the beginning of the laboratory exercise.

Spinal Cord and Meninges

Laboratory Report 26

Part A Assessments

Complete the following statements:

1. The spinal cord gives rise to 31 pairs of ________________. [1]
2. The bulge in the spinal cord that gives off nerves to the upper limbs is called the ________________. [1]
3. The bulge in the spinal cord that gives off nerves to the lower limbs is called the ________________. [1]
4. The ________________ is a groove that extends the length of the spinal cord posteriorly. [1]
5. In a spinal cord cross section, the posterior ________________ of the gray matter resemble the upper wings of a butterfly. [1]
6. The cell bodies of motor neurons are found in the ________________ horns of the spinal cord. [1]
7. The ________________ connects the gray matter on the left and right sides of the spinal cord. [1]
8. The ________________ in the gray commissure of the spinal cord contains cerebrospinal fluid and is continuous with the ventricles of the brain. [1]
9. The white matter of the spinal cord is divided into anterior, lateral, and posterior ________________. [1]
10. The longitudinal bundles of nerve fibers within the spinal cord comprise major nerve pathways called ________________. [1]

Part B Assessments

Match the nerve tracts in column A with the descriptions in column B. Place the letter of your choice in the space provided. [2]

Column A

a. Corticospinal
b. Fasciculus gracilis
c. Lateral spinothalamic
d. Posterior spinocerebellar
e. Reticulospinal

Column B

_______ 1. Ascending tract to the brain to interpret touch, pressure, and body movements

_______ 2. Descending tract whose fibers conduct motor impulses to sweat glands and muscles to control tone

_______ 3. Descending tract whose fibers conduct motor impulses to skeletal muscles

_______ 4. Ascending tract to the cerebellum necessary for coordination of skeletal muscles

_______ 5. Ascending tract to the brain to give rise to sensations of temperature and pain

Part C Assessments

Identify the features indicated in the spinal cord cross section of figure 26.4. 1

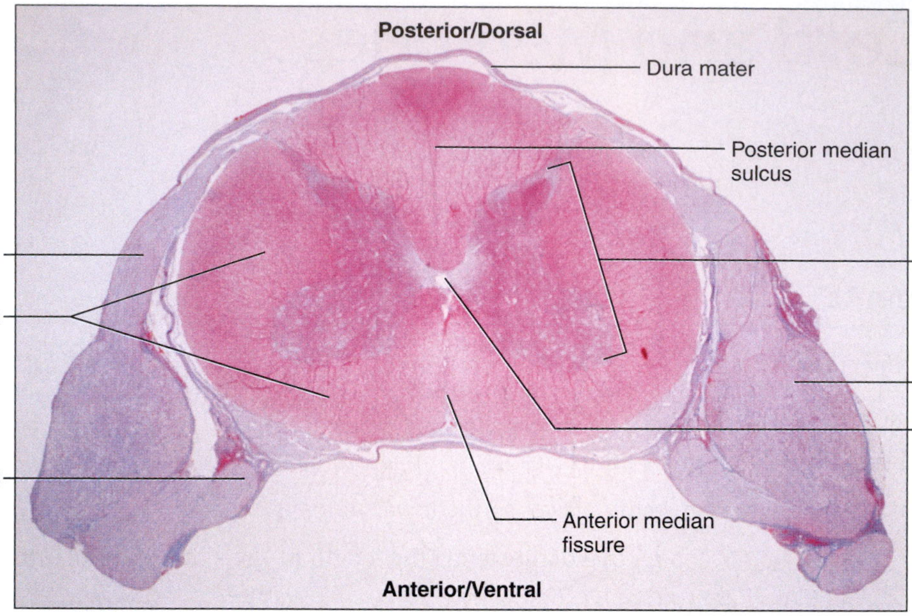

_____ Central canal
_____ Dorsal root ganglion
_____ Dorsal root of spinal nerve
_____ Gray matter
_____ Ventral root of spinal nerve
_____ White matter

Figure 26.4 Micrograph of a spinal cord cross section with spinal nerve roots (35×). Label the features by placing the correct numbers in the spaces provided.

Part D Assessments

Match the terms in column A with the descriptions in column B. Place the letter of your choice in the space provided. 3

Column A	Column B
a. Arachnoid mater	_______ **1.** Band of pia mater that anchors dura mater to cord
b. Denticulate ligament	_______ **2.** Channel through which venous blood flows
c. Dural sinus	_______ **3.** Outermost layer of meninges
d. Dura mater	_______ **4.** Follows irregular contours of spinal cord surface
e. Epidural space	_______ **5.** Contains cerebrospinal fluid
f. Pia mater	_______ **6.** Thin, weblike middle membrane
g. Subarachnoid space	_______ **7.** Separates dura mater from bone of vertebra

Laboratory Exercise 27

Reflex Arc and Reflexes

Materials Needed

Textbook
Rubber percussion hammer

A reflex arc represents the simplest type of nerve pathway found in the nervous system. This pathway begins with a receptor at the dendrite end of a sensory (afferent) neuron. The sensory neuron leads into the central nervous system and may communicate with one or more interneurons. Some of these interneurons, in turn, communicate with motor (efferent) neurons, whose axons (nerve fibers) lead outward to effectors.

Thus, when a sensory receptor is stimulated by a change occurring inside or outside the body, nerve impulses may pass through a reflex arc, and, as a result, effectors may respond. Such an automatic, subconscious response is called a *reflex*.

A *stretch reflex* involves a single synapse (monosynaptic) between a sensory and a motor neuron within the gray matter of the spinal cord. Examples of stretch reflexes include the patellar, calcaneal, biceps, triceps, and plantar reflexes. Other more complex *withdrawal reflexes* involve interneurons in combination with sensory and motor neurons; thus they are polysynaptic. Examples of withdrawal reflexes include responses to touching hot objects or stepping on sharp objects.

Reflexes demonstrated in this lab are stretch reflexes. When a muscle is stretched by a tap over its tendon, stretch receptors (proprioceptors) called *muscle spindles* are stretched within the muscle, which initiates an impulse over a reflex arc. A sensory (afferent) neuron conducts an impulse from the muscle spindle into the gray matter of the spinal cord, where it synapses with a motor (efferent) neuron, which conducts the impulse to the effector muscle. The stretched muscle responds by contracting to resist or reverse further stretching. These stretch reflexes are important to maintaining proper posture, balance, and movements. Observations of many of these reflexes in clinical tests on patients may indicate damage to a level of the spinal cord or peripheral nerves of the particular reflex arc.

Purpose of the Exercise

To review the characteristics of reflex arcs and reflex behavior and to demonstrate some of the reflexes that occur in the human body.

LEARNING OUTCOMES

After completing this exercise, you should be able to

1. Describe the components of a reflex arc.
2. Demonstrate and record stretch reflexes that occur in humans.
3. Analyze the components and patterns of stretch reflexes.

EXPLORE

Procedure—Reflex Arc and Reflexes

1. Review the section entitled "Reflex Arcs" in chapter 11 of the textbook and "Proprioception" in chapter 12 of the textbook.
2. As a review activity, label figure 27.1.
3. Complete Part A of Laboratory Report 27.
4. Work with a laboratory partner to demonstrate each of the reflexes listed. (See fig. 27.2 also.) *It is important that muscles involved in the reflexes be totally relaxed to observe proper responses.* If a person is trying too hard to experience the reflex or is trying to suppress the reflex, assign a multitasking activity while the stimulus with the rubber hammer occurs. For example, assign a physical task with the upper limbs along with a complex mental activity during the patellar (knee-jerk) reflex. After each demonstration, record your observations in the table provided in Part B of the laboratory report.
   a. *Patellar (knee-jerk) reflex.* Have your laboratory partner sit on a table (or sturdy chair) with legs relaxed and hanging freely over the edge without touching the floor. Gently strike your partner's patellar ligament (just below the patella) with the blunt side of a rubber percussion hammer

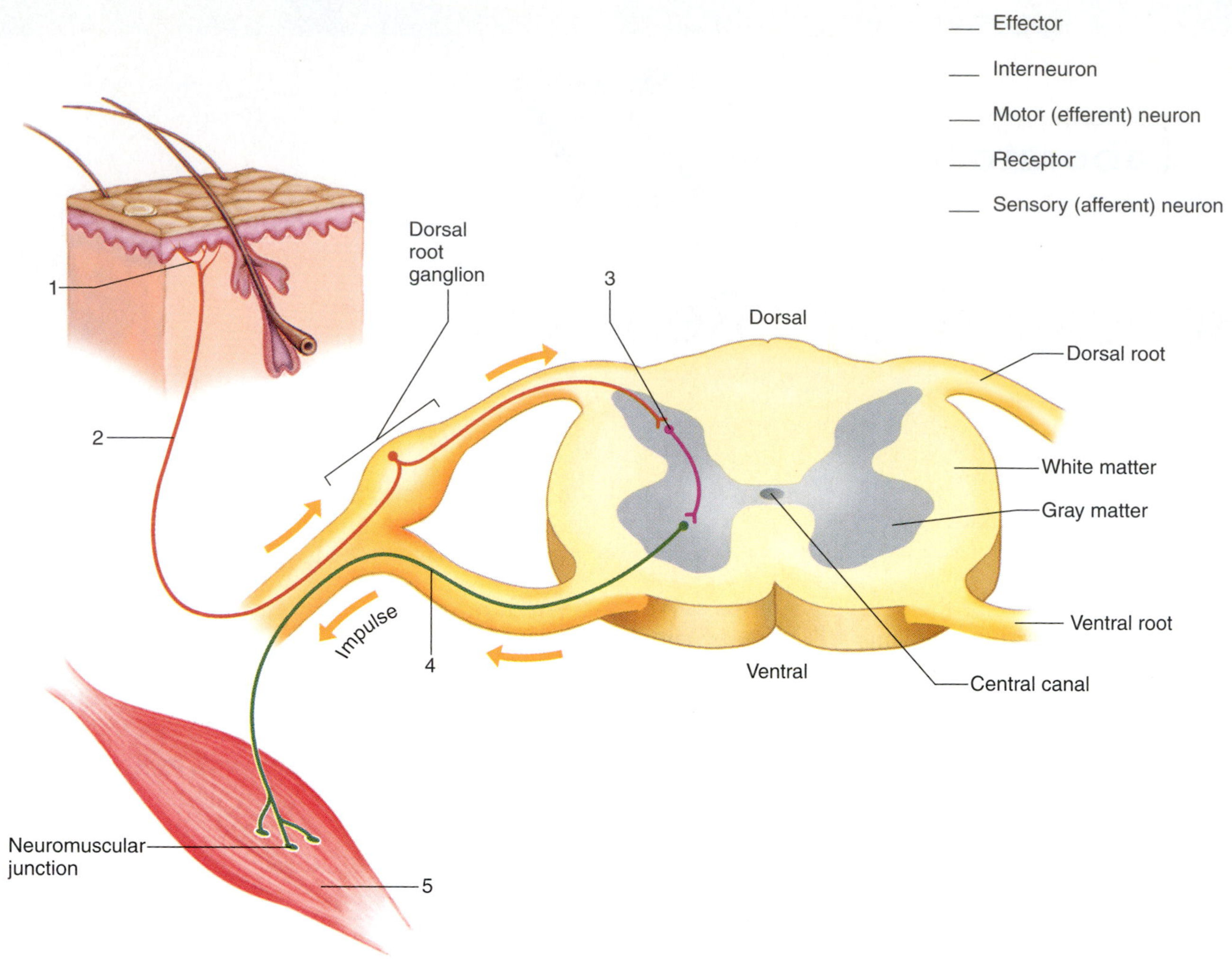

Figure 27.1 Label this diagram of a withdrawal (polysynaptic) reflex arc by placing the correct numbers in the spaces provided. Reflexes demonstrated in this lab are stretch (monosynaptic) reflex arcs and lack the interneuron. 1

(fig. 27.2*a*). The normal response is a moderate extension of the leg at the knee joint.

b. *Calcaneal (ankle-jerk) reflex.* Have your partner kneel on a chair with back toward you and with feet slightly dorsiflexed over the edge and relaxed. Gently strike the calcaneal tendon (just above its insertion on the calcaneus) with the blunt side of the rubber hammer (fig. 27.2*b*). The normal response is plantar flexion of the foot.

c. *Biceps (biceps-jerk) reflex.* Have your partner place a bare arm bent about 90° at the elbow on the table. Press your thumb on the inside of the elbow over the tendon of the biceps brachii, and gently strike your thumb with the rubber hammer (fig. 27.2*c*). Watch the biceps brachii for a response. The response might be a slight twitch of the muscle or flexion of the forearm at the elbow joint.

d. *Triceps (triceps-jerk) reflex.* Have your partner lie supine with an upper limb bent about 90° across the abdomen. Gently strike the tendon of the triceps brachii near its insertion just proximal to the olecranon process at the tip of the elbow (fig. 27.2*d*). Watch the triceps brachii for a response. The response might be a slight twitch of the muscle or extension of the forearm at the elbow joint.

e. *Plantar reflex.* Have your partner remove a shoe and sock and lie supine with the lateral surface of the foot resting on the table. Draw the metal tip of the rubber hammer, applying firm pressure, over the sole from the heel to the base of the large toe (fig. 27.2*e*). The normal response is flexion (curling) of the toes and plantar flexion of the foot. If the toes spread apart and dorsiflexion of the great toe occurs, the reflex is the abnormal *Babinski reflex* response (normal in infants until the nerve fibers have complete myelinization). If the Babinski reflex occurs later in life, it may indicate damage to the corticospinal tract of the CNS.

5. Complete Part B of the laboratory report.

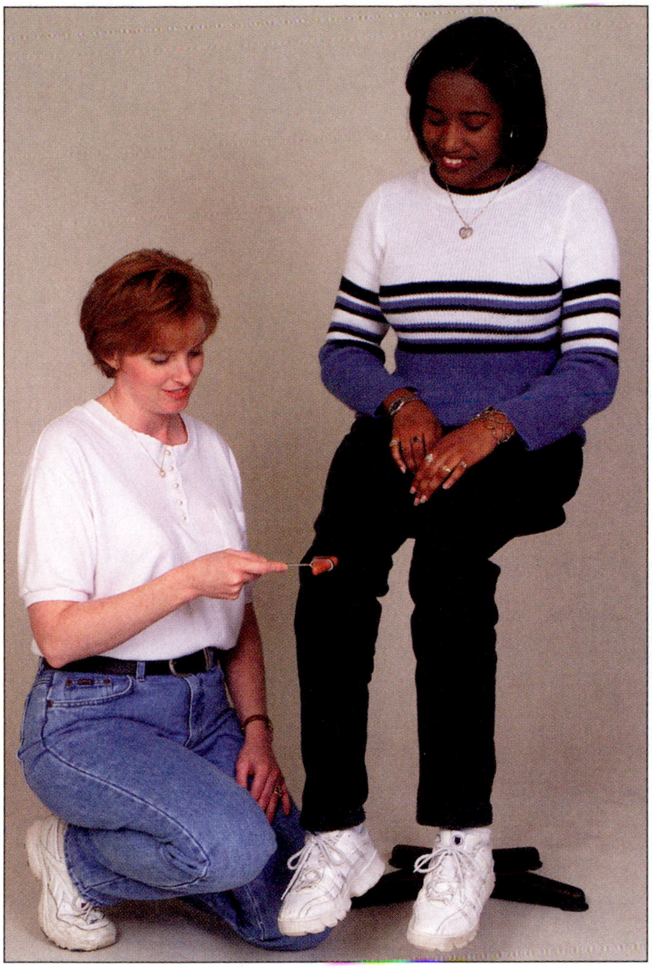

(a) Patellar (knee-jerk) reflex

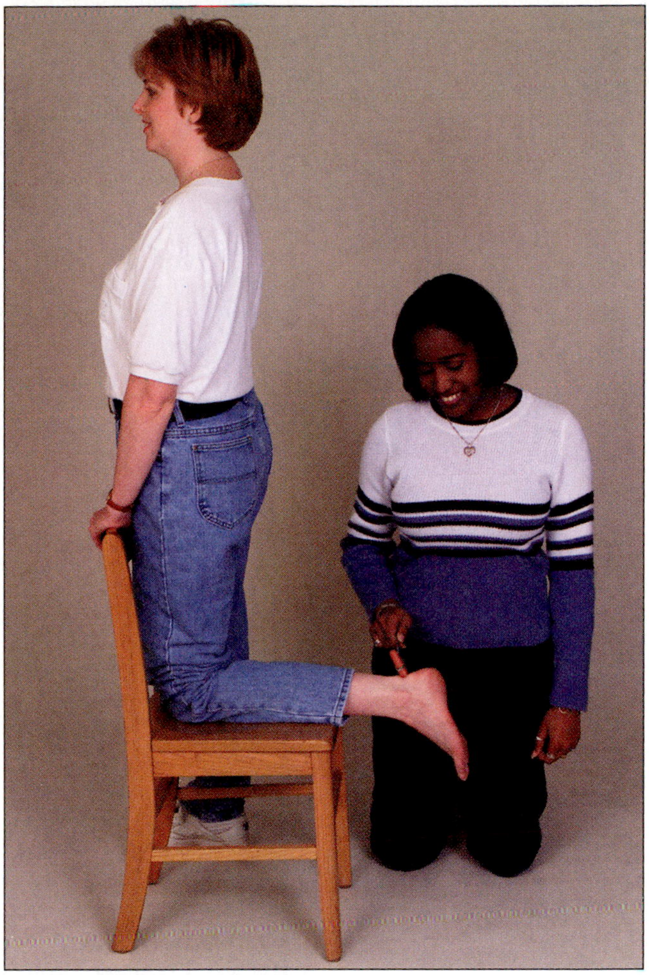

(b) Calcaneal (ankle-jerk) reflex

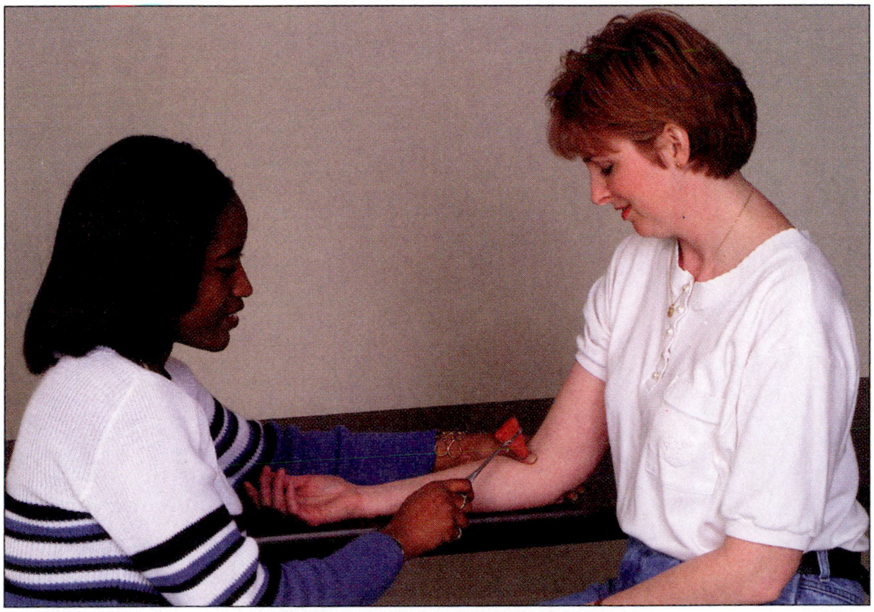

(c) Biceps reflex

Figure 27.2 Demonstrate each of the following reflexes: (*a*) patellar reflex; (*b*) calcaneal reflex; (*c*) biceps reflex; (*d*) triceps reflex; and (*e*) plantar reflex.

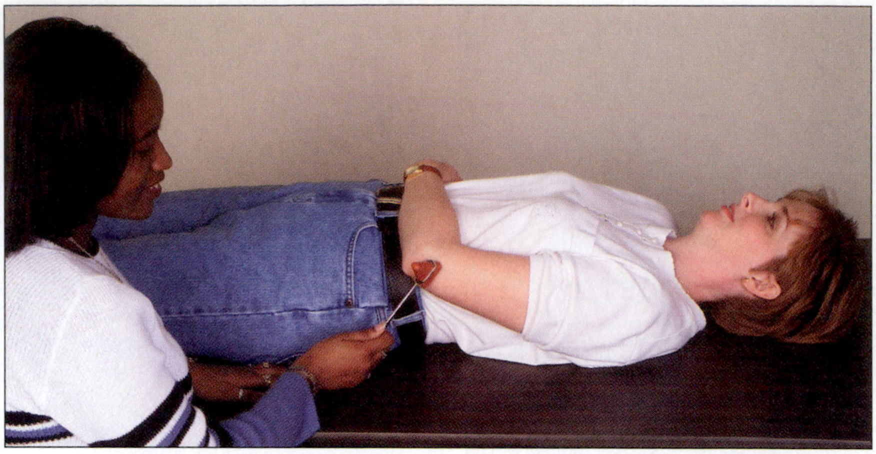

(d) Triceps reflex

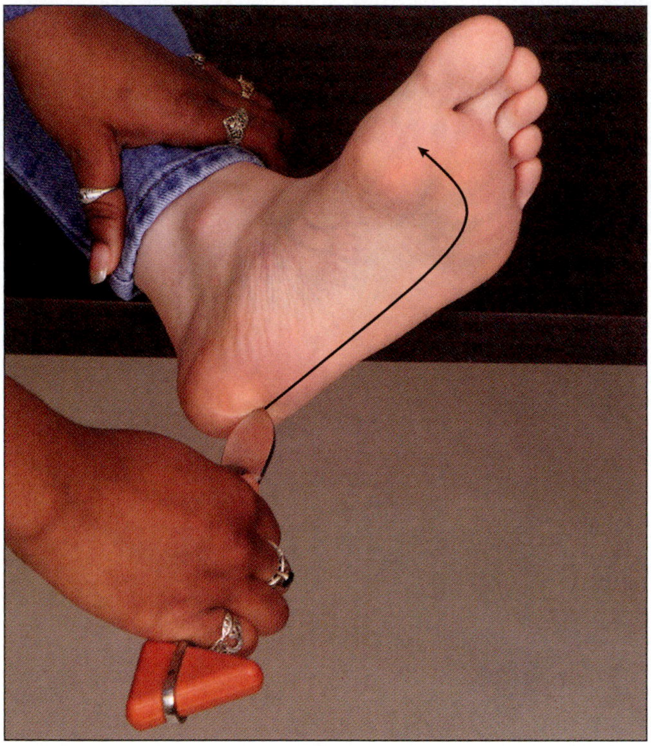

(e) Plantar reflex

Figure 27.2 *Continued.*

Name ______________________________

Date ______________________________

Section ______________________________

The [A] corresponds to the indicated Learning Outcome(s) listed at the beginning of the laboratory exercise.

Laboratory Report

27

Reflex Arc and Reflexes

Part A Assessments

Complete the following statements:

1. ______________________ are routes followed by nerve impulses as they pass through the nervous system. [1]
2. Interneurons in a withdrawal reflex are located in the ______________________ . [1]
3. ______________________ are automatic subconscious responses to external or internal stimuli. [1]
4. Effectors of a reflex arc are glands and ______________________ . [1]
5. A patellar reflex employs only ______________________ and motor neurons. [1]
6. The effector muscle of the patellar reflex is the ______________________ . [1]
7. The sensory stretch receptors (muscle spindles) of the patellar reflex are located in the ______________________ muscle. [1]
8. The patellar reflex helps the body to maintain ______________________ . [1]
9. The sensory receptors of a withdrawal reflex are located in the ______________________ . [1]
10. ______________________ muscles in the limbs are the effectors of a withdrawal reflex. [1]
11. The normal plantar reflex results in ______________________ of toes. [1]
12. Stroking the sole of the foot in infants results in dorsiflexion and toes that spread apart, called the ______________________ reflex. [1]

Part B Assessments

1. Complete the following table: 2

Reflex Tested	Response Observed	Effector Muscle Involved
Patellar (knee-jerk)		
Calcaneal (ankle-jerk)		
Biceps		
Triceps		
Plantar		

2. List the major events that occur in the patellar (knee-jerk) reflex, from the striking of the patellar ligament to the resulting response. 3

Critical Thinking Application

What characteristics do the reflexes you demonstrated have in common? 3

Laboratory Exercise 28

Brain and Cranial Nerves

Materials Needed

Textbook
Dissectible model of the human brain
Preserved human brain
Anatomical charts of the human brain

The brain, the largest and most complex part of the nervous system, contains nerve centers associated with sensory functions and is responsible for sensations and perceptions. It issues motor commands to skeletal muscles and carries on higher mental activities. It also functions to coordinate muscular movements, and it contains centers and nerve pathways necessary for the regulation of internal organs.

Twelve pairs of cranial nerves arise from the ventral surface of the brain and are designated by number and name. Although most of these nerves conduct both sensory and motor impulses, some contain only sensory fibers associated with special sense organs. Others are primarily composed of motor fibers and are involved with the activities of muscles and glands.

Purpose of the Exercise

To review the structural and functional characteristics of the human brain and cranial nerves.

LEARNING OUTCOMES

After completing this exercise, you should be able to

1. Identify the major external and internal structures in the human brain.
2. Locate the major functional regions of the brain.
3. Identify each of the cranial nerves.
4. Differentiate the functions of each cranial nerve.

EXPLORE

Procedure A—Human Brain

1. Review the section entitled "Brain" in chapter 11 of the textbook.
2. As a review activity, label figures 28.1 and 28.2.
3. Complete Part A of Laboratory Report 28.
4. Observe the anatomical charts, dissectible model, and preserved specimen of the human brain. Locate each of the following features:

cerebrum
cerebral hemispheres
corpus callosum
gyri (convolutions)
sulci
 central sulcus
 lateral sulcus
fissures
 longitudinal fissure
 transverse fissure
lobes
 frontal lobe
 parietal lobe
 temporal lobe
 occipital lobe
 insula (insular lobe)
cerebral cortex
basal nuclei (basal ganglia, a widely used clinical term)
 caudate nucleus
 putamen
 globus pallidus

ventricles
- lateral ventricles
- third ventricle
- fourth ventricle
- choroid plexuses
- cerebral aqueduct

diencephalon
- thalamus
- hypothalamus
- optic chiasma
- mammillary bodies
- pineal gland

cerebellum
- lateral (right and left) hemispheres
- falx cerebelli
- vermis
- cerebellar cortex
- arbor vitae
- cerebellar peduncles

brainstem
- midbrain (mesencephalon)
  - cerebral aqueduct
  - cerebral peduncles
  - corpora quadrigemina
- pons
- medulla oblongata

5. Locate the labeled areas in figure 28.3 that represent the following functional regions of the cerebrum:
   motor area for voluntary muscle control
   motor speech area (Broca's area)
   cutaneous sensory area
   auditory area
   visual area
   sensory speech area (Wernicke's area)
6. Complete Part B of the laboratory report.

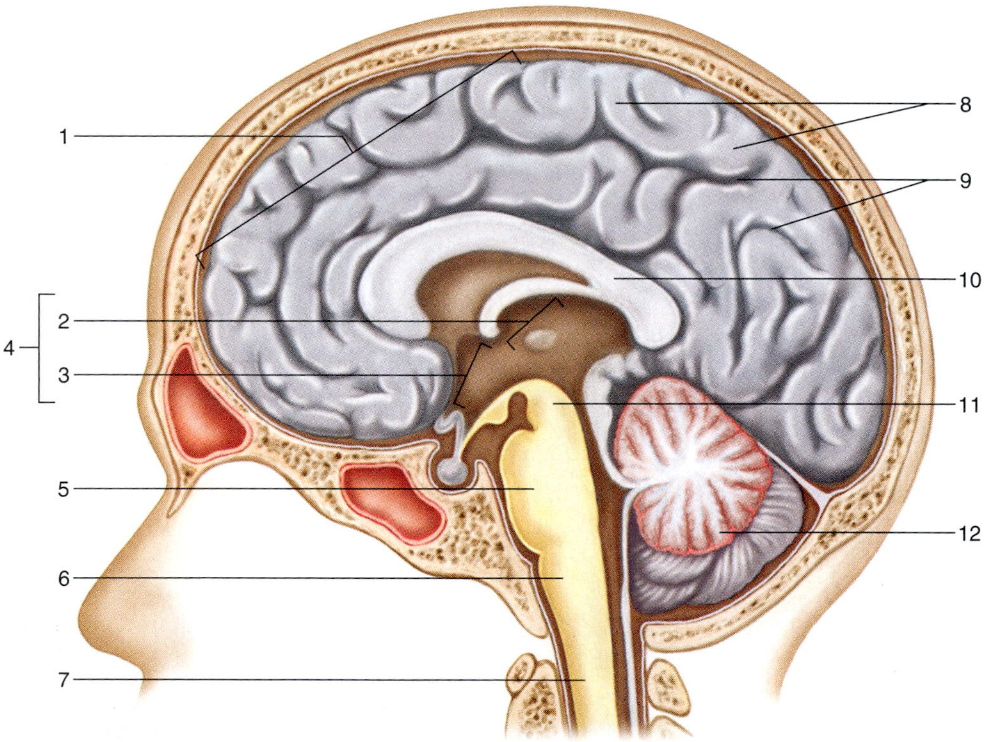

___ Cerebellum
___ Cerebrum
___ Corpus callosum
___ Diencephalon
___ Gyri/convolutions
___ Hypothalamus
___ Medulla oblongata
___ Midbrain
___ Pons
___ Spinal cord
___ Sulci
___ Thalamus

Figure 28.1 Label this diagram by placing the correct numbers in the spaces provided. 1

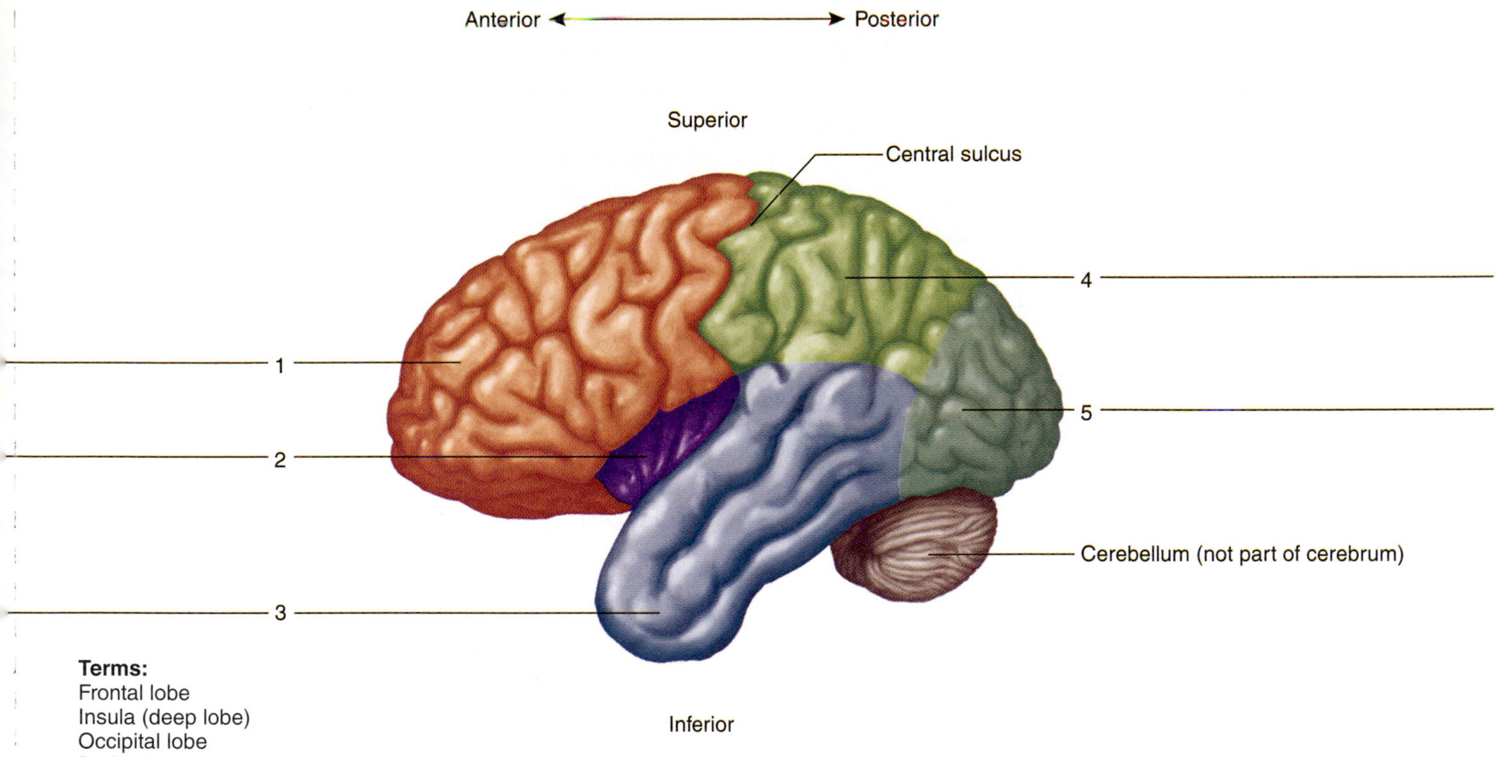

Terms:
Frontal lobe
Insula (deep lobe)
Occipital lobe
Parietal lobe
Temporal lobe

Figure 28.2 Label the five lobes of the left cerebral hemisphere, using the terms provided. Lobe 3 is retracted to expose the deep lobe 2 in this figure. 1

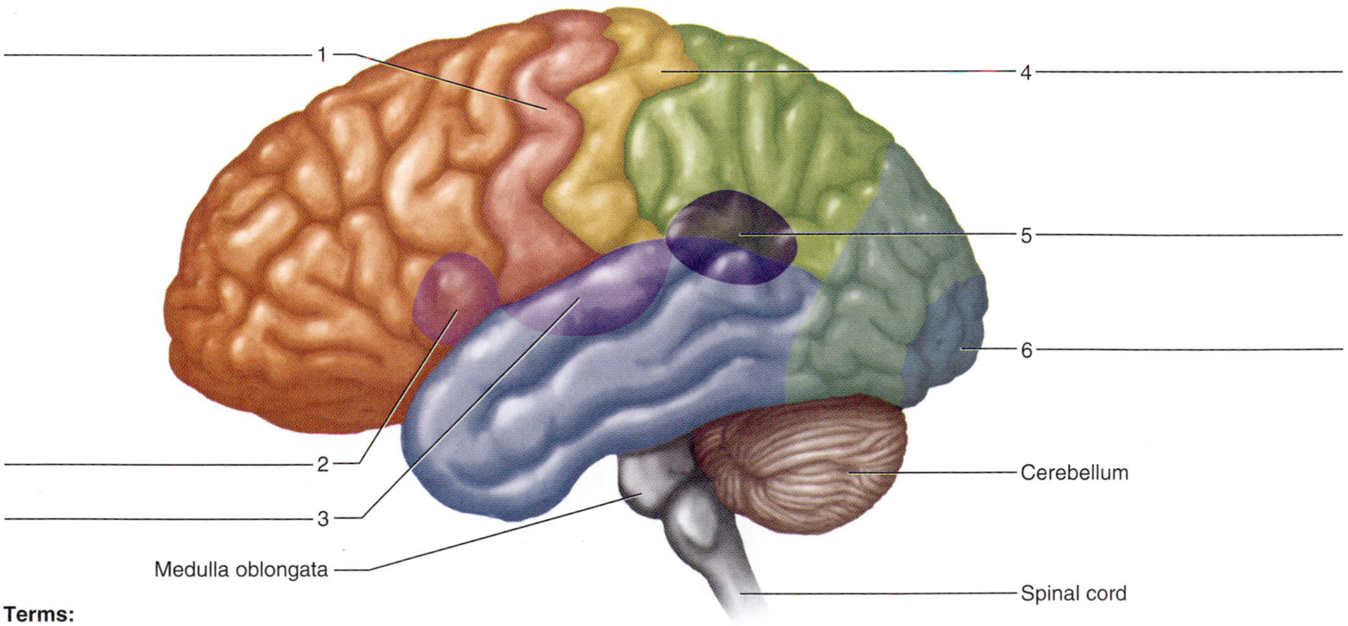

Terms:
Auditory area
Motor speech area (Broca's area)
Cutaneous sensory area
Motor area for voluntary muscle control
Visual area
Sensory speech area (Wernicke's area)

Figure 28.3 Label the functional areas of the cerebrum, using the terms provided. (*Note:* These areas are not visible as distinct parts of the brain.) 2

EXPLORE

Procedure B—Cranial Nerves

1. Review the section entitled "Cranial Nerves" in chapter 11 of the textbook.
2. As a review activity, label figure 28.4.
3. Observe the model and preserved specimen of the human brain, and locate as many of the following cranial nerves as possible as you differentiate some of their associated functions:

olfactory nerves (I)—smell

optic nerves (II)—vision

oculomotor nerves (III)—pupil constriction and open eyelid

trochlear nerves (IV)—stimulate superior oblique eye muscle

trigeminal nerves (V)—sensory from face and teeth and mastication movements

abducens nerves (VI)—lateral eye movements

facial nerves (VII)—salivation, tear secretions, and taste

vestibulocochlear nerves (VIII)—hearing and balance (equilibium)

glossopharyngeal nerves (IX)—regulate blood pressure, salivation, and swallowing

vagus nerves (X)—regulate many visceral organs including the heart rate

accessory nerves (XI)—control neck and shoulder muscles

hypoglossal nerves (XII)—control tongue movements

The following mnemonic device will help you learn the twelve pairs of cranial nerves in the proper order:

Old **Op**ie **oc**casionally **tr**ies **trig**onometry, **a**nd **f**eels **ve**ry **glo**omy, **vag**ue, **a**nd **hypo**active.[1]

4. Complete Part C of the laboratory report.

1. From *HAPS-EDucator,* Winter 2002. An official publication of the Human Anatomy & Physiology Society (HAPS).

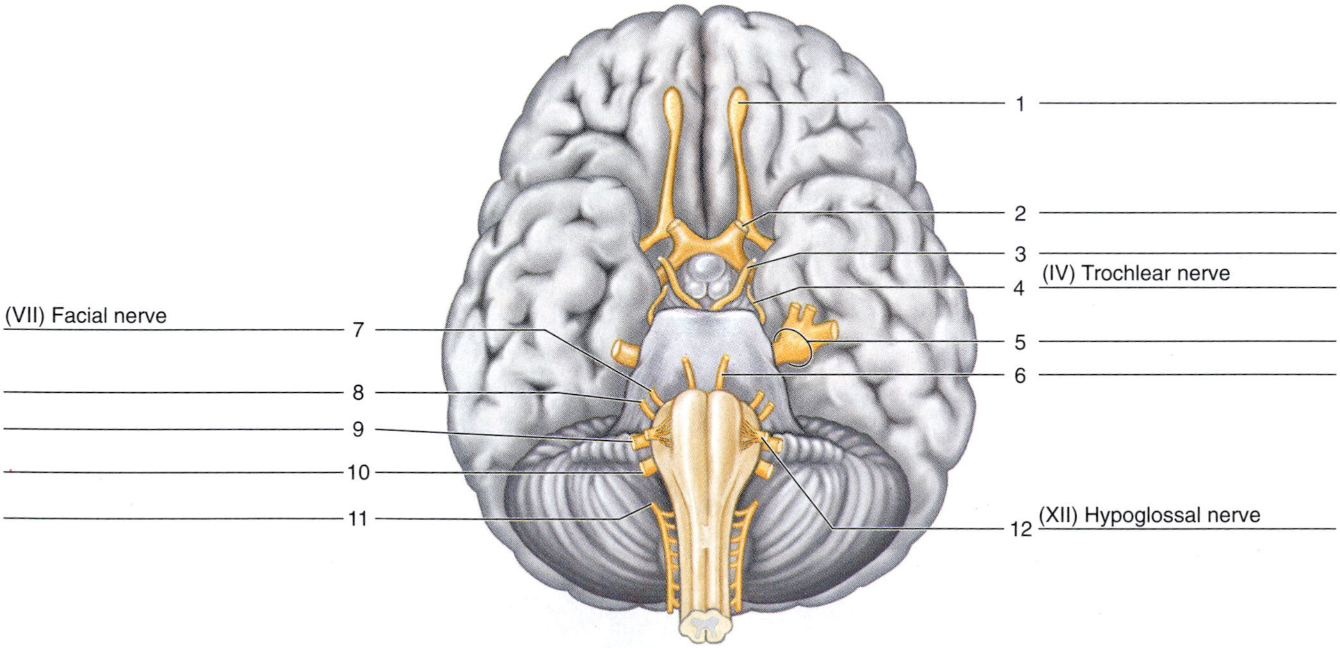

Figure 28.4 Provide the names of the cranial nerves in this ventral view. 3

Name ______________________________

Date ______________________________

Section ______________________________

The [A] corresponds to the indicated Learning Outcome(s) listed at the beginning of the laboratory exercise.

Laboratory Report 28

Brain and Cranial Nerves

Part A Assessments

Match the terms in column A with the descriptions in column B. Place the letter of your choice in the space provided. [1]

Column A		Column B
a. Central sulcus	_______	1. Structure formed by the crossing-over of the optic nerves
b. Cerebral cortex	_______	2. Part of diencephalon that forms lower walls and floor of third ventricle
c. Corpus callosum	_______	3. Cone-shaped structure attached to upper posterior portion of diencephalon
d. Falx cerebelli	_______	4. Connects cerebral hemispheres
e. Gyrus (convolution)	_______	5. Ridge on surface of cerebrum
f. Hypothalamus	_______	6. Separates frontal and parietal lobes
g. Insula	_______	7. Part of brainstem between diencephalon and pons
h. Medulla oblongata	_______	8. Rounded bulge on underside of brainstem
i. Midbrain	_______	9. Part of brainstem continuous with the spinal cord
j. Optic chiasma	_______	10. A layer of dura mater that separates cerebellar hemispheres
k. Pineal gland	_______	11. Cerebral lobe located deep within lateral sulcus
l. Pons	_______	12. Thin layer of gray matter on surface of cerebrum

Part B Assessments

Identify the features indicated in the sagittal (median) section of the right half of the human brain in figure 28.5. 1

1. ______________________

2. ______________________

3. ______________________

4. ______________________

5. ______________________

6. ______________________

7. ______________________

8. ______________________

9. ______________________

10. ______________________

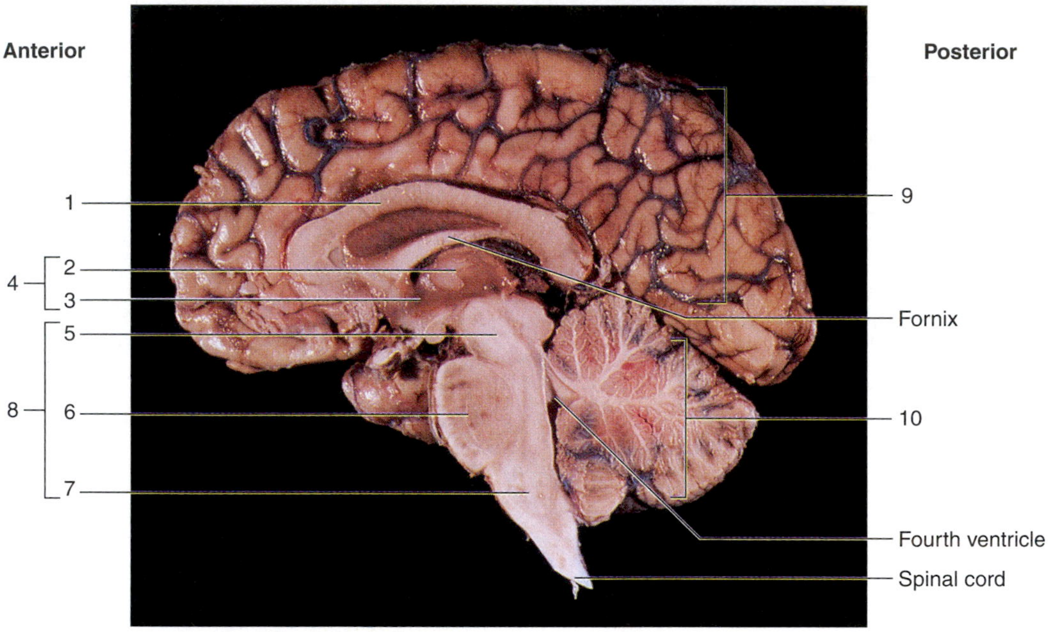

Terms:

Brainstem	Hypothalamus
Cerebellum	Medulla oblongata
Cerebrum	Midbrain
Corpus callosum	Pons
Diencephalon	Thalamus

Figure 28.5 Identify the features on this sagittal (median) section of the right half of the human brain, using the terms provided.

Part C Assessments

Indicate by name and number which cranial nerve(s) is (are) most closely associated with each of the following functions: 4

1. Sense of hearing ______________________
2. Sense of taste ______________________
3. Sense of sight ______________________
4. Sense of smell ______________________
5. Sense of equilibrium ______________________
6. Conducting sensory impulses from upper teeth ______________________
7. Conducting sensory impulses from lower teeth ______________________
8. Raising eyelids ______________________
9. Focusing lenses of eyes ______________________
10. Adjusting amount of light entering eye ______________________
11. Moving eyes ______________________
12. Stimulating salivary secretions ______________________
13. Movement of trapezius and sternocleidomastoid muscles ______________________
14. Muscular movements associated with speech ______________________
15. Muscular movements associated with swallowing ______________________

Notes

Laboratory Exercise 29

Dissection of the Sheep Brain

Materials Needed

Dissectible model of human brain
Preserved sheep brain
Dissecting tray
Dissection instruments
Long knife

For Demonstration:

Frontal sections of sheep brains

Safety

- Wear disposable gloves when handling the sheep brains.
- Save or dispose of the brains as instructed.
- Wash your hands before leaving the laboratory.

Mammalian brains have many features in common. Human brains may not be available, so sheep brains often are dissected as an aid to understanding mammalian brain structure. However, the adaptations of the sheep differ from the adaptations of the human, so comparisons of their structural features may not be precise. The sheep is a quadruped, therefore the spinal cord is horizontal, unlike the vertical orientation in a bipedal human. Preserved sheep brains have a different appearance and are firmer than those that are removed directly from the cranial cavity because of the preservatives used.

Purpose of the Exercise

To observe the major features of the sheep brain and to compare these features with those of the human brain.

LEARNING OUTCOMES

After completing this exercise, you should be able to

1. Examine the major structures of the sheep brain.
2. Locate the larger cranial nerves of the sheep brain.
3. Summarize several differences and similarities between the sheep brain and the human brain.

EXPLORE

Procedure—Dissection of the Sheep Brain

1. Obtain a preserved sheep brain and rinse it thoroughly in water to remove as much of the preserving fluid as possible.
2. Examine the surface of the brain for the presence of meninges. (The outermost layers of these membranes may have been lost during removal of the brain from the cranial cavity.) If meninges are present, locate the following:

   dura mater—the thick, opaque outer layer

   arachnoid mater—the delicate, transparent middle layer attached to the undersurface of the dura mater

   pia mater—the thin, vascular layer that adheres to the surface of the brain (should be present)

3. Remove any remaining dura mater by pulling it gently from the surface of the brain.
4. Position the brain with its ventral surface down in the dissecting tray. Study figure 29.1, and locate the following structures on the specimen:

   cerebral hemispheres
   gyri (convolutions)
   sulci
   longitudinal fissure
   frontal lobe
   parietal lobe
   temporal lobe
   occipital lobe
   cerebellum
   medulla oblongata
   spinal cord

5. Gently separate the cerebral hemispheres along the longitudinal fissure and expose the transverse band of white fibers within the fissure that connects the hemispheres. This band is the *corpus callosum.*
6. Bend the cerebellum and medulla oblongata slightly downward and away from the cerebrum (fig. 29.2).

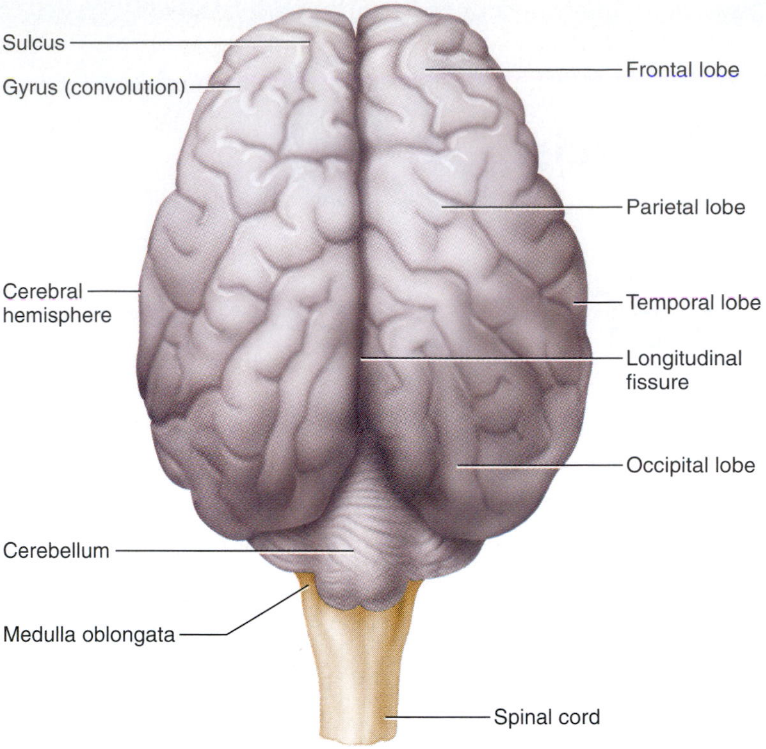

Figure 29.1 Dorsal surface of the sheep brain.

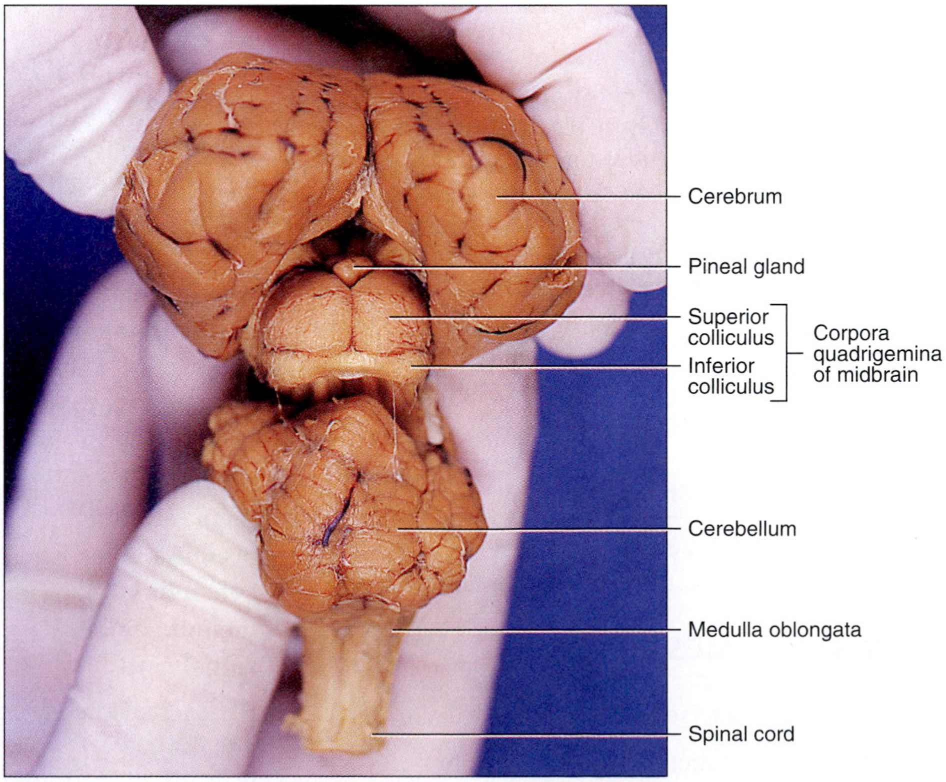

Figure 29.2 Gently bend the cerebellum and medulla oblongata away from the cerebrum to expose the pineal gland and the corpora quadrigemina.

This will expose the *pineal gland* in the upper midline and the *corpora quadrigemina,* which consists of four rounded structures, called *colliculi,* associated with the midbrain.

7. Position the brain with its ventral surface upward. Study figures 29.3 and 29.4, and locate the following structures on the specimen:

longitudinal fissure
olfactory bulbs
optic nerves
optic chiasma
optic tract
mammillary bodies
infundibulum (pituitary stalk)
midbrain
pons

8. Although some of the cranial nerves may be missing or are quite small and difficult to find, locate

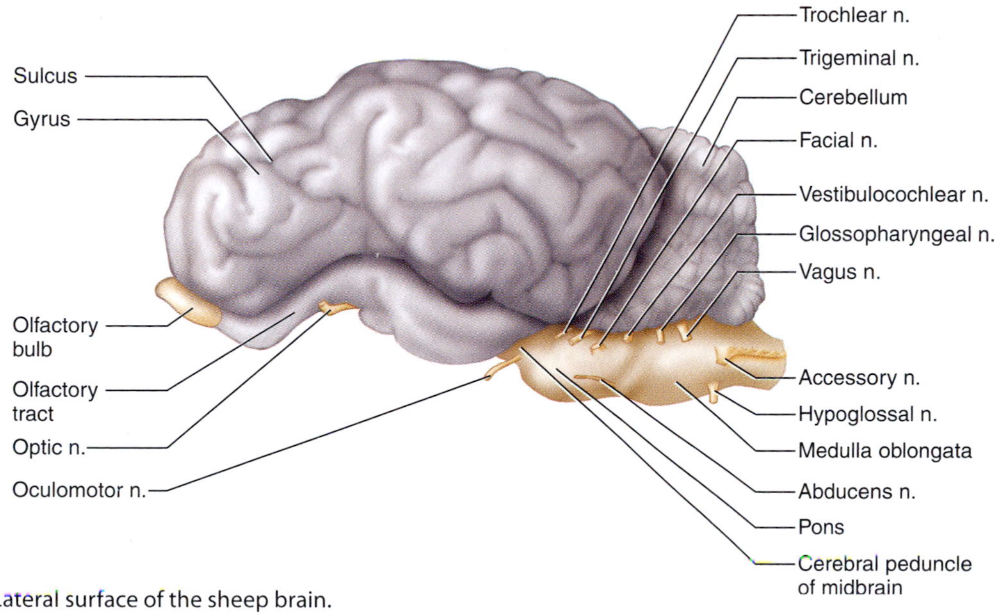

Figure 29.3 Lateral surface of the sheep brain.

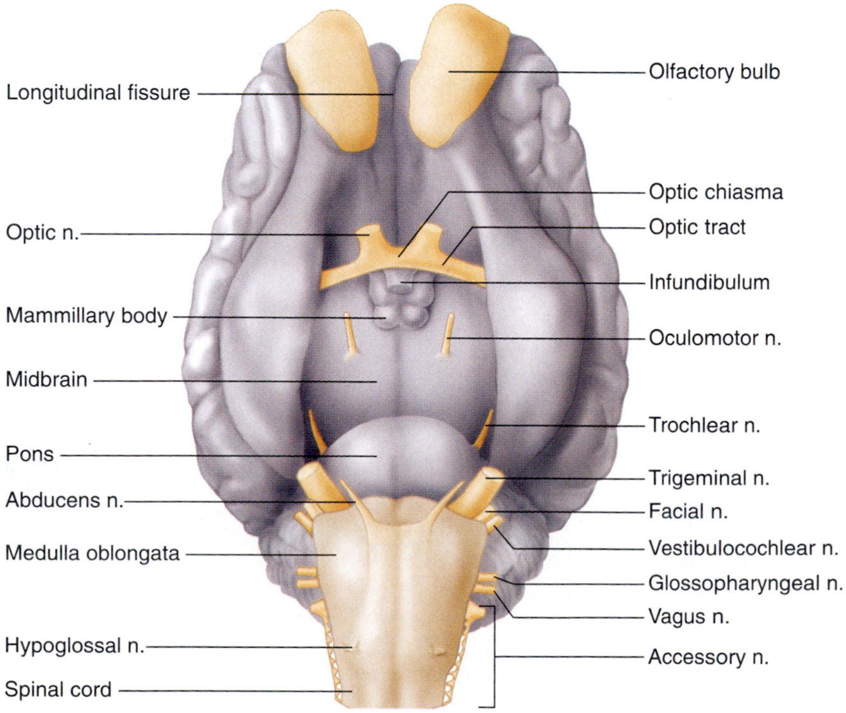

Figure 29.4 Ventral surface of the sheep brain.

as many of the following as possible, using figures 29.3 and 29.4 as references:

oculomotor nerves
trochlear nerves
trigeminal nerves
abducens nerves
facial nerves
vestibulocochlear nerves
glossopharyngeal nerves
vagus nerves
accessory nerves
hypoglossal nerves

9. Using a long, sharp knife, cut the sheep brain along the midline to produce a median section. Study figures 29.2 and 29.5, and locate the following structures on the specimen:

cerebrum
olfactory bulb
corpus callosum
cerebellum
- white matter
- gray matter

lateral ventricle (one in each cerebral hemisphere)
third ventricle (within diencephalon)
fourth ventricle (between brainstem and cerebellum)
diencephalon
- optic chiasma
- infundibulum
- pituitary gland (this structure may be missing)
- mammillary bodies
- thalamus
- hypothalamus
- pineal gland

midbrain
- corpora quadrigemina
  - superior colliculi
  - inferior colliculi

pons
medulla oblongata

Demonstration

Observe a frontal section from a sheep brain. Note the longitudinal fissure, gray matter, white matter, corpus callosum, lateral ventricles, third ventricle, and thalamus.

10. Dispose of the sheep brain as directed by the laboratory instructor.
11. Complete Parts A and B of Laboratory Report 29.

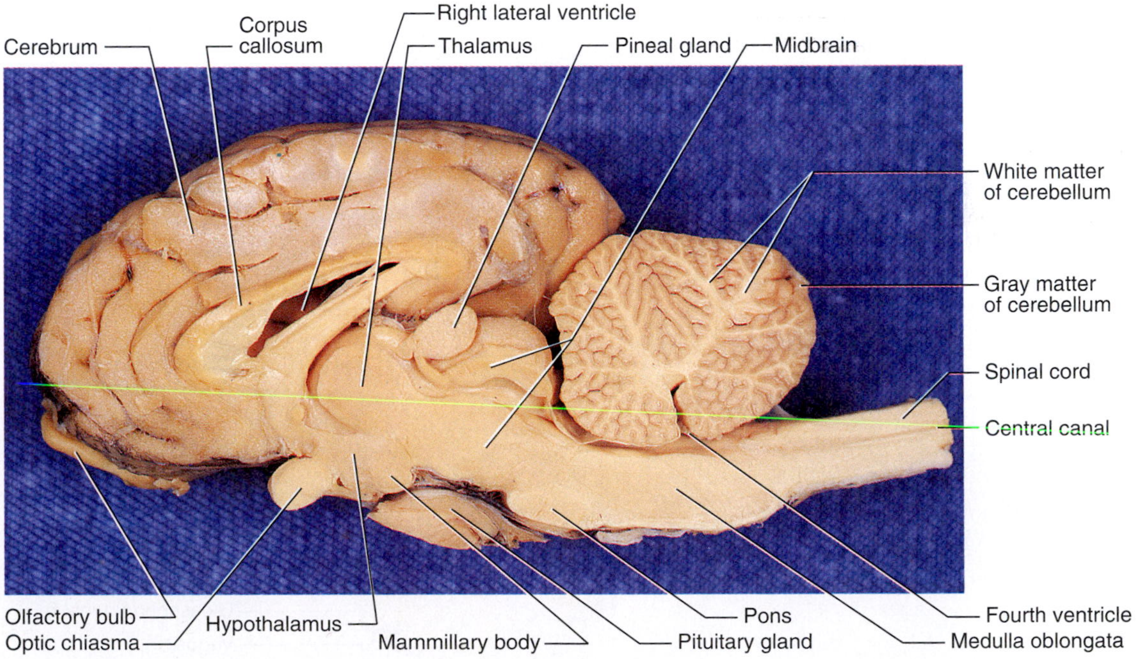

Figure 29.5 Sagittal (median) section of the right half of the sheep brain dissection.

Name ______________________

Date ______________________

Section ______________________

The [A] corresponds to the indicated Learning Outcome(s) listed at the beginning of the laboratory exercise.

Laboratory Report 29

Dissection of the Sheep Brain

Part A Assessments

Answer the following questions:

1. Compare the relative sizes of the sheep and human cerebral hemispheres. [3] ______________________

2. How do the gyri and sulci of the sheep cerebrum compare with the human cerebrum in numbers? [3] ______________________

3. What is the significance of the differences you noted in your answers for questions 1 and 2? [3] ______________________

4. What difference did you note in the structures of the sheep cerebellum and the human cerebellum? [3] ______________________

5. How do the sizes of the olfactory bulbs of the sheep brain compare with those of the human brain? [3] ______________________

6. Based on their relative sizes, which of the cranial nerves seems to be most highly developed in the sheep brain? [2] ______________________

7. What is the significance of the observations you noted in your answers for questions 5 and 6? [3] ______________________

Critical Thinking Application

Prepare a list of at least six features to illustrate ways in which the brains of sheep and humans are similar. 1 3

1. ____________________
2. ____________________
3. ____________________
4. ____________________
5. ____________________
6. ____________________

Interpret the significance of these similarities. 3 ____________________

____

____

____

Laboratory Exercise 30

Receptors and General Senses

Materials Needed

Textbook
Marking pen (washable)
Millimeter ruler
Bristle or sharp pencil
Forceps (fine points) or measuring calipers
Blunt metal probes
Three beakers (250 mL)
Warm tap water or 45°C (113°F) water bath
Cold water (ice water)
Thermometer

For Demonstration:

Prepared microscope slides of tactile (Meissner's) and lamellated (Pacinian) corpuscles
Compound light microscope

Sensory receptors are sensitive to changes that occur within the body and its surroundings. Each type of receptor is particularly sensitive to a distinct kind of environmental change and is much less sensitive to other forms of stimulation. When receptors are stimulated, they initiate nerve impulses that travel into the central nervous system. The raw form in which these receptors send information to the brain is called *sensation*. The way our brains interpret this information is called *perception*.

The sensory receptors found widely distributed throughout skin, muscles, joints, and visceral organs are associated with **general senses.** These senses include touch, pressure, temperature, pain, and the senses of muscle movement and body position.

Sensory receptors that are more specialized and confined to the head are associated with **special senses.** Laboratory Exercises 31 through 35 describe the special senses.

Purpose of the Exercise

To review the characteristics of sensory receptors and general senses and to investigate some of the general senses associated with the skin.

LEARNING OUTCOMES

After completing this exercise, you should be able to

1. Associate types of sensory receptors with general senses throughout the body.
2. Determine and record the distribution of touch, warm, and cold receptors in various regions of the skin.
3. Measure the two-point threshold of various regions of the skin.

EXPLORE

Procedure A—Receptors

1. Review the sections entitled "Receptors, Sensations, and Perception" and "General Senses" in chapter 12 of the textbook.
2. Complete Part A of Laboratory Report 30.

Demonstration

Observe the tactile (Meissner's) corpuscle with the microscope set up by the laboratory instructor. This type of receptor is abundant in the superficial dermis in outer regions of the body, such as in the fingertips, soles, lips, and external genital organs. It is responsible for the sensation of light touch. (See fig. 30.1 and chapter 6 of the textbook.)

Observe the lamellated (Pacinian) corpuscle in the second demonstration microscope. This corpuscle is composed of many layers of connective tissue cells and has a nerve fiber in its central core. Lamellated corpuscles are numerous in the hands, feet, joints, and external genital organs. They are responsible for the sense of deep pressure (fig. 30.2). How are tactile and lamellated corpuscles similar?

____

____

How are they different?

____

____

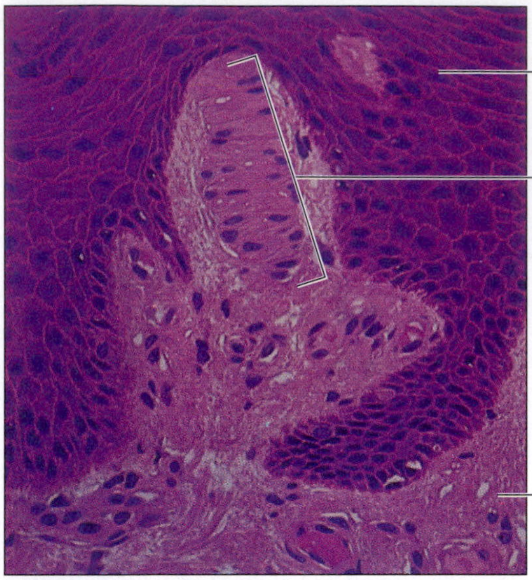

Figure 30.1 Tactile (Meissner's) corpuscles, such as this one, are responsible for the sensation of light touch (250×).

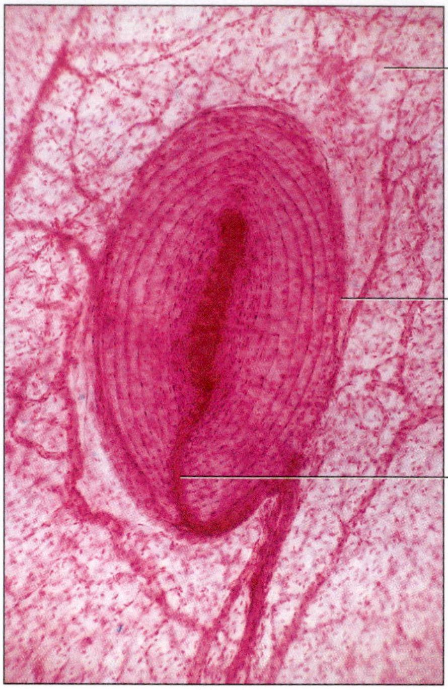

Figure 30.2 Lamellated (Pacinian) corpuscles, such as this one, are responsible for the sensation of deep pressure (25× micrograph enlarged to 100×).

EXPLORE

Procedure B—Sense of Touch

1. Investigate the distribution of touch receptors in your laboratory partner's skin. To do this, follow these steps:
   - **a.** Use a marking pen and a millimeter ruler to prepare a square with 2.5 cm on each side of the skin on your partner's inner wrist, near the palm.
   - **b.** Divide the square into smaller squares with 0.5 cm on a side, producing a small grid.
   - **c.** Ask your partner to rest the marked wrist on the tabletop and to keep his or her eyes closed throughout the remainder of the experiment.
   - **d.** Press the end of a bristle on the skin in some part of the grid, using just enough pressure to cause the bristle to bend. A sharp pencil could be used as an alternate device.
   - **e.** Ask your partner to report whenever the touch of the bristle is felt. Record the results in Part B of the laboratory report.
   - **f.** Continue this procedure until you have tested twenty-five different locations on the grid. Move randomly through the grid to help prevent anticipation of the next stimulation site.
2. Test two other areas of exposed skin in the same manner, and record the results in Part B of the laboratory report.
3. Answer the questions in Part B of the laboratory report.

EXPLORE

Procedure C—Two-Point Threshold

1. Test your partner's ability to recognize the difference between one and two points of skin being stimulated simultaneously. To do this, follow these steps:
   - **a.** Have your partner place a hand with the palm up on the table and close his or her eyes.
   - **b.** Hold the tips of a forceps tightly together and gently touch the skin of your partner's index finger.
   - **c.** Ask your partner to report if it feels like one or two points are touching the finger.
   - **d.** Allow the tips of the forceps or the points of the measuring calipers to spread so they are 1 mm apart, press both points against the skin simultaneously, and ask your partner to report as before.
   - **e.** Repeat this procedure, allowing the tips of the forceps to spread more each time until your partner can feel both tips being pressed against the skin. The minimum distance between the tips of the forceps when both can be felt is called the *two-point threshold*. As soon as you are able to distinguish two points, two separate receptors are being stimulated instead of only one receptor.
   - **f.** Record the two-point threshold for the skin of the index finger in Part C of the laboratory report.

2. Repeat this procedure to determine the two-point threshold of the palm, the back of the hand, the back of the neck, the leg, and the sole. Record the results in Part C of the laboratory report.
3. Answer the questions in Part C of the laboratory report.

EXPLORE

Procedure D—Sense of Temperature

1. Investigate the distribution of *warm receptors* in your partner's skin. To do this, follow these steps:
   a. Mark a square with 2.5 cm sides on your partner's palm.
   b. Prepare a grid by dividing the square into smaller squares, 0.5 cm on a side.
   c. Have your partner rest the marked palm on the table and close his or her eyes.
   d. Heat a blunt metal probe by placing it in a beaker of warm (hot) water (about 40–45°C/ 104–113°F) for a minute or so. *(Be sure the probe does not get so hot that it burns the skin.)* Use a thermometer to monitor the appropriate warm water from the tap or the water bath.
   e. Wipe the probe dry and touch it to the skin on some part of the grid.
   f. Ask your partner to report if the probe feels warm. Then record the results in Part D of the laboratory report.
   g. Keep the probe warm, and repeat the procedure until you have randomly tested twenty-five different locations on the grid.
2. Investigate the distribution of *cold receptors* by repeating the procedure. Use a blunt metal probe that has been cooled by placing it in ice water for a minute or so. Record the results in Part D of the laboratory report.
3. Answer the questions in Part D of the laboratory report.

Learning Extension

Prepare three beakers of water of different temperatures. One beaker should contain warm water (about 40°C/ 104°F), one should be room temperature (about 22°C/ 72°F), and one should contain cold water (about 10°C/ 50°F). Place the index finger of one hand in the warm water and, at the same time, place the index finger of the other hand in the cold water for about 2 minutes. Then, simultaneously move both index fingers into the water at room temperature. What temperature do you sense with each finger? How do you explain the resulting perceptions?

____

____

____

____

Notes

Name ______________________________

Date ______________________________

Section ______________________________

The A corresponds to the indicated Learning Outcome(s) listed at the beginning of the laboratory exercise.

Laboratory Report 30

Receptors and General Senses

Part A—Receptors and General Senses Assessments

Complete the following statements:

1. ______________________ are receptors that are sensitive to changes in the concentrations of chemicals. (1)
2. Whenever tissues are damaged, ______________________ receptors are likely to be stimulated. (1)
3. Receptors that are sensitive to temperature changes are called ______________________ . (1)
4. ______________________ are sensitive to changes in the intensity of light energy. (1)
5. A sensation may seem to fade away when receptors are continuously stimulated as a result of ________________ . (1)
6. Tactile (Meissner's) corpuscles are responsible for the sense of light ______________________ . (1)
7. Lamellated (Pacinian) corpuscles are responsible for the sense of deep ______________________ . (1)
8. Warm receptors are most sensitive to temperatures between ______________________ . (1)
9. Cold receptors are most sensitive to temperatures between ______________________ . (1)
10. Widely distributed sensory receptors throughout the body are associated with ______________________ senses in contrast to special senses. (1)

Part B—Sense of Touch Assessments

1. Record a + to indicate where the bristle was felt and a *0* to indicate where it was not felt. (2)

Skin of wrist

2. Show the distribution of touch receptors in two other regions of skin. (2)

Region tested ________________

Region tested ________________

3. Answer the following questions:

   a. How do you describe the pattern of distribution for touch receptors in the regions of the skin you tested? 2 ______________________________

   ____

   b. How does the concentration of touch receptors seem to vary from region to region? 2 ______________________________

   ____

Part C—Two-Point Threshold Assessments

1. Record the two-point threshold in millimeters for skin in each of the following regions: 3

   Index finger ______________________________

   Palm ______________________________

   Back of hand ______________________________

   Back of neck ______________________________

   Leg ______________________________

   Sole ______________________________

2. Answer the following questions:

   a. What region of the skin tested has the greatest ability to discriminate two points? 3 ______________________________

   b. What region of the skin has the least sensitivity to this test? 3 ______________________________

   c. What is the significance of these observations in questions *a* and *b*? 3 ______________________________

   ____

   ____

Part D—Sense of Temperature Assessments

1. Record a + to indicate where warm was felt and a *0* to indicate where it was not felt. 2

Skin of palm

2. Record a + to indicate where cold was felt and a *0* to indicate where it was not felt. 2

Skin of palm

3. Answer the following questions:

   a. How do temperature receptors appear to be distributed in the skin of the palm? 2 ______________________________

   ____

   b. Compare the distribution and concentration of warm and cold receptors in the skin of the palm. 2 ______________________________

   ____

   ____

Laboratory Exercise 31

Smell and Taste

Materials Needed

For Procedure A—Sense of Smell (Olfaction)

Textbook
Set of substances in stoppered bottles: cinnamon, sage, vanilla, garlic powder, oil of clove, oil of wintergreen, and perfume

For Procedure B—Sense of Taste (Gustation)

Textbook
Paper cups (small)
Cotton swabs (sterile; disposable)
5% sucrose solution (sweet)
5% NaCl solution (salt)
1% acetic acid or unsweetened lemon juice (sour)
0.5% quinine sulfate solution or 0.1% Epsom salt solution (bitter)
1% monosodium glutamate (MSG) solution (umami)

For Demonstrations:

Compound light microscope
Prepared microscope slides of olfactory epithelium and of taste buds

For Learning Extension:

Pieces of apple, potato, carrot, and onion or packages of mixed flavors of LifeSavers

Safety

- Be aware of possible food allergies when selecting test solutions and foods to taste.
- Prepare fresh solutions for use in Procedure B.
- Wash your hands before starting the taste experiment.
- Wear disposable gloves when performing taste tests on your laboratory partner.
- Use a clean cotton swab for each test. Do not dip a used swab into a test solution.
- Dispose of used cotton swabs and paper towels as directed.
- Wash your hands before leaving the laboratory.

The senses of smell (olfaction) and taste (gustation) are dependent upon chemoreceptors that are stimulated by various chemicals dissolved in liquids. When these receptors are stimulated by appropriate chemicals, it results in action potentials traveling to the brain for interpretation. The receptors of smell are found in the olfactory organs, which are located in the superior parts of the nasal cavity and in a portion of the nasal septum. The receptors of taste occur in the taste buds, which are sensory organs primarily found on the surface of the tongue, but receptor cells are also distributed in other areas of the mouth cavity and pharynx. Taste sensations are grouped into five recognized categories: sweet, sour, salt, bitter, and umami. The sweet sensation is produced from sugars, the sour sensation from acids, the salt sensation from inorganic salts, the bitter sensation from alkaloids and spoiled foods, and the umami sensation from aspartic and glutamic acids or a derivative such as monosodium glutamate.

The senses of smell and taste function closely together, because substances that are tasted often are smelled at the same moment, and they play important roles in the selection of foods. The taste and aroma of foods are also influenced by appearance, texture, temperature, and the person's mood. Chemicals are considered odorless and tasteless if receptor sites for them are absent.

Purpose of the Exercise

To review the structures of the organs of smell and taste and to investigate the abilities of smell and taste receptors to discriminate various chemical substances.

LEARNING OUTCOMES

After completing this exercise, you should be able to

1. Compare the characteristics of the smell (olfactory) receptors and taste (gustatory) receptors.
2. Explain how the senses of smell and taste function.

(3) Record recognized odors and the time needed for olfactory sensory adaptation to occur.

(4) Locate the distribution of taste receptors on the surface of the tongue and mouth cavity.

EXPLORE

Procedure A—Sense of Smell (Olfaction)

1. Review the section entitled "Sense of Smell" in chapter 12 of the textbook.
2. As a review activity, label figure 31.1.
3. Complete Part A of Laboratory Report 31.
4. Test your laboratory partner's ability to recognize the odors of the bottled substances available in the laboratory. To do this, follow these steps:
   a. Have your partner keep his or her eyes closed.
   b. Remove the stopper from one of the bottles and hold it about 4 cm under your partner's nostrils for about 2 seconds.
   c. Ask your partner to identify the odor, and then replace the stopper.
   d. Record your partner's response in Part B of the laboratory report.
   e. Repeat steps *b–d* for each of the bottled substances.
5. Repeat the preceding procedure, using the same set of bottled substances, but present them to your partner in a different sequence. Record the results in Part B of the laboratory report.
6. Wait 10 minutes and then determine the time it takes for your partner to experience olfactory sensory adaptation. To do this, follow these steps:
   a. Ask your partner to breathe in through the nostrils and exhale through the mouth.
   b. Remove the stopper from one of the bottles, and hold it about 4 cm under your partner's nostrils.
   c. Keep track of the time that passes until your partner is no longer able to detect the odor of the substance.
   d. Record the result in Part B of the laboratory report.
   e. Wait 5 minutes and repeat this procedure, using a different bottled substance.
   f. Test a third substance in the same manner.
   g. Record the results as before.
7. Complete Part B of the laboratory report.

Demonstration

Observe the olfactory epithelium in the demonstration microscope. The olfactory receptor cells are spindle-shaped, bipolar neurons with spherical nuclei. They also have six to eight cilia at their apical ends. The supporting cells are pseudostratified columnar epithelial cells. However, in this region the tissue lacks goblet cells. (See fig. 31.2 and chapter 5 of the textbook.)

___ Cilia

___ Columnar epithelial cells

___ Cribriform plate of ethmoid bone

___ Nasal cavity

___ Nerve fibers within olfactory bulb

___ Olfactory nerves

___ Olfactory receptor cells

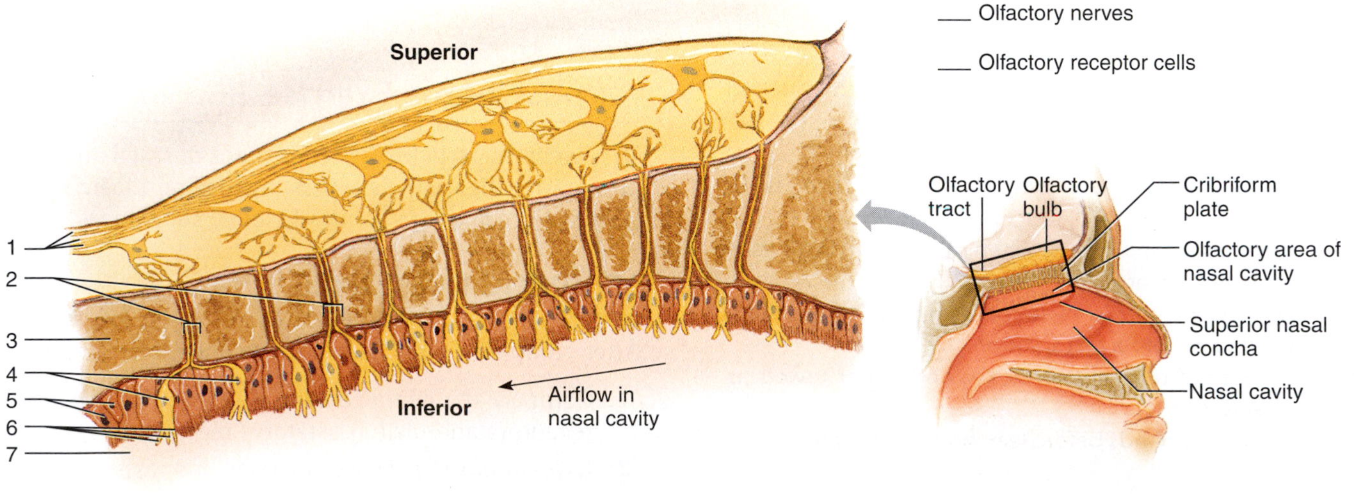

Figure 31.1 Label this diagram of the olfactory organ by placing the correct numbers in the spaces provided. (1)

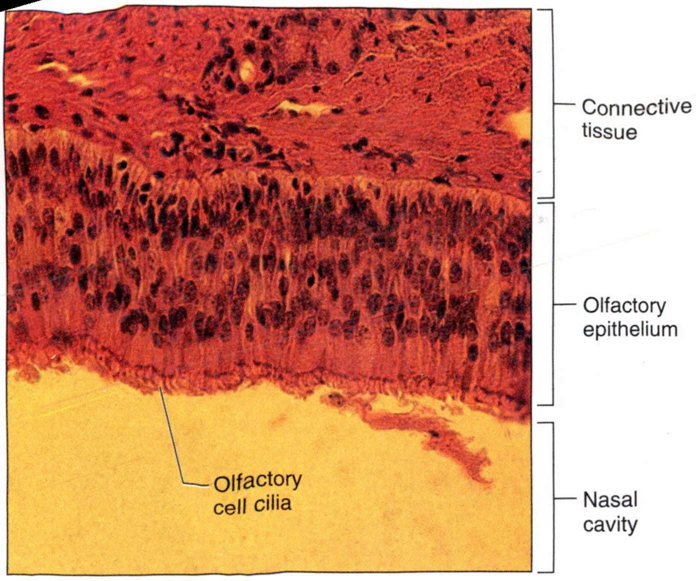

Figure 31.2 Olfactory receptors have cilia at their apical ends (250×).

EXPLORE

Procedure B—Sense of Taste (Gustation)

1. Review the section entitled "Sense of Taste" in chapter 12 of the textbook.
2. As a review activity, label figure 31.3.
3. Complete Part C of the laboratory report.
4. Map the distribution of the receptors for the primary taste sensations on your partner's tongue. To do this, follow these steps:
   a. Ask your partner to rinse his or her mouth with water and then partially dry the surface of the tongue with a paper towel.
   b. Moisten a clean cotton swab with 5% sucrose solution, and touch several regions of your partner's tongue with the swab.
   c. Each time you touch the tongue, ask your partner to report if a sweet sensation is experienced.
   d. Test the tip, sides, and back of the tongue in this manner.

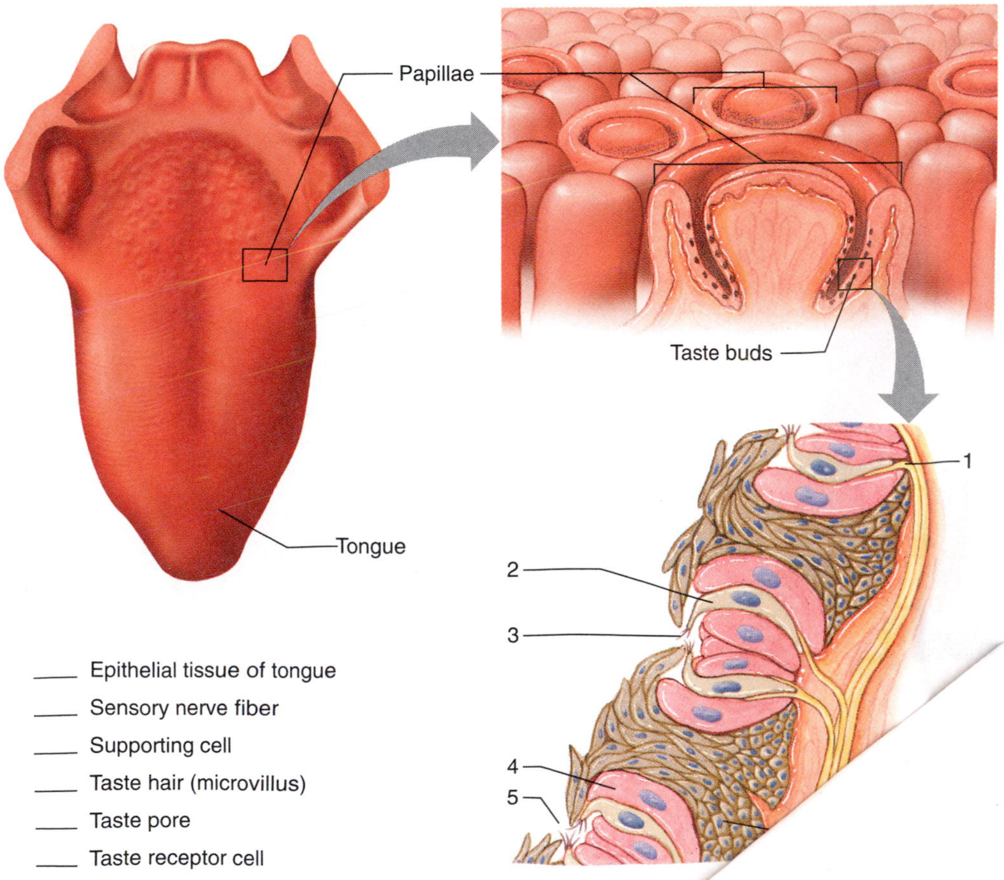

____ Epithelial tissue of tongue
____ Sensory nerve fiber
____ Supporting cell
____ Taste hair (microvillus)
____ Taste pore
____ Taste receptor cell

Figure 31.3 Label this diagram by placing the correct numbers in the spaces provi

e. Test some other representative areas inside the mouth cavity such as the cheek, gums, and roof of the mouth (hard and soft palate) for a sweet sensation.

f. Record your partner's responses in Part D of the laboratory report.

g. Have your partner rinse his or her mouth and dry the tongue again, and repeat the preceding procedure, using each of the other four test solutions—NaCl, acetic acid, quinine or Epsom salt, and MSG solution. Be sure to use a fresh swab for each test substance and dispose of used swabs and paper towels as directed.

5. Complete Part D of the laboratory report.

Demonstration

Observe the oval-shaped taste bud in the demonstration microscope. Note the surrounding epithelial cells. The taste pore, an opening into the taste bud, may be filled with taste hairs (microvilli). Within the taste bud there are supporting cells and thinner taste-receptor cells, which often have lightly stained nuclei (fig. 31.4).

Learning Extension

Test your laboratory partner's ability to recognize the tastes of apple, potato, carrot, and onion. (A package of mixed flavors of LifeSavers is a good alternative.) To do this, follow these steps:

1. Have your partner close his or her eyes and hold the nostrils shut.
2. Place a small piece of one of the test substances on your partner's tongue.
3. Ask your partner to identify the substance without chewing or swallowing it.
4. Repeat the procedure for each of the other substances.

How do you explain the results of this experiment?

____

____

____

____

____

____

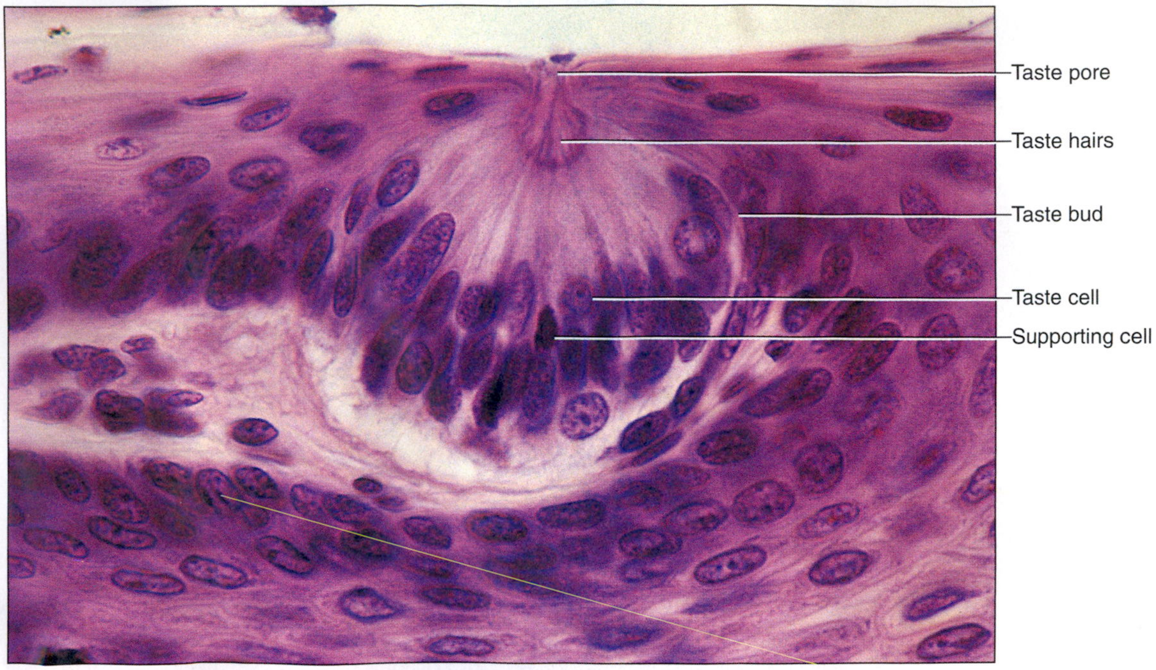

Figu

Taste receptors are found in taste buds (250× micrograph enlarged to 1,500×).

Name ______________________

Date ______________________

Section ______________________

The A corresponds to the indicated Learning Outcome(s) listed at the beginning of the laboratory exercise.

Smell and Taste

Laboratory Report 31

Part A Assessments

Complete the following statements:

1. Olfactory, or smell, receptors are ______________ neurons surrounded by columnar epithelial cells. A1
2. The distal ends of the olfactory neurons are covered with hairlike ______________. A1
3. Before gaseous substances can stimulate the olfactory receptors, they must be dissolved in ______________ that surrounds the cilia. A1
4. The axons of olfactory receptors pass through small openings in the ______________ of the ethmoid bone. A2
5. Olfactory bulbs lie on either side of the ______________ of the ethmoid bone. A2
6. The sensory impulses pass from the olfactory bulbs through the ______________ to the interpreting centers of the brain. A2
7. The olfactory interpreting centers are located deep within the temporal lobes and at the base of the ______________ lobes of the cerebrum. A2
8. Olfactory sensations usually fade rapidly as a result of ______________. A1
9. Olfactory receptor neurons are the only parts of the nervous system that come in contact with the ______________ directly. A1
10. A chemical would be considered ______________ if a person lacks a particular receptor site on the cilia of the olfactory neurons. A1

Part B—Sense of Smell Assessments

1. Record the results (as +, if recognized; as 0, if unrecognized) from the tests of odor recognition in the following table: 3

	Odor Reported	
Substance Tested	First Trial	Second Trial

2. Record the results of the olfactory sensory adaptation time in the following table: 3

Substance Tested	Adaptation Time in Seconds

3. Complete the following:

   a. How do you describe your partner's ability to recognize the odors of the substances you tested? 2 ______

   ______

   ______

   b. Compare your experimental results with those of others in the class. Did you find any evidence to indicate that individuals may vary in their abilities to recognize odors? Explain your answer. 2 ______

   ______

   ______

Critical Thinking Application

Does the time it takes for sensory adaptation to occur seem to vary with the substances tested? Explain your answer.

______

______

______

Part C Assessments

Complete the following statements:

1. Taste receptor cells are modified ______ cells. 1
2. The opening to a taste bud is called a ______. 1
3. The ______ of a taste cell are its sensitive part. 1
4. Before the taste of a substance can be detected, the substance must be dissolved in ______. 1
5. Substances that stimulate taste cells bind with ______ sites on the surfaces of taste hairs. 1
6. Sour receptors are mainly stimulated by ______. 1
7. Salt receptors are mainly stimulated by ionized inorganic ______. 1
8. Alkaloids usually have a ______ taste. 1

Part D—Sense of Taste Assessments

1. *Taste receptor distribution.* Record a + to indicate where a taste sensation seemed to originate and a *0* if no sensation occurred when the spot was stimulated. 4

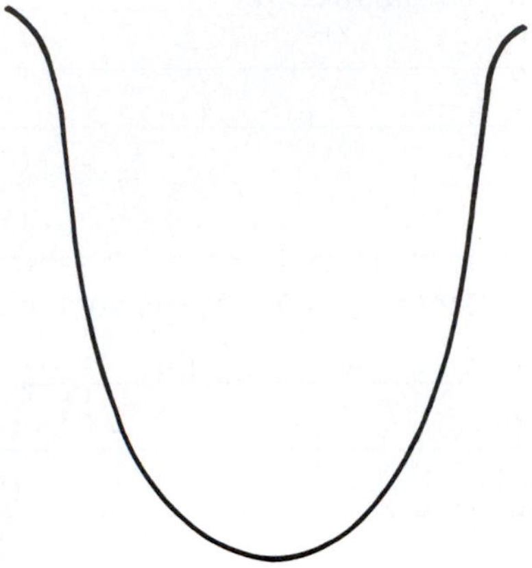

Sucrose (sweet sensation)

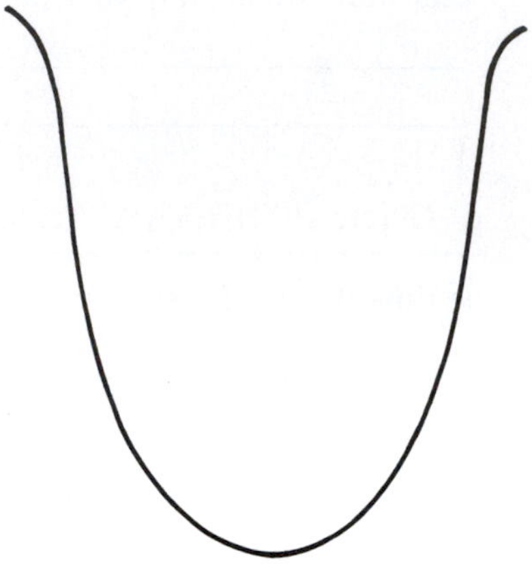

NaCl (salt sensation)

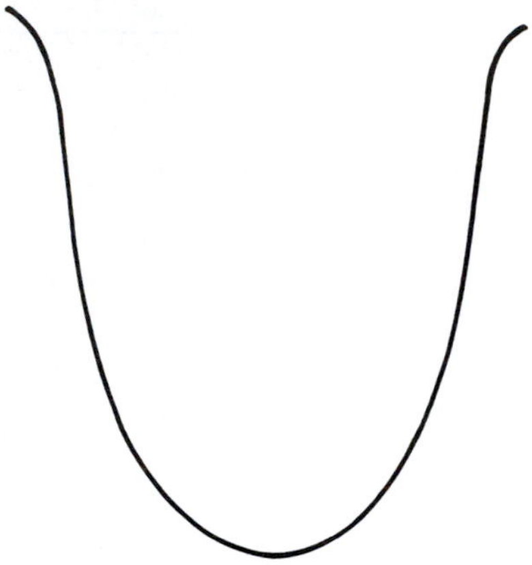

MSG (umami sensation)

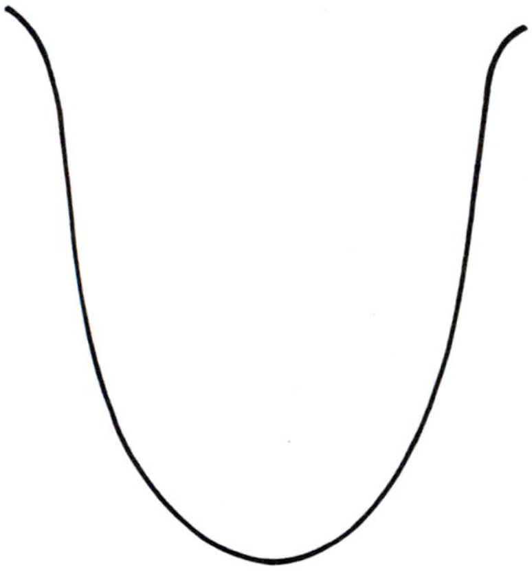

Acetic acid (sour sensation)

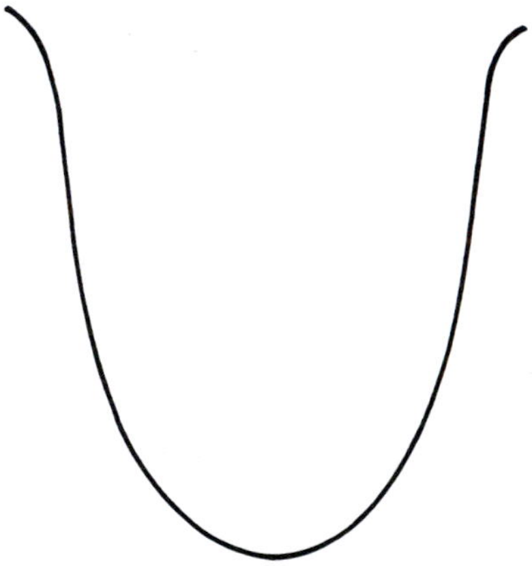

Quinine (bitter sensation)

2. Complete the following:

   a. Describe how each type of taste receptor is distributed on the surface of your partner's tongue. 4 ______

   ____

   ____

   ____

   b. Describe other locations inside the mouth where any sensations of sweet, salt, sour, bitter, or umami were located. 4 ______________________________

   ____

   c. How does your taste distribution map on the tongue compare to those of other students in the class? 4

   ____

   ____

Laboratory Exercise 32

Ear and Hearing

Materials Needed

Textbook
Dissectible ear model
Watch that ticks
Tuning fork (128 or 256 cps)
Rubber hammer
Cotton
Meterstick

For Demonstrations:

Compound light microscope
Prepared microscope slide of cochlea (section)
Audiometer

The ear is composed of outer (external), middle, and inner (internal) parts. The outer structures gather sound waves and direct them inward to the tympanic membrane. The parts of the middle ear, in turn, transmit vibrations from the tympanic membrane (eardrum) to the inner ear, where the hearing receptors are located. As they are stimulated, these receptors initiate nerve impulses to pass over the vestibulocochlear nerve into the auditory cortex of the brain, where the impulses are interpreted and the sensations of hearing are created.

Although the ear is considered here as a hearing organ, it also has important functions for equilibrium (balance). The inner ear semicircular canals, utricle, and saccule have equilibrium functions. Equilibrium will be studied in Laboratory Exercise 33.

Purpose of the Exercise

To review the structural and functional characteristics of the ear and to conduct some ordinary hearing tests.

LEARNING OUTCOMES

After completing this exercise, you should be able to

1. Locate the major structures of the ear.
2. Describe the functions of the structures of the ear.
3. Trace the pathway of sound vibrations from the tympanic membrane to the hearing receptors.
4. Conduct four ordinary hearing tests and summarize the results.

EXPLORE

Procedure A—Structure and Function of the Ear

1. Review the section entitled "Sense of Hearing" in chapter 12 of the textbook.
2. As a review activity, label figures 32.1, 32.2, and 32.3.
3. Examine the dissectible model of the ear and locate the following features:

outer (external) ear
- auricle (pinna)
- external acoustic meatus
- tympanic membrane (eardrum)

middle ear
- tympanic cavity
- auditory ossicles (fig. 32.4)
  - malleus
  - incus
  - stapes
- oval window
- tensor tympani
- stapedius

auditory tube (pharyngotympanic tube; eustachian tube)

inner (internal) ear
- osseous (bony) labyrinth
- membranous labyrinth
- cochlea
- semicircular canals and ducts
- vestibule
  - utricle
  - saccule

vestibulocochlear nerve
- vestibular nerve (balance branch)
- cochlear nerve (hearing branch)

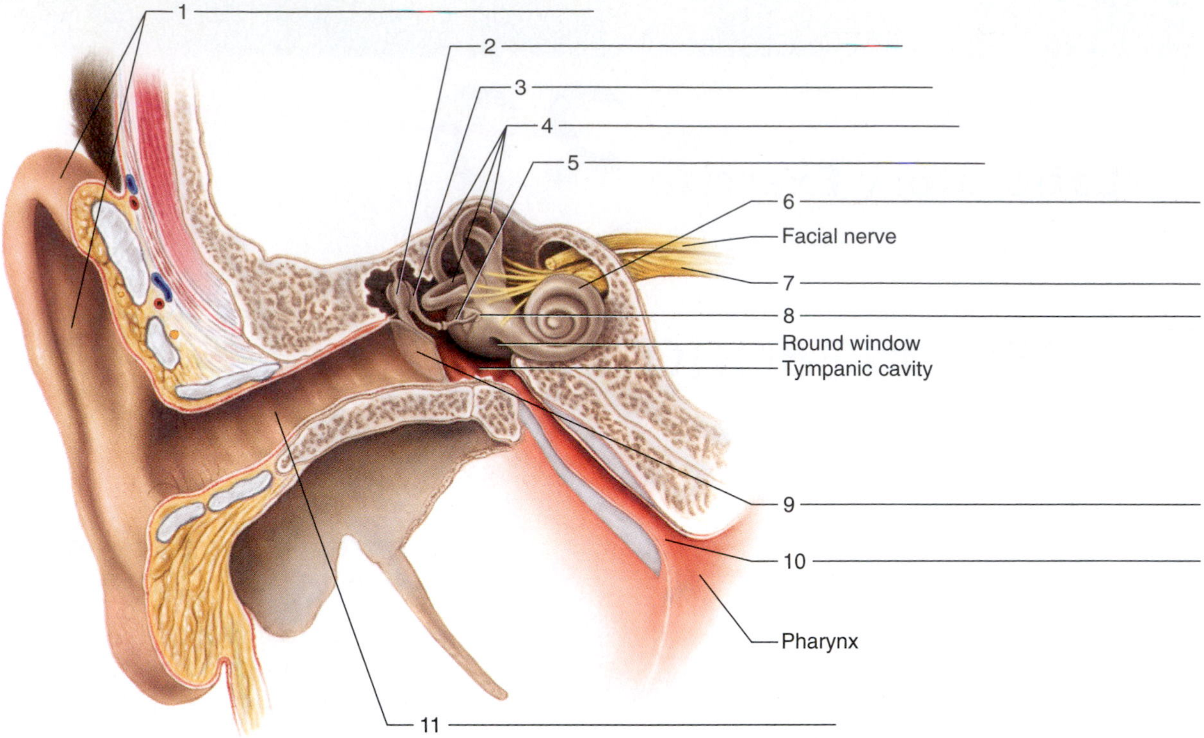

Figure 32.1 Label the major structures of the ear. 1

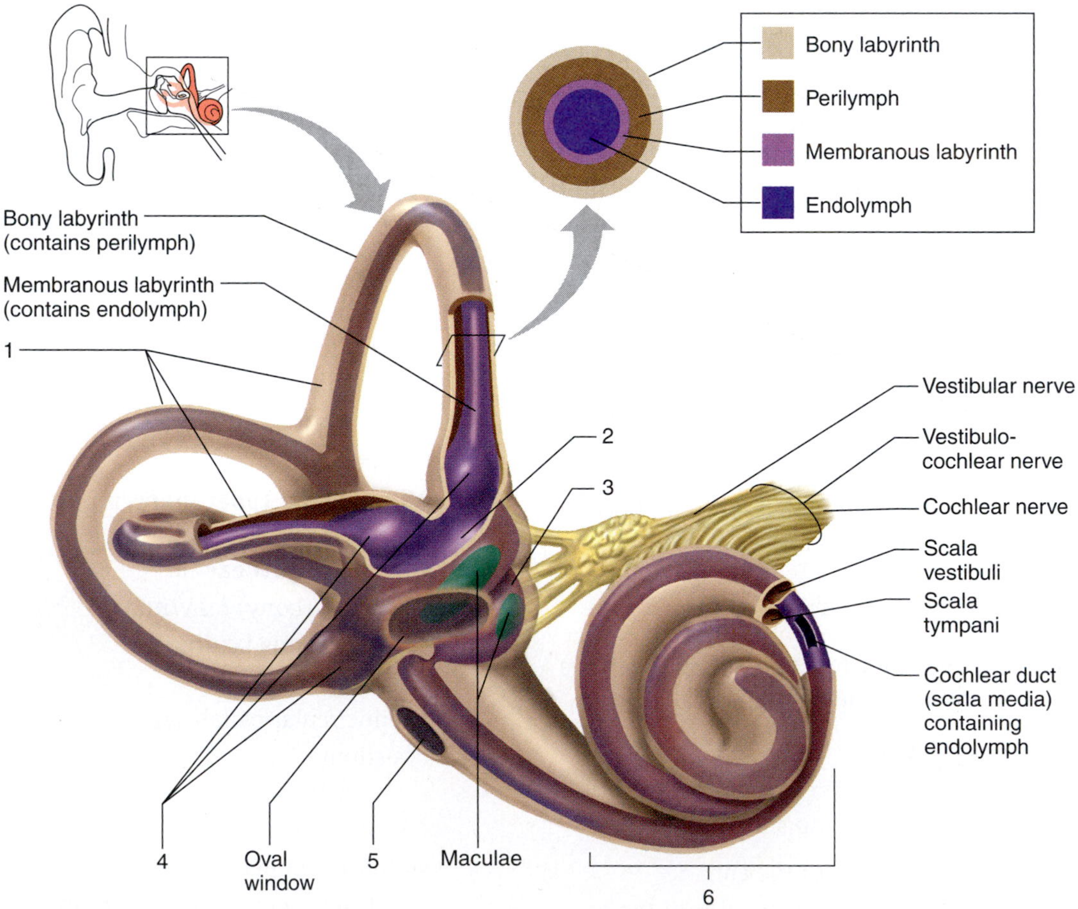

Figure 32.2 Label the structures of the inner ear by placing the correct numbers in the spaces provided. 1

___ Ampullae ___ Round window ___ Semicircular canals

___ Cochlea ___ Saccule ___ Utricle

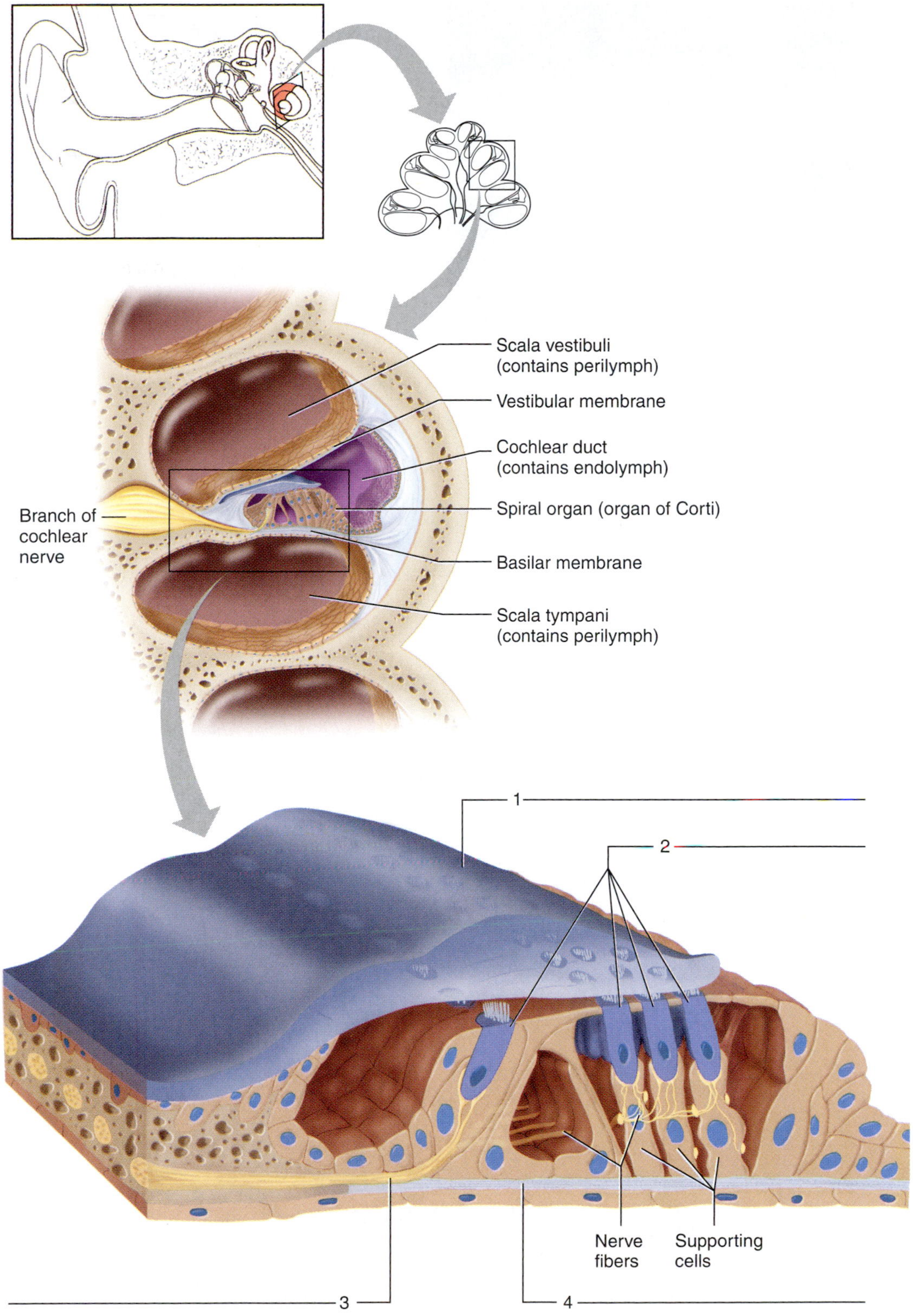

Figure 32.3 Label the structures associated with the spiral organ (organ of Corti) of a cochlea.

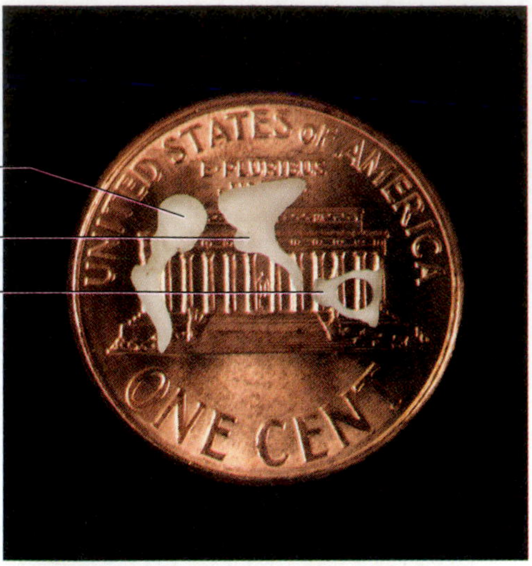

Figure 32.4 Middle ear bones (auditory ossicles) superimposed on a penny to show their relative size. The arrangement of the malleus, incus, and stapes in the enlarged photograph resembles the articulations in the middle ear.

Demonstration

Observe the section of the cochlea in the demonstration microscope. Locate one of the turns of the cochlea, and using figures 32.3 and 32.5 as a guide, identify the *scala vestibuli, cochlear duct (scala media), scala tympani, vestibular membrane, basilar membrane,* and the *spiral organ (organ of Corti).*

4. Complete Parts A and B of Laboratory Report 32.

EXPLORE

Procedure B—Hearing Tests

Perform the following tests in a quiet room, using your laboratory partner as the test subject.

1. *Auditory acuity test.* To conduct this test, follow these steps:
   a. Have the test subject sit with eyes closed.
   b. Pack one of the subject's ears with cotton.

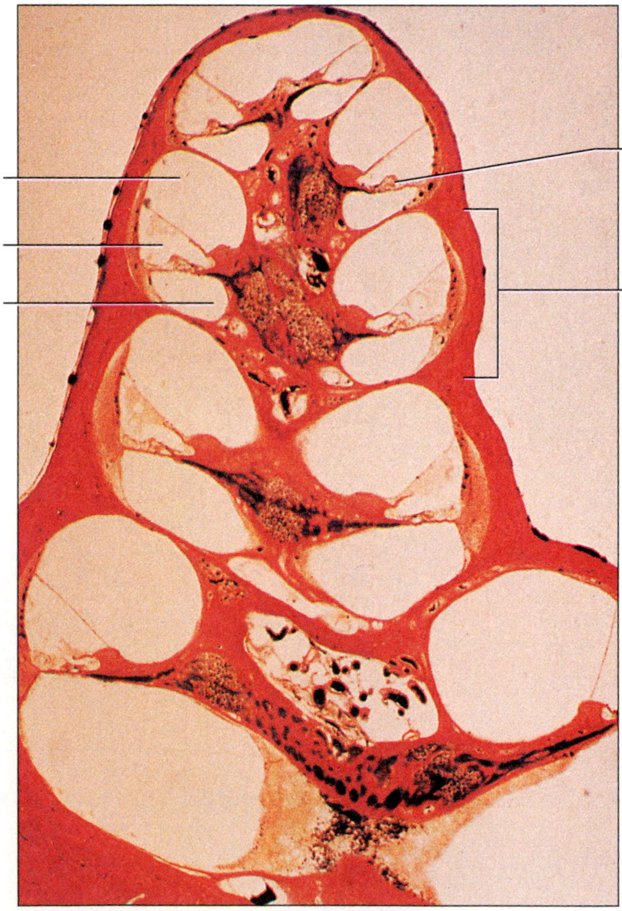

Figure 32.5 A section through the cochlea (22×).

c. Hold a ticking watch close to the open ear, and slowly move it straight out and away from the ear.
d. Have the subject indicate when the sound of the ticking can no longer be heard.
e. Use a meterstick to measure the distance in centimeters from the ear to the position of the watch.
f. Repeat this procedure to test the acuity of the other ear.
g. Record the test results in Part C of the laboratory report.

2. *Sound localization test.* To conduct this test, follow these steps:
   a. Have the subject sit with eyes closed.
   b. Hold the ticking watch somewhere within the audible range of the subject's ears, and ask the subject to point to the watch.
   c. Move the watch to another position and repeat the request. In this manner, determine how accurately the subject can locate the watch when it is in each of the following positions: in front of the head, behind the head, above the head, on the right side of the head, and on the left side of the head.
   d. Record the test results in Part C of the laboratory report.

3. *Rinne test.* This test is done to assess possible conduction deafness by comparing bone and air conduction. To conduct this test, follow these steps:
   a. Obtain a tuning fork and strike it with a rubber hammer or on the heel of your hand, causing it to vibrate.
   b. Place the end of the fork's handle against the subject's mastoid process behind one ear. Have the prongs of the fork pointed downward and away from the ear, and be sure nothing is touching them (fig. 32.6*a*). The sound sensation is that of bone conduction. If no sound is experienced, nerve deafness exists.
   c. Ask the subject to indicate when the sound is no longer heard.
   d. Then quickly remove the fork from the mastoid process and position it in the air close to the opening of the nearby external acoustic meatus (fig. 32.6*b*).

   If hearing is normal, the sound (from air conduction) will be heard again; if there is conductive impairment, the sound will not be heard. Conductive impairment involves outer or middle ear defects. Hearing aids can improve hearing for conductive deafness because bone conduction

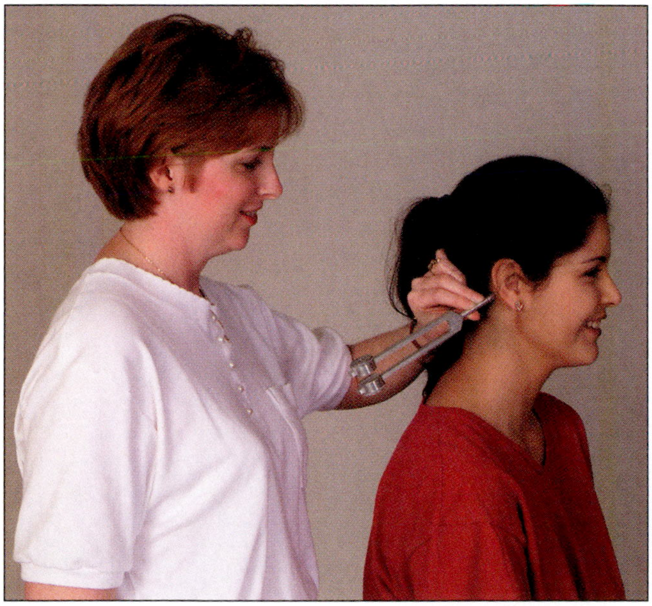
(a)

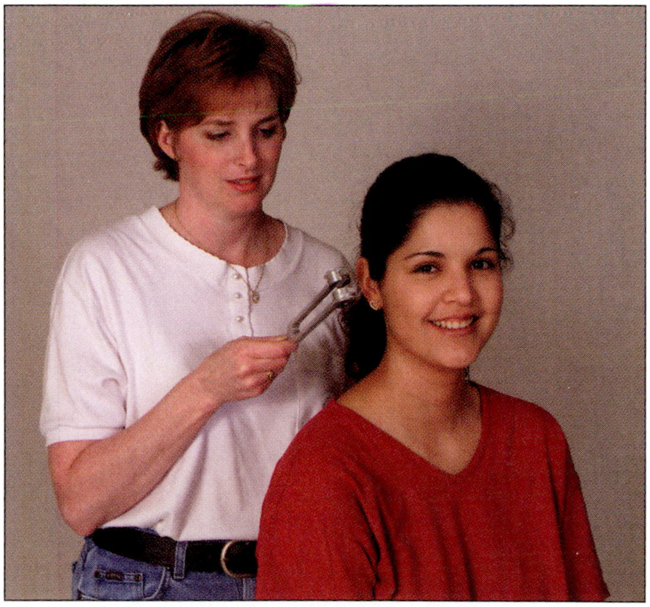
(b)

Figure 32.6 Rinne test: (*a*) first placement of vibrating tuning fork until sound is no longer heard; (*b*) second placement of tuning fork to assess air conduction.

transmits the sound into the inner ear. Surgery could possibly correct this type of defect.

e. Record the test results in Part C of the laboratory report.

4. *Weber test.* This test is used to distinguish possible conduction or sensory deafness. To conduct this test, follow these steps:

a. Strike the tuning fork with the rubber hammer.

b. Place the handle of the fork against the subject's forehead in the midline (fig. 32.7).

c. Ask the subject to indicate if the sound is louder in one ear than in the other or if it is equally loud in both ears.

If hearing is normal, the sound will be equally loud in both ears. If there is conductive impairment, the sound will appear louder in the affected ear. If some degree of sensory (nerve) deafness exists, the sound will be louder in the normal ear. The impairment involves the spiral organ (organ of Corti) or the cochlear nerve. Hearing aids will not improve sensory deafness.

d. Have the subject experience the effects of conductive impairment by packing one ear with cotton and repeating the Weber test. Usually the sound appears louder in the plugged (or impaired) ear because extraneous sounds from the room are blocked out.

e. Record the test results in Part C of the laboratory report.

5. Complete Part C of the laboratory report.

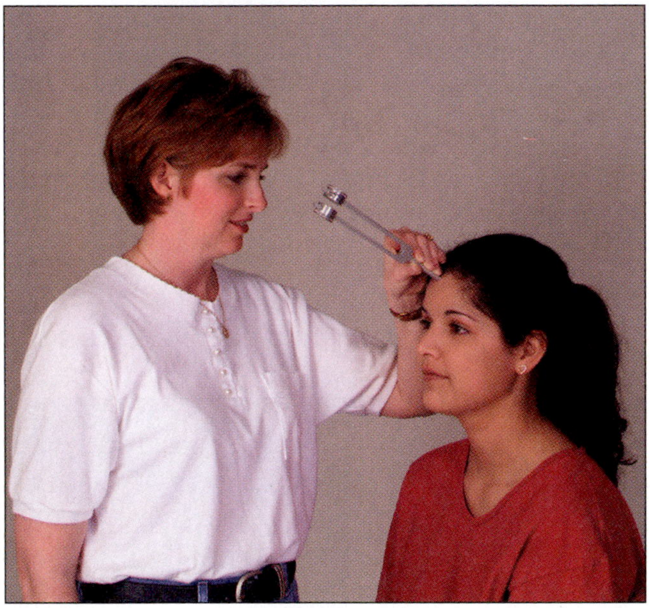

Figure 32.7 Weber test.

Critical Thinking Application

Ear structures from the outer ear into the inner ear are progressively smaller. Using results obtained from the hearing tests, explain the advantage of this arrangement. 3

____

____

____

____

Demonstration

Ask the laboratory instructor to demonstrate the use of the audiometer. This instrument produces sound vibrations of known frequencies transmitted to one or both ears of a test subject through earphones. The audiometer can be used to determine the threshold of hearing for different sound frequencies, and, in the case of hearing impairment, it can be used to determine the percentage of hearing loss for each frequency.

Name ______________________

Date ______________________

Section ______________________

The [A] corresponds to the indicated Learning Outcome(s) listed at the beginning of the laboratory exercise.

Ear and Hearing

Laboratory Report 32

Part A Assessments

Match the terms in column A with the descriptions in column B. Place the letter of your choice in the space provided. [1] [2]

Column A

- **a.** Auditory tube
- **b.** Ceruminous gland
- **c.** External acoustic meatus
- **d.** Malleus
- **e.** Membranous labyrinth
- **f.** Osseous (bony) labyrinth
- **g.** Scala tympani
- **h.** Scala vestibuli
- **i.** Stapes
- **j.** Tectorial membrane
- **k.** Tympanic cavity
- **l.** Tympanic membrane (eardrum)

Column B

_______ **1.** Auditory ossicle attached to tympanic membrane

_______ **2.** Air-filled space containing auditory ossicles

_______ **3.** Contacts hairs of hearing receptors

_______ **4.** Leads from oval window to apex of cochlea

_______ **5.** S-shaped tube leading to tympanic membrane

_______ **6.** Wax-secreting structure

_______ **7.** Cone-shaped, semitransparent membrane attached to malleus

_______ **8.** Auditory ossicle attached to oval window

_______ **9.** Contains endolymph

_______ **10.** Bony canal of inner ear in temporal bone

_______ **11.** Connects middle ear and pharynx

_______ **12.** Extends from apex of cochlea to round window

Part B Assessments

Label the structures indicated in the micrograph of the spiral organ (organ of Corti) in figure 32.8. [1]

Terms:
Basilar membrane
Cochlear duct
Hair cells
Scala tympani
Tectorial membrane

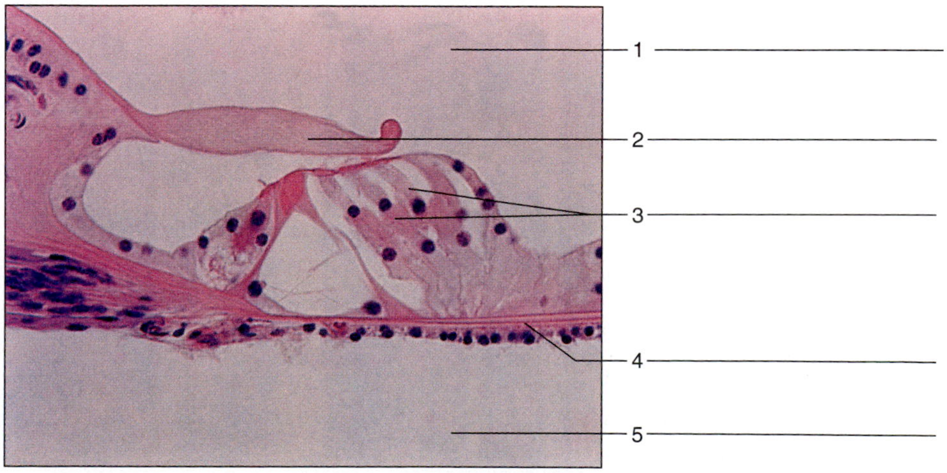

Figure 32.8 Label the structures associated with this spiral organ (organ of Corti) region of a cochlea, using the terms provided (75x micrograph enlarged to 300×).

Part C Assessments

1. Results of auditory acuity test: 4

Ear Tested	Audible Distance (cm)
Right	
Left	

2. Results of sound localization test: 4

Actual Location	Reported Location
Front of the head	
Behind the head	
Above the head	
Right side of the head	
Left side of the head	

3. Results of experiments using tuning forks: 4

Test	Left Ear (normal or impaired)	Right Ear (normal or impaired)
Rinne		
Weber		

4. Summarize the results of the hearing tests you conducted on your laboratory partner. 4

____

____

____

____

Laboratory Exercise 33

Equilibrium

Materials Needed

Textbook
Swivel chair
Bright light

For Demonstration:

Compound light microscope
Prepared microscope slide of semicircular canal (cross section through ampulla)

Safety

- Do not pick subjects who have frequent motion sickness.
- Have four people surround the subject in the swivel chair in case the person falls from vertigo or loss of balance.
- Stop your experiment if the subject becomes nauseated.

The sense of equilibrium involves two sets of sensory organs. One set helps to maintain the stability of the head and body when they are motionless or during linear acceleration and produces a sense of static (gravitational) equilibrium. The other set is concerned with balancing the head and body when they are moved suddenly or during angular acceleration and produces a sense of dynamic (rotational) equilibrium.

The organs associated with the sense of static equilibrium are located within the utricle and saccule of the vestibules of the inner ears. Those associated with the sense of dynamic equilibrium are found within the ampullae of the semicircular canals of the inner ear.

Purpose of the Exercise

To review the structure and function of the organs of equilibrium and to conduct some tests of equilibrium.

LEARNING OUTCOMES

After completing this exercise, you should be able to

1. Locate the organs of static and dynamic equilibrium and describe their functions.
2. Explain how vision contributes in the maintenance of equilibrium.
3. Conduct and record the results of the Romberg and Bárány tests of equilibrium.

EXPLORE

Procedure A—Organs of Equilibrium

1. Review the section entitled "Sense of Equilibrium" in chapter 12 of the textbook.
2. Complete Part A of Laboratory Report 33.

Demonstration

Observe the cross section of the semicircular canal through the ampulla in the demonstration microscope. Note the crista ampullaris (figs. 33.1 and 33.2) projecting into the lumen of the membranous labyrinth, which in a living person is filled with endolymph. The space between the membranous and osseous (bony) labyrinths is filled with perilymph.

EXPLORE

Procedure B—Tests of Equilibrium

Perform the following tests, using a person as a test subject who is not easily disturbed by dizziness or rotational movement. Also have some other students standing close by to help prevent the test subject from falling during the tests. *The tests should be stopped*

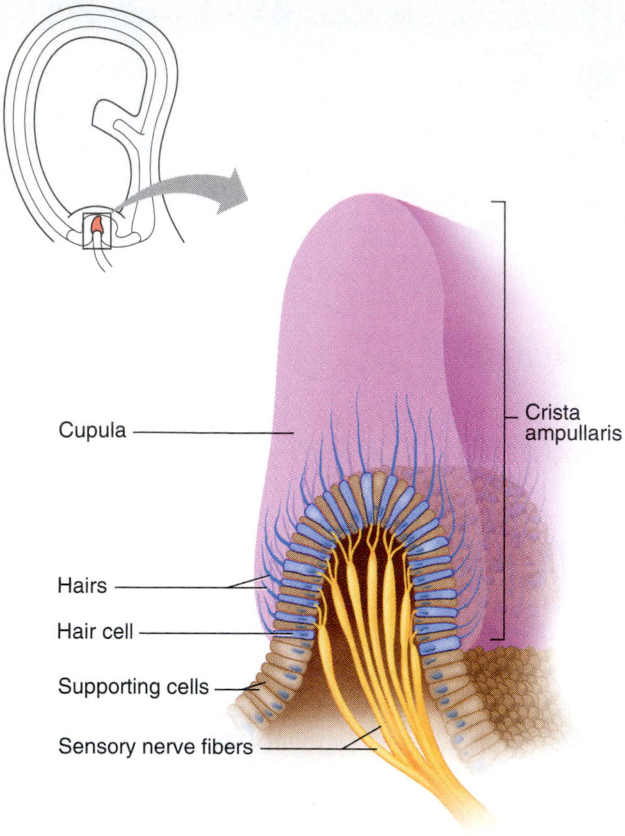

Figure 33.1 The crista ampullaris is located within the ampulla of each semicircular canal.

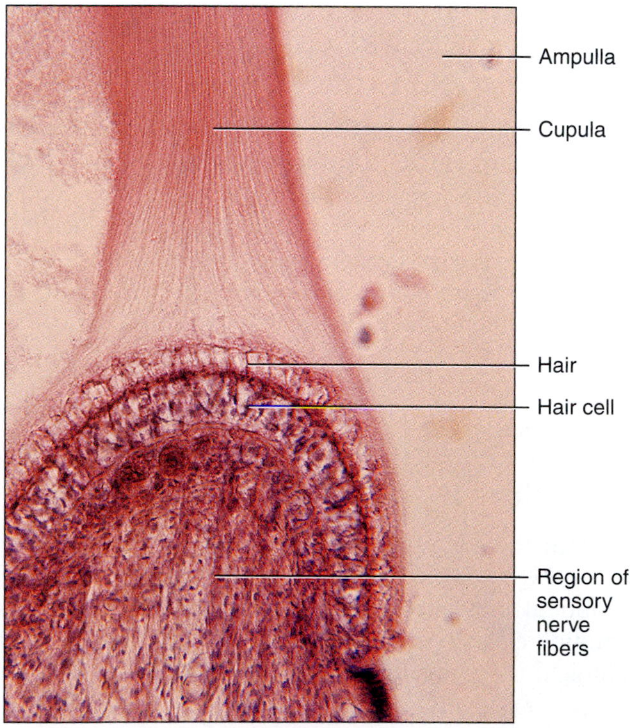

Figure 33.2 A micrograph of a crista ampullaris (1,400x).

immediately if the test subject begins to feel uncomfortable or nauseated.

1. *Vision and equilibrium.* To demonstrate the importance of vision in the maintenance of equilibrium, follow these steps:
   - **a.** Have the test subject stand erect on one foot for 1 minute with his or her eyes open.
   - **b.** Observe the subject's degree of unsteadiness.
   - **c.** Repeat the procedure with the subject's eyes closed. *Be prepared to prevent the subject from falling.*
2. *Romberg test.* To conduct this test, follow these steps:
   - **a.** Position the test subject close to a chalkboard with the back toward the board.
   - **b.** Place a bright light in front of the subject so that a shadow of the body is cast on the board.
   - **c.** Have the subject stand erect with feet close together and eyes staring straight ahead for a period of 3 minutes.
   - **d.** During the test, make marks on the chalkboard along the edge of the shadow of the subject's shoulders to indicate the range of side-to-side swaying.
   - **e.** Measure the maximum sway in centimeters and record the results in Part B of the laboratory report.
   - **f.** Repeat the procedure with the subject's eyes closed.
   - **g.** Position the subject so one side is toward the chalkboard.
   - **h.** Repeat the procedure with the eyes open.
   - **i.** Repeat the procedure with the eyes closed.

   The Romberg test is used to evaluate a person's ability to integrate sensory information from proprioceptors and receptors within the organs of static equilibrium and to relay appropriate motor impulses to postural muscles. A person who shows little unsteadiness when standing with feet together and eyes open, but who becomes unsteady when the eyes are closed, has a positive Romberg test.
3. *Bárány test.* To conduct this test, follow these steps:
   - **a.** Have the test subject sit on a swivel chair with his or her eyes open and focused on a distant object, the head tilted forward about 30°, and the hands gripped firmly to the seat. Position four people around the chair for safety. *Be prepared to prevent the subject and the chair from tipping over.*
   - **b.** Rotate the chair ten rotations within 20 seconds.
   - **c.** Abruptly stop the movement of the chair. The subject will still have the sensation of continuous movement and might experience some dizziness (vertigo).

d. Have the subject look forward and immediately note the nature of the eye movements and their direction. (Such reflex eye twitching movements are called *nystagmus.*) Also note the time it takes for the nystagmus to cease. Nystagmus will continue until the cupula is returned to an original position.

e. Record your observations in Part B of the laboratory report.

f. Allow the subject several minutes of rest, then repeat the procedure with the subject's head tilted nearly 90° onto one shoulder.

g. After another rest period, repeat the procedure with the subject's head bent forward so that the chin is resting on the chest.

In this test, when the head is tilted about 30°, the lateral semicircular canals receive maximal stimulation, and the nystagmus is normally from side to side. When the head is tilted at 90°, the superior canals are stimulated, and the nystagmus is up and down. When the head is bent forward with the chin on the chest, the posterior canals are stimulated, and the nystagmus is rotary. Observe a semicircular canal located in a temporal bone (fig. 33.3).

4. Complete Part B of the laboratory report.

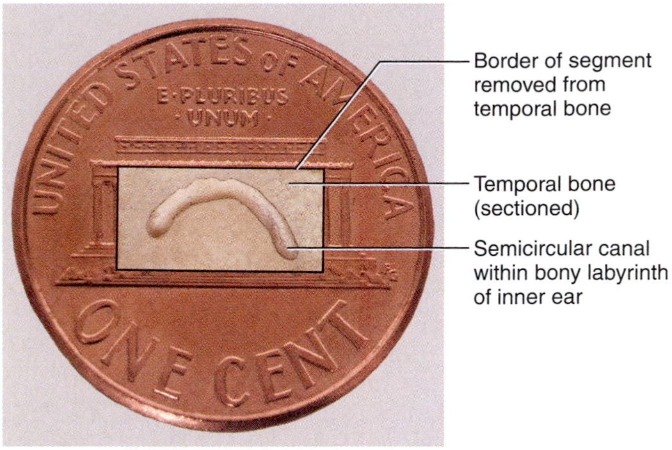

Figure 33.3 Semicircular canal superimposed on a penny to show its relative size.

Notes

Name ____________________

Date ____________________

Section ____________________

The [A] corresponds to the indicated Learning Outcome(s) listed at the beginning of the laboratory exercise.

33 Laboratory Report

Equilibrium

Part A Assessments

Complete the following statements:

1. The organs of static equilibrium are located within two expanded chambers called the ____________ and the saccule. [1]
2. All of the balance organs are located within the ____________ bone of the skull. [1]
3. The receptor cells of these organs are found in the wall of the membranous labyrinth in a structure called the ____________ . [1]
4. Otoliths are small grains composed of ____________ . [1]
5. Sensory impulses travel from the organs of equilibrium to the brain on the ____________ nerve. [1]
6. The sensory organ of a semicircular canal lies within a swelling of the canal called the ____________ . [1]
7. The sensory organ within the ampulla of a semicircular canal is called a ____________ . [1]
8. The ____________ of this sensory organ consists of a dome-shaped gelatinous mass. [1]
9. When the head is moved, the fluid inside the membranous portion of a semicircular canal tends to remain stationary because of the ____________ of the fluid. [1]
10. Parts of the ____________ of the brain interpret impulses from the equilibrium receptors. [1]

Part B—Tests of Equilibrium Assessments

1. Vision and equilibrium test results:
   a. When the eyes are open, what sensory organs provide information needed to maintain equilibrium? [2]

   ____

   ____

   b. When the eyes are closed, what sensory organs provide such information? [2] ____________

   ____

   ____

2. Romberg test results:

a. Record the test results in the following table. 3

Conditions	Maximal Movement (cm)
Back toward board, eyes open	
Back toward board, eyes closed	
Side toward board, eyes open	
Side toward board, eyes closed	

b. Did the test subject's unsteadiness increase when the eyes were closed? __________ What is the significance of this observation? 2 __________

c. Why would you expect a person with impairment of the organs of equilibrium to become more unsteady when the eyes are closed? 2 __________

3. Bárány test results:

a. Record the test results in the following table. 3

Position of Head	Description of Eye Movements	Time for Movement to Cease
Tilted 30° forward		
Tilted 90° onto shoulder		
Tilted forward, chin on chest		

b. Summarize the results of this test. 3 __________

Critical Thinking Application

What additional sensory information would you expect a person with impairment of organs of equilibrium to use to supplement their relative lack of some sensory information?

____

Laboratory Exercise 34

Eye Structure

Materials Needed

Textbook
Dissectible eye model
Compound light microscope
Prepared microscope slide of a mammalian eye (sagittal section)
Sheep or beef eye (fresh or preserved)
Dissecting tray
Dissecting instruments—forceps, sharp scissors, and dissecting needle

For Learning Extension:
Ophthalmoscope

Safety

- Wear disposable gloves when working on the eye dissection.
- Dispose of tissue remnants and gloves as instructed.
- Wash the dissecting tray and instruments as instructed.
- Wash your laboratory table.
- Wash your hands before leaving the laboratory.

The eye contains photoreceptors, modified neurons located on its inner wall. Other parts of the eye provide protective functions or make it possible to move the eyeball. Still other structures serve to focus light entering the eye so that a sharp image is projected onto the receptor cells. Nerve impulses generated when the receptors are stimulated travel along the optic nerves to the brain, which interprets the impulses and creates the sensation of sight.

Purpose of the Exercise

To review the structure and function of the eye and to dissect a mammalian eye.

LEARNING OUTCOMES

After completing this exercise, you should be able to

1. Locate the major structures of an eye.
2. Describe the functions of the structures of an eye.
3. Trace the structures through which light passes as it travels from the cornea to the retina.
4. Dissect a mammalian eye and locate its major features.

EXPLORE

Procedure A—Structure and Function of the Eye

1. Review the sections entitled "Visual Accessory Organs," "Structure of the Eye," and "Visual Receptors" in chapter 12 of the textbook.
2. As a review activity, label figures 34.1, 34.2, and 34.3.
3. Complete Part A of Laboratory Report 34.
4. Examine the dissectible model of the eye and locate the following features:

eyelid
conjunctiva
orbicularis oculi
levator palpebrae superioris
lacrimal apparatus
- lacrimal gland
- canaliculi
- lacrimal sac
- nasolacrimal duct

extrinsic muscles
- superior rectus
- inferior rectus
- medial rectus

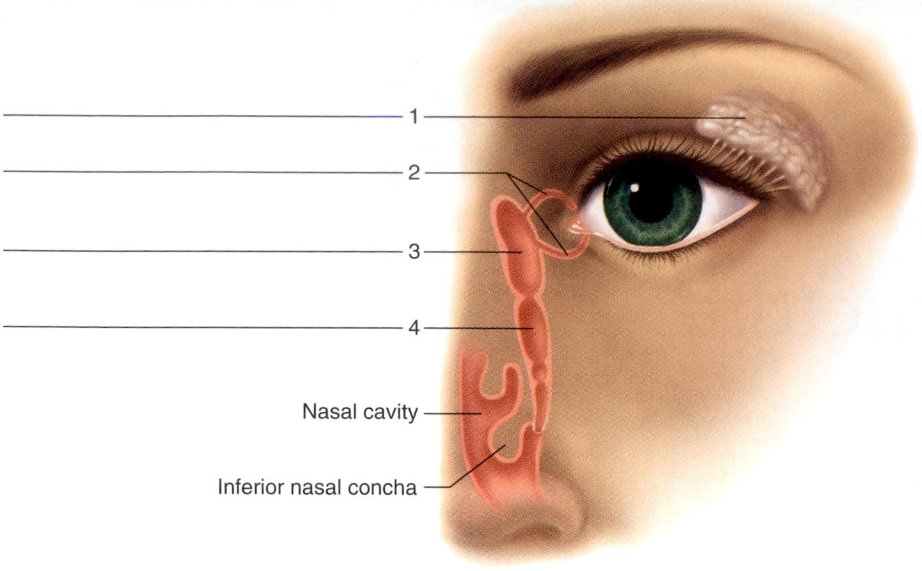

Figure 34.1 Label the structures of the lacrimal apparatus.

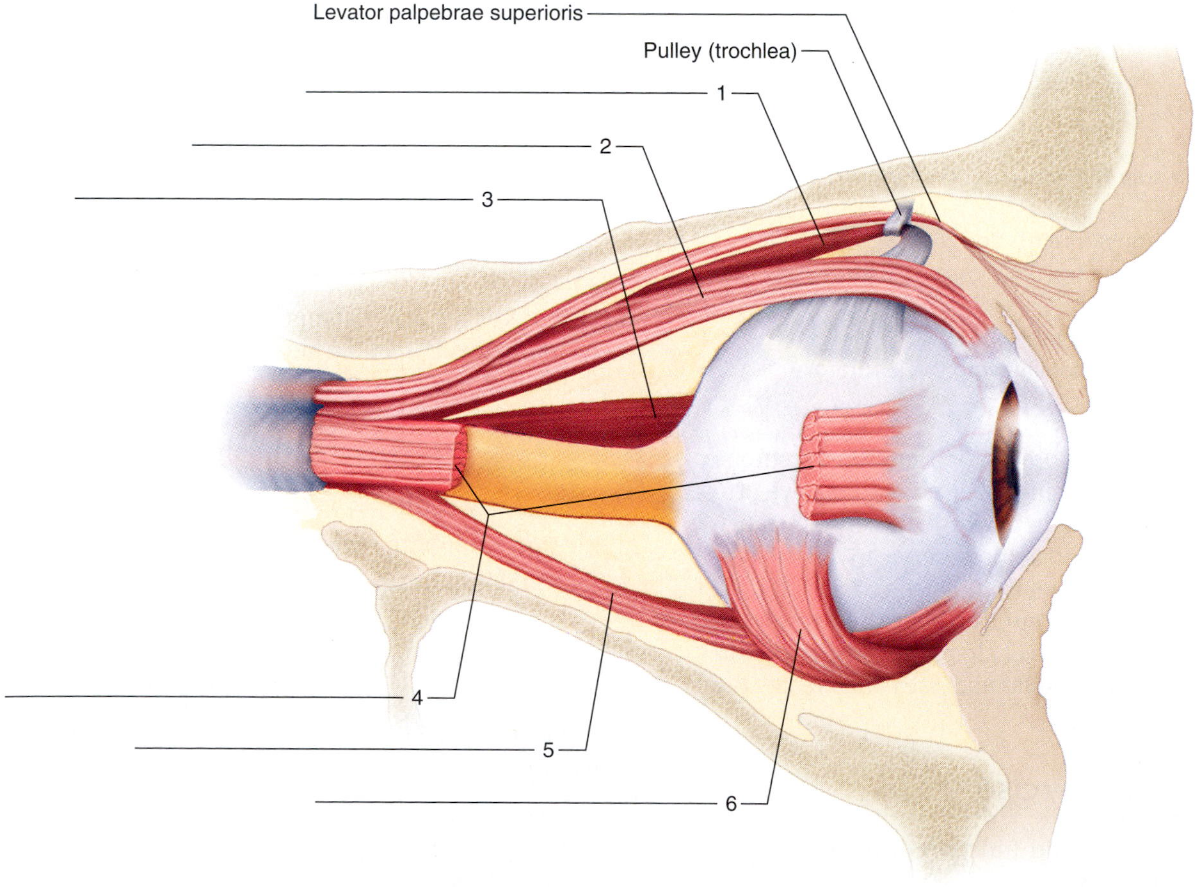

Figure 34.2 Label the extrinsic muscles of the right eye (lateral view).

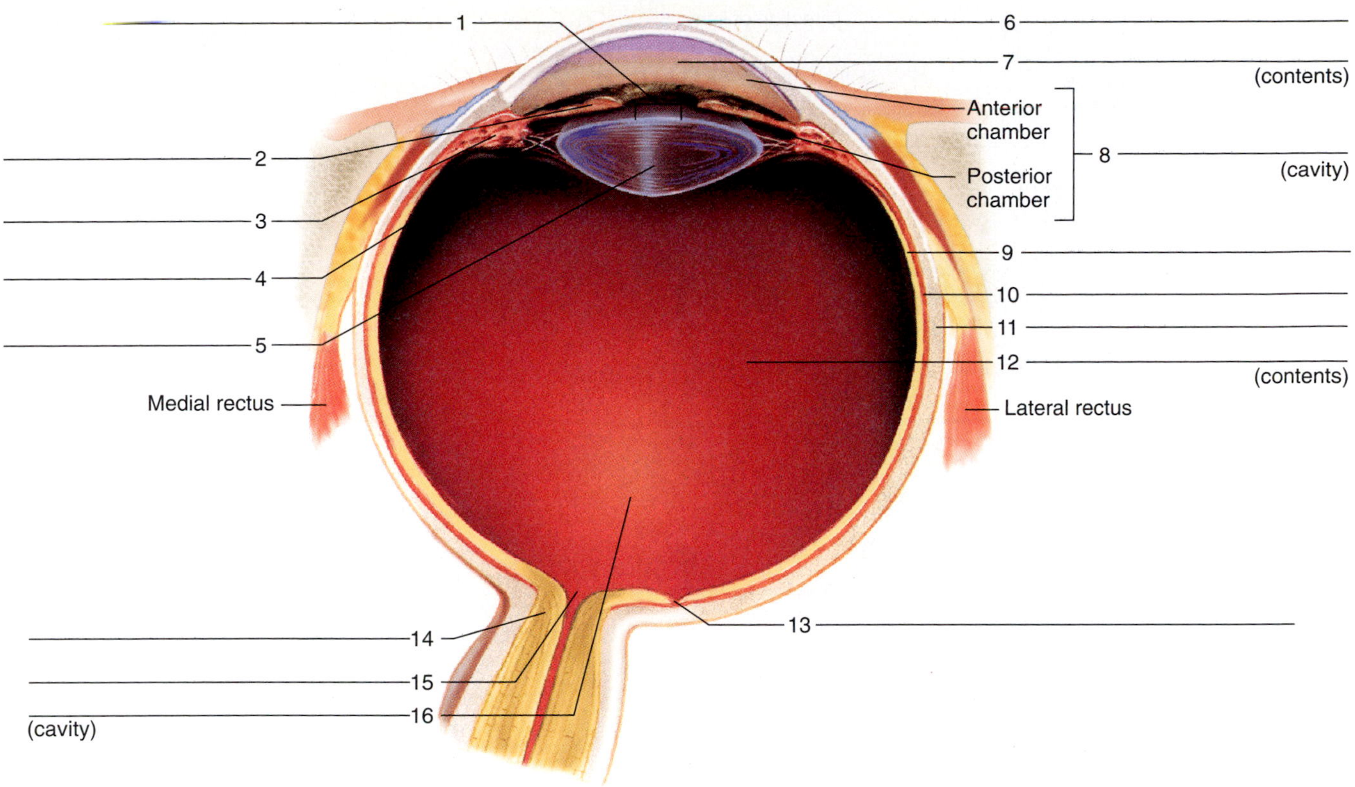

Figure 34.3 Label the structures indicated in this transverse section of the right eye (superior view). 1

lateral rectus
superior oblique
inferior oblique

trochlea (pulley)

outer (fibrous) tunic (layer)
- sclera
- cornea

middle (vascular) tunic (layer)
- choroid coat
- ciliary body
  - ciliary processes
  - ciliary muscles
  - suspensory ligaments
- iris
- pupil (opening in center of iris)

inner (nervous) tunic (layer)
- retina
  - macula lutea
  - fovea centralis
  - optic disc
- optic nerve

lens

anterior cavity
- anterior chamber
- posterior chamber
- aqueous humor

posterior cavity
- vitreous humor

5. Obtain a microscope slide of a mammalian eye section, and locate as many of the features in step 4 as possible.
6. Using high-power magnification, observe the posterior portion of the eye wall, and locate the sclera, choroid coat, and retina.
7. Using high-power magnification, examine the retina, and note its layered structure (fig. 34.4). Locate the following:

nerve fibers leading to the optic nerve (innermost layer of the retina)
layer of ganglion cells
layer of bipolar neurons
nuclei of rods and cones
receptor ends of rods and cones
pigmented epithelium (outermost layer of the retina)

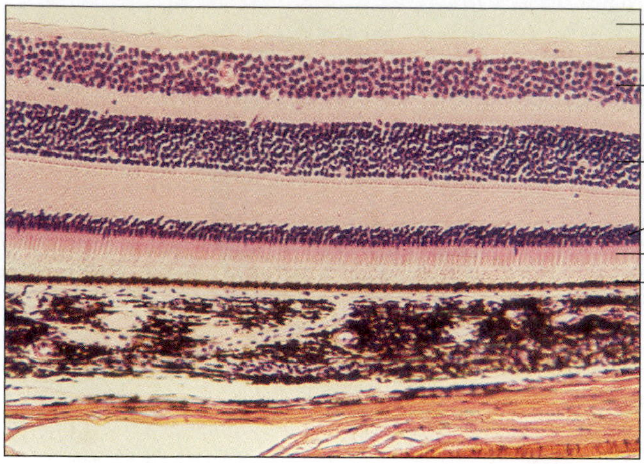

Figure 34.4 The cells of the retina are arranged in distinct layers (75×).

Learning Extension

Use an ophthalmoscope to examine the interior of your laboratory partner's eye. This instrument consists of a set of lenses held in a rotating disc, a light source, and some mirrors that reflect the light into the test subject's eye.

The examination should be conducted in a dimly lit room. Have your partner seated and staring straight ahead at eye level. Move the rotating disc of the ophthalmoscope so that the *O* appears in the lens selection window. Hold the instrument in your right hand with the end of your index finger on the rotating disc (fig. 34.5). Direct the light at a slight angle from a distance of about 15 cm into the subject's right eye. The light beam should pass along the inner edge of the pupil. Look through the instrument, and you should see a reddish, circular area—the interior of the eye. Rotate the disc of lenses to higher values until sharp focus is achieved.

Move the ophthalmoscope to within about 5 cm of the eye being examined *being very careful that the instrument does not touch the eye,* and again rotate the lenses to sharpen the focus (fig. 34.6). Locate the optic disc and the blood vessels that pass through it. Also locate the yellowish macula lutea by having your partner stare directly into the light of the instrument (fig. 34.7).

Examine the subject's iris by viewing it from the side and by using a lens with a +15 or +20 value.

Figure 34.5 The ophthalmoscope is used to examine the interior of the eye.

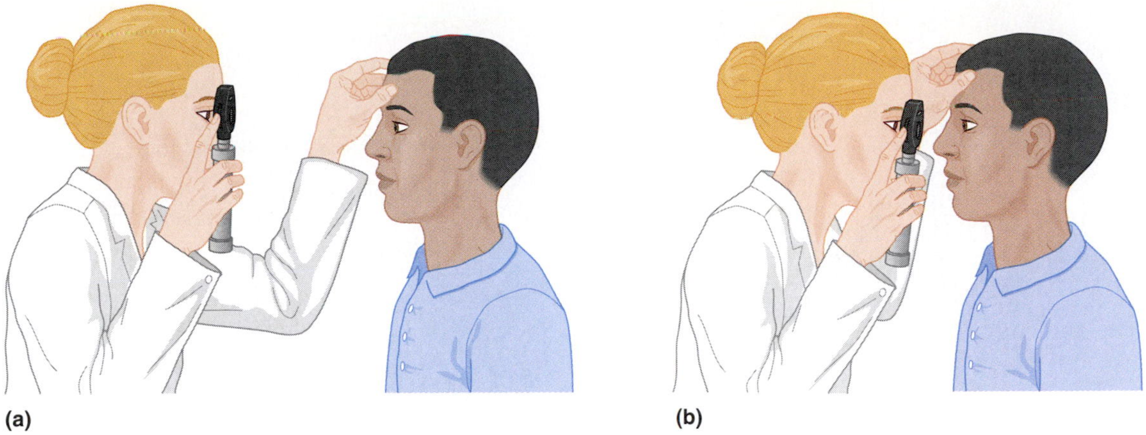

Figure 34.6 (*a*) Rotate the disc of lenses until sharp focus is achieved. (*b*) Move the ophthalmoscope to within 5 cm of the eye to examine the optic disc.

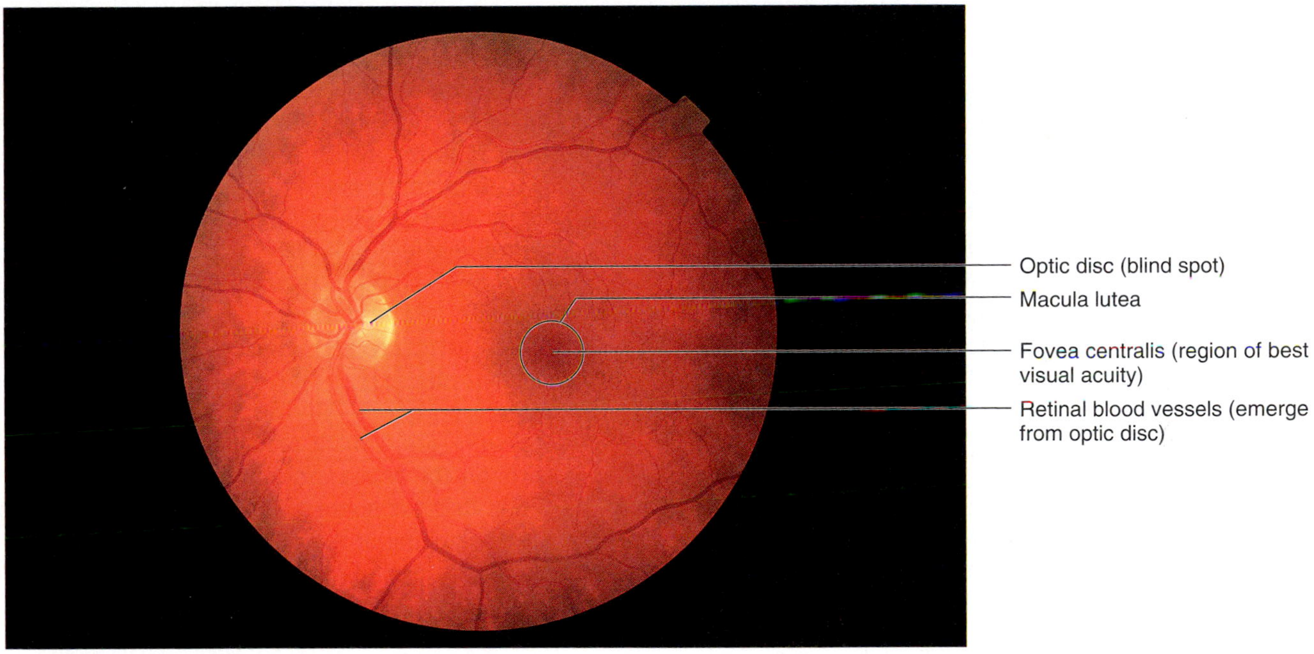

Figure 34.7 Retina of the eye as seen through the pupil using an ophthalmoscope.

EXPLORE

Procedure B—Eye Dissection

1. Obtain a mammalian eye, place it in a dissecting tray, and dissect it as follows:
   a. Trim away the fat and other connective tissues but leave the stubs of the *extrinsic muscles* and of the *optic nerve.* This nerve projects outward from the posterior region of the eyeball.
   b. The *conjunctiva* lines the inside of the eyelid and is reflected over the anterior surface of the eye, except for the cornea. Lift some of this thin membrane away from the eye with forceps and examine it.
   c. Locate and observe the *cornea, sclera,* and *iris.* Also note the *pupil* and its shape. The cornea from a fresh eye will be transparent; when preserved, it becomes opaque.

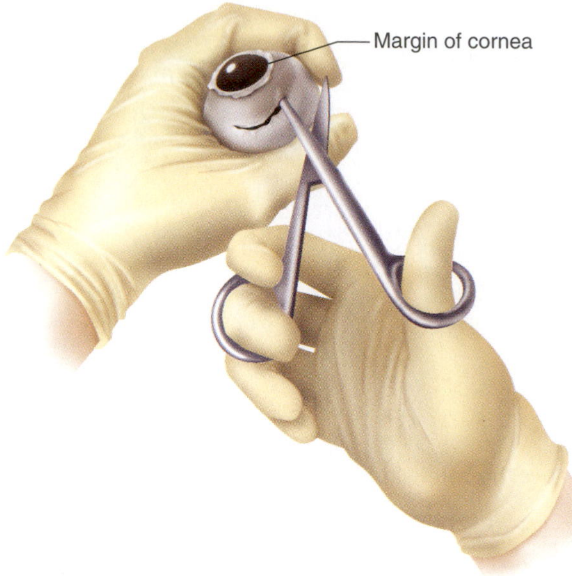

Figure 34.8 Prepare a frontal section of the eye.

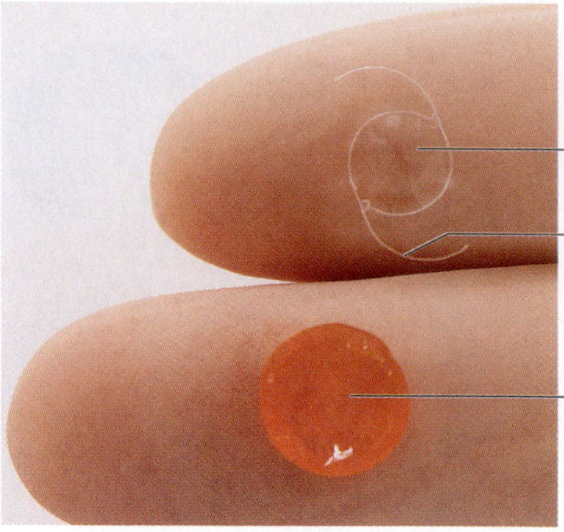

Figure 34.9 Human lens showing a cataract and one type of lens implant (intraocular lens replacement) on tips of fingers to show their relative size. A normal lens is transparent.

d. Use sharp scissors to make a frontal section of the eye. To do this, cut through the wall about 1 cm from the margin of the cornea and continue all the way around the eyeball. Try not to damage the internal structures of the eye (fig. 34.8).

e. Gently separate the eyeball into anterior and posterior portions. Usually the jellylike vitreous humor will remain in the posterior portion, and the lens may adhere to it. Place the parts in the dissecting tray with their contents facing upward.

f. Examine the anterior portion of the eye, and locate the *ciliary body,* which appears as a dark, circular structure. Also note the *iris* and the *lens* if it remained in the anterior portion. The lens is normally attached to the ciliary body by many *suspensory ligaments,* which appear as delicate, transparent threads.

g. Use a dissecting needle to gently remove the lens, and examine it. If the lens is still transparent, hold it up and look through it at something in the distance and note that the lens inverts the image. The lens of a preserved eye is usually too opaque for this experience. If the lens of the human eye becomes opaque, the defect is called a cataract (fig. 34.9).

h. Examine the posterior portion of the eye. Note the *vitreous humor.* This jellylike mass helps to hold the lens in place anteriorly and helps to hold the *retina* against the choroid coat.

i. Carefully remove the vitreous humor and examine the retina. This layer will appear as a thin, nearly colorless to cream-colored membrane that detaches easily from the choroid coat. Compare the structures identified to figure 34.10.

j. Locate the *optic disc*—the point where the retina is attached to the posterior wall of the eyeball and where the optic nerve originates. There are no receptor cells in the optic disc, so this region is also called the "blind spot."

k. Note the iridescent area of the choroid coat beneath the retina. This colored surface in ungulates (mammals having hoofs) is called the *tapetum fibrosum.* It serves to reflect light back through the retina, an action thought to aid the night vision of some animals. The tapetum fibrosum is lacking in the human eye.

2. Discard the tissues of the eye as directed by the laboratory instructor.
3. Complete Parts B and C of the laboratory report.

Critical Thinking Application

A strong blow to the head might cause the retina to detach. From observations made during the eye dissection, explain why this could happen.

____

____

____

____

____

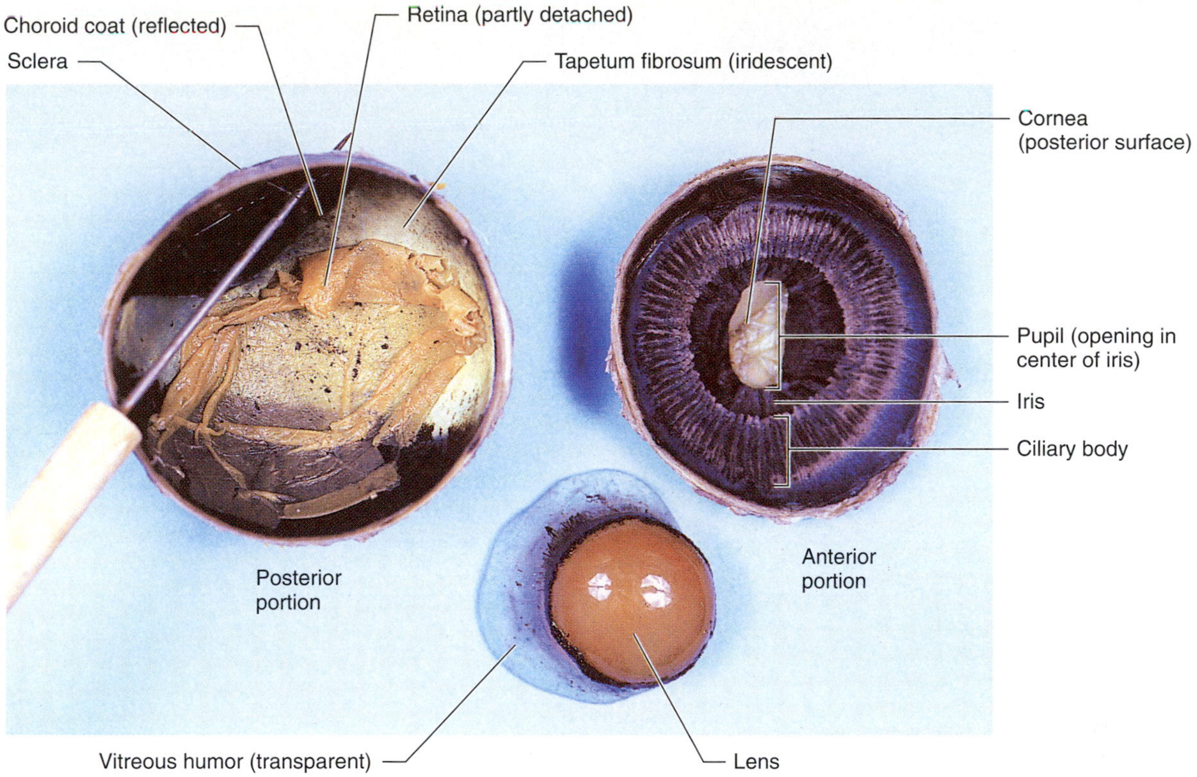

Figure 34.10 Internal structures of the beef eye dissection.

Notes

Name ______________________________

Date ______________________________

Section ______________________________

The [A] corresponds to the indicated Learning Outcome(s) listed at the beginning of the laboratory exercise.

Eye Structure

Laboratory Report 34

Part A Assessments

Match the terms in column A with the descriptions in column B. Place the letter of your choice in the space provided. [1] [2]

Column A		Column B
a. Aqueous humor	_____	1. Posterior five-sixths of middle (vascular) tunic
b. Choroid coat	_____	2. White part of outer (fibrous) tunic
c. Ciliary muscles	_____	3. Transparent anterior portion of outer tunic
d. Conjunctiva	_____	4. Inner lining of eyelid
e. Cornea	_____	5. Secretes tears
f. Iris	_____	6. Fills posterior cavity of eye
g. Lacrimal gland	_____	7. Area where optic nerve exits the eye
h. Optic disc	_____	8. Smooth muscle that controls light entering the eye
i. Retina	_____	9. Fills anterior and posterior chambers of the anterior cavity of the eye
j. Sclera	_____	10. Contains visual receptors called rods and cones
k. Suspensory ligament	_____	11. Connects lens to ciliary body
l. Vitreous humor	_____	12. Cause lens to change shape

Complete the following:

13. List the structures and fluids through which light passes as it travels from the cornea to the retina. [3] ______________________________

____

____

____

14. List three ways in which rods and cones differ in structure or function. [2] ______________________________

____

____

____

Part B Assessments

Complete the following:

1. Which tunic/layer of the eye was the most difficult to cut? 4 ______
2. What kind of tissue do you think is responsible for this quality of toughness? 4 ______
3. How do you compare the shape of the pupil in the dissected eye with your pupil? 4 ______
4. Where do you find aqueous humor in the dissected eye? 4 ______
5. What is the function of the dark pigment in the choroid coat? 2 ______
6. Describe the lens of the dissected eye. 4 ______
7. Describe the vitreous humor of the dissected eye. 4 ______

Part C Assessments

Identify the features of the eye indicated in figure 34.11. 1

a. Anterior portion of eye:

1. ______
2. ______
3. ______
4. ______
5. ______

b. Posterior portion of eye:

6. ______
7. ______
8. ______
9. ______
10. ______
11. ______

Terms:

Aqueous humor	Optic disc
Choroid coat/middle tunic	Optic nerve
Conjunctiva	Retina/inner tunic
Cornea	Sclera/outer tunic
Iris	Vitreous humor
Lens	

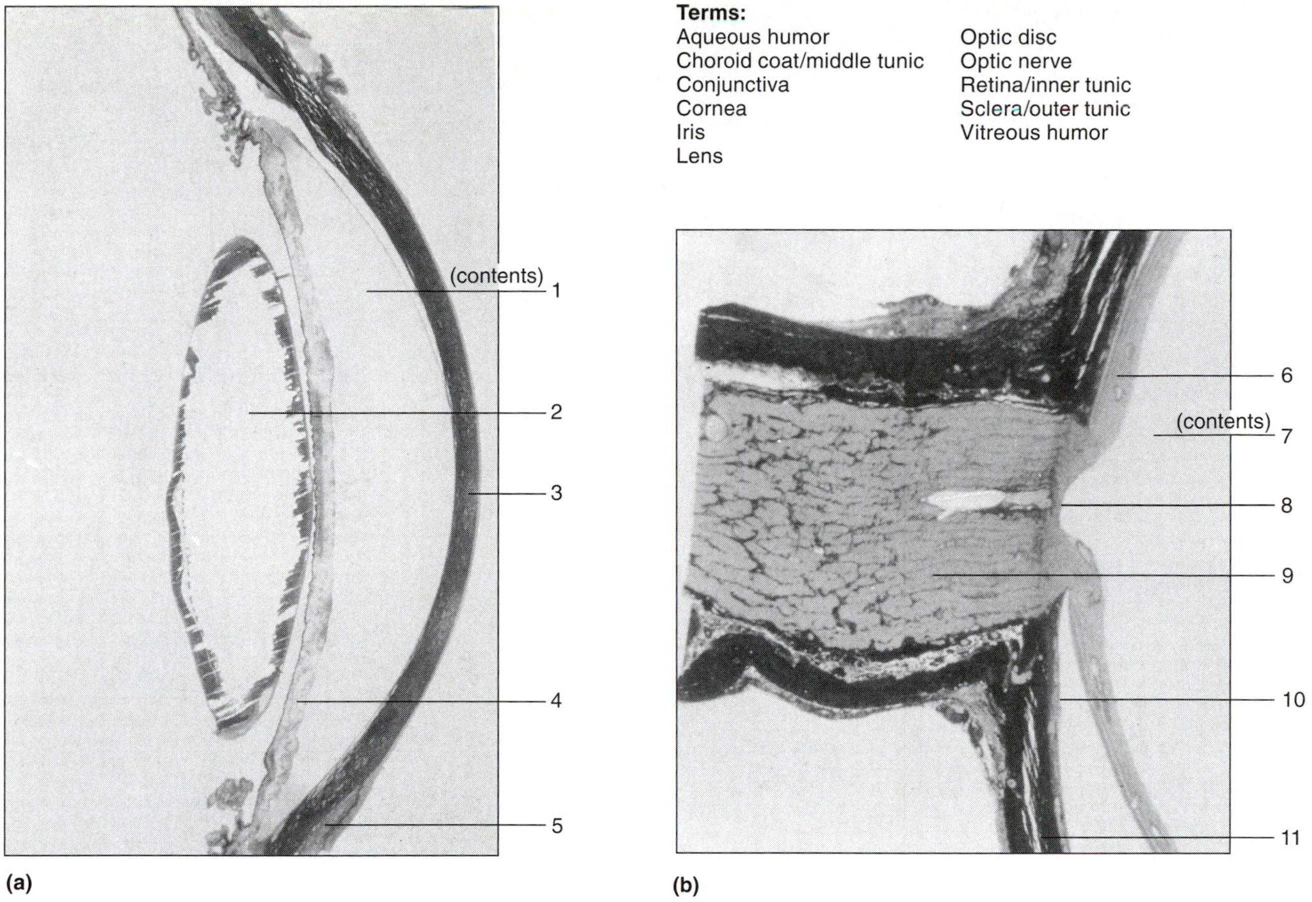

Figure 34.11 Sections of the eye: (*a*) anterior portion (10×); (*b*) posterior portion (53×). Identify the numbered features, using the terms provided.

Notes

Laboratory Exercise 35

Visual Tests and Demonstrations

Materials Needed

Snellen eye chart
3" × 5" card (plain)
3" × 5" card with word typed in center
Astigmatism chart
Meterstick
Metric ruler
Pen flashlight
Ichikawa's or Ishihara's color plates for colorblindness test

Normal vision (emmetropia) results when light rays from objects in the external environment are refracted by the cornea and lens of the eye and focused onto the photoreceptors of the retina. If a person has a condition of nearsightedness (myopia), the image focuses in front of the retina because the eyeball is too long; nearby objects are clear, but distant objects are blurred. If a person has a condition of farsightedness (hyperopia), the image focuses behind the retina because the eyeball is too short; distant objects are clear, but nearby objects are blurred. A concave lens is required to correct vision for nearsightedness; a convex lens is required to correct vision for farsightedness. If a defect occurs in the curvature of the cornea or lens, a condition called astigmatism results. Focusing in different planes cannot occur simultaneously. Corrective cylindrical lenses will allow proper refraction of light onto the retina to compensate for the unequal curvatures.

As part of a natural aging process, the lens's elasticity decreases, which degrades our ability to focus on nearby objects. The near point of accommodation test is used to determine the ability to accommodate. Often a person will use "reading glasses" to compensate for this condition. Hereditary defects in color vision result from a lack of certain cones needed to absorb certain wavelengths of light. Special color test plates are used to diagnose a condition of colorblindness.

As you conduct this laboratory exercise, it does not matter in what order the visual tests or visual demonstrations are performed. None of the tests and demonstrations depends upon the results of any of the others. If you wear glasses, perform the tests with and without the corrective lenses. If you wear contact lenses, it is not necessary to run the tests under both conditions, but indicate in the laboratory report that all tests were performed with contact lenses in place.

Purpose of the Exercise

To conduct and interpret the tests for visual acuity, astigmatism, accommodation, color vision, the blind spot, and certain reflexes of the eye.

LEARNING OUTCOMES

After completing this exercise, you should be able to

1. Conduct and interpret the tests used to evaluate visual acuity, astigmatism, the ability to accommodate for close vision, and color vision.
2. Describe four conditions that can lead to defective vision.
3. Demonstrate and describe the blind spot, photopupillary reflex, accommodation pupillary reflex, and convergence reflex.

EXPLORE

Procedure A—Visual Tests

Perform the following visual tests, using your laboratory partner as a test subject. If your partner usually wears glasses, test each eye with and without the glasses.

1. *Visual acuity test.* Visual acuity (sharpness of vision) can be measured by using a Snellen eye chart (fig. 35.1). This chart consists of several sets of letters in different sizes printed on a white card. The letters near the top of the chart are relatively large, and those in each lower set become smaller. At one end of each set of letters is an acuity value in the form of a fraction. One of the sets near the bottom of the chart, for example, is marked 20/20. The normal eye can clearly see these letters from

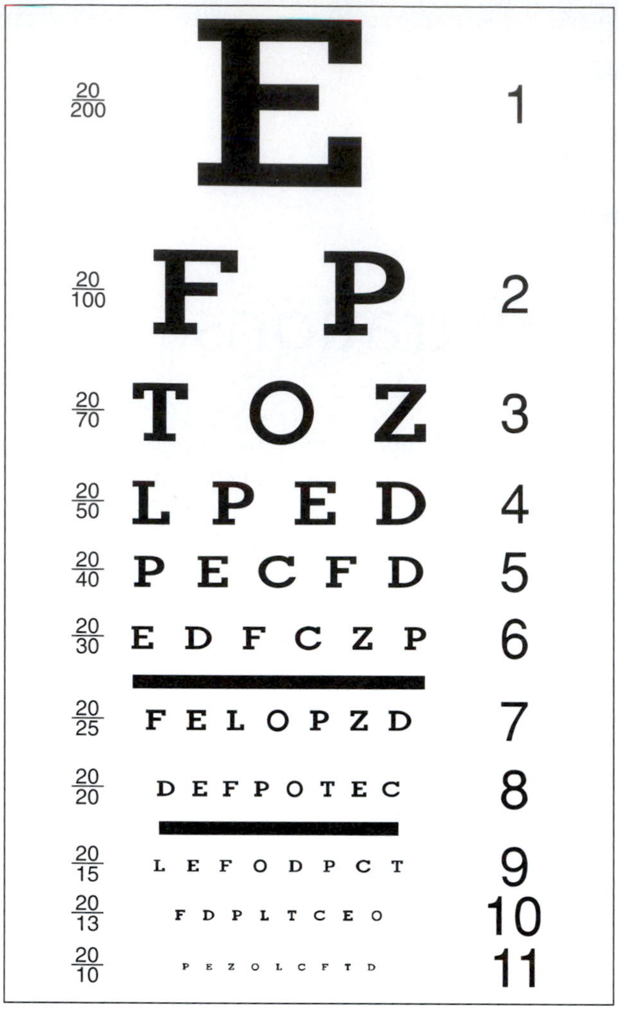

Figure 35.1 The Snellen eye chart looks similar to this but is somewhat larger.

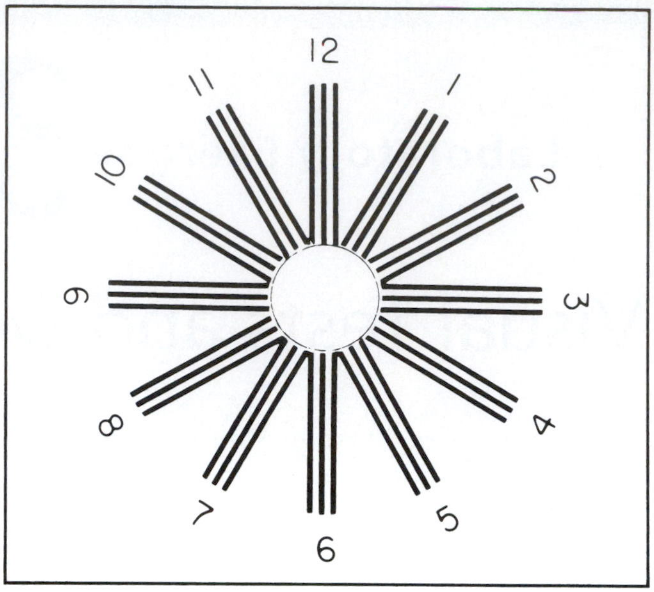

Figure 35.2 Astigmatism is evaluated using a chart such as this one.

the standard distance of 20 feet and thus is said to have 20/20 vision. The letter at the top of the chart is marked 20/200. The normal eye can read letters of this size from a distance of 200 feet. Thus, an eye able to read only the top letter of the chart from a distance of 20 feet is said to have 20/200 vision. This person has less than normal vision. A line of letters near the bottom of the chart is marked 20/15. The normal eye can read letters of this size from a distance of 15 feet, but a person might be able to read it from 20 feet. This person has better than normal vision.

To conduct the visual acuity test, follow these steps:

a. Hang the Snellen eye chart on a well-illuminated wall at eye level.

b. Have your partner stand 20 feet in front of the chart, gently cover the left eye with a 3" × 5" card, and read the smallest set of letters possible using your right eye.

c. Record the visual acuity value for that set of letters in Part A of Laboratory Report 35.

d. Repeat the procedure, using the left eye.

2. *Astigmatism test.* Astigmatism is a condition that results from a defect in the curvature of the cornea or lens. As a consequence, some portions of the image projected on the retina are sharply focused, and other portions are blurred. Astigmatism can be evaluated by using an astigmatism chart (fig. 35.2). This chart consists of sets of black lines radiating from a central spot like the spokes of a wheel. To a normal eye, these lines appear sharply focused and equally dark; however, if the eye has an astigmatism, some sets of lines appear sharply focused and dark, while others are blurred and less dark.

To conduct the astigmatism test, follow these steps:

a. Hang the astigmatism chart on a well-illuminated wall at eye level.

b. Stand 20 feet in front of the chart, and gently cover the left eye with a 3" × 5" card. Using your right eye, focus on the spot in the center of the radiating lines, and report which lines, if any, appear more sharply focused and darker.

c. Repeat the procedure, using the left eye.

d. Record the results in Part A of the laboratory report.

3. *Accommodation test.* Accommodation is the changing of the shape of the lens that occurs when the normal eye is focused for close vision. It involves a reflex in which muscles of the ciliary body are stimulated to contract, releasing tension on the suspensory ligaments that are fastened to the lens

Figure 35.3 To determine the near point of accommodation, slide the 3" × 5" card along the meterstick toward your partner's open eye until it reaches the closest location where your partner can still see the word sharply focused.

capsule. This allows the capsule to rebound elastically, causing the surface of the lens to become more convex. The ability to accommodate is likely to decrease with age because the tissues involved tend to lose their elasticity.

To evaluate the ability to accommodate, follow these steps:

a. Hold the end of a meterstick against your partner's chin so that the stick extends outward at a right angle to the plane of the face (fig. 35.3).
b. Have your partner close the left eye.
c. Hold a 3" × 5" card with a word typed in the center at the distal end of the meterstick.
d. Slide the card along the stick toward your partner's open right eye, and locate the *point closest to the eye* where your partner can still see the letters of the word sharply focused. This distance is called the *near point of accommodation,* and it tends to increase with age (table 35.1).

Table 35.1 Near Point of Accommodation

Age (years)	Average Near Point (cm)
10	7
20	10
30	13
40	20
50	45
60	90

e. Repeat the procedure with the right eye closed.
f. Record the results in Part A of the laboratory report.

4. *Color vision test.* Some people exhibit defective color vision because they lack certain cones, usually those sensitive to the reds or greens. This trait is an X-linked (sex-linked) inheritance, so the condition is more prevalent in males (7%) than in females (0.4%). People who lack or possess decreased sensitivity to the red-sensitive cones possess protanopia colorblindness; those who lack or possess decreased sensitivity to green-sensitive cones possess deuteranopia colorblindness. The colorblindness condition is often more of a deficiency or a weakness than one of blindness. Laboratory Exercise 53 describes the genetics for colorblindness.

   To conduct the color vision test, follow these steps:

   a. Examine the color test plates in Ichikawa's or Ishihara's book to test for any red-green color vision deficiency. Also examine figure 35.4.
   b. Hold the test plates approximately 30 inches from the subject in bright light. All responses should occur within 3 seconds.
   c. Compare your responses with the correct answers in Ichikawa's or Ishihara's book. Determine the percentage of males and females in your class who exhibit any color-deficient vision. If an individual exhibits color-deficient vision, determine if the condition is protanopia or deuteranopia.
   d. Record the class results in Part A of the laboratory report.

5. Complete Part A of the laboratory report.

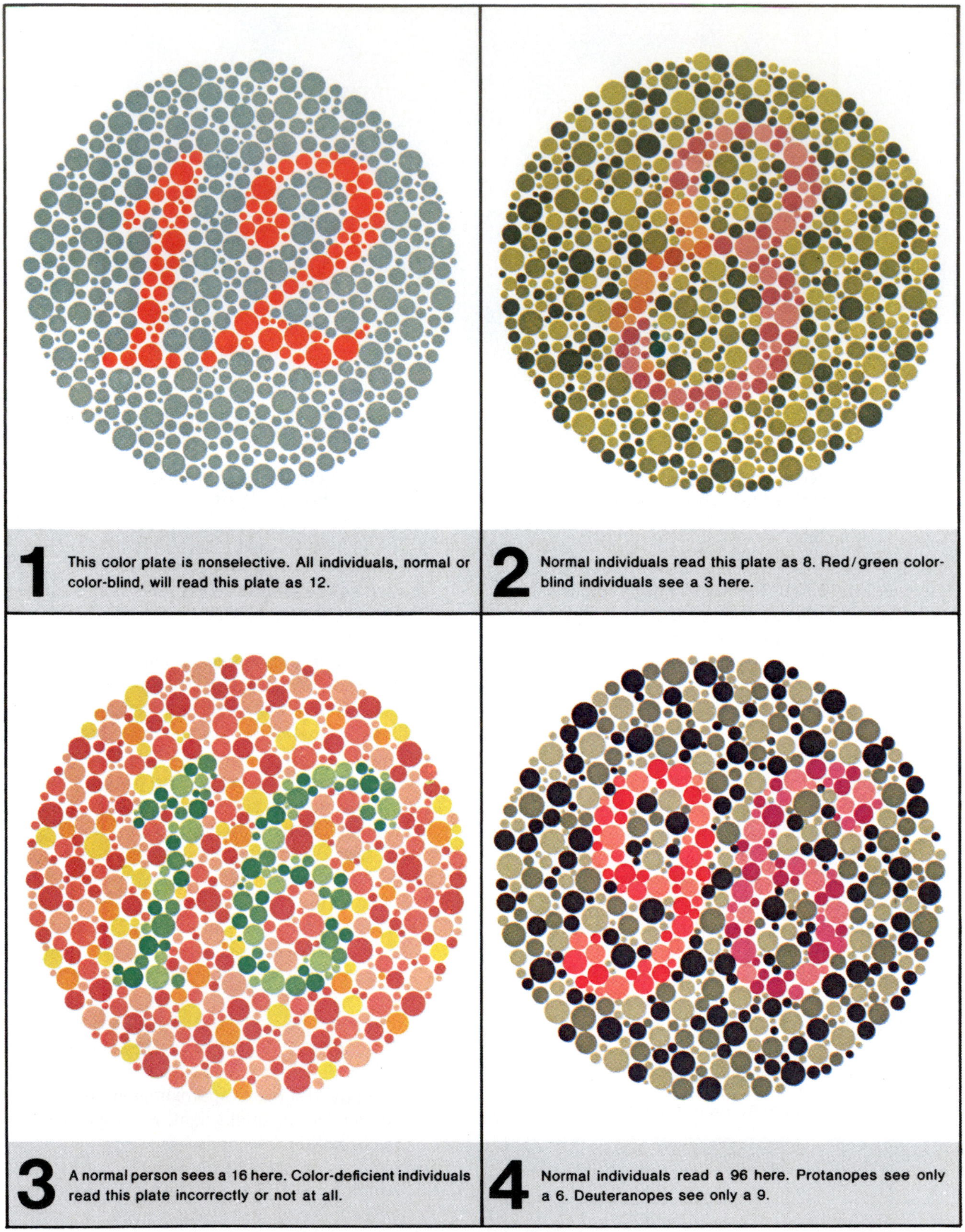

Figure 35.4 Samples of Ishihara's color plates. These plates are reproduced from *Ishihara's Tests for Colour Blindness* published by Kanehara & Co., Ltd., Tokyo, Japan, but tests for colorblindness cannot be conducted with this material. For accurate testing, the original plates should be used.

EXPLORE

Procedure B—Visual Demonstrations

Perform the following demonstrations with the help of your laboratory partner.

1. *Blind-spot demonstration.* There are no photoreceptors in the optic disc, located where the nerve fibers of the retina leave the eye and enter the optic nerve. Consequently, this region of the retina is commonly called the *blind spot.*

   To demonstrate the blind spot, follow these steps:

   a. Close your left eye, hold figure 35.5 about 35 cm away from your face, and stare at the + sign in the figure with your right eye.

   b. Move the figure closer to your face as you continue to stare at the + until the dot on the figure suddenly disappears. This happens when the image of the dot is focused on the optic disc. Measure the right eye distance using a metric ruler or a meterstick.

   c. Repeat the procedures with your right eye closed. This time stare at the dot using your left eye, and the + will disappear when the image falls on the optic disc. Measure the distance.

   d. Record the results in Part B of the laboratory report.

Critical Thinking Application

Under normal visual circumstances, explain why small objects are not lost from our vision.

2. *Photopupillary reflex.* The smooth muscles of the iris control the size of the pupil. For example, when the intensity of light entering the eye increases, a photopupillary reflex is triggered, and the circular muscles of the iris are stimulated to contract. As a result, the size of the pupil decreases, and less light enters the eye.

   To demonstrate this photopupillary reflex, follow these steps:

   a. Ask your partner to sit with his or her hands thoroughly covering his or her eyes for 2 minutes.

   b. Position a pen flashlight close to one eye with the light shining on the hand that covers the eye.

   c. Ask your partner to remove the hand quickly.

   d. Observe the pupil and note any change in its size.

   e. Have your partner remove the other hand, but keep that uncovered eye shielded from extra light.

   f. Observe both pupils and note any difference in their sizes.

Figure 35.5 Blind-spot demonstration.

3. *Accommodation pupillary reflex.* The pupil constricts as a normal accommodation reflex to focusing on close objects. To demonstrate the accommodation pupillary reflex, follow these steps:

   a. Have your partner stare for several seconds at some dimly illuminated object in the room that is more than 20 feet away.

   b. Observe the size of the pupil of one eye. Then hold a pencil about 25 cm in front of your partner's face, and have your partner stare at it.

   c. Note any change in the size of the pupil.

4. *Convergence reflex.* The eyes converge as a normal convergence reflex to focusing on close objects. To demonstrate the convergence reflex, follow these steps:

   a. Repeat the procedure outlined for the accommodation pupillary reflex.

   b. Note any change in the position of the eyeballs as your partner changes focus from the distant object to the pencil.

5. Complete Part B of the laboratory report.

Notes

Name ______________________

Date ______________________

Section ______________________

The [A] corresponds to the indicated Learning Outcome(s) listed at the beginning of the laboratory exercise.

Visual Tests and Demonstrations

Laboratory Report 35

Part A Assessments

1. Visual acuity test results: [A] 1

Eye Tested	Acuity Values
Right eye	
Right eye with glasses (if applicable)	
Left eye	
Left eye with glasses (if applicable)	

2. Astigmatism test results: [A] 1

Eye Tested	Darker Lines
Right eye	
Right eye with glasses (if applicable)	
Left eye	
Left eye with glasses (if applicable)	

3. Accommodation test results: [A] 1

Eye Tested	Near Point (cm)
Right eye	
Right eye with glasses (if applicable)	
Left eye	
Left eye with glasses (if applicable)	

4. Color vision test results: 1

Condition	Males			Females		
	Class Number	Class Percentage	Expected Percentage	Class Number	Class Percentage	Expected Percentage
Normal color vision			93			99.6
Deficient red-green color vision			7			0.4
Protanopia (lack red-sensitive cones)			less-frequent type			less-frequent type
Deuteranopia (lack green-sensitive cones)			more-frequent type			more-frequent type

5. Complete the following: 2

a. What is meant by 20/70 vision? ______________________________

____

b. What is meant by 20/10 vision? ______________________________

____

c. What visual problem is created by astigmatism? ______________________________

____

d. Why does the near point of accommodation often increase with age? ______________________________

____

e. Describe the eye defect that causes color-deficient vision. ______________________________

____

Part B Assessments

1. Blind-spot results: 3

a. Right eye distance ______________________________

b. Left eye distance ______________________________

2. Complete the following:

a. Explain why an eye has a blind spot. 3 ______________________________

____

b. Describe the photopupillary reflex. 3 ______________________________

____

c. What difference did you note in the size of the pupils when one eye was exposed to bright light and the other eye was shielded from the light? 3 ______________________________

____

d. Describe the accommodation pupillary reflex. 3 ______________________________

____

e. Describe the convergence reflex. 3 ______________________________

____

Laboratory Exercise 36

Endocrine Histology and Diabetic Physiology

Materials Needed

For Procedure A—Endocrine Histology:
Textbook
Human torso model
Compound light microscope
Prepared microscope slides of the following:
- Pituitary gland
- Thyroid gland
- Parathyroid gland
- Adrenal gland
- Pancreas

For Procedure B—Insulin Shock:
500 mL beaker
Live small fish (1" to 1.5" goldfish, guppy, rosy red feeder, or other)
Small fish net
Insulin (regular U-100) (HumulinR in 10 mL vials has 100 units/mL; store in refrigerator—do not freeze)
Syringes for U-100 insulin
10% glucose solution
Clock with second hand or timer

For Learning Extension:
Compound light microscope
Prepared microscope slides of the following:
- Normal human pancreas (stained for alpha and beta cells)
- Human pancreas of diabetic

Safety

- Review all the safety guidelines inside the front cover.
- Wear disposable gloves when handling the fish and syringes.
- Dispose of the used syringe and needle in the puncture-resistant container.
- Wash your hands before leaving the laboratory.

The endocrine system consists of ductless glands that act together with parts of the nervous system to help control body activities. The endocrine glands secrete hormones that are transported in body fluids and affect cells that have appropriate receptor molecules. In this way, hormones influence the rate of metabolic reactions, the transport of substances through cell membranes, the regulation of water and electrolyte balances, and many other functions. By controlling cellular activities, endocrine glands play important roles in the maintenance of homeostasis.

The primary "fuel" for the mitochondria of cells is sugar (glucose). Insulin produced by the beta cells of the pancreatic islets has a primary regulation role in the transport of blood sugar across the cell membranes. Some individuals do not produce enough insulin; others might not have adequate transport of glucose into the body cells when cells lose insulin receptors. This rather common disease is called *diabetes mellitus.* Diabetes mellitus is a functional disease that might have its onset either during childhood or later in life.

The protein hormone insulin has several functions: it stimulates the liver and skeletal muscles to form glycogen from glucose; it inhibits the conversion of noncarbohydrates into glucose; it promotes the facilitated diffusion of glucose across cell membranes of cells possessing insulin receptors (adipose tissue, cardiac muscle, and resting skeletal muscle); it decreases blood sugar (glucose); it increases protein synthesis; and it promotes fat storage in adipose cells. In a normal person, as blood sugar increases during nutrient absorption after a meal, the rising glucose levels directly stimulate beta cells of the pancreas to secrete insulin. This increase in insulin prevents blood sugar from sudden surges (hyperglycemia) by promoting glycogen production in the liver and an increase in the entry of glucose into muscle and adipose cells.

Between meals and during sleep, blood glucose levels decrease and insulin also decreases. When insulin concentration decreases, more glucose is available to enter cells that lack insulin receptors. Neurons, liver cells, kidney cells, and red blood cells either lack insulin receptors or express limited insulin receptors, and they obtain their glucose by facilitated diffusion. These cells depend upon blood glucose concentrations to enable cellular respiration and ATP production. When insulin is given as a medication for diabetes mellitus, blood glucose levels may drop drastically (hypoglycemia) if the individual does not eat. This can have significant effects on the nervous system, resulting in insulin shock.

Type 1 diabetes mellitus usually has a rapid onset relatively early in life, but it can occur in adults. It occurs when insulin is no longer produced; hence sometimes it is referred to as *insulin-dependent diabetes mellitus (IDDM)*. This autoimmune disorder occurs when the immune system destroys beta cells, resulting in a decrease in insulin production. This disorder affects about 10 to 15% of diabetics. The treatment for type 1 diabetes mellitus is to administer insulin, usually by injection.

Type 2 diabetes mellitus has a gradual onset, usually in those over age 40. It is sometimes called *non-insulin-dependent diabetes mellitus (NIDDM)*. This disease results as body cells become less sensitive to insulin or lose insulin receptors and thus cannot respond to insulin even though insulin levels may remain normal. This situation is called *insulin resistance*. Risk factors for the onset of type 2 diabetes mellitus, which afflicts about 85 to 90% of diabetics, include heredity, obesity, and lack of exercise. Obesity is a major indicator for predicting type 2 diabetes; a weight loss of a mere 10 pounds will increase the body's sensitivity to insulin as a means of prevention. The treatments for type 2 diabetes mellitus include a diet that avoids foods that stimulate insulin production, weight control, exercise, and medications. (The causes and treatments for type 2 diabetes closely resemble those for coronary artery disease!)

Purpose of the Exercise

To examine microscopically the tissues of endocrine glands and to observe behavior changes that occur during insulin shock.

LEARNING OUTCOMES

After completing this exercise, you should be able to

1. Name and locate the major endocrine glands.
2. Associate the principal hormones and functions of the major endocrine glands.
3. Sketch and label tissue sections from the pituitary gland, thyroid gland, parathyroid glands, adrenal glands, and pancreas.
4. Compare the causes, symptoms, and treatments for type 1 and type 2 diabetes mellitus.
5. Observe and record the behavioral changes caused by insulin shock.
6. Observe and record the recovery from insulin shock when sugar is provided to cells.

EXPLORE

Procedure A—Endocrine Histology

1. Review the sections entitled "Pituitary Gland," "Thyroid Gland," "Parathyroid Glands," "Adrenal Glands," "Pancreas," and "Other Endocrine Glands" in chapter 13 of the textbook.
2. As a review activity, label figure 36.1.
3. Complete Part A of Laboratory Report 36.
4. Examine the human torso model and locate the following:

   hypothalamus
   pituitary stalk (infundibulum)
   pituitary gland (hypophysis)
   - anterior lobe (adenohypophysis)
   - posterior lobe (neurohypophysis)

   thyroid gland
   parathyroid glands
   adrenal glands (suprarenal glands)
   - adrenal medulla
   - adrenal cortex

   pancreas
   pineal gland
   thymus
   ovaries
   testes

5. Examine the microscopic tissue sections of the following glands, and identify the features described: **Pituitary gland.** To examine the pituitary tissue, follow these steps:
   a. Observe the tissues using low-power magnification (fig. 36.2).
   b. Locate the *pituitary stalk (infundibulum)*, the *anterior lobe* (the largest part of the gland), and the *posterior lobe*. The anterior lobe makes hormones and appears glandular; the posterior lobe stores and releases hormones made by the hypothalamus and is nervous in appearance.
   c. Observe an area of the anterior lobe with high-power magnification. Locate a cluster of relatively large cells and identify some *acidophil cells*, which contain pink-stained granules, and some *basophil cells*, which contain blue-stained

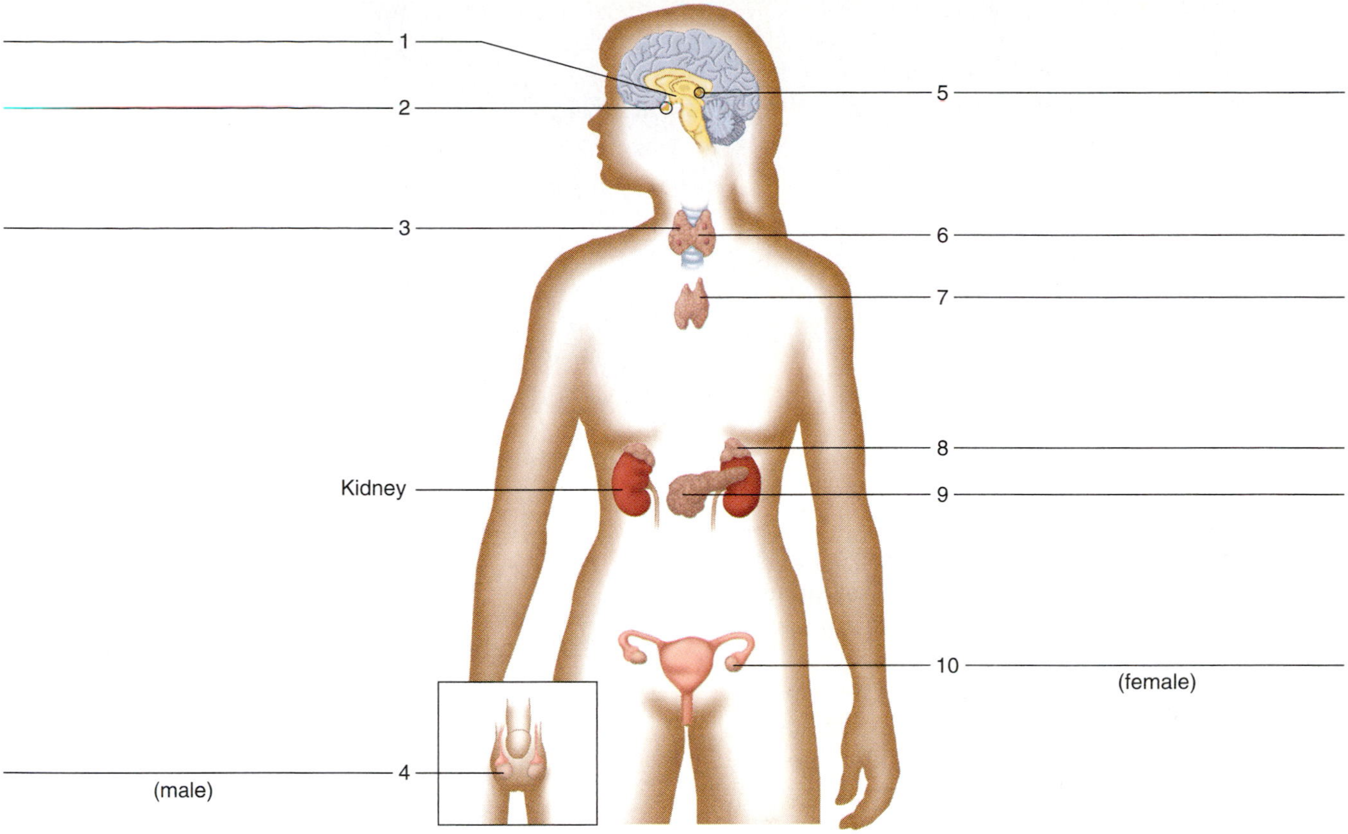

Figure 36.1 Label the major endocrine glands.

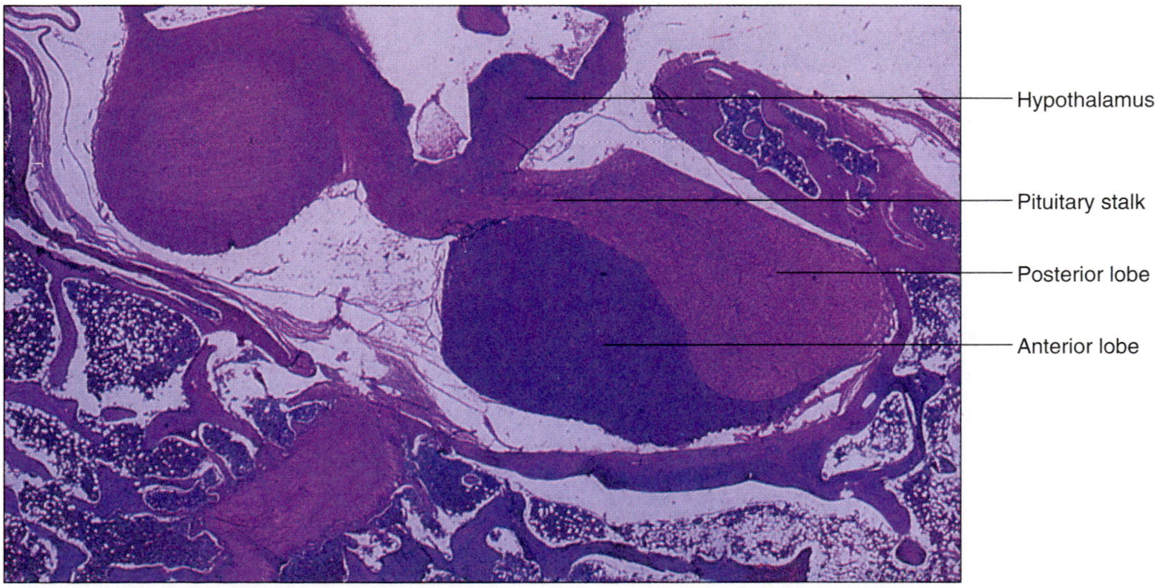

Figure 36.2 Micrograph of the pituitary gland (6×).

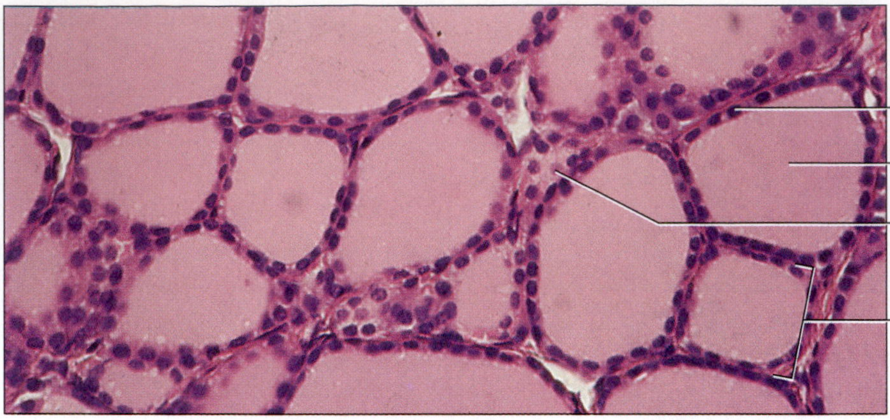

Figure 36.3 Micrograph of the thyroid gland (100x micrograph enlarged to 300×).

granules. These acidophil and basophil cells are the hormone-secreting cells. The hormones secreted include TSH, ACTH, prolactin, growth hormone, FSH, and LH.

d. Observe an area of the posterior lobe with high-power magnification. Note the numerous unmyelinated nerve fibers present in this lobe. Also locate some *pituicytes,* a type of neuroglia, scattered among the nerve fibers. The posterior lobe secretes ADH and oxytocin.

e. Prepare labeled sketches of representative portions of the anterior and posterior lobes of the pituitary gland in Part B of the laboratory report.

Thyroid gland. To examine the thyroid tissue, follow these steps:

a. Use low-power magnification to observe the tissue (fig. 36.3). Note the numerous *follicles,* each of which consists of a layer of cells surrounding a colloid-filled cavity.

b. Observe the tissue using high-power magnification. The cells forming the wall of a follicle are simple cuboidal epithelial cells. These cells secrete thyroid hormone (T_3 and T_4). The extrafollicular (parafollicular) cells secrete calcitonin.

c. Prepare a labeled sketch of a representative portion of the thyroid gland in Part B of the laboratory report.

Parathyroid gland. To examine the parathyroid tissue, follow these steps:

a. Use low-power magnification to observe the tissue (fig. 36.4). The gland consists of numerous tightly packed secretory cells.

b. Switch to high-power magnification and locate two types of cells—a smaller form (chief cells) arranged in cordlike patterns and a larger form (oxyphil cells) that have distinct cell boundaries

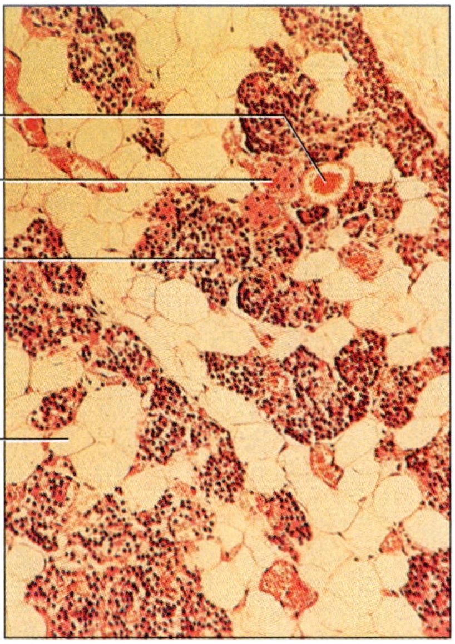

Figure 36.4 Micrograph of the parathyroid gland (65×).

and are present in clusters. *Chief cells* secrete parathyroid hormone, whereas the function of *oxyphil cells* is not clearly understood.

c. Prepare a labeled sketch of a representative portion of the parathyroid gland in Part B of the laboratory report.

Adrenal gland. To examine the adrenal tissue, follow these steps:

a. Use low-power magnification to observe the tissue (fig. 36.5). Note the thin capsule of connective tissue that covers the gland. Just beneath the capsule, there is a relatively thick *adrenal cortex.* The central portion of the gland is the *adrenal medulla.* The cells of the cortex are in three poorly defined layers.. Those of the outer layer (zona glomerulosa) are arranged irregularly; those of the middle layer (zona fasciculata) are in long cords;

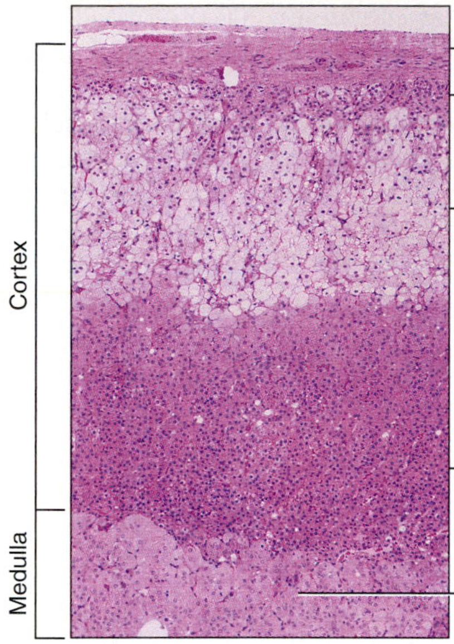

Figure 36.5 Micrograph of the adrenal cortex and the adrenal medulla (75×).

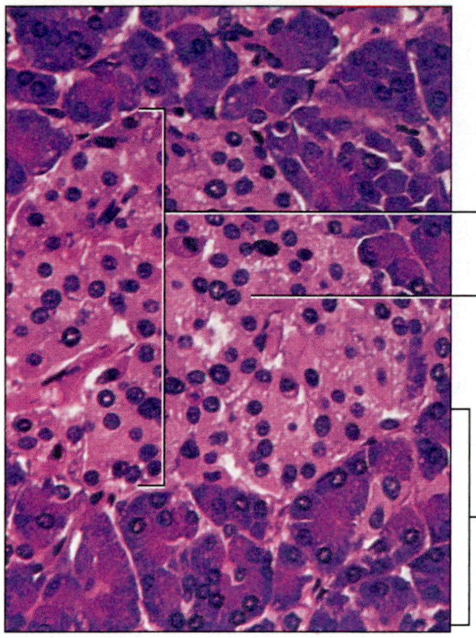

Figure 36.6 Micrograph of the pancreas (100×).

and those of the inner layer (zona reticularis) are arranged in an interconnected network of cords. The most noted hormones secreted from the cortex include aldosterone from the zona glomerulosa, cortisol from the zona fasciculata, and estrogens and androgens from the zona reticularis. The cells of the medulla are relatively large and irregularly shaped, and they often occur in clusters. The medulla cells secrete epinephrine and norepinephrine.

b. Using high-power magnification, observe each of the layers of the cortex and the cells of the medulla.

c. Prepare labeled sketches of representative portions of the adrenal cortex and medulla in Part B of the laboratory report.

Pancreas. To examine the pancreas tissue, follow these steps:

a. Use low-power magnification to observe the tissue (fig. 36.6). The gland largely consists of deeply stained exocrine cells arranged in clusters around secretory ducts. These exocrine cells (acinar cells) secrete pancreatic juice rich in digestive enzymes. There are circular masses of lightly stained cells scattered throughout the gland. These clumps of cells constitute the *pancreatic islets (islets of Langerhans)*, and they represent the endocrine portion of the pancreas. The beta cells secrete insulin and the alpha cells secrete glucagon.

b. Examine an islet, using high-power magnification. Unless special stains are used, it is not possible to distinguish alpha and beta cells.

c. Prepare a labeled sketch of a representative portion of the pancreas in Part B of the laboratory report.

EXPLORE

Procedure B—Insulin Shock

1. Review the sections entitled "Pancreas" and "Diabetes Mellitus—Clinical Application" in chapter 13 of the textbook, and the introduction to this lab.
2. Complete the table in Part C of the laboratory report.
3. A fish can be used to observe normal behavior, induced insulin shock, and recovery from insulin shock. The pancreatic islets are located in the pyloric ceca or other scattered regions in fish. The behavioral changes of fish when they experience insulin shock and recovery mimic those of humans. Observe a fish in 200 mL of aquarium water in a 500 mL beaker. Make your observations of swimming, gill cover (operculum), and mouth movements for 5 minutes. Record your observations in Part D of the laboratory report.
4. Add 200 units of room-temperature insulin slowly, using the syringe, into the water with the fish to induce insulin shock. Record the time or start the timer when the insulin is added. Watch for changes of the swimming, gill cover, and mouth movements as insulin diffuses into the blood at the gills. Signs of insulin shock might include rapid and irregular

movements. As hypoglycemia occurs from the increased insulin absorption, the brain cells obtain less of the much-needed glucose, and epinephrine is secreted. Consequently, a rapid heart rate, anxiety, sweating, mental disorientation, impaired vision, dizziness, convulsions, and possible unconsciousness are complications during insulin shock. Record your observations of any behavioral changes and the time of the onset in Part D of the laboratory report.

5. After definite behavioral changes are observed, use the small fish net to move the fish into another 500 mL beaker containing 200 mL of a 10% room-temperature glucose solution. Record the time when the fish is transferred into this container. Make observations of any recovery of normal behaviors, including the amount of time involved. Record your observations in Part D of the laboratory report.
6. When the fish appears to be fully recovered, return the fish to its normal container. Dispose of the syringe according to directions from your laboratory instructor.
7. Complete Part D of the laboratory report.

Learning Extension

Examine the normal pancreas under high-power magnification (fig. 36.6). Locate the clumps of cells that constitute the *pancreatic islets (islets of Langerhans)* that represent the endocrine portions of the organ. A pancreatic islet contains four distinct cells that secrete different hormones: *alpha cells,* which secrete glucagon, *beta cells,* which secrete insulin, *delta cells,* which secrete somatostatin, and *F cells,* which secrete pancreatic polypeptide. Some of the specially stained pancreatic islets allow for distinguishing the different types of cells (figs. 36.7 and 36.8). The normal pancreatic islet has about 80% beta cells concentrated in the central region of the islet, about 15% alpha cells near the periphery of the islet, about 5% delta cells, and a small number of F cells. There are complex interactions among the four hormones affecting blood sugar, but in general, glucagon will increase blood sugar while insulin will lower blood sugar. Somatostatin acts as a paracrine secretion to inhibit alpha and beta cell secretions, and pancreatic polypeptide inhibits secretions from delta cells and digestive enzyme secretions from the pancreatic acini.

Examine the pancreas of a diabetic under high-power magnification (fig. 36.9). Locate a pancreatic islet and note the number of beta cells as compared to a normal number of beta cells in the normal pancreatic islet. Prepare a labeled sketch of a pancreatic islet showing indications of diabetes mellitus in Part B of the laboratory report.

Describe the differences that you observed between a normal pancreatic islet and a pancreatic islet of a person with diabetes mellitus.

____

____

____

____

____

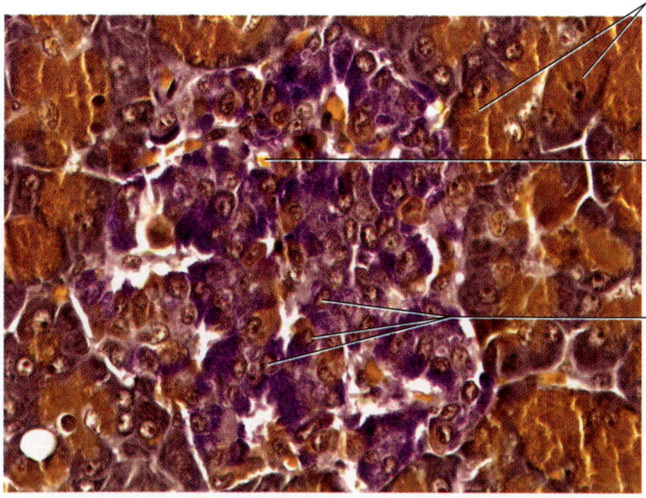

Figure 36.7 Micrograph of specially stained pancreatic islet surrounded by exocrine (acinar) cells (500×). The insulin-containing beta cells are stained dark purple.

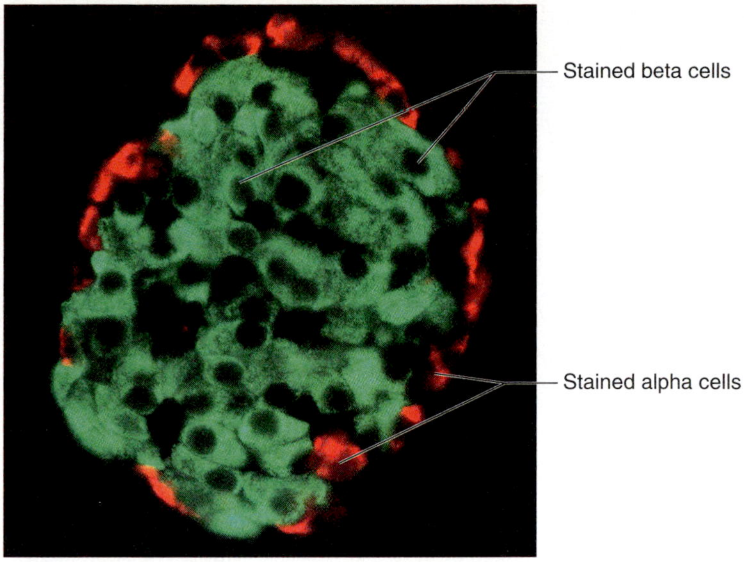

Figure 36.8 Micrograph of a pancreatic islet using immunofluorescence stains to visualize alpha and beta cells. Only the areas of interest are visible in the fluorescence microscope.

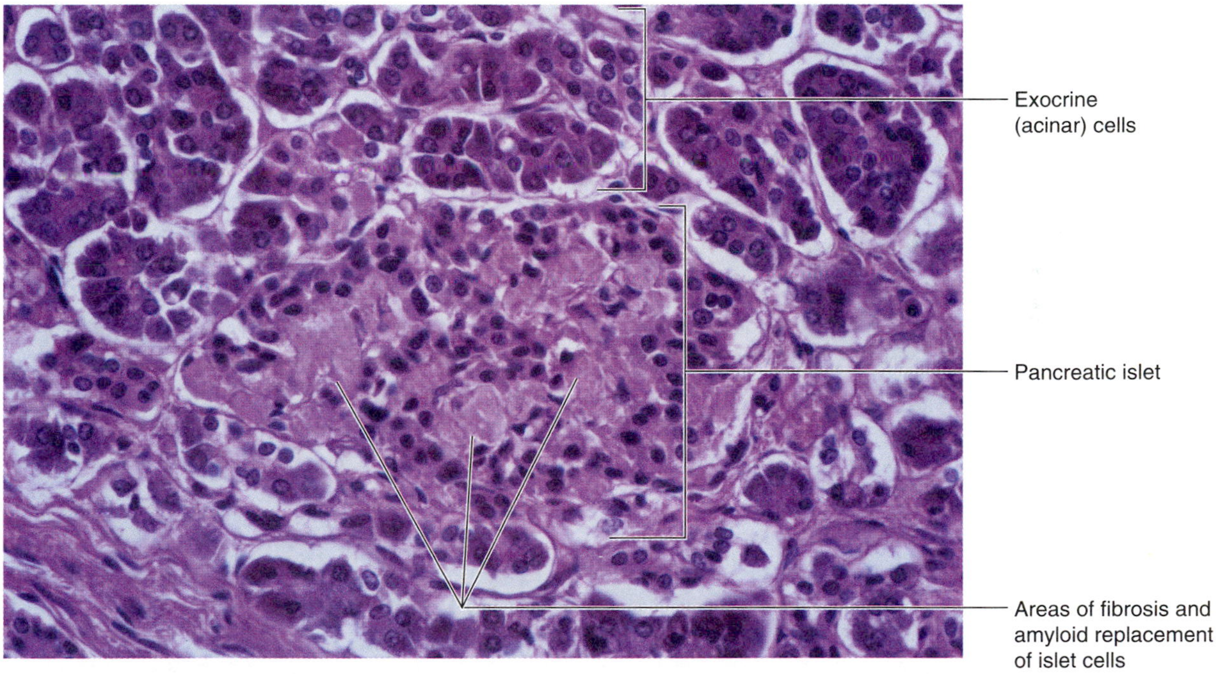

Figure 36.9 Micrograph of pancreas showing indications of changes from diabetes mellitus (100×).

Notes

Name ______________________________

Date ______________________________

Section ______________________________

The [A] corresponds to the indicated Learning Outcome(s) listed at the beginning of the laboratory exercise.

Laboratory Report

36

Endocrine Histology and Diabetic Physiology

Part A Assessments

Complete the following:

1. Name six hormones secreted by the anterior lobe of the pituitary gland. [2] ______________________________

____

2. Name two hormones secreted by the posterior lobe of the pituitary gland. [2] ______________________________

3. Name the pituitary hormone responsible for the following actions: [2]

   a. causes kidneys to conserve water ______________________________

   b. stimulates cells to increase in size and divide more rapidly ______________________________

   c. stimulates secretion from thyroid gland ______________________________

   d. causes contraction of uterine wall muscles ______________________________

   e. stimulates secretion from adrenal cortex ______________________________

   f. stimulates milk production ______________________________

4. Name two thyroid hormones that affect metabolic rate. [2] ______________________________

5. Name a hormone secreted by the thyroid gland that acts to lower blood calcium. [2] ______________________________

6. Name a hormone that acts to raise blood calcium. [2] ______________________________

7. Name three target organs of parathyroid hormone. [2] ______________________________

8. Name two hormones secreted by the adrenal medulla. [2] ______________________________

9. List five different effects produced by these medullary hormones. [2] ______________________________

____

10. Name the most important mineralocorticoid secreted by the adrenal cortex. [2] ______________________________

11. List two actions of this mineralocorticoid. [2] ______________________________

____

12. Name the most important glucocorticoid secreted by the adrenal cortex. [2] ______________________________

13. List three actions of this glucocorticoid. [2] ______________________________

____

14. Distinguish the two hormones secreted by the alpha and beta cells of the pancreatic islets. [2] ______________________________

____

Part B Assessments

Sketch and label representative portions of the following endocrine glands: 3

Pituitary gland (____×) (anterior and posterior lobes)	Thyroid gland (____×)
Parathyroid gland (____×)	Adrenal gland (____×) (cortex and medulla)
Pancreatic islet (____×) (normal)	Pancreatic islet (____×) (showing changes from diabetes mellitus)

Part C Assessments

Complete the missing parts of the table of diabetes mellitus: 4

Characteristic	Type 1 Diabetes	Type 2 Diabetes
Onset age	Early age or adult	
Onset of symptoms		Slow
Percentage of diabetics		85–90%
Natural insulin levels	Below normal	
Beta cells of pancreatic islets		Not destroyed
Pancreatic islet cell antibodies	Present	
Risk factors of having the disease	Heredity	
Typical treatments	Insulin administration	
Untreated blood sugar levels		Hyperglycemia

Part D Assessments

1. Record your observations of normal behavior of the fish.
   a. Swimming movements: ______________________________
   ____
   b. Gill cover movements: ______________________________
   ____
   c. Mouth movements: ______________________________
   ____
2. Record your observations of the fish after insulin was added to the water. 5
   a. Recorded time that insulin was added: ______________________________
   b. Swimming movements: ______________________________
   ____
   c. Gill cover movements: ______________________________
   ____
   d. Mouth movements: ______________________________
   ____
   e. Elapsed time until the insulin shock symptoms occurred: ______________________________

3. Record your observations after the fish was transferred into a glucose solution. 6

   a. Recorded time that fish was transferred into the glucose solution: ______________________

   b. Describe any changes in the behavior of the fish that indicates a recovery from insulin shock. ______________________

   c. Elapsed time until indications of a recovery from insulin shock occurred: ______________________

Critical Thinking Application

Justify the importance for a type 1 diabetic of regulating insulin administration, meals, and exercise. 4

Laboratory Exercise 37

Blood Cells and Blood Typing

Materials Needed

Textbook
Compound light microscope
Prepared microscope slides of human blood (Wright's stain)
Colored pencils
ABO blood-typing kit
Simulated blood-typing kits are suggested as a substitute for collected blood

For Demonstration:

Microscope slide
Alcohol swabs
Sterile blood lancet
Toothpicks
Anti-D serum
Slide warming box (Rh blood-typing box or Rh view box)

For Learning Extension:

Prepared slides of pathological blood, such as eosinophilia, leukocytosis, leukopenia, and lymphocytosis

Safety

- It is important that students learn and practice correct procedures for handling body fluids. Consider using simulated blood-typing kits, mammal blood other than human, or contaminant-free blood that has been tested and is available from various laboratory supply houses. Some of the procedures might be accomplished as demonstrations only. If student blood is used, it is important that students handle only their own blood.
- Use an appropriate disinfectant to wash the laboratory tables before and after the procedures.
- Wear disposable gloves when handling blood samples.
- Clean the end of a finger with 70% alcohol before the puncture is performed.
- The sterile blood lancet should be used only once.
- Dispose of used lancets and blood-contaminated items in an appropriate container (never use the wastebasket).
- Wash your hands before leaving the laboratory.

Blood is a type of connective tissue whose cells are suspended in a liquid extracellular matrix. These cells are mainly formed in red bone marrow, and they include *red blood cells, white blood cells,* and some cellular fragments called *platelets.*

Red blood cells contain hemoglobin and transport gases between the body cells and the lungs, white blood cells defend the body against infections, and platelets play an important role in stoppage of bleeding (hemostasis).

Blood typing involves identifying protein substances called *antigens* present on red blood cell membranes. Although there are many different antigens associated with human red blood cells, only a few of them are of clinical importance. These include the antigens of the ABO group and those of the Rh group.

To determine which antigens are present, a blood sample is mixed with blood-typing sera that contain known types of antibodies. If a particular antibody contacts a corresponding antigen, a reaction occurs, and the red blood cells clump together (agglutination). Thus, if blood cells are mixed with serum containing antibodies that react with antigen A and the cells clump together, antigen A must be present in those cells.

Purpose of the Exercise

To review the characteristics of blood cells, to examine them microscopically, to perform a differential white blood cell count, to determine the ABO blood type of a blood sample, and to observe an Rh blood-typing test.

LEARNING OUTCOMES

After completing this exercise, you should be able to

1. Describe the structure and function of red blood cells, white blood cells, and platelets.
2. Identify and sketch red blood cells, five types of white blood cells, and platelets.
3. Perform and interpret the results of a differential white blood cell count.
4. Analyze the basis of ABO blood typing.
5. Match the ABO type of a blood sample.
6. Explain the basis of Rh blood typing.
7. Interpret how the Rh type of a blood sample is determined.

WARNING: Because of the possibility of blood infections being transmitted from one student to another if blood testing is performed in the classroom, it is suggested that commercially prepared blood slides and blood-typing kits containing virus-free human blood be used for ABO blood typing. The instructor may wish to demonstrate Rh blood typing. Observe all of the safety procedures listed for this lab.

EXPLORE

Procedure A—Types of Blood Cells

1. Review the section entitled "Blood and Blood Cells" in chapter 14 of the textbook.
2. Complete Part A of Laboratory Report 37.
3. Refer to figure 37.1 as an aid in identifying the various types of blood cells. Study the functions of the blood cells listed in table 37.1. Use the prepared slide of blood and locate each of the following:

   red blood cell (erythrocyte)
   white blood cell (leukocyte)
   - granulocytes
     - neutrophil
     - eosinophil
     - basophil
   - agranulocytes
     - lymphocyte
     - monocyte

   platelet (thrombocyte)

Table 37.1 Cellular Components of Blood

Component	Function
Red blood cell (erythrocyte)	Transports oxygen and carbon dioxide
White blood cell (leukocyte)	Destroys pathogenic microorganisms and parasites and removes worn cells
Granulocytes	
1. Neutrophil	Phagocytizes small particles
2. Eosinophil	Kills parasites and helps control inflammation and allergic reactions
3. Basophil	Releases heparin and histamine
Agranulocytes	
1. Lymphocyte	Provides immunity
2. Monocyte	Phagocytizes large particles
Platelet (thrombocyte)	Helps control blood loss from broken vessels

4. In Part B of the laboratory report, prepare sketches of single blood cells to illustrate each type. Pay particular attention to relative size, nuclear shape, and color of granules in the cytoplasm (if present). The sketches should be accomplished using either the high-power objective or the oil immersion objective of the compound light microscope.

EXPLORE

Procedure B—Differential White Blood Cell Count

A differential white blood cell count is performed to determine the percentage of each of the various types of white blood cells present in a blood sample. The test is useful because the relative proportions of white blood cells may change in particular diseases as indicated in table 37.2. Neutrophils, for example, usually increase during bacterial infections, whereas eosinophils may increase during certain parasitic infections and allergic reactions.

1. To make a differential white blood cell count, follow these steps:
   a. Using high-power magnification or an oil immersion objective, focus on the cells at one end of a prepared blood slide where the cells are well distributed.
   b. Slowly move the blood slide back and forth, following a path that avoids passing over the same cells twice (fig. 37.2).
   c. Each time you encounter a white blood cell, identify its type and record it in Part C of the laboratory report.
   d. Continue searching for and identifying white blood cells until you have recorded 100 cells in the data table. *Percent* means "parts of 100" for each type of white blood cell, the total number observed is equal to its percentage in the blood sample.
2. Complete Part C of the laboratory report.

Table 37.2 Differential White Blood Cell Count

Cell Type	Normal Value (percent)	Elevated Levels May Indicate
Neutrophil	54–62	Bacterial infections, stress
Lymphocyte	25–33	Mononucleosis, whooping cough, viral infections
Monocyte	3–9	Malaria, tuberculosis, fungal infections
Eosinophil	1–3	Allergic reactions, autoimmune diseases, parasitic worms
Basophil	<1	Cancers, chicken pox, hypothyroidism

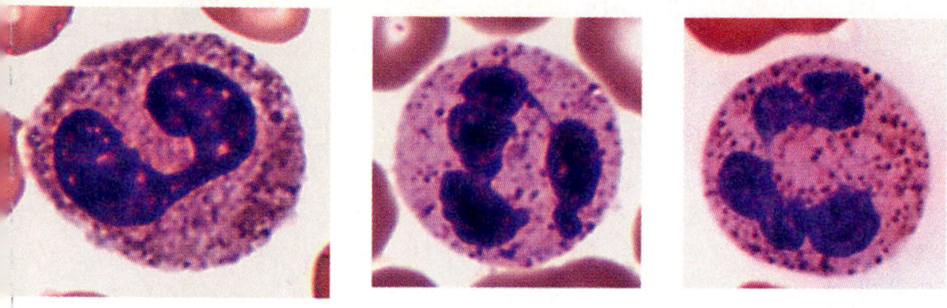

Neutrophils (3 of many variations)
•Fine light-purple granules
•Nucleus single to five lobes (highly variable)
•Immature neutrophils, called bands, have a single C-shaped nucleus
•Mature neutrophils, called segs, have a lobed nucleus
•Often called polymorphonuclear leukocytes when older

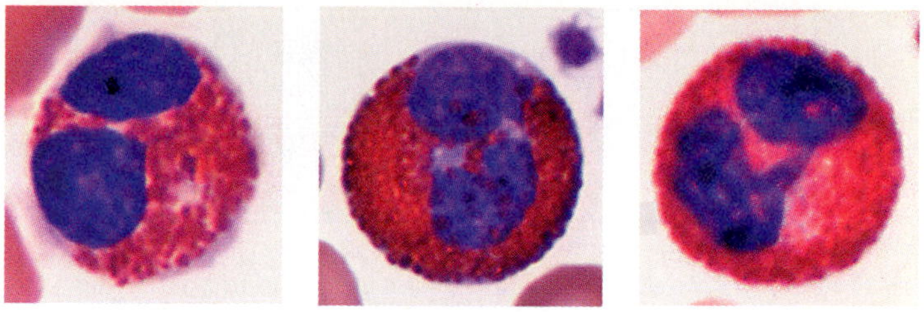

Eosinophils (3 of many variations)
•Coarse reddish granules
•Nucleus usually bilobed

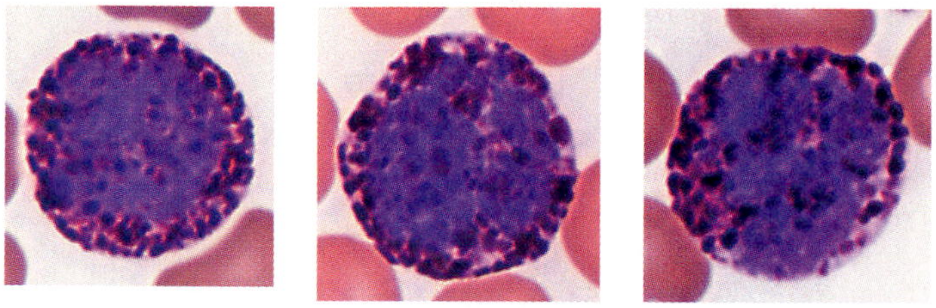

Basophils (3 of many variations)
•Coarse deep blue to almost black granules
•Nucleus often almost hidden by granules

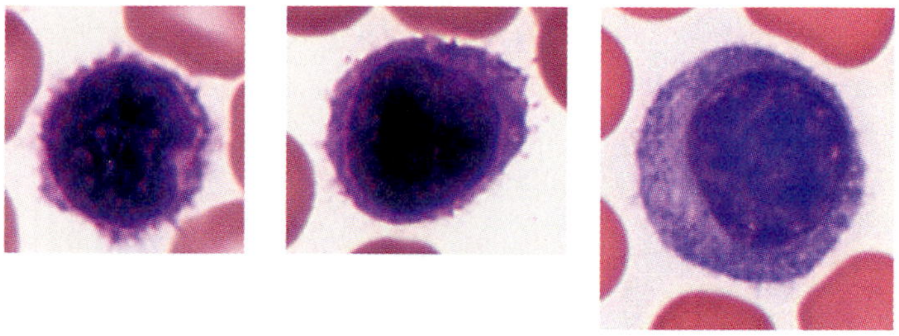

Lymphocytes (3 of many variations)
•Slightly larger than RBCs
•Thin rim of nearly clear cytoplasm
•Nearly round nucleus appears to fill most of cell in smaller lymphocytes
•Larger lymphocytes hard to distinguish from monocytes

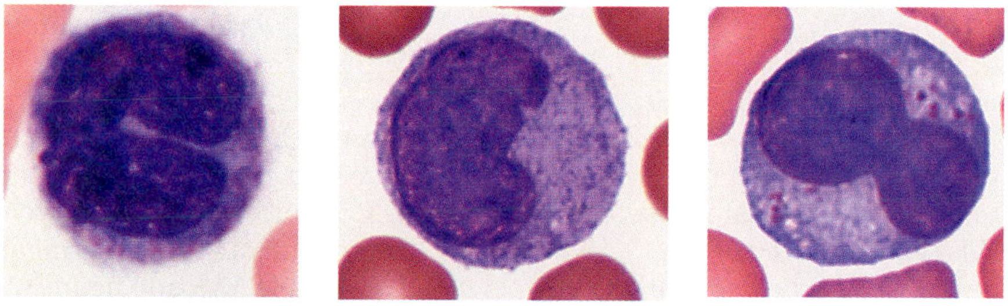

Monocytes (3 of many variations)
•Largest WBC; 2–3x larger than RBCs
•Cytoplasm nearly clear
•Nucleus round, kidney-shaped, oval, or lobed

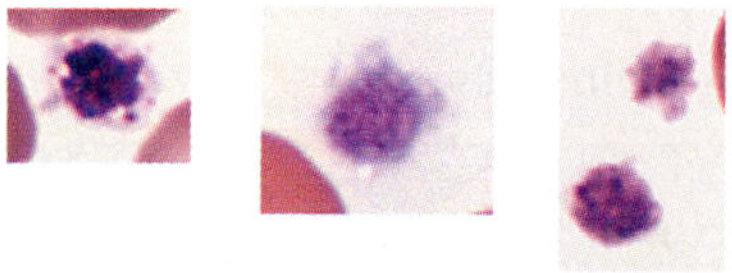

Platelets (several variations)
•Cell fragments
•Single to small clusters

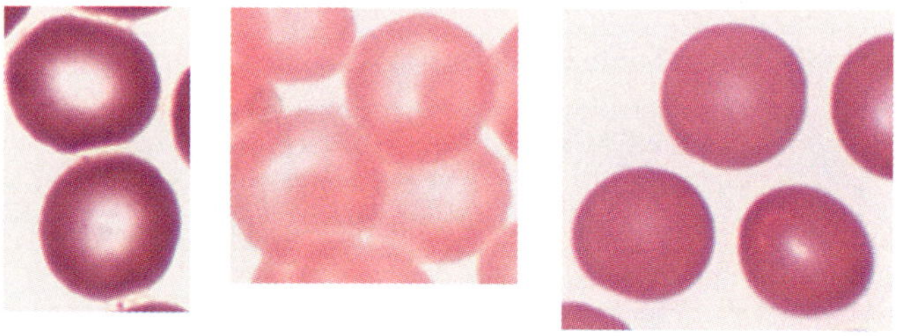

Erythrocytes (several variations)
•Lack nucleus (mature cell)
•Biconcave discs
•Thin centers appear almost hollow

Figure 37.1 Micrographs of blood cells illustrating some of the numerous variations of each type. Appearance characteristics for each cell pertain to a thin blood film using Wright's stain (1,000 ×).

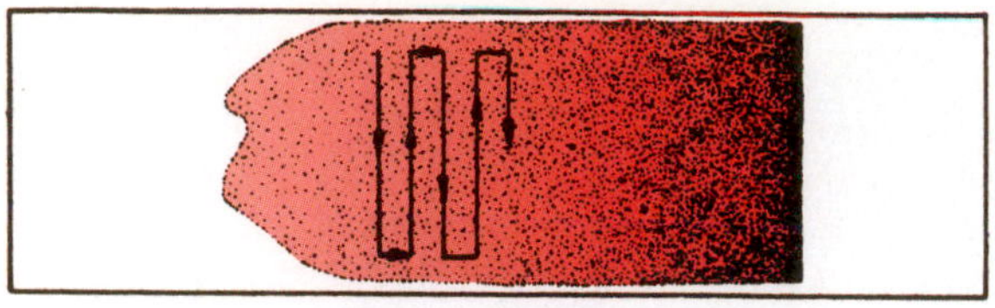

Figure 37.2 Move the blood slide back and forth to avoid passing the same cells twice.

Learning Extension

Obtain a prepared slide of pathological blood that has been stained with Wright's stain. Perform a differential white blood cell count using this slide, and compare the results with the values for normal blood listed in table 37.2. What differences do you note?

____

____

____

____

____

____

EXPLORE

Procedure C—ABO Blood Typing

1. Review the section entitled "ABO Blood Group" in chapter 14 of the textbook.
2. Complete Part D of the laboratory report.
3. Perform the ABO blood type test using the blood-typing kit. To do this, follow these steps:
   a. Obtain a clean microscope slide and mark across its center with a wax pencil to divide it into right and left halves. Also write "Anti-A" near the edge of the left half and "Anti-B" near the edge of the right half (fig. 37.3).

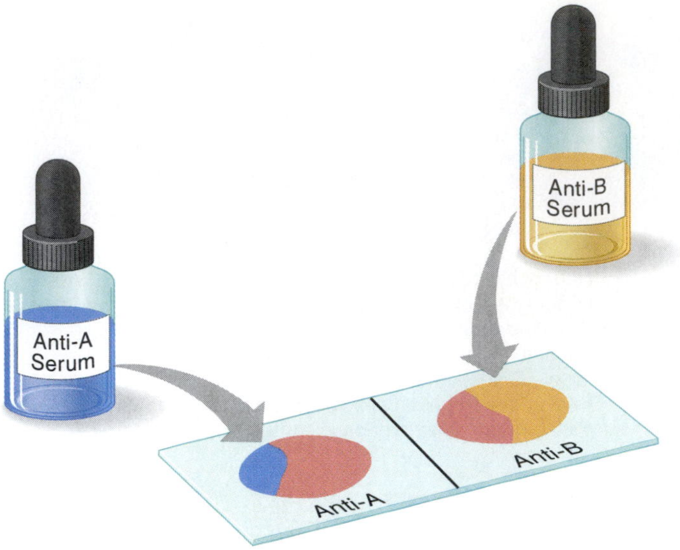

Figure 37.3 Slide prepared for ABO blood typing.

   b. Place a small drop of blood on each half of the microscope slide. Work quickly so that the blood will not have time to clot.
   c. Add a drop of anti-A serum to the blood on the left half and a drop of anti-B serum to the blood on the right half. Note the color coding of the anti-A and anti-B typing sera. To avoid contaminating the serum, avoid touching the blood with the serum while it is in the dropper; instead allow the serum to fall from the dropper onto the blood.
   d. Use separate toothpicks to stir each sample of serum and blood together, and spread each over an area about as large as a quarter. Dispose of toothpicks in an appropriate container.
   e. Examine the samples for clumping of blood cells (agglutination) after 2 minutes.
   f. See table 37.3 and figure 37.4 for aid in interpreting the test results.
   g. Discard contaminated materials as directed by the laboratory instructor.
4. Complete Part E of the laboratory report.

Table 37.3 Possible Reactions of ABO Blood-Typing Sera and Donor Compatibilities

Reactions						
Anti-A serum	**Anti-B serum**	**Blood Type**	**Average U.S. Frequency**	**Preferred Donor Type**	**Permissible Donor in Limited Amounts**	**Incompatible Donor**
Clumping	No clumping	Type A	38%	A	O	B, AB
No clumping	Clumping	Type B	11%	B	O	A, AB
Clumping	Clumping	Type AB (universal recipient)	4%	AB	A, B, O	None
No clumping	No clumping	Type O (universal donor)	47%	O	No alternative types	A, B, AB

Note: The inheritance of the ABO blood groups is described in Laboratory Exercise 53.

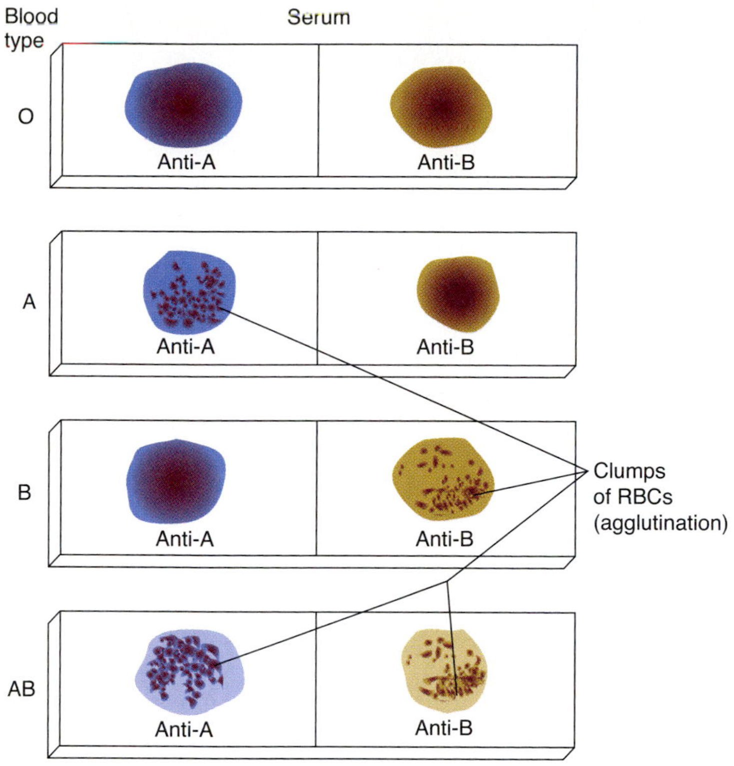

Figure 37.4 Four possible results of the ABO test.

Critical Thinking Application

Judging from the observations of the blood-typing results, suggest which components in the anti-A and anti-B sera caused clumping (agglutination).

____

____

____

____

Demonstration—Rh Blood Typing

1. Review the section entitled "Rh Blood Group" in chapter 14 of the textbook.
2. Complete Part F of the laboratory report.
3. To determine the Rh blood type of a blood sample, follow these steps:
   a. Thoroughly wash hands with soap and water, and dry them with paper towels.
   b. Cleanse the end of the middle finger with an alcohol swab, and let the finger dry in the air.
   c. Lance the tip of a finger using a sterile disposable blood lancet. Place a small drop of blood in the center of a clean microscope slide. Cover the lanced finger location with a bandage.
   d. Add a drop of anti-D serum to the blood and mix them together with a clean toothpick.
   e. Place the slide on the plate of a warming box (Rh blood-typing box or Rh view box) prewarmed to 45°C (113°F)(fig. 37.5).
   f. Slowly rock the box back and forth to keep the mixture moving, and watch for clumping (agglutination) of the blood cells. When clumping occurs in anti-D serum, the clumps usually are smaller than those that appear in anti-A or anti-B sera, so they may be less obvious. However, if clumping occurs, the blood is called Rh positive; if no clumping occurs *within 2 minutes,* the blood is called Rh negative.
   g. Discard all contaminated materials in appropriate containers.
4. Complete Part G of the laboratory report.

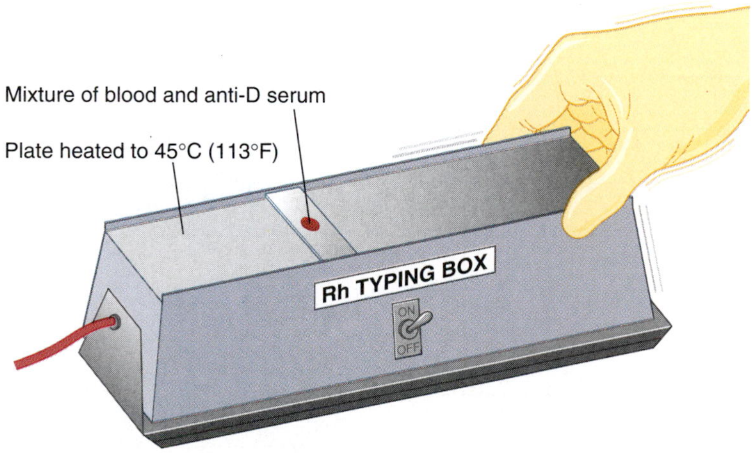

Figure 37.5 Slide warming box used for Rh blood typing.

Notes

Name ______________________________

Date ______________________________

Section ______________________________

The A corresponds to the indicated Learning Outcome(s) listed at the beginning of the laboratory exercise.

Laboratory Report 37

Blood Cells and Blood Typing

Part A Assessments

Complete the following:

1. Mature red blood cells are also called ______________________. A1
2. The functions of red blood cells are ______________________________________. A1
3. ______________________ is the oxygen-carrying substance in a red blood cell. A1
4. White blood cells are also called ______________________. A1
5. White blood cells with granular cytoplasm are called ______________________. A1
6. White blood cells lacking granular cytoplasm are called ______________________. A1
7. Normally, the most numerous white blood cells are ______________________. A1
8. A platelet, a fragment of a cell, lacks a ______________________. A1

Part B Assessments

Sketch a single blood cell of each type in the following spaces. Use colored pencils to represent the stained colors of the cells. Label any features that can be identified. A2

Red blood cell and platelet (_______ ×)

Neutrophil (_______ ×)

Lymphocyte (_______ ×)

Monocyte (_______ ×)

Eosinophil (_______ ×)

Basophil (_______ ×)

Part C Assessments

1. *Differential White Blood Cell Count Data Table.* As you identify white blood cells, record them on the table by using a tally system, such as ||||| ||. Place tally marks in the "Number Observed" column, and total each of the five WBCs when the differential count is completed. Obtain a total of all five WBCs counted to determine the percent of each WBC type.

Type of WBC	Number Observed	Total	Percent
Neutrophil			
Lymphocyte			
Monocyte			
Eosinophil			
Basophil			
		Total of column	

2. How do the results of your differential white blood cell count compare with the normal values listed in table 37.2? ____

Critical Thinking Application

What is the difference between a differential white blood cell count and a total white blood cell count?

____

Part D Assessments

Complete the following statements:

1. The antigens of the ABO blood group are located in the ____________________.
2. The blood of every person contains one of (how many possible?) ____________________ combinations of antigens.
3. Type A blood contains antigen ____________________.
4. Type B blood contains antigen ____________________.
5. Type A blood contains ____________________ antibody in the plasma.
6. Type B blood contains ____________________ antibody in the plasma.
7. Persons with ABO blood type ____________________ are often called universal recipients.
8. Persons with ABO blood type ____________________ are often called universal donors.

Part E Assessments

Complete the following:

1. What was the ABO type of the blood tested? 5 ____________________
2. What ABO antigens are present in the red blood cells of this type of blood? 5 ____________________.
3. What ABO antibodies are present in the plasma of this type of blood? 5 ____________________.
4. If a person with this blood type needed a blood transfusion, what ABO type(s) of blood could be received safely? 5 ____________________.
5. If a person with this blood type was serving as a blood donor, what ABO blood type(s) could receive the blood safely? 5 ____________________.

Part F Assessments

Complete the following statements:

1. The Rh blood group was named after the ____________________. 6
2. Of the antigens in the Rh group, the most important is ____________________. 6
3. If red blood cells lack Rh antigens, the blood is called ____________________. 6
4. If an Rh-negative person sensitive to Rh-positive blood receives a transfusion of Rh-positive blood, the donor's cells are likely to ____________________. 6
5. An Rh-negative woman who might be carrying an ____________________ fetus is given an injection of RhoGAM to prevent hemolytic disease of the fetus and newborn (erythroblastosis fetalis). 6

Part G Assessments

Complete the following:

1. What was the Rh type of blood tested in the demonstration? 7 ____________________.
2. What Rh antigen is present in the red blood cells of this type of blood? 7 ____________________.
3. If a person with this blood type needed a blood transfusion, what type of blood could be received safely? 7

   ____________________.
4. If a person with this blood type was serving as a blood donor, a person with what type of blood could receive the blood safely? 7 ____________________.

Notes

Laboratory Exercise 38

Heart Structure

Materials Needed

Textbook
Dissectible human heart model
Preserved sheep or other mammalian heart
Dissecting tray
Dissecting instruments

For Learning Extension:
Colored pencils

Safety

- Wear disposable gloves when working on the heart dissection.
- Save or dispose of the dissected heart as instructed.
- Wash the dissecting tray and instruments as instructed.
- Wash your laboratory table.
- Wash your hands before leaving the laboratory.

The heart is a muscular pump located within the mediastinum and resting upon the diaphragm. It is enclosed by the lungs, thoracic vertebrae, and sternum, and attached at its superior end (the base) are several large blood vessels. Its inferior end extends downward to the left and terminates as a bluntly pointed apex.

The heart and the proximal ends of the attached blood vessels are enclosed by a double-layered pericardium. The innermost layer of this membrane (visceral pericardium) consists of a thin covering closely applied to the surface of the heart, whereas the outer layer (parietal pericardium with fibrous pericardium) forms a tough, protective sac surrounding the heart. Between the parietal and visceral layers of the pericardium is a space, the pericardial cavity, that contains a small volume of serous (pericardial) fluid.

Purpose of the Exercise

To review the structural characteristics of the human heart and to examine the major features of a mammalian heart.

LEARNING OUTCOMES

After completing this exercise, you should be able to

1. Identify and label the major structural features of the human heart and closely associated blood vessels.
2. Match heart structures with appropriate locations and functions.
3. Compare the features of the human heart with those of another mammal.

EXPLORE

Procedure A—The Human Heart

1. Review the section entitled "Structure of the Heart" in chapter 15 of the textbook.
2. As a review activity, label figures 38.1, 38.2, and 38.3.
3. Complete Part A of Laboratory Report 38.
4. Examine the human heart model, and locate the following features:

heart
- base (superior region where blood vessels emerge)
- apex (inferior, rounded end)

pericardium (pericardial sac)
- fibrous pericardium (outer layer)
- parietal pericardium (inner lining of fibrous pericardium)
- visceral pericardium (epicardium)

pericardial cavity (between parietal and visceral pericardial membranes)

myocardium (cardiac muscle)

endocardium (lines heart chambers)

atria
- right atrium
- left atrium
- auricles

ventricles
- right ventricle
- left ventricle

atrioventricular orifices

atrioventricular valves (AV valves)

tricuspid valve (right atrioventricular valve)

mitral valve (bicuspid valve; left atrioventricular valve)

semilunar valves

pulmonary valve

aortic valve

chordae tendineae

papillary muscles

atrioventricular sulcus

interventricular sulci

anterior sulcus

posterior sulcus

superior vena cava

inferior vena cava

pulmonary trunk

pulmonary arteries

pulmonary veins

aorta

left coronary artery

circumflex artery

anterior interventricular (descending) artery

right coronary artery

posterior interventricular artery

marginal artery

cardiac veins

great cardiac vein

middle cardiac vein

small cardiac vein

coronary sinus (for return of blood from cardiac veins into right atrium)

5. Label the human heart model in figure 38.4.

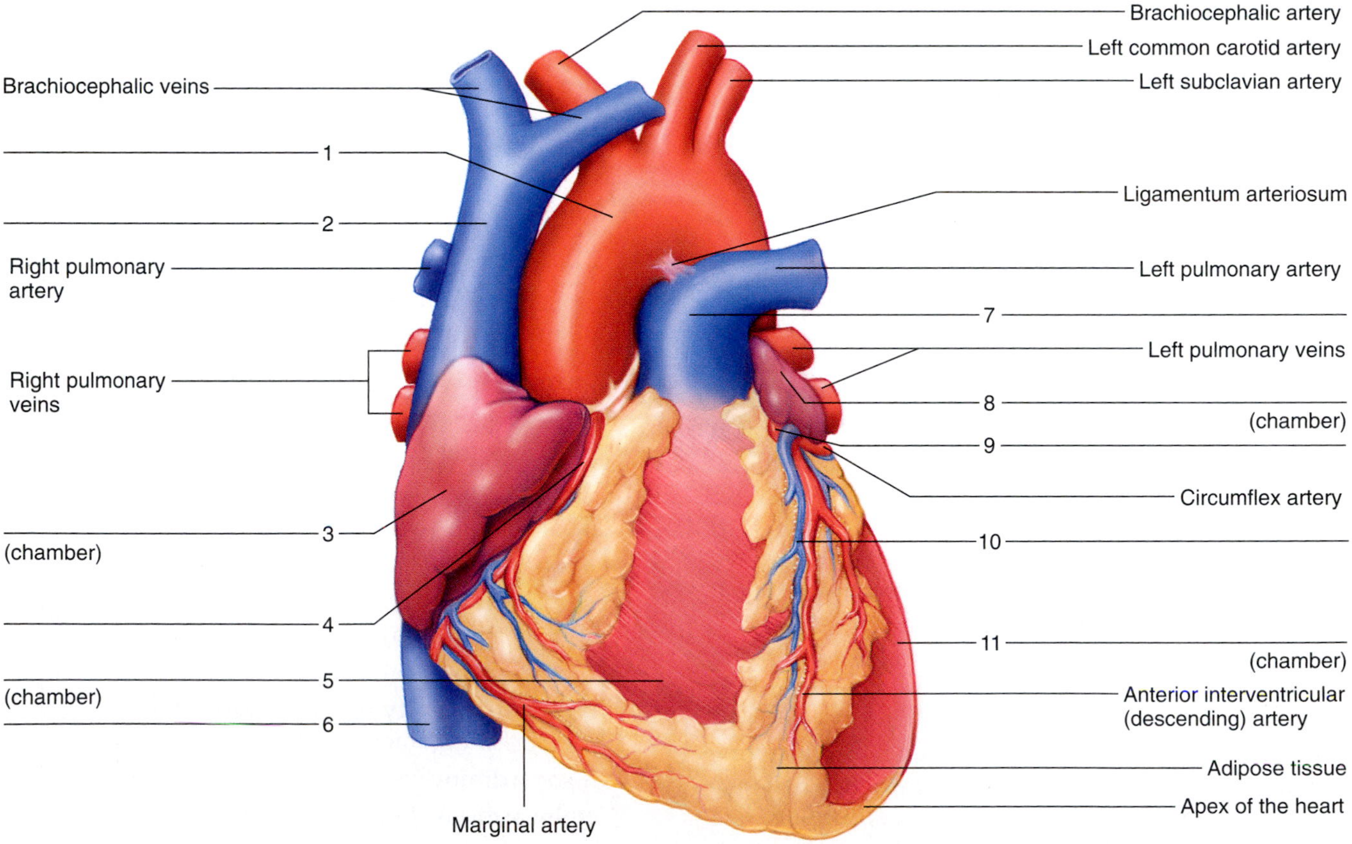

Figure 38.1 Label this anterior view of the human heart. 1

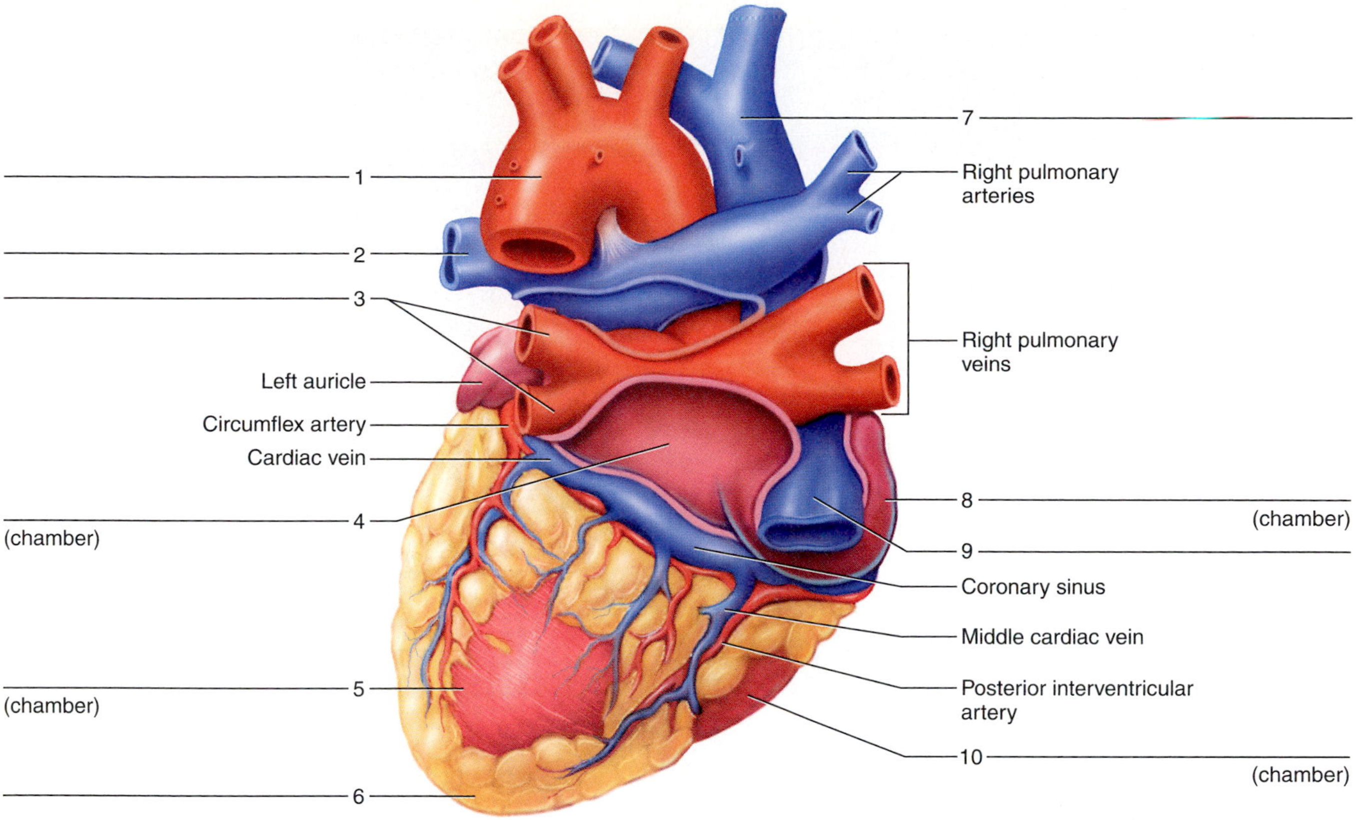

Figure 38.2 Label this posterior view of the human heart.

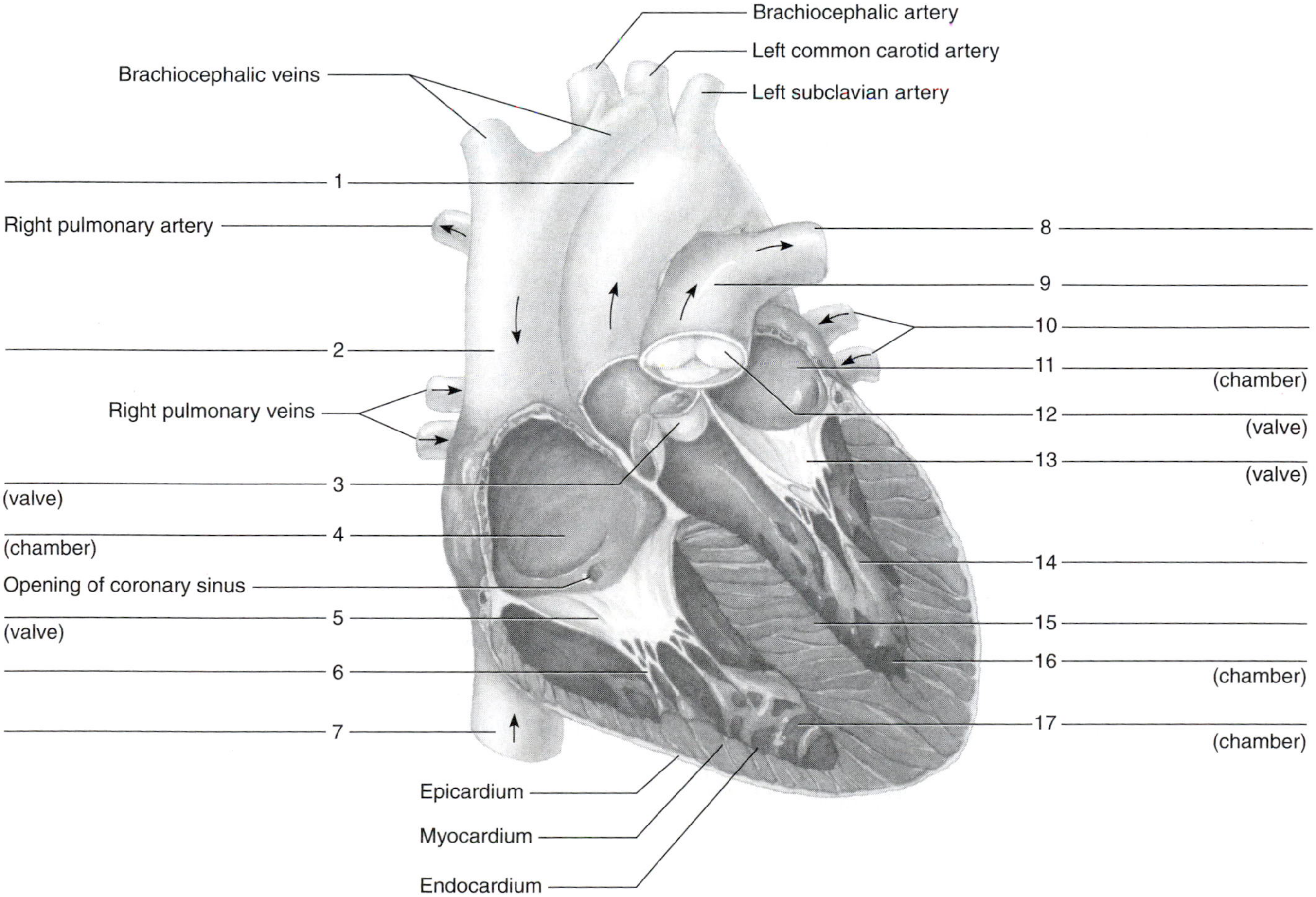

Figure 38.3 Label this frontal section of the human heart. The arrows indicate the direction of blood flow.

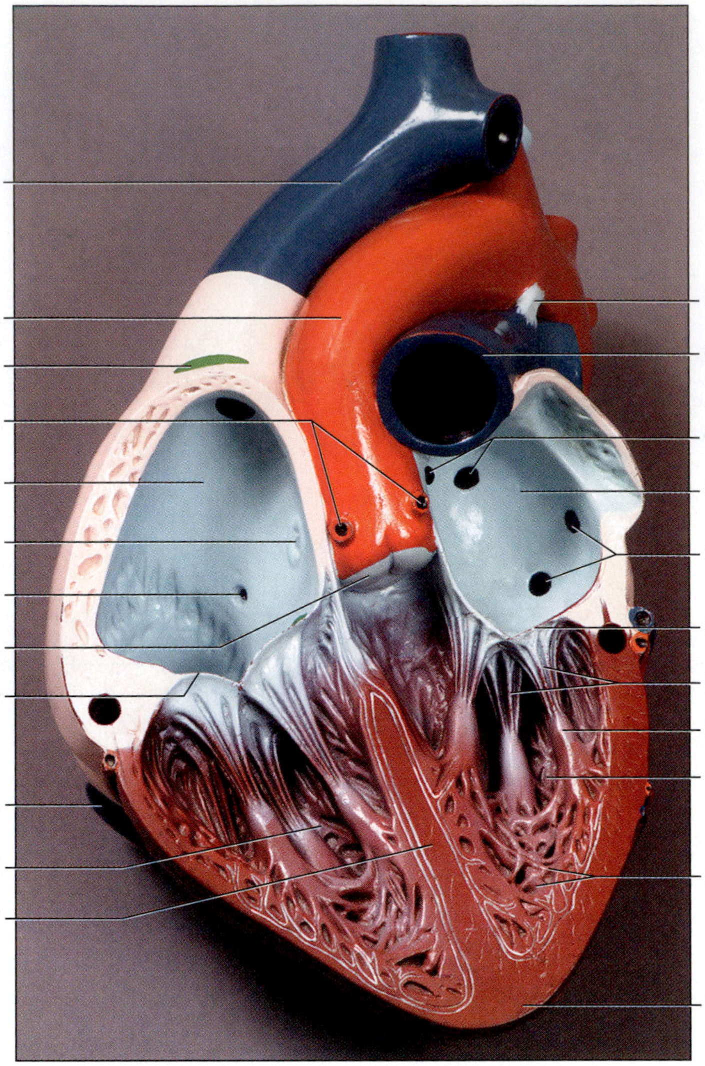

Terms:

Aorta	Mitral valve
Aortic valve	Right atrium
Apex	Right ventricle
Interventricular septum	Superior vena cava
Left atrium	Tricuspid valve
Left ventricle	

Figure 38.4 Identify the features indicated on this anterior view of a frontal section of a human heart model, using the terms provided. (*Note:* The pulmonary valve is not shown on the portion of the model photographed.) 1

Learning Extension

Use red and blue colored pencils to color the blood vessels in figure 38.3. Use red to illustrate a blood vessel high in oxygen, and use blue to illustrate a blood vessel low in oxygen. You can check your work by referring to the corresponding figures in the textbook, presented in full color.

EXPLORE

Procedure B—Dissection of a Sheep Heart

1. Obtain a preserved sheep heart. Rinse it in water thoroughly to remove as much of the preservative as possible. Also run water into the large blood vessels to force any blood clots out of the heart chambers.

2. Place the heart in a dissecting tray with its ventral surface up (fig. 38.5), and proceed as follows:
   a. Although the relatively thick *pericardial sac* probably is missing, look for traces of this membrane around the origins of the large blood vessels.
   b. Locate the *visceral pericardium,* which appears as a thin, transparent layer on the surface of the heart. Use a scalpel to remove a portion of this layer and expose the *myocardium* beneath. Also note the abundance of fat along the paths of various blood vessels. This adipose tissue occurs in the loose connective tissue that underlies the visceral pericardium.
   c. Identify the following:

      right atrium
      right ventricle
      left atrium
      left ventricle
      atrioventricular sulcus
      anterior interventricular sulcus

   d. Carefully remove the fat from the anterior interventricular sulcus, and expose the blood vessels that pass along this groove. They include a branch of the *left coronary artery* (anterior interventricular artery) and a *cardiac vein.*
3. Examine the dorsal surface of the heart (fig. 38.6), and proceed as follows:
   a. Identify the *atrioventricular sulcus* and the *posterior interventricular sulcus.*
   b. Locate the stumps of two relatively thin-walled veins that enter the right atrium. Demonstrate this connection by passing a slender probe through them. The upper vessel is the *superior vena cava,* and the lower one is the *inferior vena cava.*
4. Open the right atrium. To do this, follow these steps:
   a. Insert a blade of the scissors into the superior vena cava and cut downward through the atrial wall (fig. 38.6).
   b. Open the chamber, locate the *right atrioventricular valve (tricuspid valve),* and examine its cusps.

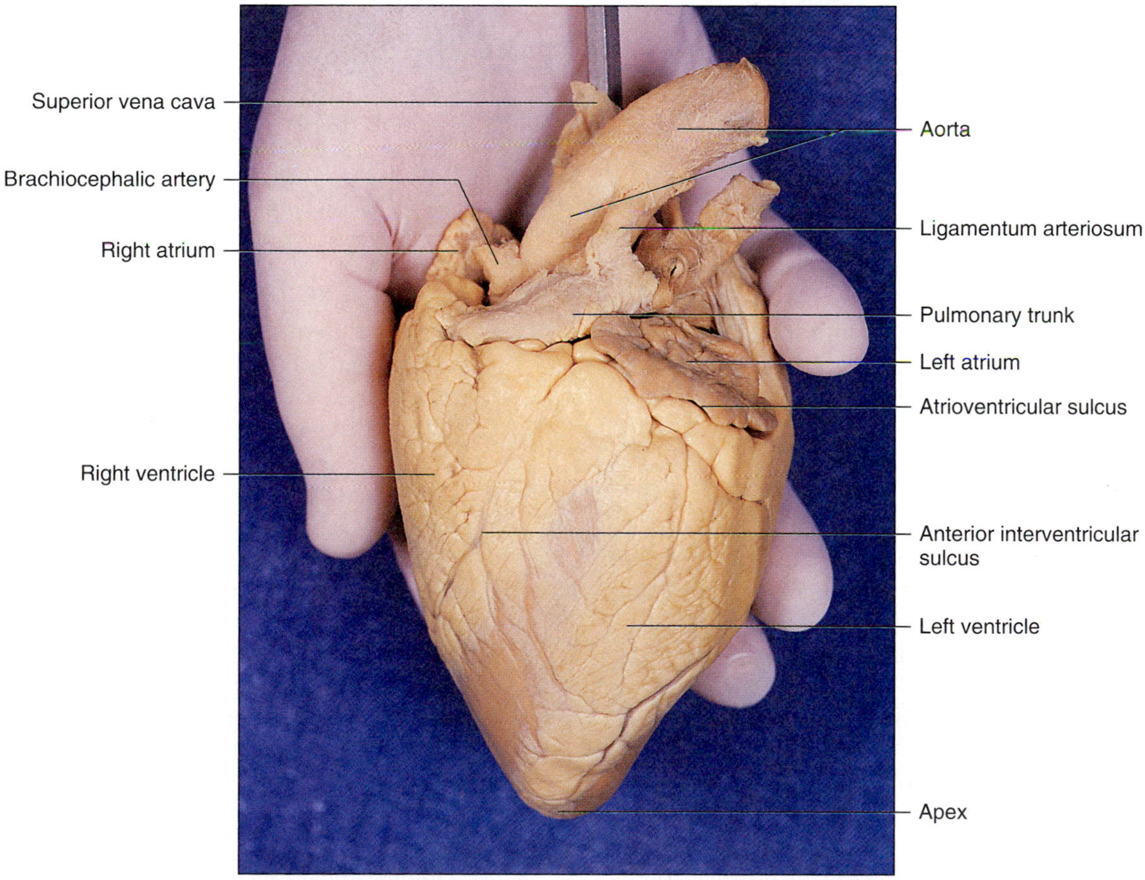

Figure 38.5 Ventral surface of sheep heart.

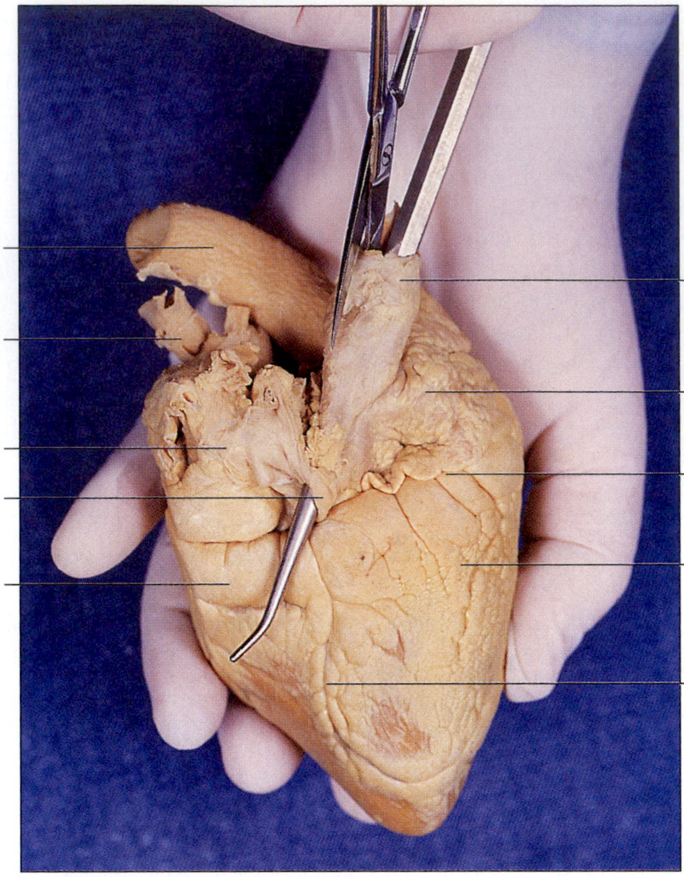

Figure 38.6 Dorsal surface of sheep heart. To open the right atrium, insert a blade of the scissors into the superior (anterior in sheep) vena cava and cut downward.

   c. Also locate the opening to the *coronary sinus* between the valve and the inferior vena cava.
   d. Run some water through the right atrioventricular valve to fill the chamber of the right ventricle.
   e. Gently squeeze the ventricles, and watch the cusps of the valve as the water moves up against them.
5. Open the right ventricle as follows:
   a. Continue cutting downward through the right atrioventricular valve and the right ventricular wall until you reach the apex of the heart.
   b. Locate the *chordae tendineae* and the *papillary muscles.*
   c. Find the opening to the *pulmonary trunk,* and use the scissors to cut upward through the wall of the right ventricle. Follow the pulmonary trunk until you have exposed the *pulmonary valve.*
   d. Examine the valve and its cusps.
6. Open the left side of the heart. To do this, follow these steps:
   a. Insert the blade of the scissors through the wall of the left atrium and cut downward to the apex of the heart.
   b. Open the left atrium, and locate the four openings of the *pulmonary veins.* Pass a slender probe through each opening, and locate the stump of its vessel.
   c. Examine the *left atrioventricular valve (bicuspid valve)* and its cusps.
   d. Also examine the left ventricle, and compare the thickness of its wall with that of the right ventricle.
7. Locate the aorta, which leads away from the left ventricle, and proceed as follows:
   a. Compare the thickness of the aortic wall with that of a pulmonary artery.
   b. Use scissors to cut along the length of the aorta to expose the *aortic valve* at its base.
   c. Examine the cusps of the valve, and locate the openings of the *coronary arteries* just distal to them.
8. As a review, locate and identify the stumps of each of the major blood vessels associated with the heart.
9. Discard or save the specimen as directed by the laboratory instructor.
10. Complete Part B of the laboratory report.

Name ______________________________

Date ______________________________

Section ______________________________

The [A] corresponds to the indicated Learning Outcome(s) listed at the beginning of the laboratory exercise.

Laboratory Report 38

Heart Structure

Part A Assessments

Match the terms in column A with the descriptions in column B. Place the letter of your choice in the space provided. [2]

Column A		Column B
a. Aorta	_______	1. Structure from which chordae tendineae originate
b. Cardiac vein	_______	2. Prevents blood movement from right ventricle to right atrium
c. Coronary artery	_______	3. Membranes around heart
d. Coronary sinus	_______	4. Prevents blood movement from left ventricle to left atrium
e. Endocardium	_______	5. Gives rise to left and right pulmonary arteries
f. Mitral valve	_______	6. Drains blood from myocardium into right atrium
g. Myocardium	_______	7. Inner lining of heart chamber
h. Papillary muscle	_______	8. Layer largely composed of cardiac muscle tissue
i. Pericardial cavity	_______	9. Space containing serous fluid to reduce friction during heartbeats
j. Pericardial sac	_______	10. Drains blood from myocardial capillaries
k. Pulmonary trunk	_______	11. Supplies blood to heart muscle
l. Tricuspid valve	_______	12. Distributes blood to body organs (systemic circuit) except lungs

Part B Assessments

Complete the following:

1. Compare the structure of the right atrioventricular valve with that of the pulmonary valve. [3]

____

____

____

____

2. Describe the action of the right atrioventricular valve when you squeezed the water-filled right ventricle. [3]

____

____

____

____

3. Describe the function of the chordae tendineae and the papillary muscles. 3

4. What is the significance of the difference in thickness between the wall of the aorta and the wall of the pulmonary trunk? 3

5. List in order the major blood vessels, chambers, and valves through which blood must pass in traveling from a vena cava to the aorta. 3

Critical Thinking Application

What is the significance of the difference in thickness of the ventricular walls? 3

Laboratory Exercise 39

Cardiac Cycle

Materials Needed

Textbook

For Procedure A—Heart Sounds:

Stethoscope
Alcohol swabs

For Procedure B—Electrocardiogram:

Electrocardiograph (or other instrument for recording an ECG)
Cot or table
Alcohol swabs
Electrode cream (paste)
Plate electrodes and cables
Self-sticking leads (optional)
Lead selector switch

A set of atrial contractions while the ventricular walls relax, followed by ventricular contractions while the atrial walls relax, constitutes a cardiac cycle. Such a cycle is accompanied by blood pressure changes within the heart chambers, movement of blood in and out of the chambers, and opening and closing of heart valves. These valves closing produce vibrations in the tissues and thus create the sounds associated with the heartbeat. If backflow of blood occurs when a heart valve is closed, creating a turbulence noise, the condition is called a murmur.

The regulation and coordination of the cardiac cycle involves the cardiac conduction system. The pathway of electrical signals originates from the SA (sinoatrial) node located in the right atrium near the entrance of the superior vena cava. The stimulation of the heartbeat and the heart rate originate from the SA node, so it is often called the pacemaker. As the signals pass through the atrial walls toward the AV (atrioventricular) node, contractions of the atria take place. When the AV node has been signaled, the rapid continuation of the electrical signals occurs through the AV bundle, passes through the right and left bundle branches within the interventricular septum, and terminates via the Purkinje fibers throughout the ventricular walls. Once the myocardium has been stimulated, ventricular contractions happen, completing one cardiac cycle. The sympathetic and parasympathetic subdivisions of the autonomic nervous system influence the activity of the pacemaker under various conditions. An increased rate results from sympathetic responses; a decreased rate results from parasympathetic responses.

A number of electrical changes also occur in the myocardium as it contracts and relaxes. These changes can be detected by using metal electrodes and an instrument called an *electrocardiograph.* The recording produced by the instrument is an *electrocardiogram,* or *ECG (EKG).*

Purpose of the Exercise

To review the events of a cardiac cycle, to become acquainted with normal heart sounds, and to record an electrocardiogram.

LEARNING OUTCOMES

After completing this exercise, you should be able to

1. Interpret the major events of a cardiac cycle.
2. Associate the sounds produced during a cardiac cycle with the valves closing.
3. Correlate the components of a normal ECG pattern with the phases of a cardiac cycle.
4. Record and interpret an electrocardiogram.

EXPLORE

Procedure A—Heart Sounds

1. Review the sections entitled "Cardiac Cycle" and "Heart Sounds" in chapter 15 of the textbook.
2. Complete Part A of Laboratory Report 39.
3. Listen to your heart sounds. To do this, follow these steps:
   a. Obtain a stethoscope, and clean its earpieces and the diaphragm by using alcohol swabs.
   b. Fit the earpieces into your ear canals so that the angles are positioned in the forward direction.

c. Firmly place the diaphragm (or bell) of the stethoscope on the chest over the fifth intercostal space near the apex of the heart (fig. 39.1) and listen to the sounds. This is a good location to hear the first sound (*lubb*) of a cardiac cycle when the AV valves are closing, which occurs during ventricular systole (contraction).

d. Move the diaphragm to the second intercostal space, just to the left of the sternum, and listen to the sounds from this region. You should be able to hear the second sound (*dupp*) of the cardiac cycle clearly when the semilunar valves are closing, which occurs during ventricular diastole (relaxation).

e. It is possible to hear sounds associated with the aortic and pulmonary valves by listening from the second intercostal space on either side of the sternum. The aortic valve sound comes from the right and the pulmonary valve sound from the left. The sound associated with the mitral valve can be heard from the fifth intercostal space at the nipple line on the left. The sound of the tricuspid valve can be heard at the fifth intercostal space just to the right of the sternum (fig. 39.1).

4. Inhale slowly and deeply, and exhale slowly while you listen to the heart sounds from each of the locations as before. Note any changes that have occurred in the sounds.
5. Exercise moderately outside the laboratory for a few minutes so that other students listening to heart sounds will not be disturbed. After the exercise period, listen to the heart sounds and note any changes that have occurred in them. *This exercise should be avoided by anyone with health risks.*
6. Complete Part B of the laboratory report.

EXPLORE

Procedure B—Electrocardiogram

1. Review sections entitled "Cardiac Conduction System," "Electrocardiogram," and "Arrhythmias—Clinical Application" in chapter 15 of the textbook.
2. Complete Part C of the laboratory report.
3. The laboratory instructor will demonstrate the proper adjustment and use of the instrument available to record an electrocardiogram.
4. Record your laboratory partner's ECG. To do this, follow these steps:
   a. Have your partner lie on a cot or table close to the electrocardiograph, remaining as relaxed and still as possible.

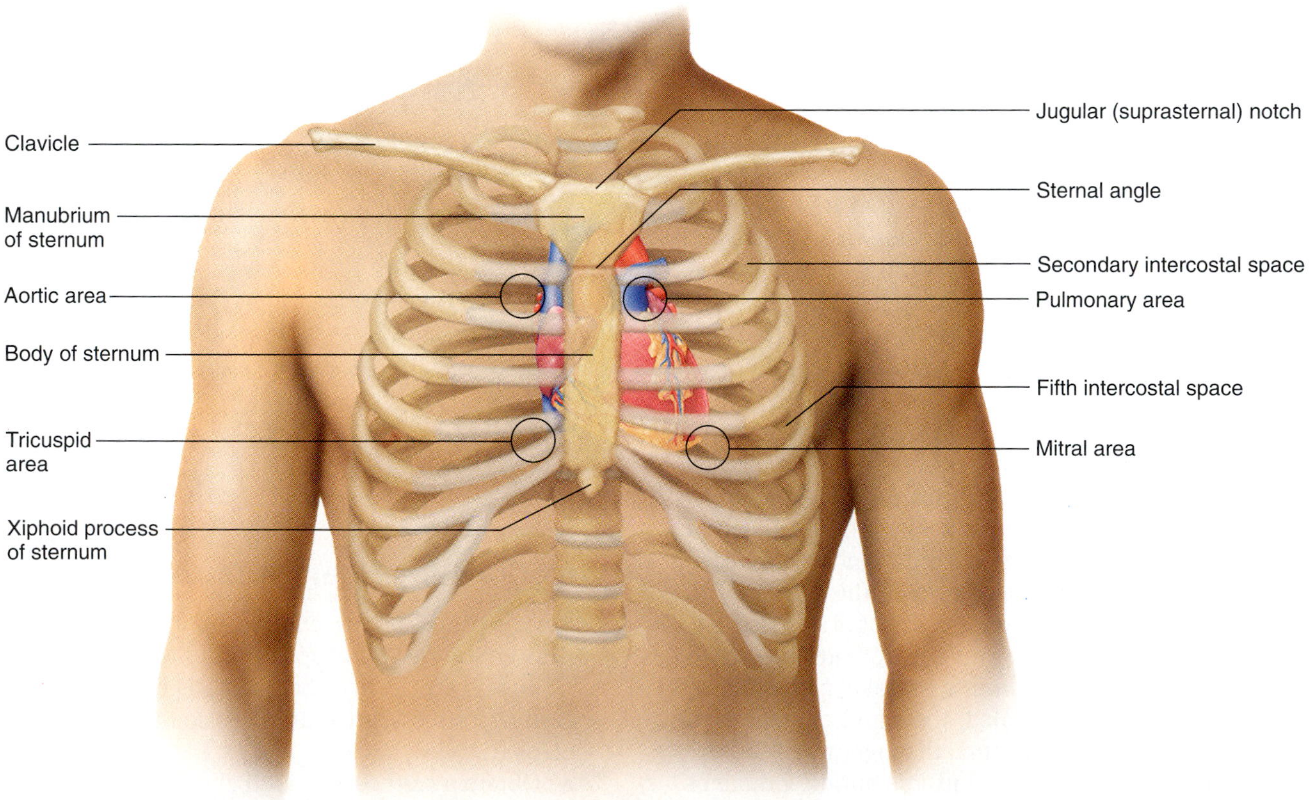

Figure 39.1 The first sound (lubb) of a cardiac cycle can be heard by placing the diaphragm of a stethoscope over the fifth intercostal space near the apex of the heart. The second sound (dupp) can be heard over the second intercostal space, just left of the sternum. The thoracic regions circled indicate where the sounds of each heart valve are most easily heard.

b. Scrub the electrode placement locations with alcohol swabs (fig. 39.2). Apply a small quantity of electrode cream to the skin on the insides of the wrists and ankles. (Any jewelry on the wrists or ankles should be removed.)

c. Spread some electrode cream over the inner surfaces of four plate electrodes and attach one to each of the prepared skin areas (fig. 39.2). Make sure there is good contact between the skin and the metal of the electrodes. The electrode plate on the right ankle is the grounding system.

d. Attach the plate electrodes to the corresponding cables of a lead selector switch. When an ECG recording is made, only two electrodes are used at a time, and the selector switch allows various combinations of electrodes (leads) to be activated. Three standard limb leads placed on the two wrists and the left ankle are used for an ECG. This arrangement has become known as *Einthoven's triangle,*[1] which permits the recording of the potential difference between any two of the electrodes.

The standard leads I, II, and III are called bipolar leads because they are the potential difference between two electrodes (a positive and a negative). Lead I measures the potential difference between the right wrist (negative) and the left wrist (positive). Lead II measures the potential difference between the right wrist and the left ankle, and lead III measures the potential difference between the left wrist and the left ankle. The right ankle is always the ground.

e. Turn on the recording instrument and adjust it as previously demonstrated by the laboratory instructor. The paper speed should be set at 2.5 cm/second. This is the standard speed for ECG recordings.

f. Set the lead selector switch to lead I (right wrist, left wrist electrodes), and record the ECG for 1 minute.

g. Set the lead selector switch to lead II (right wrist, left ankle electrodes), and record the ECG for 1 minute.

h. Set the lead selector switch to lead III (left wrist, left ankle electrodes), and record the ECG for 1 minute.

i. Remove the electrodes and clean the cream from the metal and skin.

j. Use figure 39.3 to label the ECG components of the results from leads I, II, and III. The P-Q interval is often called the P-R interval because the Q wave is often small or absent. The normal P-Q interval is 0.12–0.20 seconds. The normal QRS complex duration is less than 0.10 seconds.

5. Complete Part D of the laboratory report.

[1]Willem Einthoven (1860–1927), a Dutch physiologist, received the Nobel prize for physiology or medicine for his work with electrocardiograms.

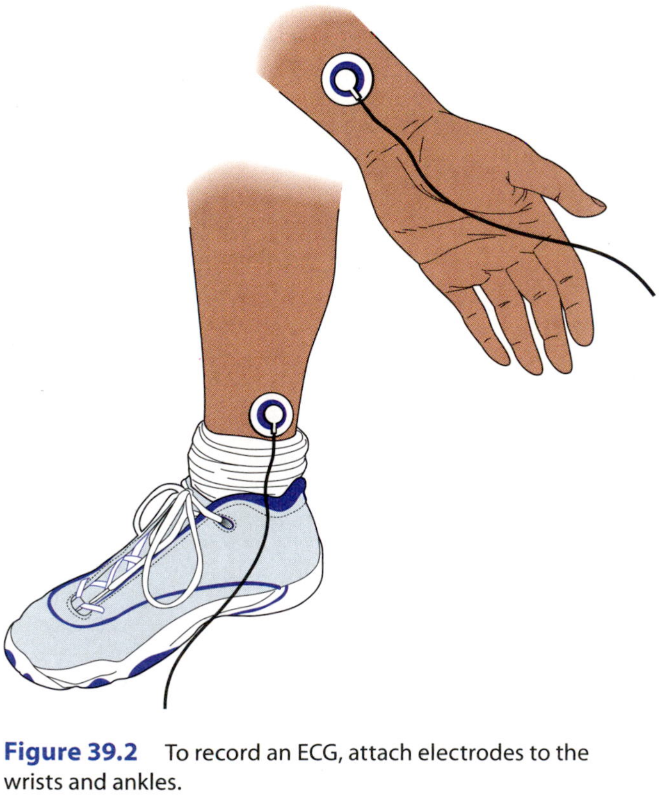

Figure 39.2 To record an ECG, attach electrodes to the wrists and ankles.

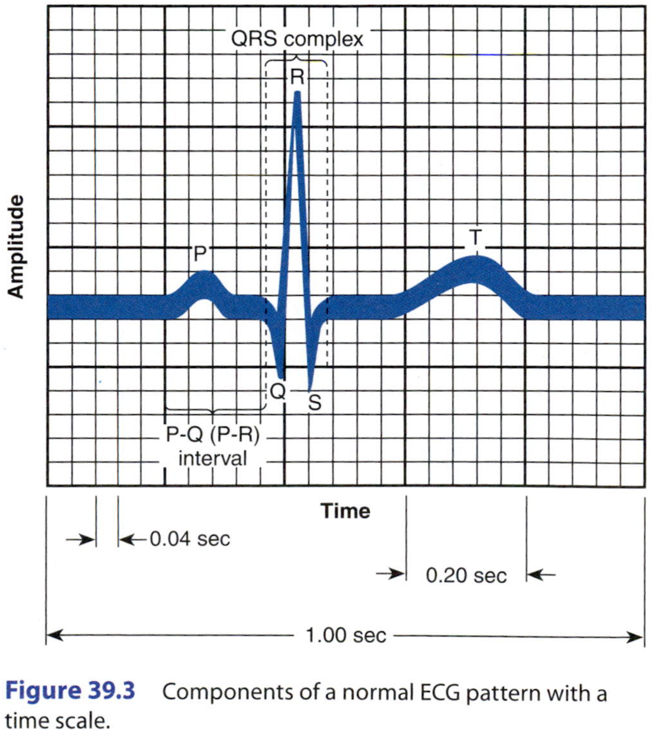

Figure 39.3 Components of a normal ECG pattern with a time scale.

Notes

Name ______________________________

Date ______________________________

Section ______________________________

The [A] corresponds to the indicated Learning Outcome(s) listed at the beginning of the laboratory exercise.

Cardiac Cycle

Laboratory Report 39

Part A Assessments

Complete the following statements:

1. About ______________ % of the blood from the atria passes into the ventricles before the atrial walls contract. [1]
2. The period during which a heart chamber is contracting is called ______________ . [1]
3. The period during which a heart chamber is relaxing is called ______________ . [1]
4. During ventricular contraction, the AV valves (tricuspid and mitral valves) are ______________ . [2]
5. During ventricular relaxation, the AV valves are ______________ . [2]
6. The pulmonary and aortic valves open when the pressure in the ______________ exceeds the pressure in the pulmonary trunk and aorta. [1]
7. Heart sounds are due to ______________ in heart tissues created by changes in blood flow. [2]
8. The first sound of a cardiac cycle occurs when the ______________ are closing. [2]
9. The second sound of a cardiac cycle occurs when the ______________ are closing. [2]
10. The sound created when blood leaks back through an incompletely closed valve is called a ______________ . [2]

Part B Assessments

Complete the following:

1. What changes did you note in the heart sounds when you inhaled deeply? [2]

____

____

2. What changes did you note in the heart sounds following the exercise period? [2]

____

____

Part C Assessments

Complete the following statements:

1. The fibers of the cardiac conduction system are specialized ______________________ tissue. 3
2. Normally, the ______________________ node serves as the pacemaker of the heart. 3
3. The ______________________ node is located in the inferior portion of the interatrial septum. 3
4. The large fibers on the distal side of the AV node make up the ______________________ . 3
5. The fibers that carry cardiac impulses from the interventricular septum into the myocardium are called ______________________ . 3
6. A(n) ______________________ is a recording of electrical changes occurring in the myocardium during the cardiac cycle. 3
7. Between cardiac cycles, cardiac muscle fibers remain ____________ with no detectable electrical changes. 3
8. The P wave corresponds to depolarization of the muscle fibers of the ______________________ . 3
9. The QRS complex corresponds to depolarization of the muscle fibers of the ______________________ . 3
10. The T wave corresponds to repolarization of the muscle fibers of the ______________________ . 3
11. Why is atrial repolarization not observed in the ECG? 3 ____

____
12. A rapid heartbeat is called ______________________; a slow heartbeat is called ______________________ . 4

Part D Assessments

1. Attach a short segment of the ECG recording from each of the three leads you used, and label the waves of each. 4

   Lead I

   Lead II

Lead III

2. What differences do you find in the ECG patterns of these leads? ____________________

3. How much time passed from the beginning of the P wave to the beginning of the QRS complex (P-Q interval, or P-R interval) in the ECG from lead I? ____________________

4. What is the significance of this P-Q (P-R) interval? ____________________

5. How can you determine the heart rate from an electrocardiogram? ____________________

6. What was your heart rate as determined from the ECG? ____________________

Critical Thinking Application

If a person's heart rate is 72 beats per minute, determine the number of QRS complexes that would have appeared on an ECG during the first 30 seconds.

Notes

Laboratory Exercise 40

Blood Vessel Structure, Arteries, and Veins

Materials Needed

Textbook
Human torso model
Anatomical charts of the cardiovascular system
Compound light microscope
Prepared microscope slides:
  Artery cross section
  Vein cross section
Live frog or goldfish
Frog Ringer's solution
Paper towel
Rubber bands
Frog board or heavy cardboard (with a 1-inch hole cut in one corner)
Dissecting pins
Thread
Masking tape

For Learning Extension:

Ice
Hot plate
Thermometer

Safety

- Wear disposable gloves when handling the live frogs.
- Return the frogs to the location indicated after the experiment.
- Wash your hands before leaving the laboratory.

The blood vessels form a closed system of tubes that carry blood to and from the heart, lungs, and body cells. These tubes include arteries and arterioles that conduct blood away from the heart; capillaries in which exchanges of substances occur between the blood and surrounding tissues; and venules and veins that return blood to the heart.

The blood vessels of the cardiovascular system can be divided into two major pathways—the pulmonary circuit and the systemic circuit. Within each circuit, arteries transport blood away from the heart. After exchanges of gases, nutrients, and wastes have occurred between the blood and the surrounding tissues, veins return the blood to the heart.

Frequent variations exist in anatomical structures among humans, especially in arteries and veins. The illustrations in this laboratory manual represent normal (normal means the most common variation) anatomy.

Purpose of the Exercise

To review the structure and functions of blood vessels, trace major circulatory pathways, and locate the major arteries and veins.

LEARNING OUTCOMES

After completing this exercise, you should be able to

1. Distinguish structures and functions of arteries, veins, and capillaries.
2. Identify the types of blood vessels in the web of a frog's foot.
3. Distinguish and trace the pulmonary and systemic circuits.
4. Locate the major arteries in these circuits on a diagram, chart, or model.
5. Locate the major veins in these circuits on a diagram, chart, or model.

EXPLORE

Procedure A—Blood Vessel Structure

1. Review the section entitled "Blood Vessels" in chapter 15 of the textbook.
2. As a review activity, label figure 40.1.
3. Complete Part A of Laboratory Report 40.
4. Obtain a microscope slide of an artery cross section, and examine it using low-power and high-power

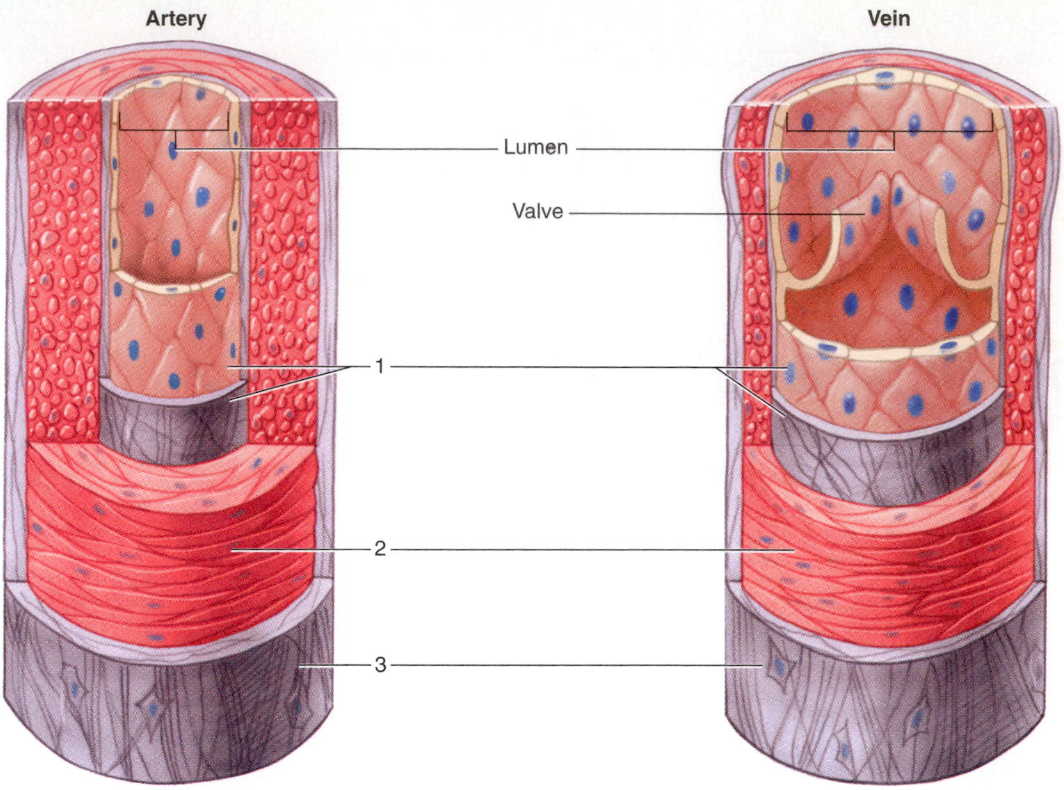

Figure 40.1 Label the tunics of the walls of this artery and vein.

magnification. Identify the three distinct layers (tunics) of the arterial wall. The inner layer **(tunica interna)** is composed of an endothelium (simple squamous epithelium) and appears as a wavy line due to an abundance of elastic fibers that have recoiled just beneath it. The middle layer **(tunica media)** consists of numerous concentrically arranged smooth muscle cells with elastic fibers scattered among them. The outer layer **(tunica externa)** contains connective tissue rich in collagenous fibers (fig. 40.2).

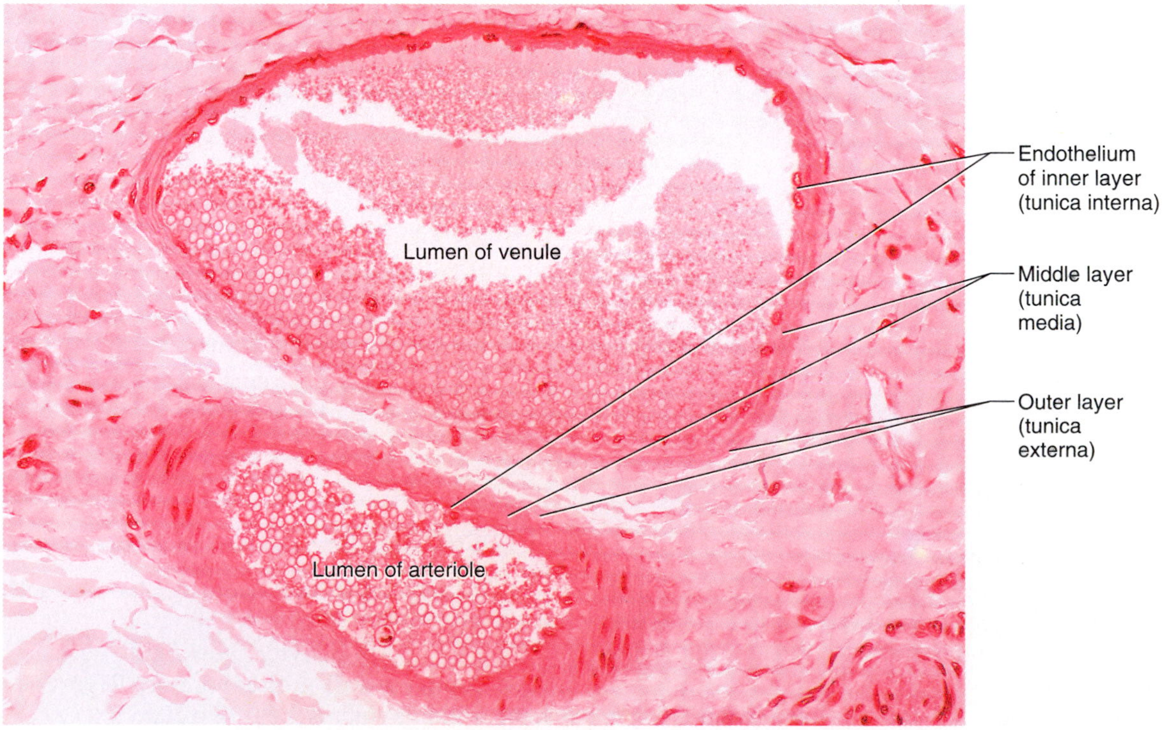

Figure 40.2 Cross section of an artery and a vein (200×).

5. Prepare a labeled sketch of the arterial wall in Part B of the laboratory report.
6. Obtain a slide of a vein cross section and examine it as you did the artery cross section. Note the thinner wall and larger lumen relative to an artery of comparable locations. Identify the three layers of the wall, and prepare a labeled sketch in Part B of the laboratory report.
7. Complete Part B of the laboratory report.
8. Observe the blood vessels in the webbing of a frog's foot. (As a substitute for a frog, a live goldfish could be used to observe circulation in the tail.) To do this, follow these steps:
   - **a.** Obtain a live frog. Wrap its body in a moist paper towel, leaving one foot extending outward. Secure the towel with rubber bands, but be careful not to wrap the animal so tightly that it could be injured. Try to keep the nostrils exposed.
   - **b.** Place the frog on a frog board or on a piece of heavy cardboard with the foot near the hole in one corner.
   - **c.** Fasten the wrapped body to the board with masking tape.
   - **d.** Carefully spread the web of the foot over the hole and secure it to the board with dissecting pins and thread (fig. 40.3). Keep the web moist with frog Ringer's solution.
   - **e.** Secure the board on the stage of a microscope with heavy rubber bands, and position it so that the web is beneath the objective lens.
   - **f.** Focus on the web, using low-power magnification, and locate some blood vessels. Note the movement of the blood cells and the direction of the blood flow. You might notice that red blood cells of frogs are nucleated. Identify an arteriole, a capillary, and a venule.

Learning Extension

Investigate the effect of temperature change on the blood vessels of the frog's foot by flooding the web with a small quantity of ice water. Observe the blood vessels with low-power magnification and note any changes in their diameters or the rate of blood flow. Remove the ice water and replace it with water heated to about 35°C (95°F). Repeat your observations. What do you conclude from this experiment?

____

____

____

   - **g.** Examine each of these vessels with high-power magnification.
   - **h.** When finished, return the frog to the location indicated by your instructor. The microscope lenses and stage will likely need cleaning after the experiment.
9. Complete Part C of the laboratory report.

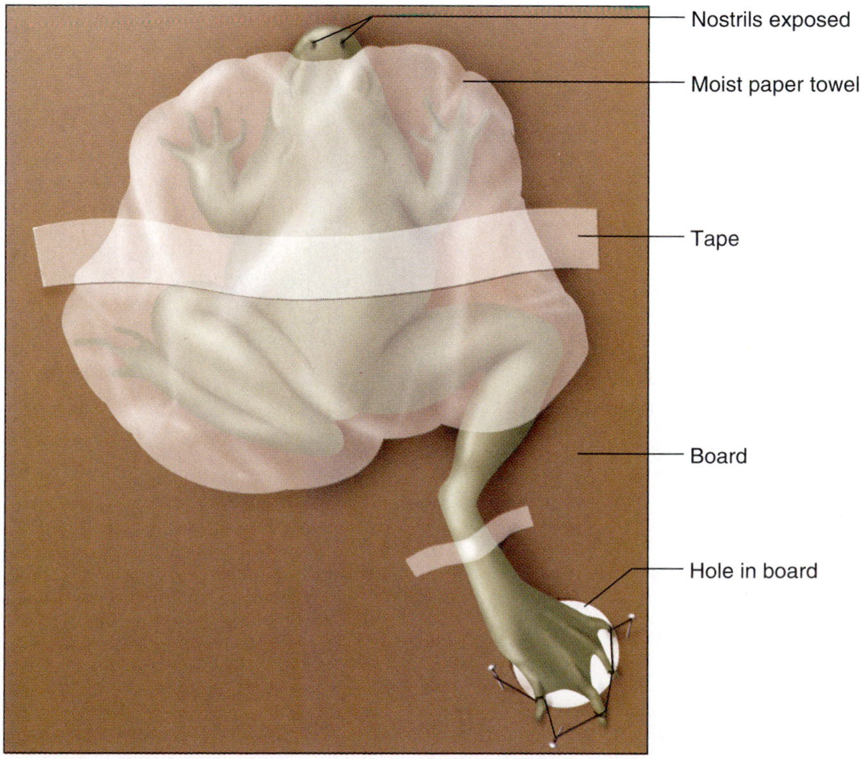

Figure 40.3 Spread the web of the foot over the hole and secure it to the board with pins and thread.

EXPLORE

Procedure B—Paths of Circulation

1. Review the sections entitled "Pulmonary Circuit" and "Systemic Circuit" in chapter 15 of the textbook.
2. As a review activity, label figure 40.4.
3. Locate the following blood vessels on the available anatomical charts and the human torso model:

pulmonary trunk 1
pulmonary arteries 2
pulmonary veins 4
aorta 1
superior vena cava 1
inferior vena cava 1

Critical Thinking Application

Why is the left ventricle wall thicker than the right ventricle wall?

____

____

____

____

EXPLORE

Procedure C—Arterial System

1. Review the section entitled "Arterial System" in chapter 15 of the textbook.
2. As a review activity, label figures 40.5, 40.6, 40.7, and 40.8.

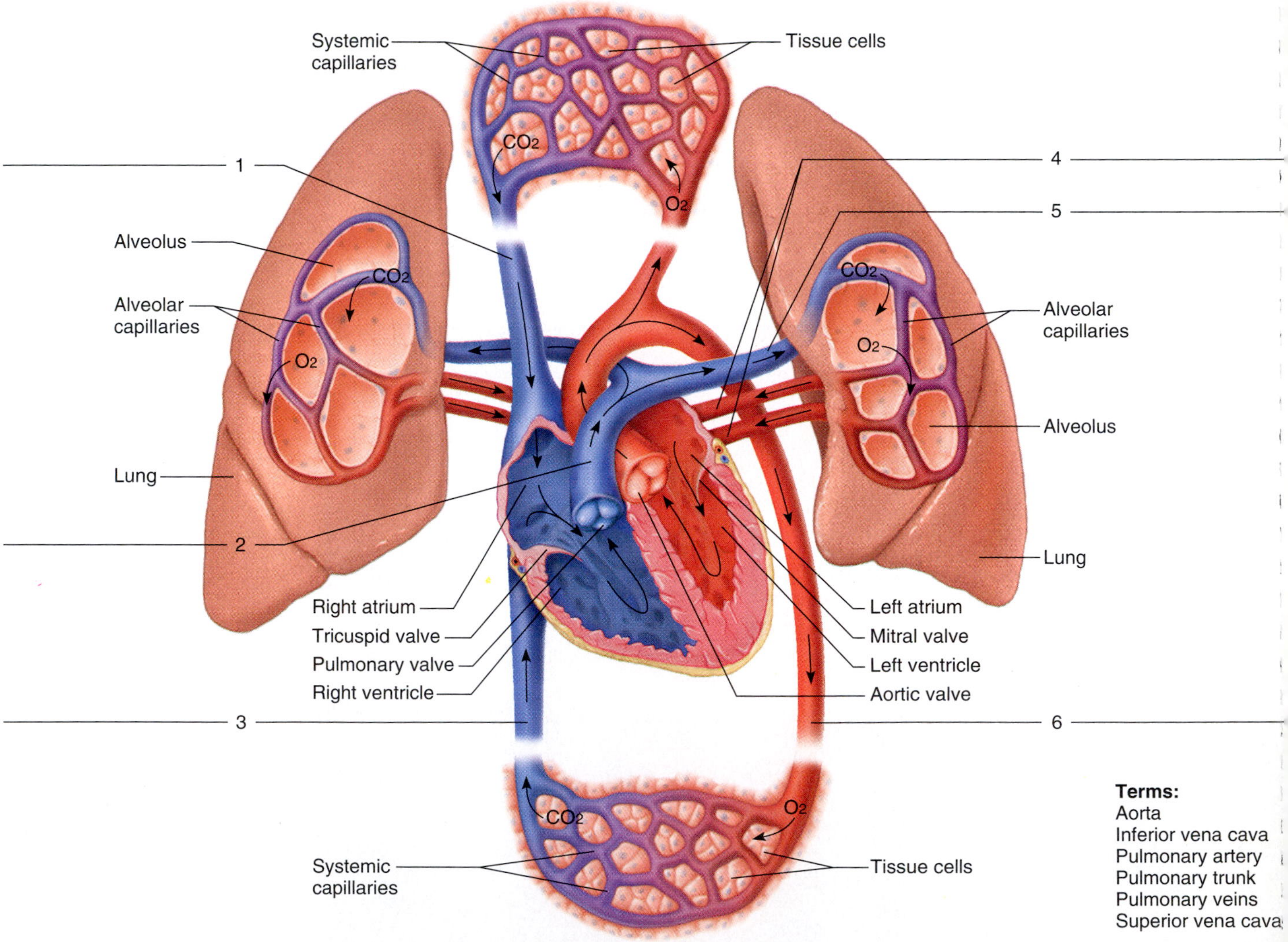

Figure 40.4 Label the major blood vessels associated with the pulmonary and systemic circuits, using the terms provided. 3

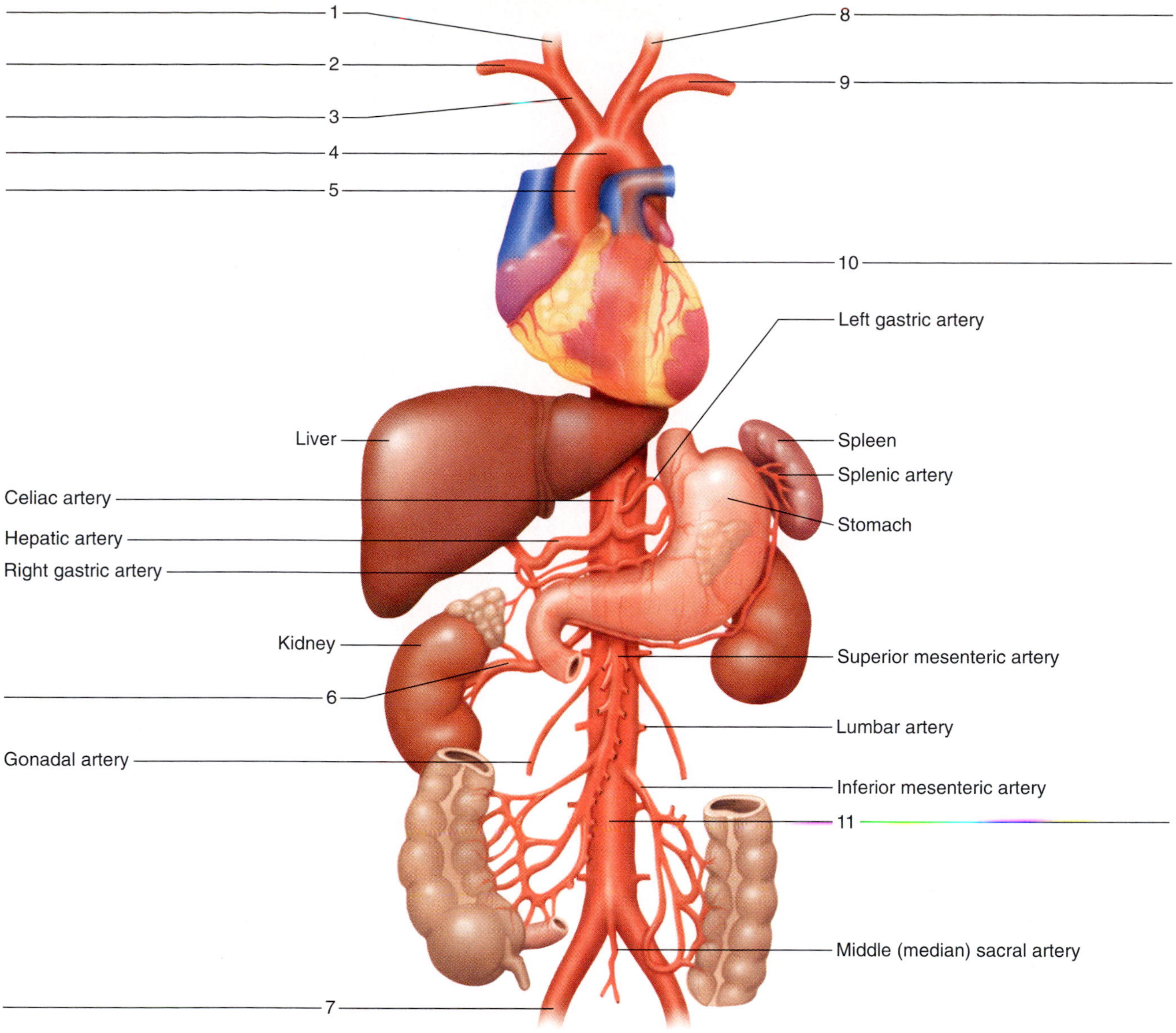

Figure 40.5 Label the portions of the aorta and its principal branches. 4

3. Locate the following arteries of the systemic circuit on the charts and human torso model:

aorta

- ascending aorta
- aortic arch (arch of aorta)
- thoracic aorta
- abdominal aorta

branches of the aorta

- coronary artery
- brachiocephalic trunk (artery)
- left common carotid artery
- left subclavian artery
- celiac trunk (artery)
- superior mesenteric artery
- renal artery
- inferior mesenteric artery
- common iliac artery

arteries to neck, head, and brain

- vertebral artery
- common carotid artery
- external carotid artery
- internal carotid artery
- facial artery

arteries to shoulder and upper limb

- subclavian artery
- axillary artery

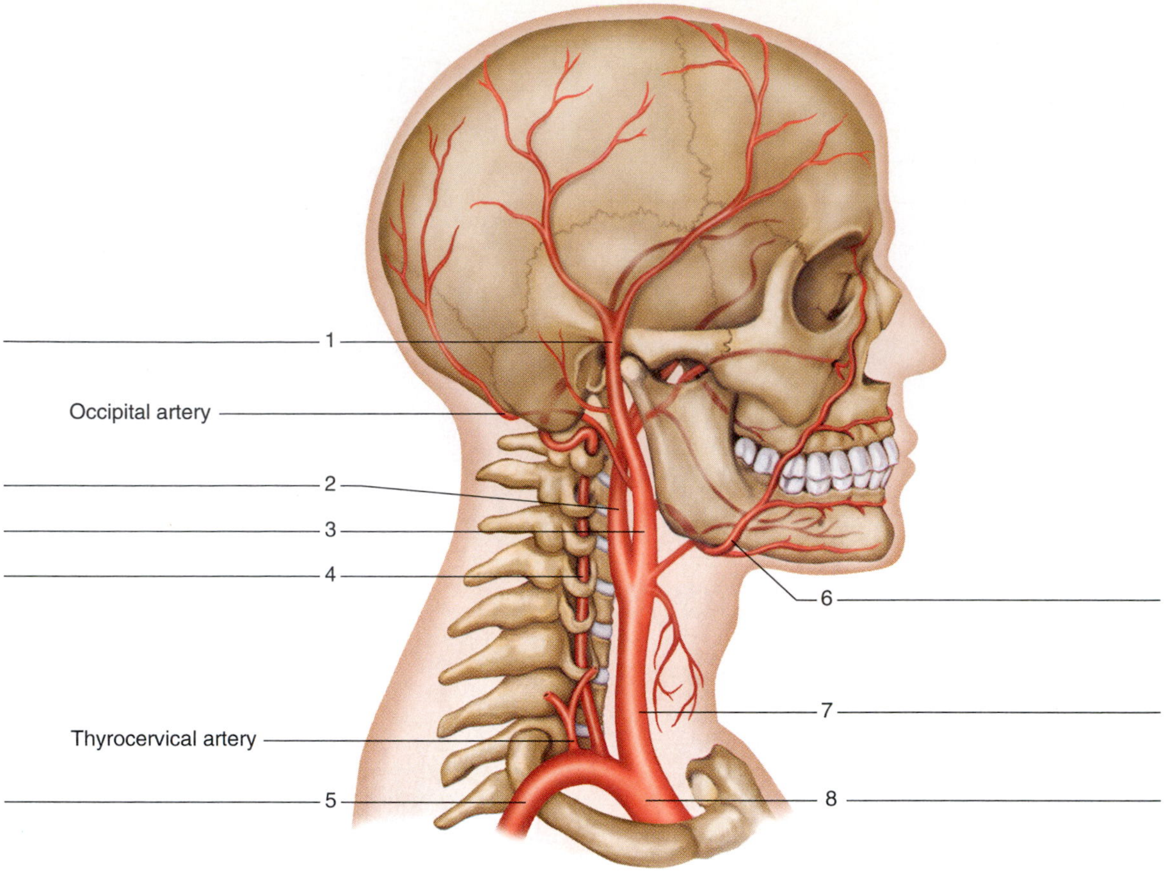

Figure 40.6 Label the arteries supplying the right side of the neck and head. (The clavicle has been removed.)

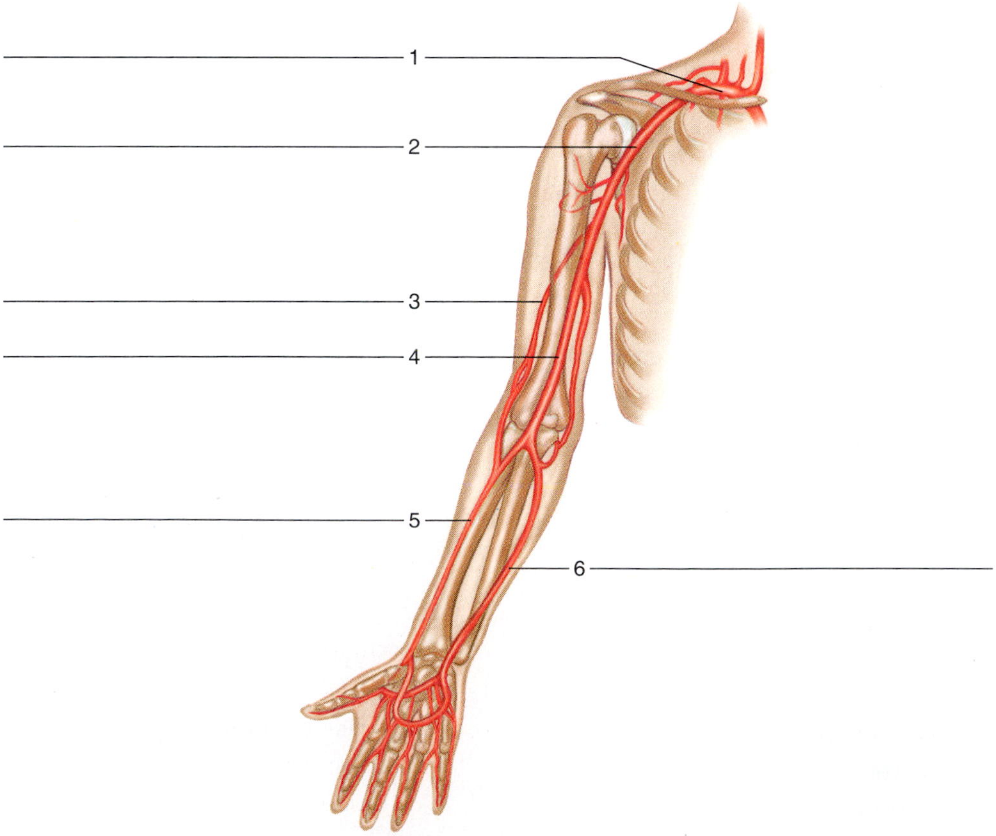

Figure 40.7 Label the major arteries of the shoulder and upper limb.

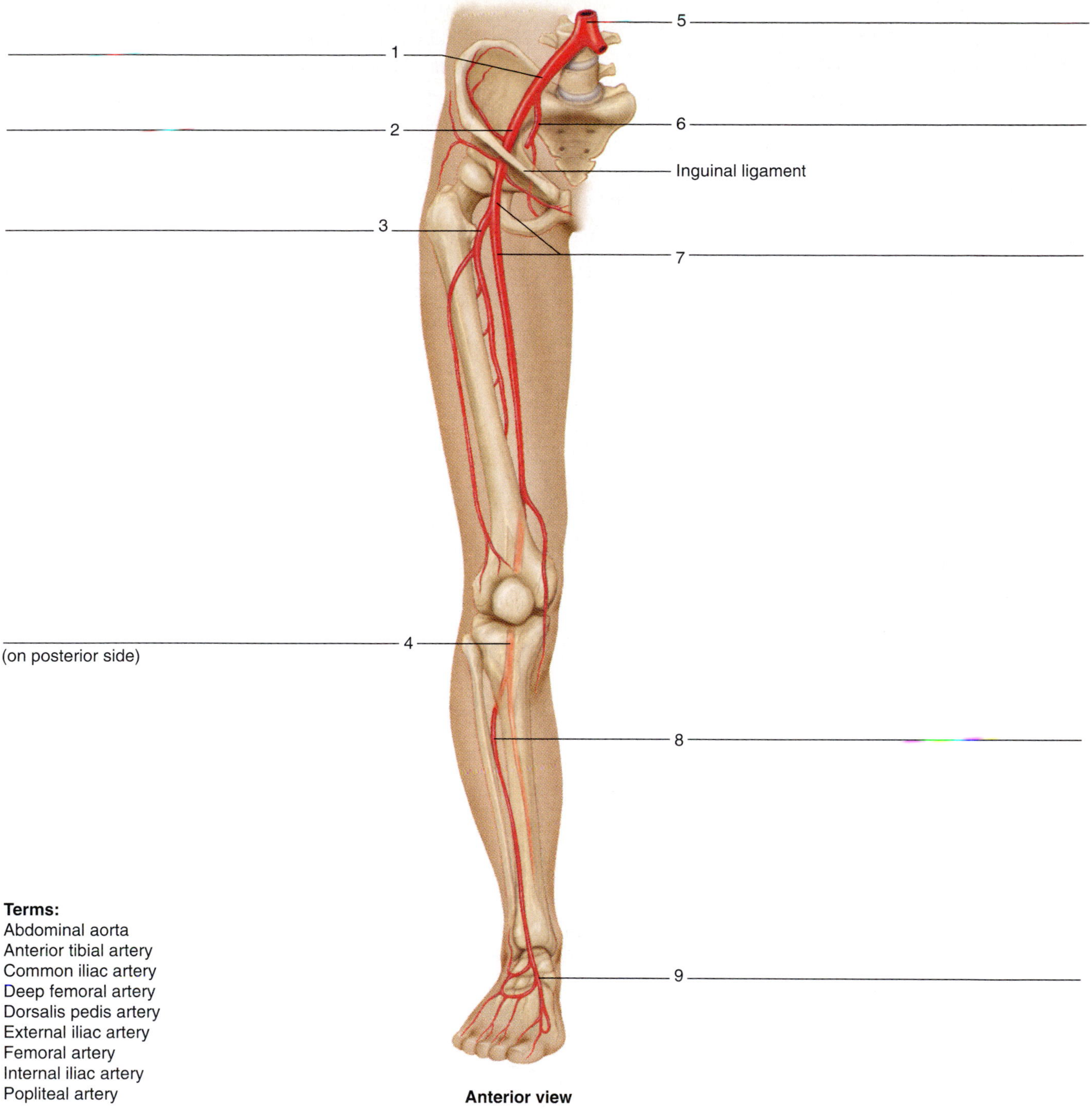

Figure 40.8 Label the major arteries supplying the pelvis and lower limb, using the terms provided. 4

- brachial artery
- deep brachial artery
- ulnar artery
- radial artery

arteries to pelvis and lower limb

- common iliac artery
- internal iliac artery
- external iliac artery
- femoral artery
- popliteal artery
- anterior tibial artery
- dorsalis pedis artery (dorsal artery of foot)
- posterior tibial artery

4. Complete Part D of the laboratory report.

EXPLORE

Procedure D—Venous System

1. Review the section entitled "Venous System" in chapter 15 of the textbook.
2. As a review activity, label figures 40.9, 40.10, 40.11, and 40.12.
3. Locate the following veins of the systemic circuit on the charts and human torso model:

veins from brain, head, and neck
- external jugular vein
- internal jugular vein
- vertebral vein
- subclavian vein
- brachiocephalic vein
- superior vena cava

veins from upper limb and shoulder
- radial vein
- ulnar vein
- brachial vein
- basilic vein
- cephalic vein
- median cubital vein (antecubital vein)
- axillary vein
- subclavian vein

veins of abdominal viscera
- hepatic portal vein
- gastric vein
- superior mesenteric vein
- splenic vein
- inferior mesenteric vein
- hepatic vein
- renal vein

veins from lower limb and pelvis
- anterior tibial vein
- posterior tibial vein
- popliteal vein
- femoral vein
- great (long) saphenous vein
- small (short) saphenous vein
- external iliac vein
- internal iliac vein
- common iliac vein
- inferior vena cava

4. Complete Parts E and F of the laboratory report.

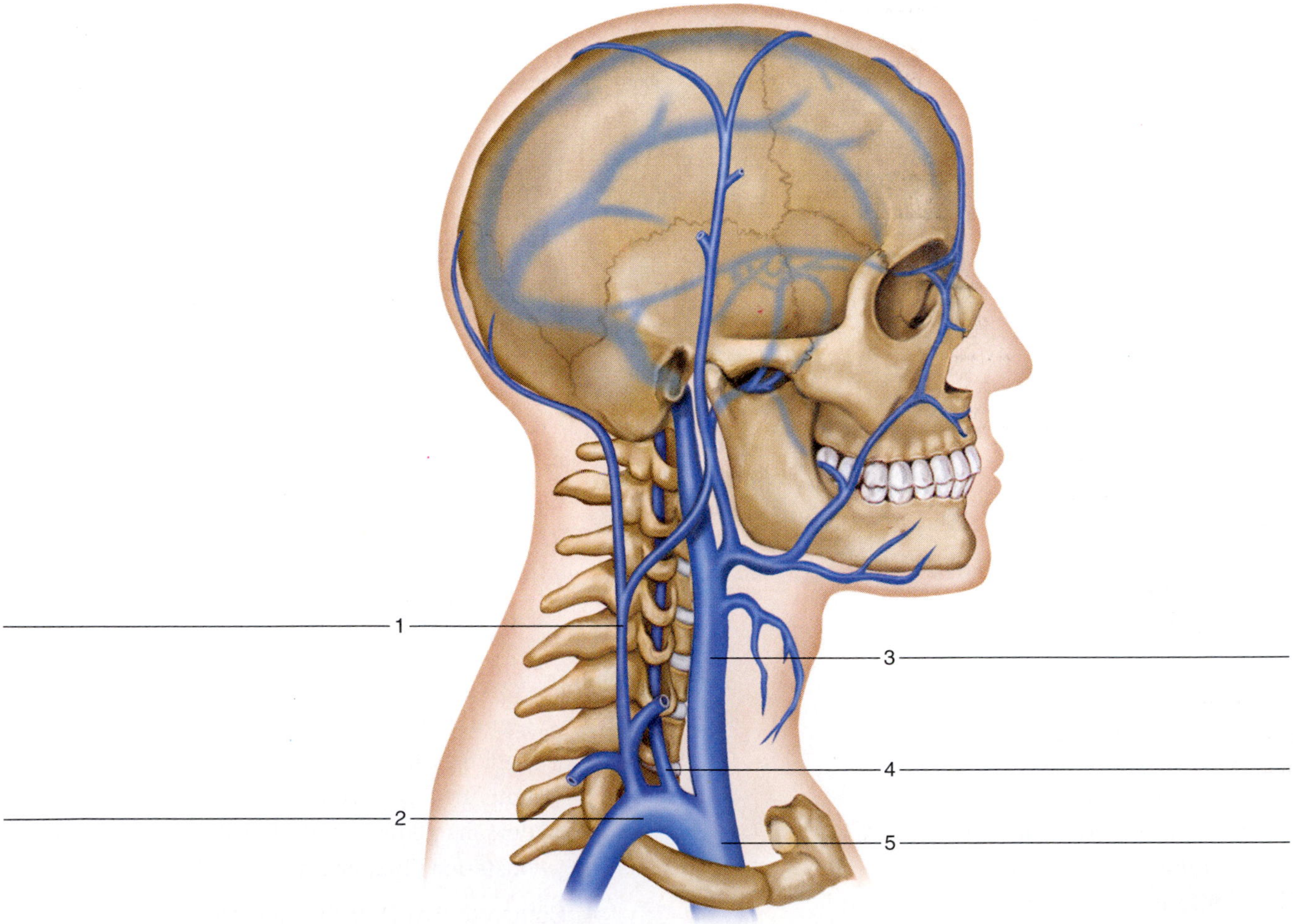

Figure 40.9 Label the major veins associated with the head and neck. (The clavicle has been removed.) 5

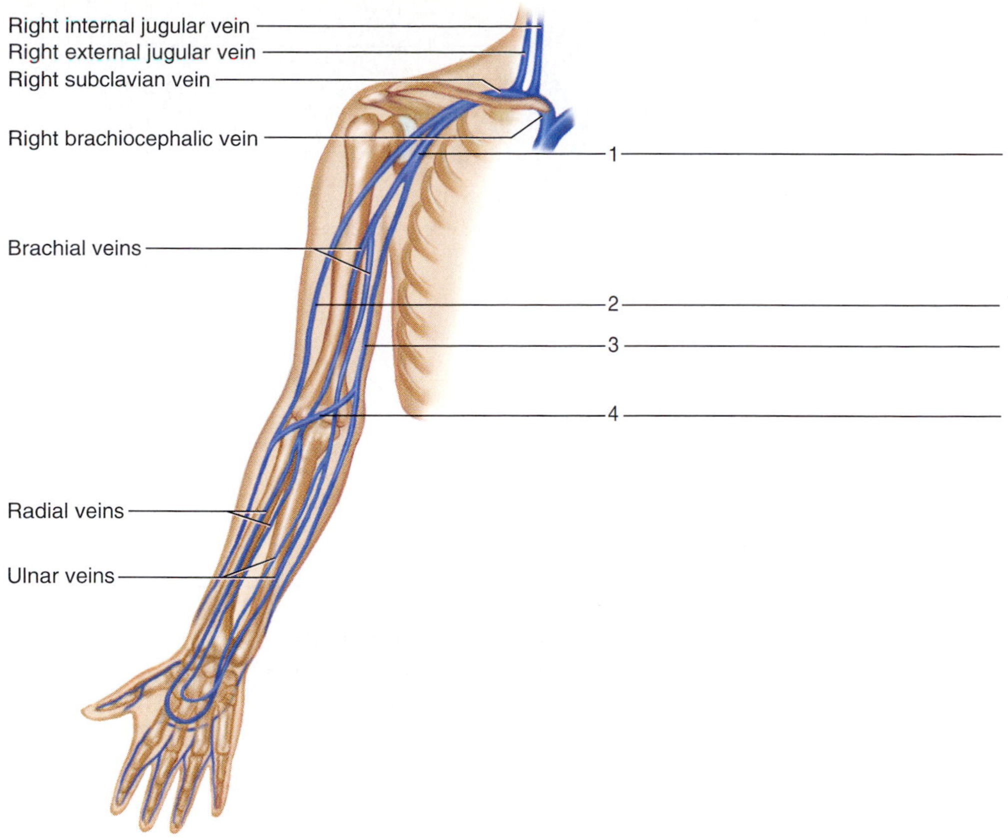

Figure 40.10 Label the veins of the upper limb and shoulder. 5

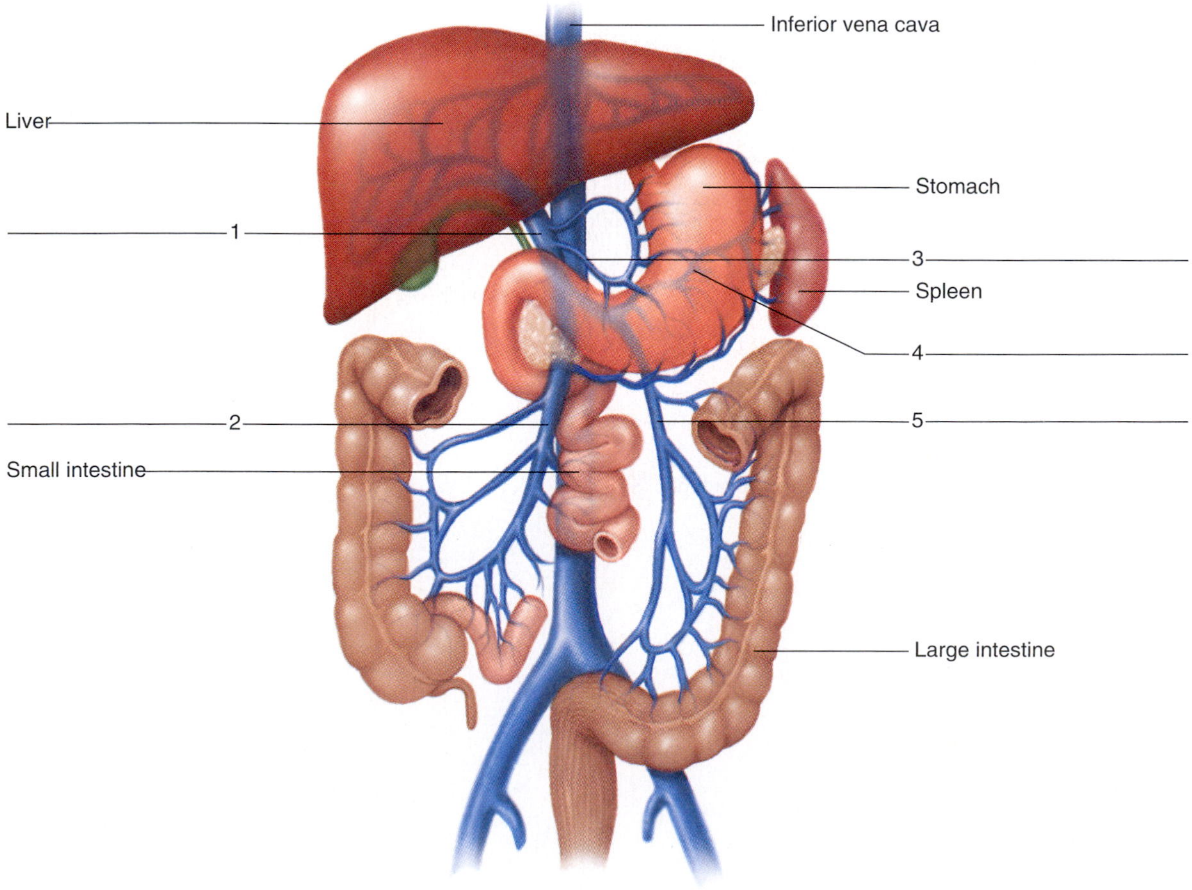

Figure 40.11 Label the veins that drain the abdominal viscera (hepatic portal system). 5

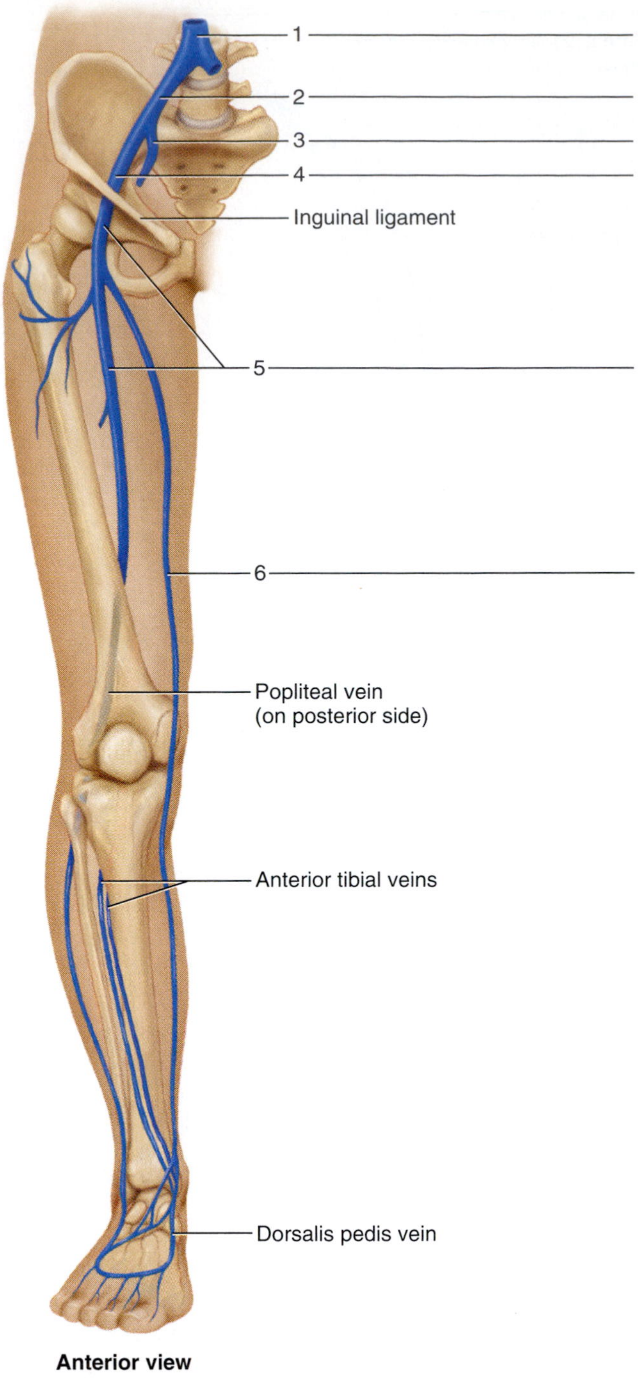

Figure 40.12 Label the veins of the pelvis and lower limb. 5

Name ______________________________

Date ______________________________

Section ______________________________

The [A] corresponds to the indicated Learning Outcome(s) listed at the beginning of the laboratory exercise.

Blood Vessel Structure, Arteries, and Veins

Laboratory Report 40

Part A Assessments

Complete the following statements:

1. Simple squamous epithelial tissue called ______________________ forms the inner linings of the tunica interna of blood vessels. [1]
2. The ______________________ of an arterial wall contains many smooth muscle cells. [1]
3. The ______________________ of an arterial wall is largely composed of connective tissue. [1]
4. When contraction of the smooth muscle in a blood vessel wall occurs, the vessel is referred to as being in a condition of ______________________. [1]
5. Relaxation of the smooth muscle in a blood vessel wall results in the vessel being in a condition of ______________________. [1]
6. The smallest blood vessels are called ______________________. [1]
7. Filtration results when substances are forced through capillary walls by ______________________ pressure. [1]
8. The presence of plasma proteins in the blood increases its ______________________ pressure as compared to tissue fluids. [1]
9. ______________________ in certain veins close if the blood begins to back up in that vein. [1]

Part B Assessments

1. Sketch and label a section of an artery wall and a vein wall. 1

Artery wall **Vein wall**

2. Describe the differences you noted in the structures of the wall of an artery and a vein. Mention each of the three layers of the wall. 1 ______

______

______

Critical Thinking Application

Explain the functional significance of the differences you noted in the structures of the arterial and venous walls. 1

______

______

Part C Assessments

1. How did you distinguish between arterioles and venules when you observed the vessels in the web of the frog's foot? 2 ______

______

2. How did you recognize capillaries in the web? 2 ______

______

3. What differences did you note in the rate of blood flow through the arterioles, capillaries, and venules? 2

______

______

Part D Assessments

Match the arteries in column A with the descriptions in column B. Place the letter of your choice in the space provided. 4

Column A		Column B
a. Anterior tibial	_______	1. Jaw, teeth, and face
b. Brachial	_______	2. Kidney
c. Celiac	_______	3. Upper digestive tract, spleen, and liver
d. External carotid	_______	4. Foot and toes
e. Internal carotid	_______	5. Gluteal muscles
f. Internal iliac	_______	6. Biceps muscle
g. Phrenic	_______	7. Knee joint
h. Popliteal	_______	8. Adrenal gland
i. Renal	_______	9. Diaphragm
j. Suprarenal	_______	10. Brain
k. Ulnar	_______	11. Forearm muscles

Part E Assessments

Match each vein in column A with the vein it drains into from column B. Place the letter of your choice in the space provided. 5

Column A		Column B
a. Anterior tibial	_______	1. Popliteal
b. Basilic	_______	2. Axillary
c. Brachiocephalic	_______	3. Inferior vena cava
d. Common iliac	_______	4. Subclavian
e. External jugular	_______	5. Brachial
f. Femoral	_______	6. Superior vena cava
g. Popliteal	_______	7. Femoral
h. Radial	_______	8. External iliac

Part F Assessments

Label the major arteries and veins indicated in figure 40.13. 3 4 5

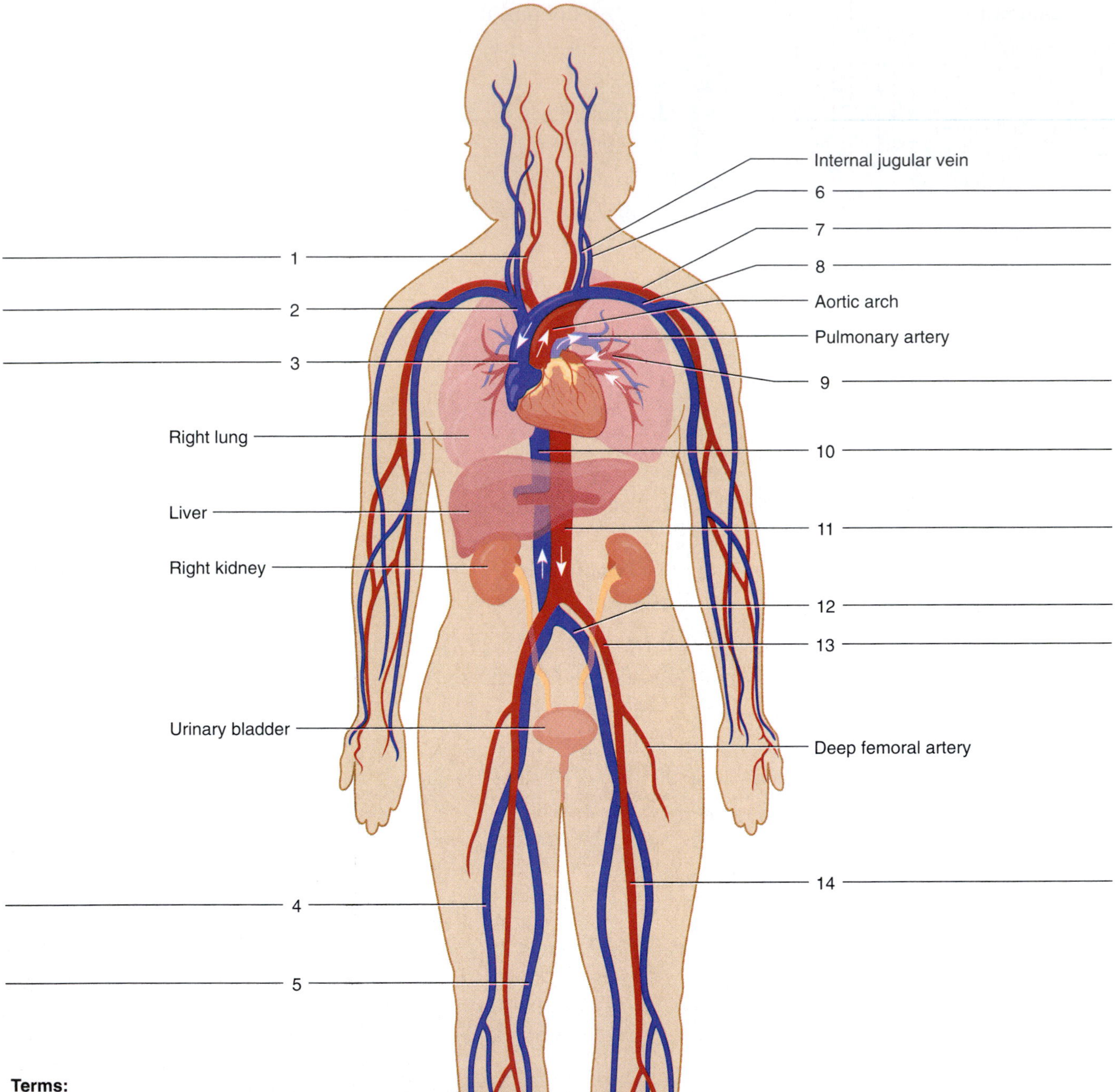

Terms:

Aorta
Brachiocephalic vein
Common carotid artery
Common iliac artery
Common iliac vein
External jugular vein
Femoral artery
Femoral vein
Great saphenous vein
Inferior vena cava
Pulmonary vein
Subclavian artery
Subclavian vein
Surperior vena cava

Figure 40.13 Label the major arteries and veins of the systemic and pulmonary circuits, using the terms provided.

Laboratory Exercise 41

Pulse Rate and Blood Pressure

Materials Needed

Textbook
Clock with second hand
Sphygmomanometer
Stethoscope
Alcohol swabs
Absorbent cotton

The surge of blood that enters the arteries each time the ventricles of the heart contract causes the elastic walls of these vessels to swell. Then, as the ventricles relax, the walls recoil. This alternate expanding and recoiling of an arterial wall can be felt as a pulse in vessels that run close to the surface of the body.

The force exerted by the blood pressing against the inner walls of arteries also creates blood pressure. This pressure reaches a maximum during ventricular contraction and then drops to its lowest level while the ventricles are relaxed.

Purpose of the Exercise

To examine the pulse, determine the pulse rate, measure blood pressure, and investigate the effects of body position and exercise on pulse rate and blood pressure.

LEARNING OUTCOMES

After completing this exercise, you should be able to

1. Determine pulse rate and pulse characteristics.
2. Test the effects of various factors on pulse rate.
3. Describe and measure blood pressure using a sphygmomanometer.
4. Test the effects of various factors on blood pressure.

EXPLORE

Procedure A—Pulse Rate

1. Review the sections entitled "Arterial Blood Pressure" and "Measurement of Arterial Blood Pressure—Clinical Application" in chapter 15 of the textbook.
2. Complete Part A of Laboratory Report 41.
3. Examine your laboratory partner's radial pulse. To do this, follow these steps:
   a. Have your partner sit quietly, remaining as relaxed as possible.
   b. Locate the pulse by placing your index and middle fingers over the radial artery on the anterior surface of the wrist. Do not use your thumb for sensing the pulse because you may feel a pulse coming from an artery in the thumb. The radial artery is most often used because it is easy and convenient to locate.
   c. Note the characteristics of the pulse. That is, could it be described as regular or irregular, strong or weak, hard or soft? The pulse should be regular and the amplitude (magnitude) will decrease as the distance from the left ventricle increases. The strength of the pulse reveals some indications of blood pressure. Under high blood pressure, the pulse feels very hard and strong; under low blood pressure, the pulse feels weak and can be easily compressed.
   d. To determine the pulse rate, count the number of pulses that occur in 1 minute. This can be accomplished by counting pulses in 30 seconds and multiplying that number by 2. (A pulse count for a full minute, although taking a little longer, would give a better opportunity to detect irregularities.)
   e. Record the pulse rate in Part B of the laboratory report.
4. Repeat the procedure and determine the pulse rate in each of the following conditions:
   a. immediately after lying down;
   b. 3–5 minutes after lying down;
   c. immediately after standing;
   d. 3–5 minutes after standing quietly;
   e. immediately after 3 minutes of moderate exercise *(omit if the person has health problems)*;
   f. 3–5 minutes after exercise has ended.
5. Complete Part B of the laboratory report.

Learning Extension

Determine the pulse rate and pulse characteristics in two additional locations on your laboratory partner. Locate and record the pulse rate from the common carotid artery and the dorsalis pedis artery. 1

Common carotid artery pulse rate ______________

Dorsalis pedis artery pulse rate ______________

Compare the amplitude of the pulse characteristics between these two pulse locations and interpret any of the variations noted. ______________

____

____

____

____

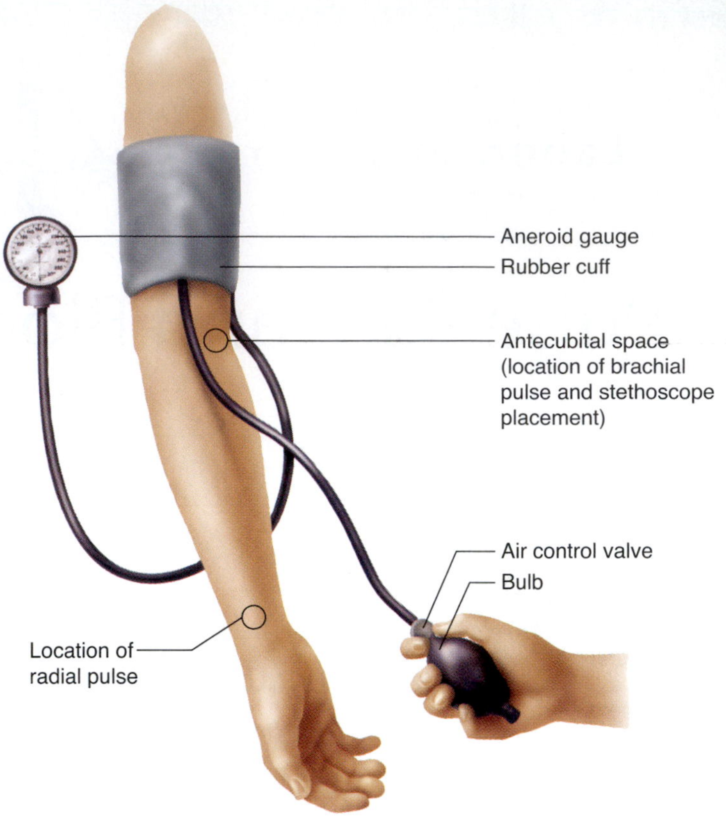

Figure 41.1 Blood pressure is commonly measured by using a sphygmomanometer (blood pressure cuff). The use of a column of mercury is the most accurate measurement, but due to environmental concerns, it has been replaced by alternative gauges and digital readouts.

EXPLORE

Procedure B—Blood Pressure

1. Although blood pressure is present in all blood vessels, the standard location to record blood pressure is the brachial artery.
2. Measure your laboratory partner's arterial blood pressure. To do this, follow these steps:
   a. Obtain a sphygmomanometer and a stethoscope.
   b. Clean the earpieces and the diaphragm of the stethoscope with cotton moistened with an alcohol swab.
   c. Have your partner sit quietly with bare upper limb resting on a table at heart level. Have the person remain as relaxed as possible.
   d. Locate the brachial artery at the antecubital space. Wrap the cuff of the sphygmomanometer around the arm so that its lower border is about 2.5 cm above the bend of the elbow. Center the bladder of the cuff in line with the *brachial pulse* (fig. 41.1).
   e. Palpate the *radial pulse.* Close the valve on the neck of the rubber bulb connected to the cuff, and pump air from the bulb into the cuff. Inflate the cuff while watching the sphygmomanometer, and note the pressure when the pulse disappears. (This is a rough estimate of the systolic pressure.) Immediately deflate the cuff.
   f. Position the stethoscope over the brachial artery. Reinflate the cuff to a level 30 mm Hg higher than the point where the pulse disappeared during palpation.
   g. Slowly open the valve of the bulb until the pressure in the cuff drops at a rate of about 2 or 3 mm Hg per second.
   h. Listen for sounds (Korotkoff sounds) from the brachial artery. When the first loud tapping sound is heard, record the reading as the systolic pressure. This indicates the pressure exerted against the arterial wall during systole.
   i. Continue to listen to the sounds as the pressure drops, and note the level when the last sound is heard. Record this reading as the diastolic pressure, which measures the constant arterial resistance.
   j. Release all of the pressure from the cuff.
   k. Repeat the procedure until you have two blood pressure measurements from each arm, allowing 2–3 minutes of rest between readings.
   l. Average your readings and enter them in the table in Part C of the laboratory report.
3. Measure and record your partner's blood pressure in each of the following conditions:
   a. 3–5 minutes after lying down;
   b. 3–5 minutes after standing quietly;
   c. immediately after 3 minutes of moderate exercise *(omit if the person has health problems);*
   d. 3–5 minutes after exercise has ended.
   e. Record the blood pressures in Part C of the laboratory report.
4. Complete Part C of the laboratory report.

Name ______________________________

Date ______________________________

Section ______________________________

The [A] corresponds to the indicated Learning Outcome(s) listed at the beginning of the laboratory exercise.

Pulse Rate and Blood Pressure

Laboratory Report 41

Part A Assessments

Complete the following statements:

1. The term *blood pressure* most commonly is used to refer to systemic ______________ pressure. [3]
2. The maximum pressure achieved during ventricular contraction is called ______________ pressure. [3]
3. The lowest pressure that remains in the arterial system during ventricular relaxation is called ______________ pressure. [3]
4. The pulse rate is equal to the ______________ rate. [1]
5. A pulse that feels full and is not easily compressed is produced by an elevated ______________. [1]
6. The instrument commonly used to measure systemic arterial blood pressure is called a (an) ______________. [3]
7. Blood pressure is expressed in units of ______________. [3]
8. The upper number of the fraction used to record blood pressure indicates the ______________ pressure. [3]
9. The difference between the systolic and diastolic pressure is called the ______________. [3]
10. The ______________ artery in the arm is the standard systemic artery in which blood pressure is measured. [3]

Part B Assessments

1. Enter your observations of pulse characteristics and pulse rates in the table. [1] [2]

Test Subject	Pulse Characteristics	Pulse Rate (beats/min)
Sitting		
Lying down		
3–5 minutes later		
Standing		
3–5 minutes later		
After exercise		
3–5 minutes later		

2. Summarize the effects of body position and exercise on the characteristics and rates of the pulse. 2

Part C Assessments

1. Enter the initial measurements of blood pressure in the table. 3

Reading	Blood Pressure in Right Arm	Blood Pressure in Left Arm
First		
Second		
Average		

2. Enter your test results in the table. 4

Test Subject	Blood Pressure
3–5 minutes after lying down	
3–5 minutes after standing	
After 3 minutes of moderate exercise	
3–5 minutes later	

3. Summarize the effects of body position and exercise on blood pressure. 4

4. Summarize any correlations between pulse rate and blood pressure from any of the experimental conditions. 2 4

Critical Thinking Application

When a pulse is palpated and counted, which pressure (systolic or diastolic) would be characteristic at that moment? Explain your answer. 1

Laboratory Exercise 42

Lymphatic System

Materials Needed

Textbook
Human torso model
Anatomical chart of the lymphatic system
Compound light microscope
Prepared microscope slides:
- Lymph node section
- Human thymus section
- Human spleen section

The lymphatic system is a vast collection of cells and biochemicals that travel in lymphatic capillaries and vessels and the organs and glands that produce them. The system is closely associated with the cardiovascular system and includes a network of vessels that assist in the circulation of body fluids. These vessels provide pathways through which excess fluid can be transported away from interstitial spaces within most tissues and returned to the bloodstream. Without the lymphatic system, this fluid would accumulate in tissue spaces, producing edema.

The organs of the lymphatic system also help defend the tissues against infections by filtering particles from lymph and by supporting the activities of lymphocytes that furnish immunity against specific disease-causing agents or pathogens.

Purpose of the Exercise

To review the structure of the lymphatic system and to observe the microscopic structure of a lymph node, thymus, and spleen.

LEARNING OUTCOMES

After completing this exercise, you should be able to

1. Trace the major lymphatic pathways in an anatomical chart or model.
2. Locate and identify the major clusters of lymph nodes (lymph glands) in an anatomical chart or model.
3. Describe the structure and function of a lymph vessel, a lymph node, the thymus, and the spleen.
4. Locate and sketch the major microscopic structures of a lymph node, the thymus, and the spleen.

EXPLORE

Procedure A—Lymphatic Pathways

1. Review the section entitled "Lymphatic Pathways" in chapter 16 of the textbook.
2. As a review activity, label figure 42.1.
3. Complete Part A of Laboratory Report 42.
4. Observe the human torso model and the anatomical chart of the lymphatic system, and locate the following features:

lymphatic vessels
lymph nodes
lymphatic trunks
- lumbar trunk
- intestinal trunk
- intercostal trunk
- bronchomediastinal trunk
- subclavian trunk
- jugular trunk

cisterna chyli
collecting ducts
- thoracic (left lymphatic) duct
- right lymphatic duct

internal jugular veins
subclavian veins

EXPLORE

Procedure B—Lymph Nodes

1. Review the section entitled "Lymph Nodes" in chapter 16 of the textbook.
2. As a review activity, label figure 42.2.

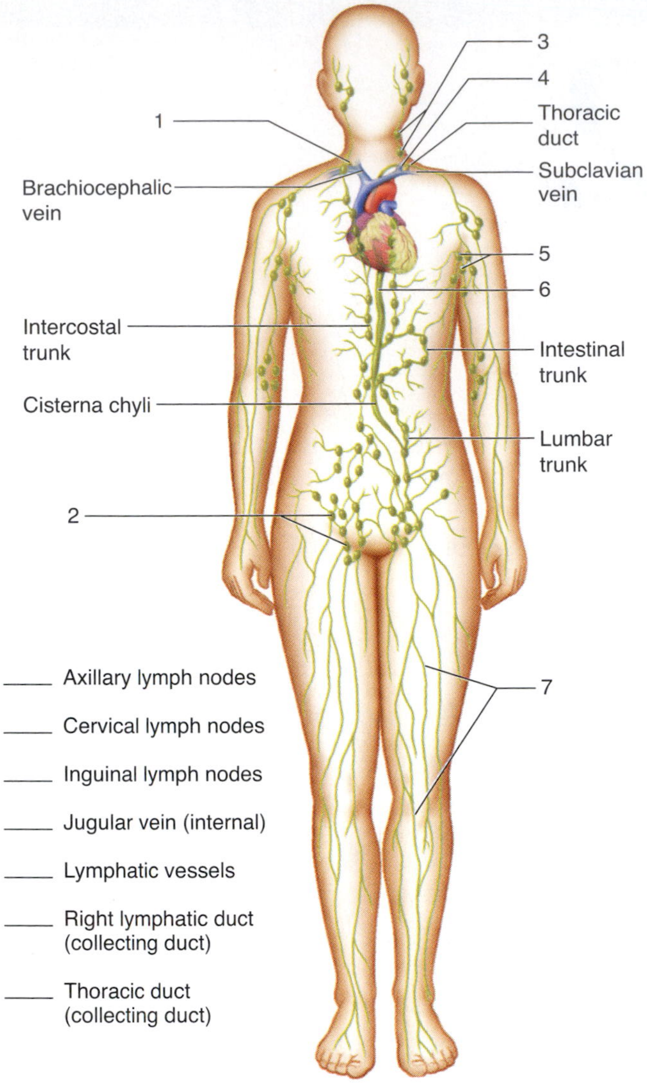

Figure 42.1 Label the diagram by placing the correct numbers in the spaces provided. 1

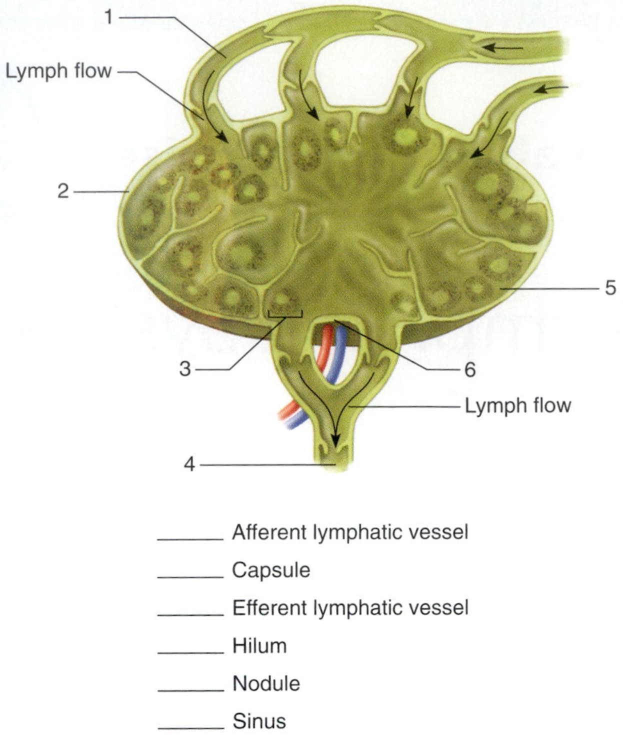

Figure 42.2 Label this diagram of a lymph node by placing the correct numbers in the spaces provided. 3

3. Complete Part B of the laboratory report.
4. Observe the anatomical chart of the lymphatic system and the human torso model, and locate the clusters of lymph nodes in the following regions:

   cervical region
   axillary region
   inguinal region
   pelvic cavity
   abdominal cavity
   thoracic cavity

5. Palpate the lymph nodes in your cervical region. They are located along the lower border of the mandible and between the ramus of the mandible and the sternocleidomastoid muscle. They feel like small, firm lumps.
6. Study figure 42.3.
7. Obtain a prepared microscope slide of a lymph node and observe it, using low-power magnification (fig. 42.3*b*). Identify the *capsule* that surrounds the node and is mainly composed of collagenous fibers, the *lymph nodules* that appear as dense masses near the surface of the node, and the *lymph sinuses* that appear as narrow spaces between the nodules and the capsule.
8. Using high-power magnification, examine a nodule within the lymph node. The nodule contains densely packed *lymphocytes.*
9. Prepare a labeled sketch of a representative section of a lymph node in Part D of the laboratory report.

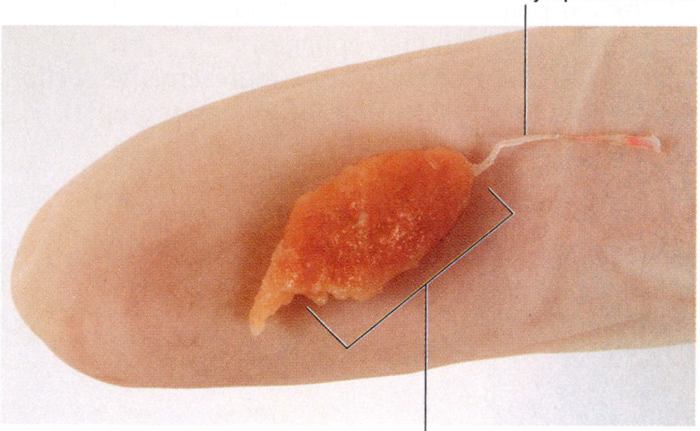

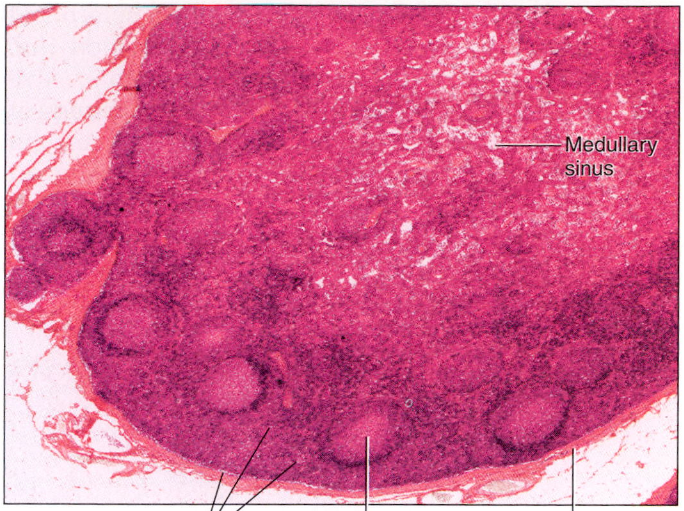

Figure 42.3 Lymph nodes: (*a*) photograph of a human lymph node on tip of finger; (*b*) micrograph of a lymph node (20×).

EXPLORE

Procedure C—Thymus and Spleen

1. Review the section entitled "Thymus and Spleen" in chapter 16 of the textbook.
2. Locate the thymus and spleen in the anatomical chart of the lymphatic system and on the human torso model.
3. Complete Part C of the laboratory report.
4. Obtain a prepared microscope slide of human thymus and observe it, using low-power magnification (fig. 42.4). Note how the thymus is subdivided into *lobules* by *septa* of connective tissue that contain blood vessels. Identify the *capsule* of loose connective tissue that surrounds the thymus, the outer *cortex* of a lobule composed of densely packed cells and deeply stained, and the inner *medulla* of a lobule composed of loosely packed *lymphocytes* and epithelial cells and lightly stained.
5. Examine the cortex tissue of a lobule using high-power magnification. The cells of the cortex are composed of densely packed lymphocytes among some epithelial cells and *macrophages*. Some of these cortical cells may be undergoing mitosis, so their chromosomes may be visible.
6. Prepare a labeled sketch of a representative section of the thymus in Part D of the laboratory report.

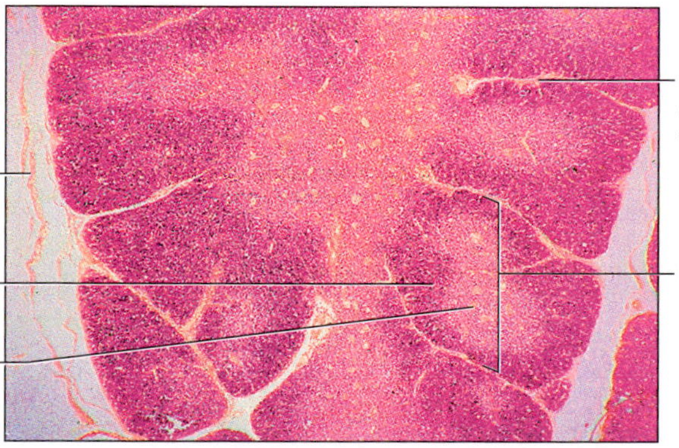

Figure 42.4 Micrograph of a section of the thymus (15×).

7. Obtain a prepared slide of the human spleen and observe it, using low-power magnification (fig. 42.5). Identify the *capsule* of dense connective tissue that surrounds the spleen. The tissues of the spleen include circular *nodules* of *white* (in unstained tissue) *pulp* enclosed in a matrix of *red pulp*.
8. Using high-power magnification, examine a nodule of white pulp and red pulp. The cells of the white pulp are mainly *lymphocytes*. Also, there may be an arteriole centrally located in the nodule. The cells of the red pulp are mostly red blood cells with many lymphocytes and macrophages.
9. Prepare a labeled sketch of a representative section of the spleen in Part D of the laboratory report.

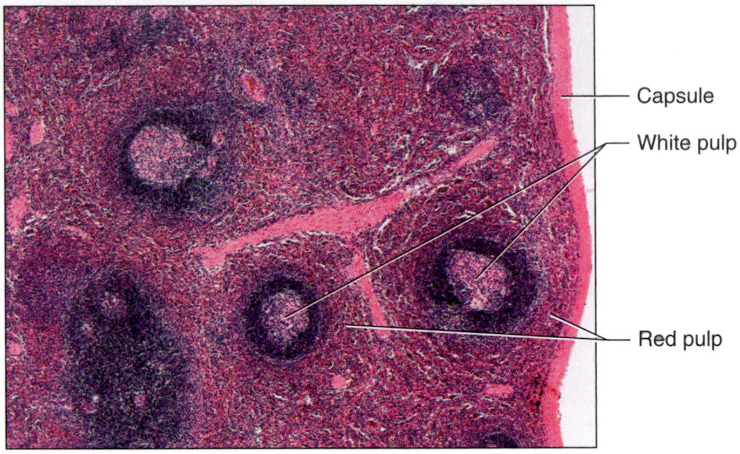

Figure 42.5 Micrograph of a section of the spleen (40×).

Name ______________________________

Date ______________________________

Section ______________________________

The [A] corresponds to the indicated Learning Outcome(s) listed at the beginning of the laboratory exercise.

Lymphatic System

Laboratory Report 42

Part A Assessments

Complete the following statements:

1. Lymphatic pathways begin as ______________________ that merge to form lymphatic vessels. [1]
2. The wall of a lymphatic capillary consists of a single layer of ______________________ cells. [3]
3. Once tissue (interstitial) fluid is inside a lymph capillary, the fluid is called ______________________ . [3]
4. Lymphatic vessels have walls similar to those of ______________________ but thinner. [3]
5. Lymphatic vessels contain ______________________ that help prevent the backflow of lymph. [1]
6. Lymphatic vessels usually lead to ______________________ that filter the fluid being transported. [1]
7. The lymphatic trunk that drains the abdominal viscera is called the ______________________ trunk. [1]
8. The lymphatic trunk that drains the head and neck is called the ______________________ trunk. [1]
9. The ______________________ is the larger and longer of the two lymphatic collecting ducts. [1]

Part B Assessments

Complete the following statements:

1. Lymph nodes contain large numbers of white blood cells called ______________________ and macrophages that fight invading microorganisms. [3]
2. The indented region where blood vessels and nerves join a lymph node is called the ______________________ . [3]
3. ______________________ that contain germinal centers are the structural units of a lymph node. [3]
4. The spaces within a lymph node are called ______________________ through which lymph circulates. [3]
5. Lymph enters a node through a(an) ______________________ lymphatic vessel. [3]
6. The partially encapsulated lymph nodes in the pharynx are called ______________________ . [3]
7. The aggregations of lymph nodules found within the mucosal lining of the small intestine are called ______________________ . [2]
8. The lymph nodes in the cervical region are associated with the lymphatic vessels that drain the ______________________ . [1]
9. The lymph nodes associated with the lymphatic vessels that drain the lower limbs are located in the ______________________ region. [2]

Part C Assessments

Complete the following statements:

1. The thymus is located in the ________________________ , anterior to the aortic arch. 3
2. The thymus reaches its greatest size during infancy and early childhood and tends to decrease in size following ________________________ . 3
3. As a person ages, the thymus tissue is often replaced by ________________________ . 3
4. The lymphocytes of the thymus develop from precursor cells that originated in the ____________________ . 3
5. The hormones secreted by the thymus are called ____________________ . 3
6. The ________________________ is the largest of the lymphatic organs. 3
7. Blood vessels enter the spleen through the region called the ________________________ . 3
8. The sinuses within the spleen contain ________________________ . 3
9. The tiny islands (nodules) of tissue within the spleen that contain many lymphocytes comprise the ________________ pulp. 3
10. The ____________________ pulp of the spleen contains large numbers of red blood cells, lymphocytes, and macrophages. 3
11. ________________________ within the spleen engulf and destroy foreign particles and cellular debris. 3

Part D Assessments

Sketch and label structural features of the following lymphatic organs: 4

Lymph node (_______×)	Thymus (_______×)
Spleen (_______×)	

Laboratory Exercise 43

Digestive Organs

Materials Needed

Textbook
Human torso model
Head model, sagittal section
Skull with teeth
Teeth, sectioned
Tooth model, sectioned
Paper cup
Compound light microscope
Prepared microscope slides of the following:
- Salivary gland
- Esophagus
- Stomach (fundus)
- Pancreas (exocrine portion)
- Small intestine (jejunum)
- Large intestine

The digestive system includes the organs associated with the alimentary canal and several accessory structures. The alimentary canal, a muscular tube, passes through the body from the opening of the mouth to the anus. It includes the mouth, pharynx, esophagus, stomach, small intestine, and large intestine. The function of the canal is to move substances throughout its length. It is specialized in various regions to store, digest, and absorb food materials and to eliminate the residues. The accessory organs, which include the salivary glands, liver, gallbladder, and pancreas, secrete products into the alimentary canal that aid digestive functions.

Purpose of the Exercise

To review the structure and function of the digestive organs and to examine the tissues of these organs.

LEARNING OUTCOMES

After completing this exercise, you should be able to

1. Locate and label the major digestive organs and their major structures.
2. Describe the functions of these organs.
3. Examine and sketch the structures of a section of the small intestine.

EXPLORE

Procedure A—Mouth and Salivary Glands

1. Review the sections entitled "Mouth" and "Salivary Glands" in chapter 17 of the textbook.
2. As a review activity, label figures 43.1, 43.2, and 43.3.
3. Examine the head model (sagittal section) and a skull. Locate the following structures:

oral cavity (mouth)
vestibule
tongue
- lingual frenulum
- papillae

palate
- hard palate
- soft palate
- uvula

palatine tonsils
gums (gingivae)
teeth
- incisors
- canines (cuspids)
- premolars (bicuspids)
- molars

4. Examine a sectioned tooth and a tooth model. Locate the following features:

crown
- enamel
- dentin

neck
root
- pulp cavity
- cementum
- root canal

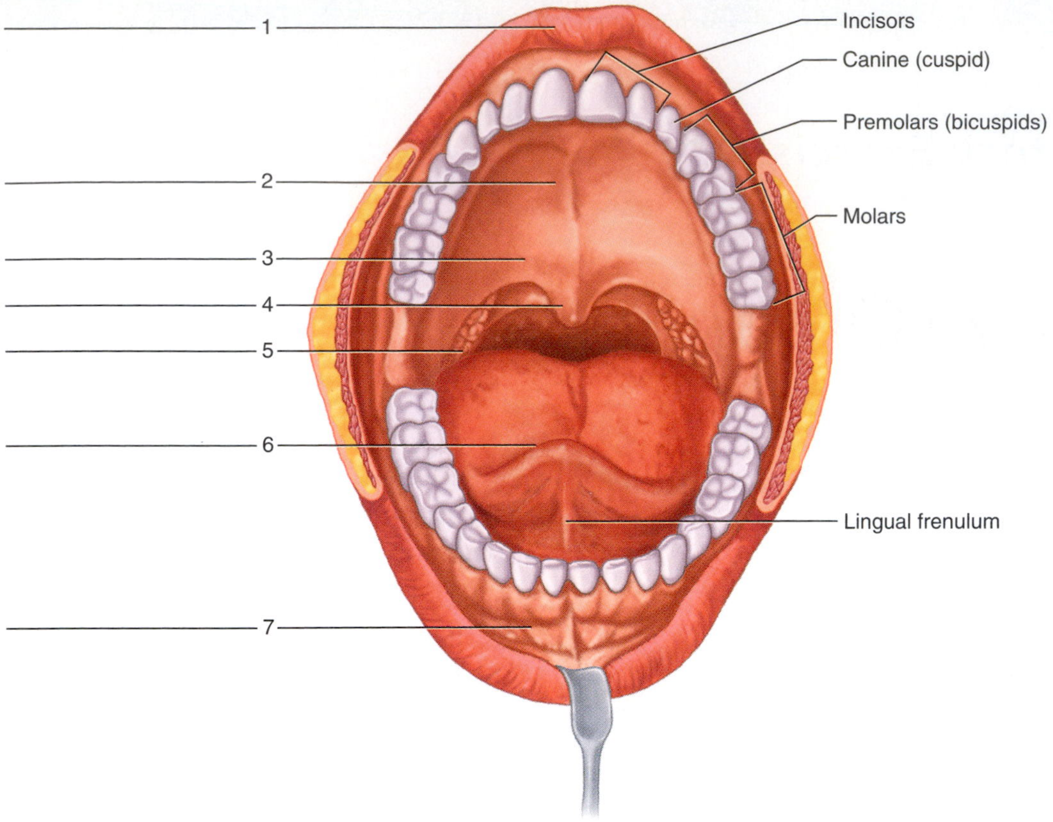

Figure 43.1 Label the features of the mouth. 1

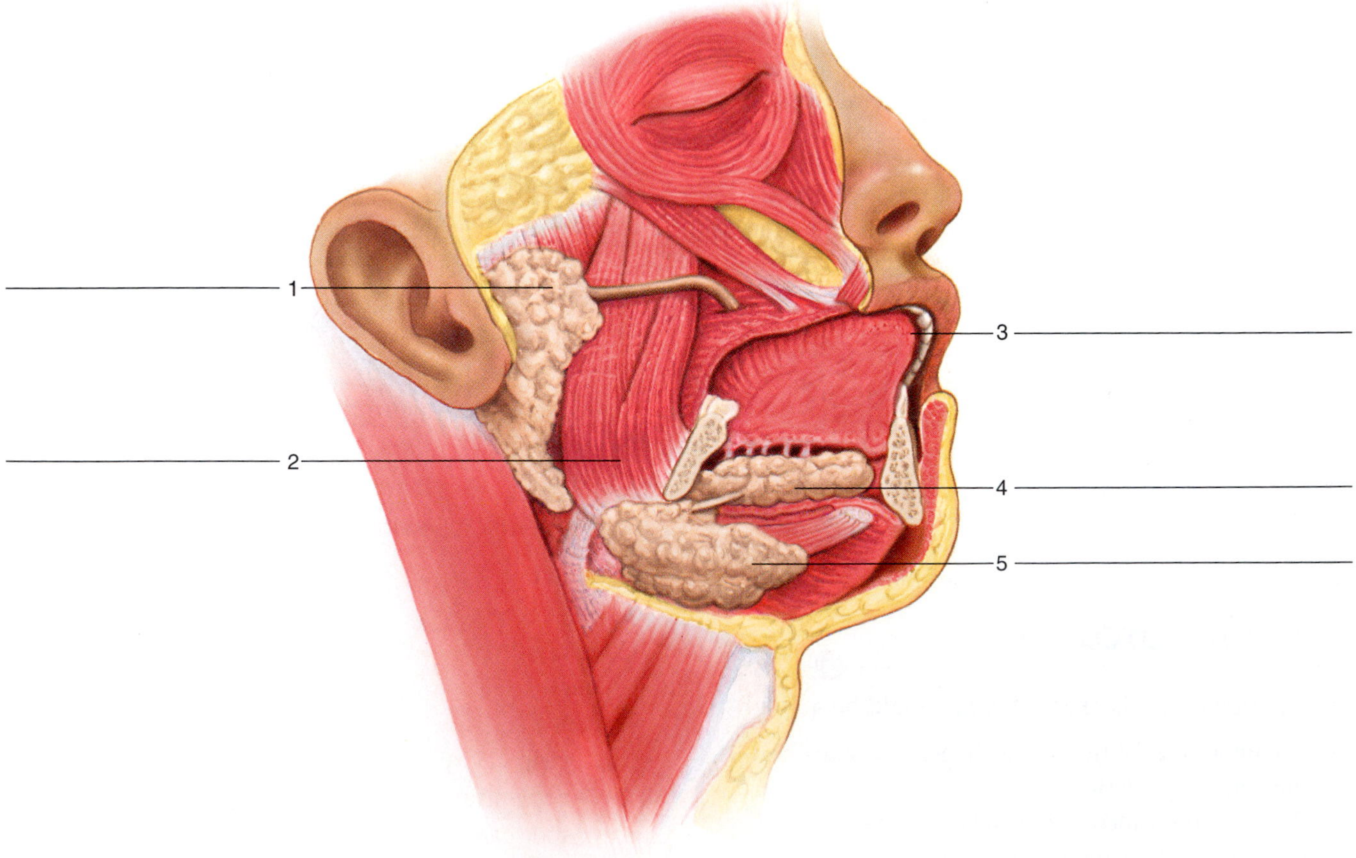

Figure 43.2 Label the features associated with the major salivary glands. 1

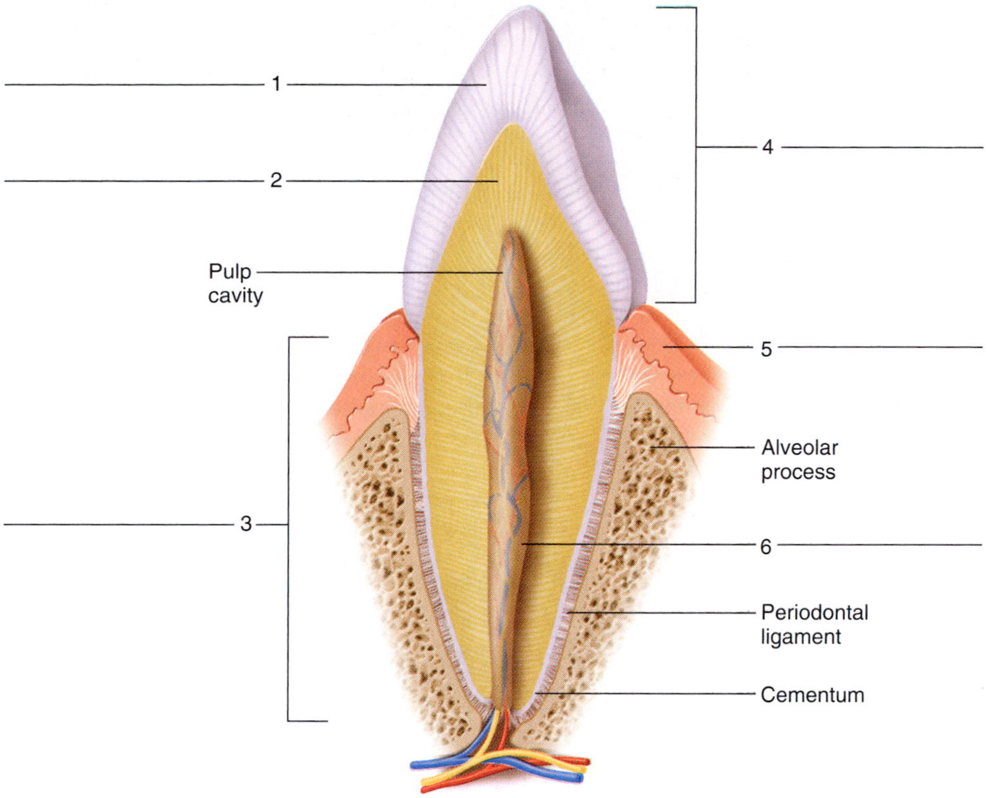

Figure 43.3 Label the features of this canine (cuspid) tooth. 1

5. Observe the head of the human torso model, and locate the following:

   parotid salivary gland
   parotid duct (Stensen's duct)
   submandibular salivary gland
   submandibular duct (Wharton's duct)
   sublingual salivary gland
   sublingual ducts (10–12)

6. Examine a microscopic section of a salivary gland, using low- and high-power magnification. Note the mucous cells that produce mucus and serous cells that produce enzymes are clustered around small ducts. Also note a larger secretory duct surrounded by lightly stained cuboidal epithelial cells (fig. 43.4).
7. Complete Part A of Laboratory Report 43.

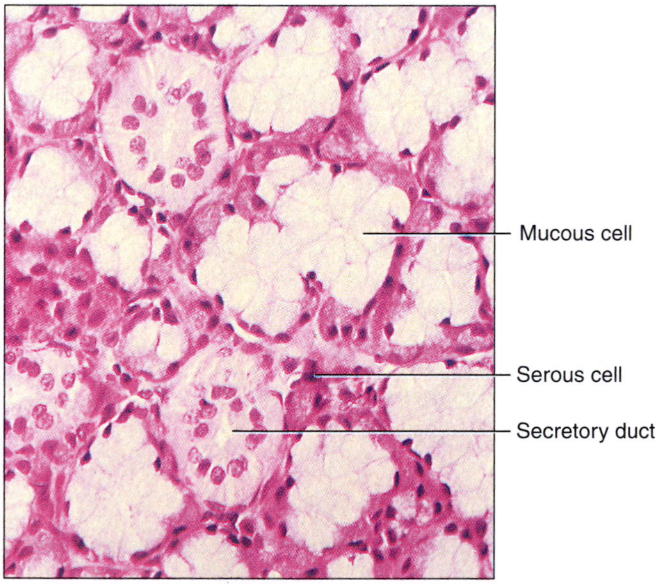

Figure 43.4 Micrograph of the salivary gland (300×).

EXPLORE

Procedure B—Pharynx and Esophagus

1. Review the section entitled "Pharynx and Esophagus" in chapter 17 of the textbook.
2. As a review activity, label figure 43.5.

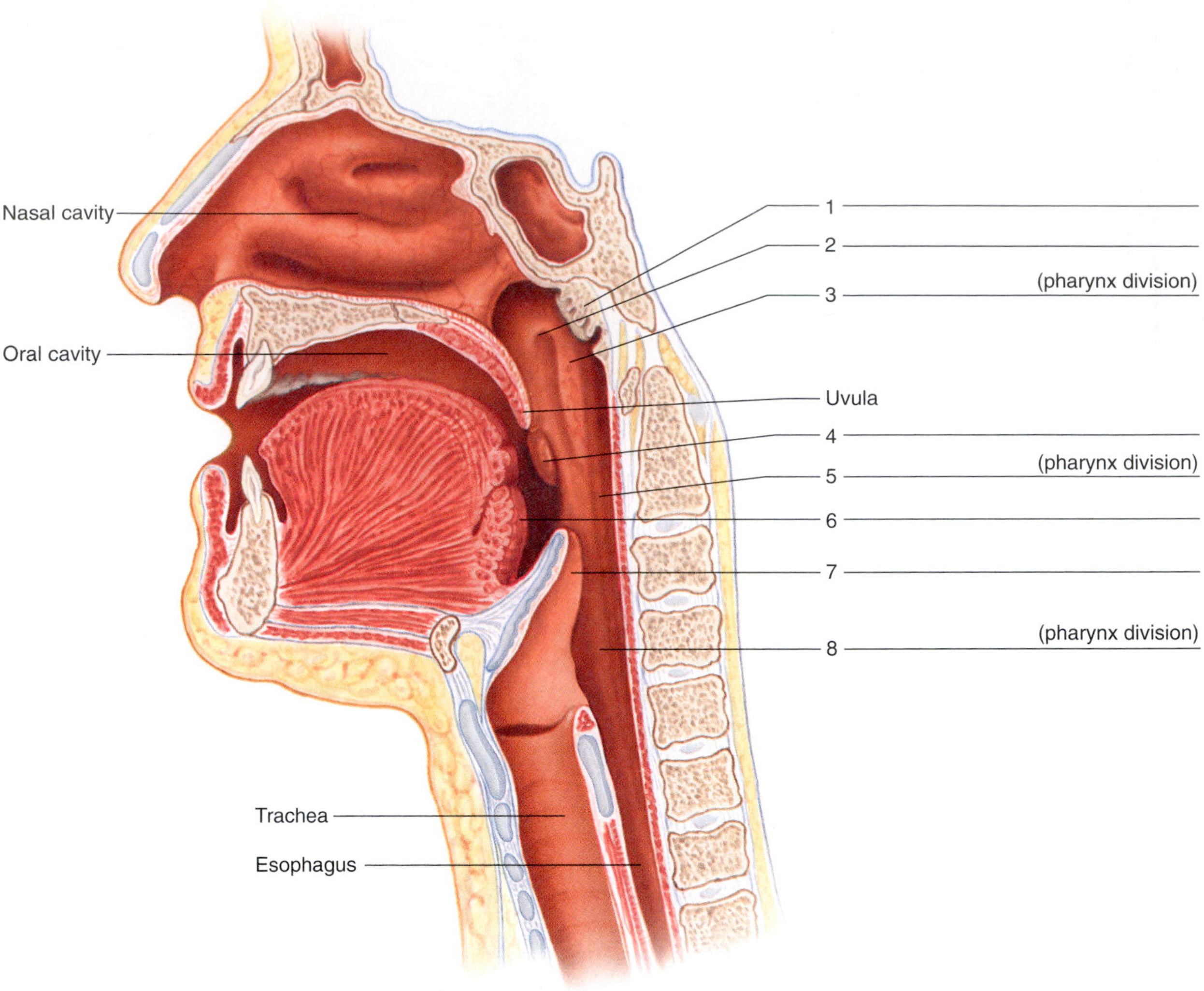

Figure 43.5 Label the features associated with the pharynx. 1

3. Observe the human torso model, and locate the following features:

   pharynx
   - nasopharynx
   - oropharynx
   - laryngopharynx
   - constrictor muscles

   epiglottis
   esophagus
   lower esophageal sphincter (cardiac sphincter)

4. Have your partner take a swallow from a cup of water. Carefully watch the movements in the anterior region of the neck. What steps in the swallowing process did you observe?

   ____

   ____

   ____

5. Examine a microscopic section of esophagus wall, using low-power magnification (fig. 43.6). The

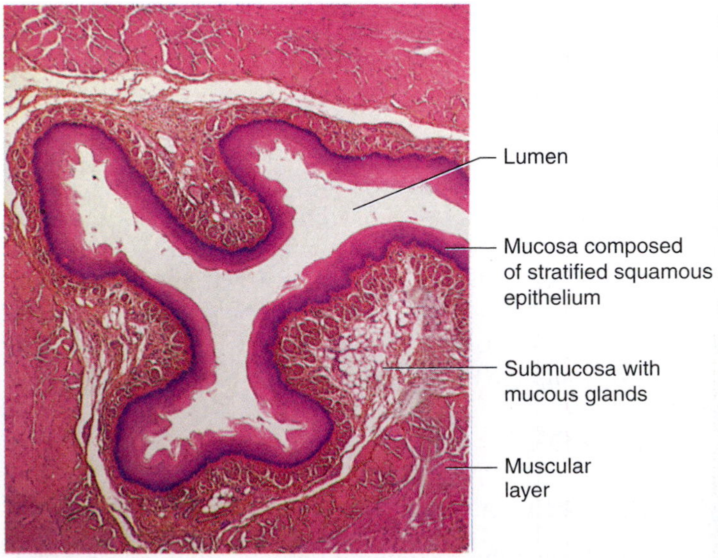

Figure 43.6 Micrograph of a cross section of the esophagus (10×).

inner lining is composed of stratified squamous epithelium, and there are layers of muscle tissue in the wall. Locate some mucous glands in the submucosa. They appear as clusters of lightly stained cells.

6. Complete Part B of the laboratory report.

EXPLORE

Procedure C—The Stomach

1. Review the section entitled "Stomach" in chapter 17 of the textbook.
2. As a review activity, label figures 43.7 and 43.8.

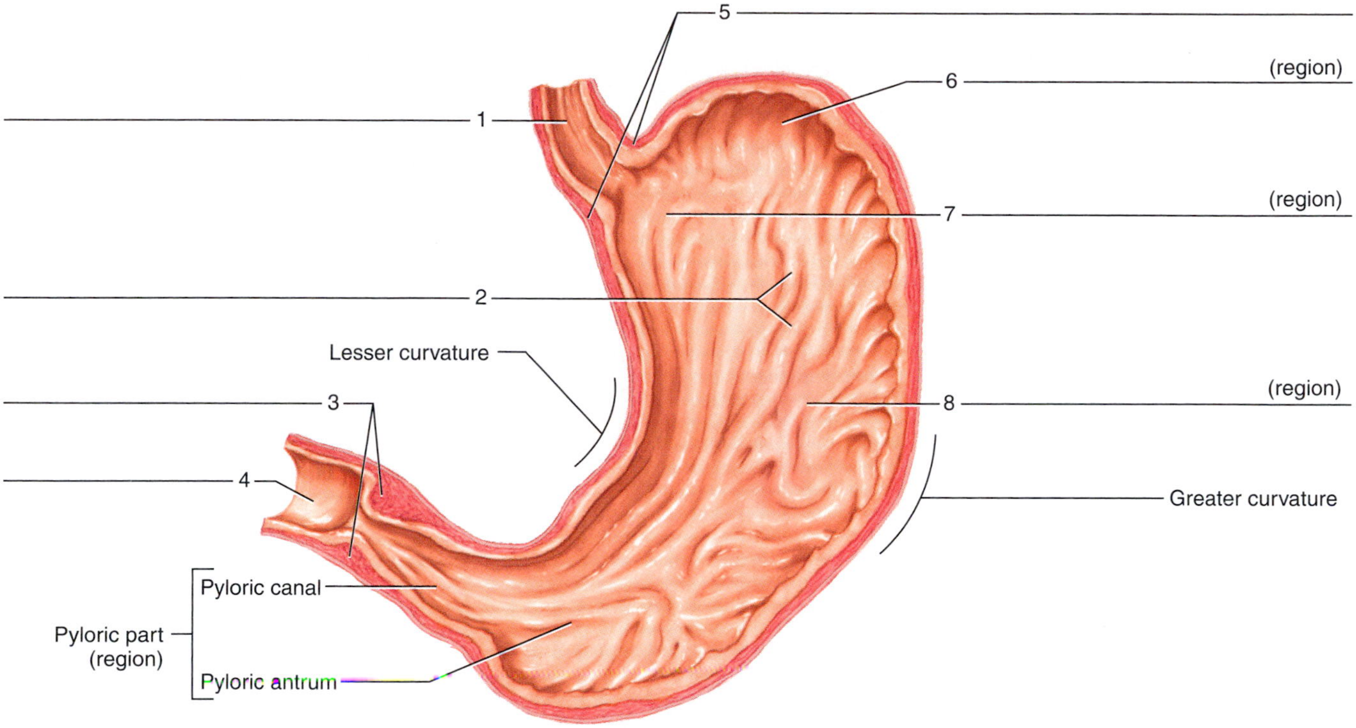

Figure 43.7 Label the stomach and associated structures.

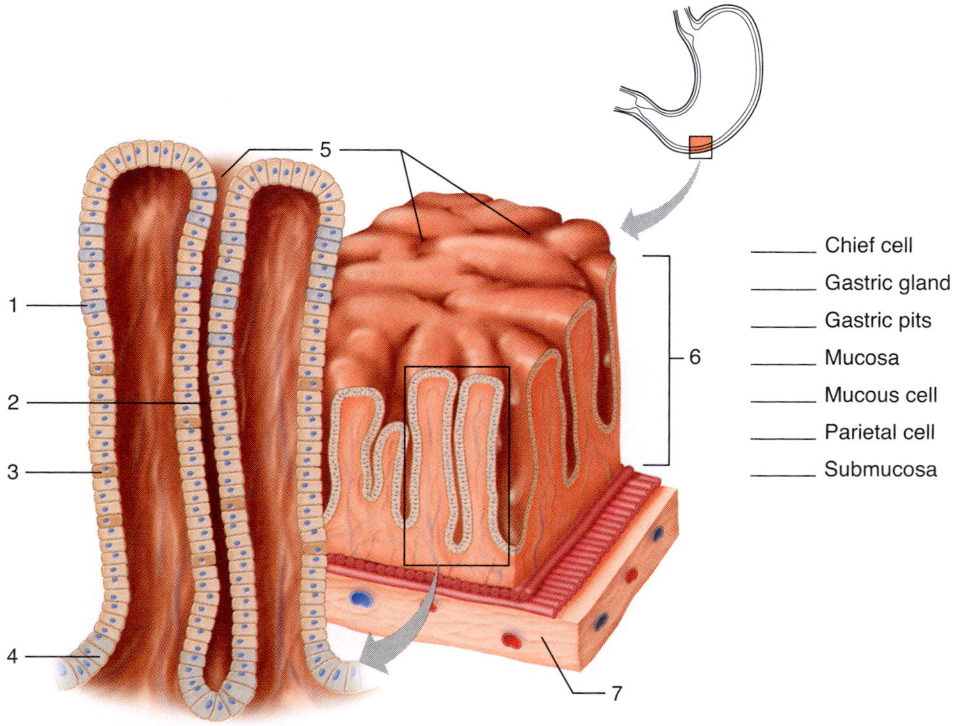

Figure 43.8 Label the lining of the stomach by placing the correct numbers in the spaces provided.

3. Observe the human torso model, and locate the following features of the stomach:

cardia
fundus
body
pyloric part (region)
pyloric antrum
pyloric canal
pyloric sphincter (valve)
lesser curvature
greater curvature
gastric folds (rugae)

4. Examine a microscopic section of stomach wall, using low-power magnification. Note how the inner lining of simple columnar epithelium dips inward to form gastric pits. The gastric glands are tubular structures that open into the gastric pits. Near the deep ends of these glands, you should be able to locate some intensely stained (bluish) chief cells and some lightly stained (pinkish) parietal cells (fig. 43.9).

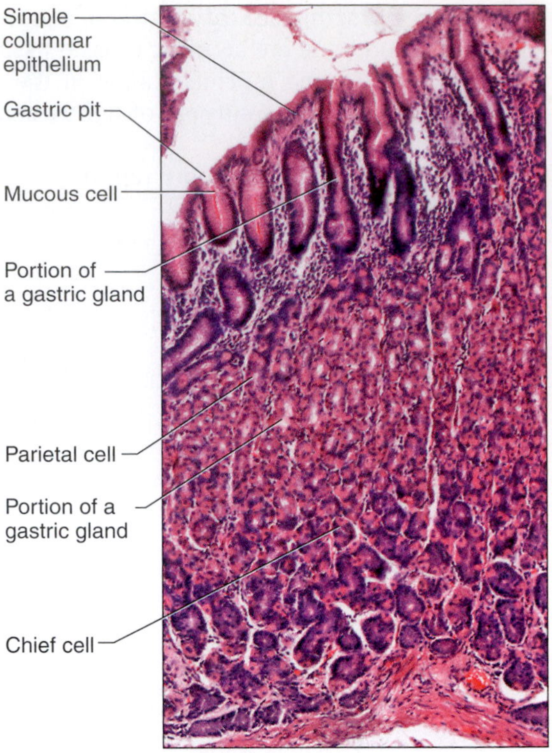

Figure 43.9 Micrograph of the mucosa of the stomach wall (60×).

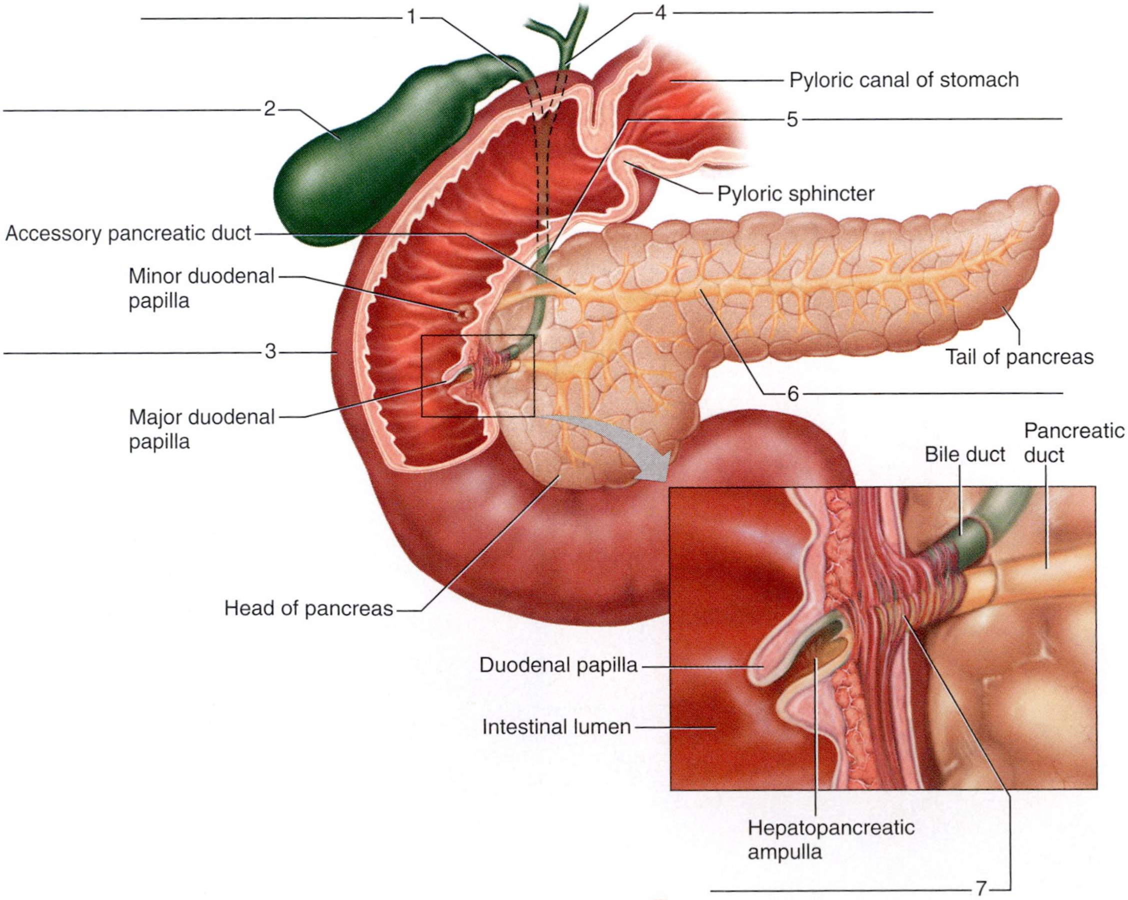

Figure 43.10 Label the features associated with the liver and pancreas. 1

What are the functions of these cells?

___

___

___

___

5. Complete Part C of the laboratory report.

EXPLORE

Procedure D—Pancreas and Liver

1. Review the sections entitled "Pancreas," "Liver," and "Small Intestine" in chapter 17 of the textbook.
2. As a review activity, label figure 43.10.
3. Observe the human torso model, and locate the following structures:

pancreas
pancreatic duct
liver
  right lobe
  quadrate lobe
  caudate lobe
  left lobe
falciform ligament
gallbladder
hepatic ducts
common hepatic duct
cystic duct
bile duct
hepatopancreatic sphincter (sphincter of Oddi)

4. Examine the pancreas slide, using low-power magnification. Observe the exocrine (acinar) cells that secrete pancreatic juice. See figure 36.6 in Laboratory Exercise 36 for a micrograph of the pancreas.

EXPLORE

Procedure E—Small and Large Intestines

1. Review the sections entitled "Small Intestine" and "Large Intestine" in chapter 17 of the textbook.
2. As a review activity, label figure 43.11 and examine figure 43.12.

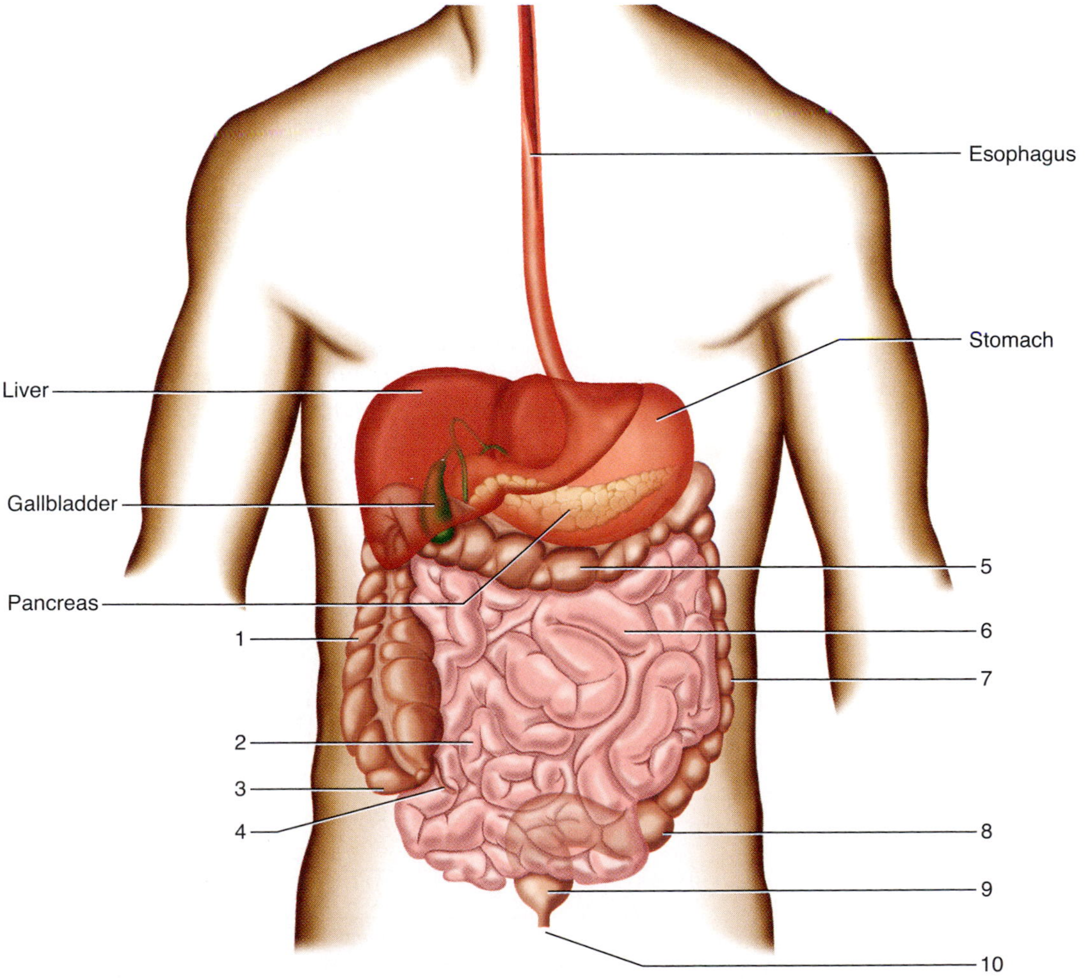

___ Anus
___ Appendix
___ Ascending colon
___ Cecum
___ Descending colon
___ Ileum
___ Jejunum
___ Rectum
___ Sigmoid colon
___ Transverse colon

Figure 43.11 Label this diagram of the small and large intestines by placing the correct numbers in the spaces provided. 1

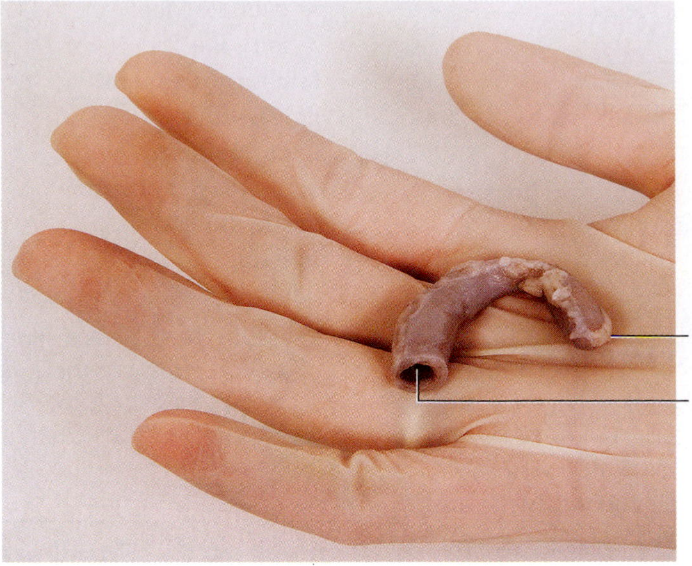

Figure 43.12 Normal appendix.

3. Observe the human torso model, and locate each of the following features:

   small intestine
   - duodenum
   - jejunum
   - ileum

   mesentery

   ileocecal sphincter (valve)

   large intestine
   - large intestinal wall
     - haustra
     - teniae coli
     - epiploic (fatty) appendages
   - cecum
   - appendix (vermiform appendix)
   - ascending colon
   - right colic (hepatic) flexure
   - transverse colon
   - left colic (splenic) flexure
   - descending colon
   - sigmoid colon
   - rectum
   - anal canal
   - anal columns

   anal sphincter muscles
   - external anal sphincter
   - internal anal sphincter

   anus

4. Using low-power magnification, examine a microscopic section of small intestine wall. Identify the mucosa, submucosa, muscular layer, and serosa. Note the villi that extend into the lumen of the tube. Study a single villus, using high-power magnification. Note the core of connective tissue and the covering of simple columnar epithelium that contains some lightly stained goblet cells (fig. 43.13). What is the function of these villi?

   ____

   ____

   ____

5. Prepare a labeled sketch of the wall of the small intestine in Part D of the laboratory report.
6. Examine a microscopic section of large intestine wall. Note the lack of villi. Also note the tubular mucous glands that open on the surface of the inner lining and the numerous lightly stained goblet cells. Locate the four layers of the wall (fig. 43.14). What is the function of the mucus secreted by these glands?

   ____

   ____

   ____

7. Complete Part E of the laboratory report.

Critical Thinking Application

How is the structure of the small intestine better adapted for absorption than the large intestine?

____

____

____

____

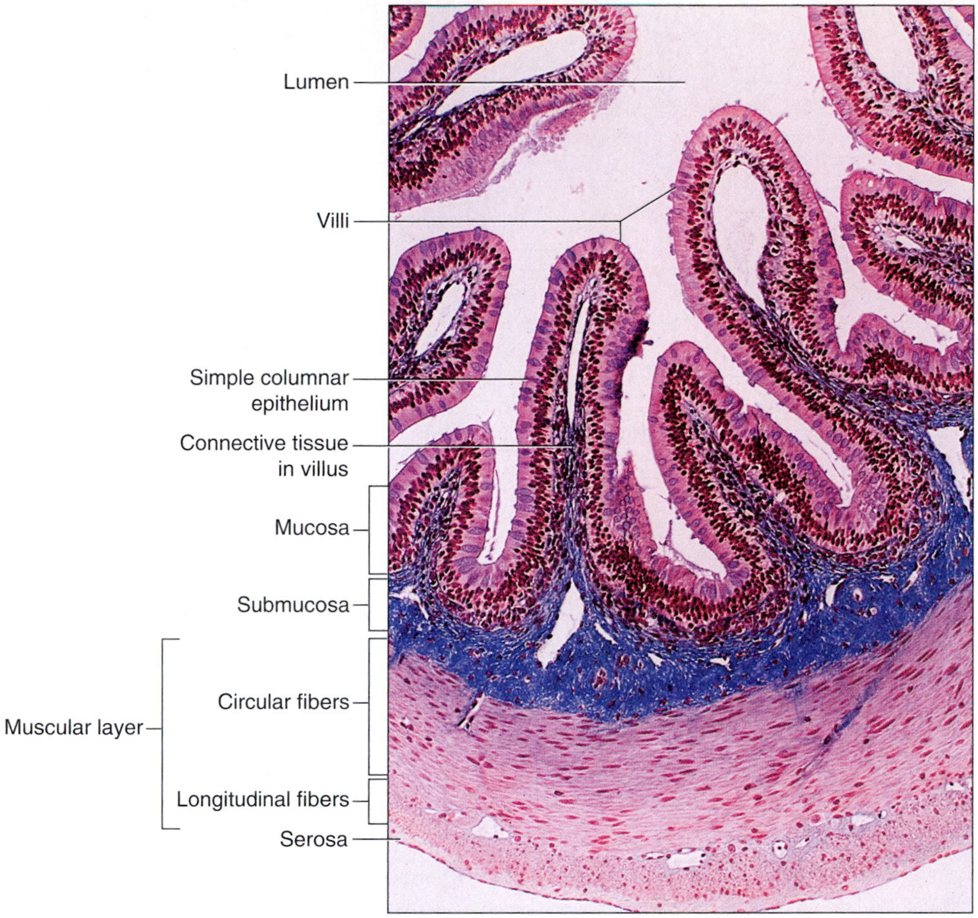

Figure 43.13 Micrograph of the small intestine wall (40×).

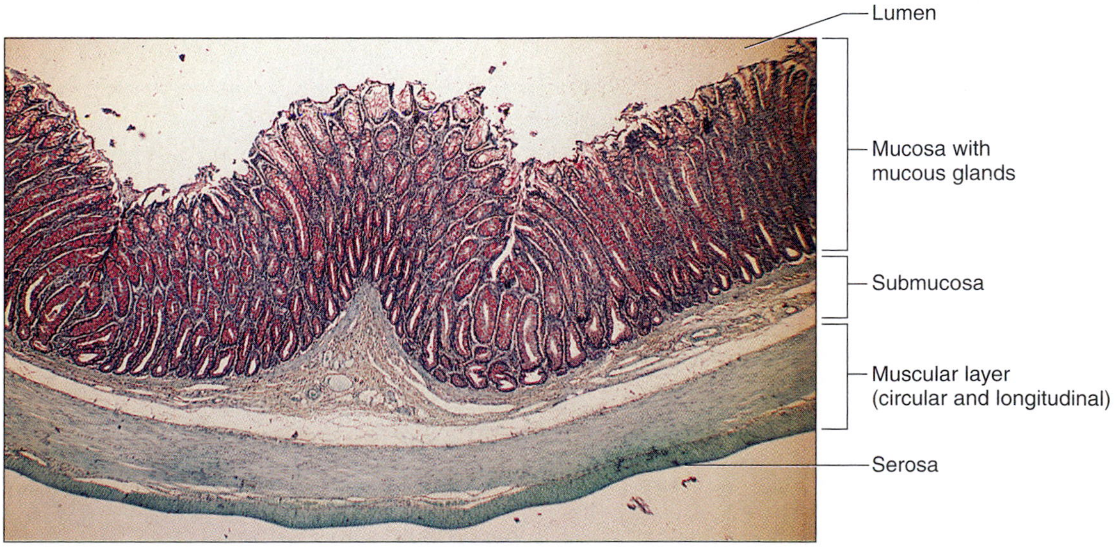

Figure 43.14 Micrograph of the large intestine wall (64×).

Notes

Name ______________________________

Date ______________________________

Section ______________________________

The [A] corresponds to the indicated Learning Outcome(s) listed at the beginning of the laboratory exercise.

Digestive Organs

Laboratory Report 43

Part A Assessments

Match the terms in column A with the descriptions in column B. Place the letter of your choice in the space provided. [1]

Column A		Column B
a. Crown	_______	**1.** Bonelike substance beneath tooth enamel
b. Dentin	_______	**2.** Smallest of major salivary glands
c. Incisor	_______	**3.** Tooth specialized for grinding
d. Lingual frenulum	_______	**4.** Chamber between tongue and palate
e. Molar	_______	**5.** Projections on tongue surface
f. Oral cavity	_______	**6.** Cone-shaped projection of soft palate
g. Papillae	_______	**7.** Attaches tooth to jaw
h. Periodontal ligament	_______	**8.** Chisel-shaped tooth
i. Sublingual gland	_______	**9.** Space between the teeth, cheeks, and lips
j. Uvula	_______	**10.** Anchors tongue to floor of mouth
k. Vestibule	_______	**11.** Portion of tooth projecting beyond gum

Part B Assessments

Complete the following:

1. The part of the pharynx superior to the soft palate is called the ______________________________. [1]
2. The middle part of the pharynx is called the ______________________________. [1]
3. The inferior portion of the pharynx is called the ______________________________. [1]
4. ______________________ is the main secretion of the esophagus. [2]
5. Summarize the functions of the esophagus. [2] ______________________________

____

____

Part C Assessments

Complete the following:

1. Name the four regions of the stomach. 1 ______
2. Name the gastric cells that secrete digestive enzymes. 2 ______
3. Name the gastric cells that secrete hydrochloric acid. 2 ______
4. Name the most important digestive enzyme in gastric juice. 2 ______
5. Name the semifluid paste of food particles and gastric juice. 2 ______
6. Summarize the functions of the stomach. 2 ______

Part D Assessments

Prepare a sketch of a cross section of the small intestine. Label the four layers, villi, and lumen. Indicate the magnification used for your sketch. 3

Part E Assessments

Complete the following:

1. Name the three portions of the small intestine. 1 ______
2. Describe the function of the mesentery. 2 ______
3. Name five digestive enzymes secreted by the small intestinal mucosa. 2 ______
4. Name the valve located between the small and large intestines. 1 ______
5. Summarize the functions of the small intestine. 2 ______
6. Summarize the functions of the large intestine. 2 ______

Laboratory Exercise 44

Action of a Digestive Enzyme

Materials Needed

0.5% amylase solution*
Beakers (50 and 500 mL)
Distilled water
Funnel
Pipets (1 and 10 mL)
Pipet rubber bulbs
0.5% starch solution
Graduated cylinder (10 mL)
Test tubes
Test-tube clamps
Wax marker
Iodine-potassium-iodide solution
Medicine dropper
Ice
Water bath, 37°C (98.6°F)
Porcelain test plate
Benedict's solution
Hot plates
Test-tube rack
Thermometer

For Alternative Procedure:

Small disposable cups
Distilled water

*The amylase should be free of sugar for best results; a low-maltose solution of amylase yields good results.

Safety

- Use only a mechanical pipetting device (never your mouth). Use pipets with rubber bulbs or dropping pipets.
- Wear safety glasses when working with acids and when heating test tubes.
- Use test-tube clamps when handling hot test tubes.
- If an open flame is used for heating the test solutions, keep clothes and hair away from the flame.
- Dispose of chemicals according to appropriate directions.
- Review the safety guidelines inside the front cover.
- Use appropriate disinfectant to wash laboratory tables before and after the procedures.
- If student saliva is used as a source of amylase, it is important that students wear disposable gloves and only handle their own materials.
- Wash your hands before leaving the laboratory.

The digestive enzyme in salivary secretions is called *salivary amylase.* Pancreatic amylase is secreted among several other pancreatic enzymes. A bacterial extraction of amylase is available for laboratory experiments. This enzyme splits starch molecules into sugar (disaccharide) molecules, which is the first step in the digestion of complex carbohydrates.

As in the case of other enzymes, amylase is a protein catalyst. Its activity is affected by exposure to certain environmental factors, including various temperatures, pH, radiation, and electricity. As temperatures increase, faster chemical reactions occur as the collisions of molecules happen at a greater frequency. Eventually, temperatures increase to a point that the enzyme is denatured and the rate of the enzyme activity rapidly declines. As temperatures decrease, enzyme activity also decreases due to fewer collisions of the molecules; however, the colder temperatures do not denature the enzyme. Normal body temperature provides an environment for enzyme activity near the optimum for enzymatic reactions.

The pH of the environment where enzymes are secreted also has a major influence on the enzyme reactions. The optimum pH for amylase activity is between 6.8 and 7.0, typical of salivary secretions. When salivary amylase arrives in the stomach, hydrochloric acid deactivates the enzyme, diminishing any further chemical digestion of remaining starch in the stomach. Pancreatic amylase is secreted into the small intestine where an optimum pH for amylase is once more provided. Other enzymes in the digestive system have different optimum pH ranges of activity compared to amylase. For example, pepsin from stomach secretions has an optimum activity around pH 2, while trypsin from pancreatic secretions operates best around pH 7–8.

Purpose of the Exercise

To investigate the action of amylase and the effect of heat on its enzymatic activity.

LEARNING OUTCOMES

After completing this exercise, you should be able to

1. Explain the action of amylase.
2. Test a solution for the presence of starch or the presence of sugar.
3. Test the effects of varying temperatures on the activity of amylase.

EXPLORE

Procedure A—Amylase Activity

1. Study the lock-and-key model of amylase action on starch digestion (fig. 44.1).
2. Examine the digestive locations where starch is converted to glucose (fig. 44.2).
3. Mark three clean test tubes as *tubes 1, 2,* and *3,* and prepare the tubes as follows:

   Tube 1: **Add 6 mL of amylase solution.**

   Tube 2: **Add 6 mL of starch solution.**

   Tube 3: **Add 5 mL of starch solution and 1 mL of amylase solution.**

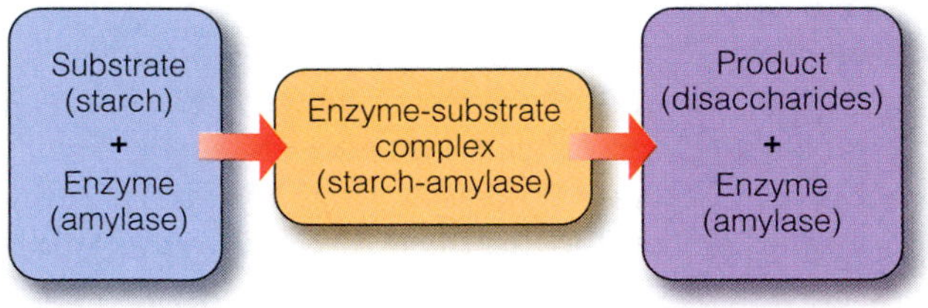

Figure 44.1 Lock-and-key model of enzyme action of amylase on starch digestion.

4. Shake the tubes well to mix the contents, and place them in a warm water bath, 37°C (98.6°F), for 10 minutes.
5. At the end of the 10 minutes, test the contents of each tube for the presence of starch. To do this, follow these steps:
   - **a.** Place 1 mL of the solution to be tested in a depression of a porcelain test plate.
   - **b.** Next add one drop of iodine-potassium-iodide solution, and note the color of the mixture. If the solution becomes blue-black, starch is present.
   - **c.** Record the results in Part A of Laboratory Report 44.
6. Test the contents of each tube for the presence of sugar (disaccharides in this instance). To do this, follow these steps:
   - **a.** Place 1 mL of the solution to be tested in a clean test tube.
   - **b.** Add 1 mL of Benedict's solution.
   - **c.** Place the test tube with a test-tube clamp in a beaker of boiling water for 2 minutes.
   - **d.** Note the color of the liquid. If the solution becomes green, yellow, orange, or red, sugar is present. Blue indicates a negative test, whereas green indicates a positive test with the least amount of sugar, and red indicates the greatest amount of sugar present.
   - **e.** Record the results in Part A of the laboratory report.
7. Complete Part A of the laboratory report.

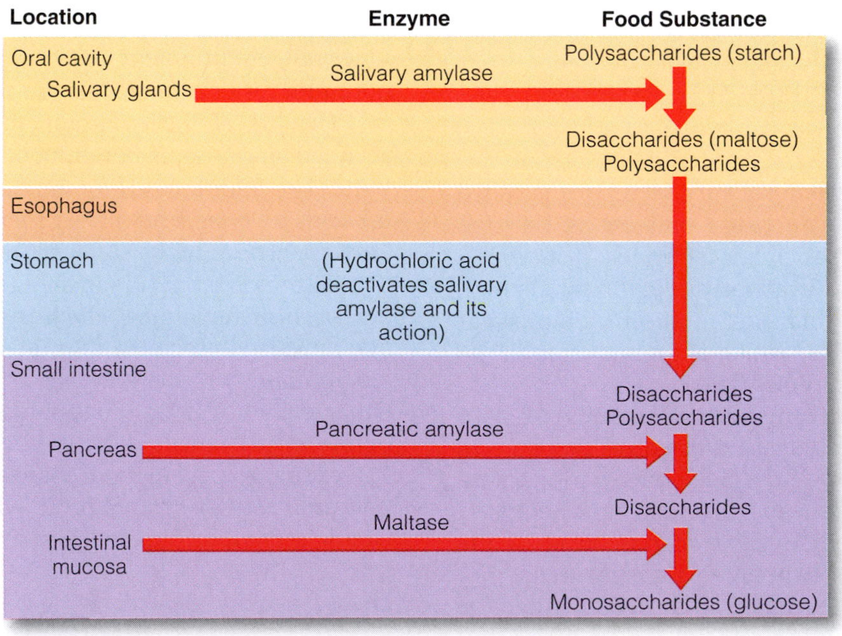

Figure 44.2 Flowchart of starch digestion.

Alternative Procedure

Human saliva could be used as a source of amylase solutions instead of bacterial amylase preparations. Collect about 5 mL of saliva in a small disposable cup. Add an equal amount of distilled water and mix together for the amylase solutions during the laboratory procedures. Be sure to follow all of the safety guidelines.

EXPLORE

Procedure B—Effect of Heat

1. Mark three clean test tubes as *tubes 4, 5,* and *6.*
2. Add 1 mL of amylase solution to each of the tubes, and expose each solution to a different test temperature for 3 minutes as follows:

   Tube 4: **Place in beaker of ice water (about 0°C [32°F]).**

   Tube 5: **Place in warm water bath (about 37°C [98.6°F]).**

   Tube 6: **Place in beaker of boiling water (about 100°C [212°F]). Use a test-tube clamp.**
3. Add 5 mL of starch solution to each tube, shake to mix the contents, and return the tubes to their respective test temperatures for 10 minutes. It is important that the 5 mL of starch solution added to tube 4 be at ice-water temperature before it is added to the 1 mL of amylase solution.
4. At the end of the 10 minutes, test the contents of each tube for the presence of starch and the presence of sugar by following the directions in Procedure A.
5. Complete Part B of the laboratory report.

Learning Extension

Devise an experiment to test the effect of some other environmental factor on amylase activity. For example, you might test the effect of a strong acid by adding a few drops of concentrated hydrochloric acid to a mixture of starch and amylase solutions. Be sure to include a control in your experimental plan. That is, include a tube containing everything except the factor you are testing. Then you will have something with which to compare your results. *Carry out your experiment only if it has been approved by the laboratory instructor.*

Notes

Name ____________________

Date ____________________

Section ____________________

The [A] corresponds to the indicated Learning Outcome(s) listed at the beginning of the laboratory exercise.

Action of a Digestive Enzyme

44 Laboratory Report

Part A—Amylase Activity Assessments

1. Test results: [2]

Tube	Starch	Sugar
1 Amylase solution		
2 Starch solution		
3 Starch-amylase solution		

2. Complete the following:

   a. Explain the reason for including tube 1 in this experiment. [1] ____________________

   b. What is the importance of tube 2? [1] ____________________

   c. What do you conclude from the results of this experiment? [1] ____________________

Part B—Effect of Heat Assessments

1. Test results:

Tube	Starch	Sugar
4 0°C (32°F)		
5 37°C (98.6°F)		
6 100°C (212°F)		

2. Complete the following:

   a. What do you conclude from the results of this experiment?

   b. If digestion failed to occur in one of the tubes in this experiment, how can you tell if the amylase was destroyed by the factor being tested or if the amylase activity was simply inhibited by the test treatment?

Critical Thinking Application

What test result would occur if the amylase you used contained sugar?
Would your results be as valid? Explain your answer.

Laboratory Exercise 45

Respiratory Organs

Materials Needed

Textbook
Human skull (sagittal section)
Human torso model
Larynx model
Thoracic organs model
Compound light microscope
Prepared microscope slides of the following:
- Trachea (cross section)
- Lung, human (normal)

For Demonstrations:

Animal lung with trachea (fresh or preserved)
Bicycle pump
Prepared microscope slides of the following:
- Lung tissue (smoker)
- Lung tissue (emphysema)

Safety

- Wear disposable gloves when working on the fresh or preserved animal lung demonstration.
- Wash your hands before leaving the laboratory.

The organs of the respiratory system include the nose, nasal cavity, sinuses, pharynx, larynx, trachea, bronchial tree, and lungs. They mainly function to process incoming air and to transport it to and from the atmosphere outside the body and the air sacs of the lungs.

In the air sacs, gas exchanges take place between the air and the blood of nearby capillaries. The blood, in turn, transports gases to and from the air sacs and the body cells. This entire process of transporting and exchanging gases between the atmosphere and the body cells is called *respiration*.

Purpose of the Exercise

To review the structure and function of the respiratory organs and to examine the tissues of some of these organs.

LEARNING OUTCOMES

After completing this exercise, you should be able to

1. Locate the major organs and structural features of the respiratory system.
2. Describe the functions of these organs.
3. Sketch and label features of tissue sections of the trachea and lung.

EXPLORE

Procedure A—Respiratory Organs

1. Review the section entitled "Organs of the Respiratory System" in chapter 19 of the textbook.
2. As a review activity, label figures 45.1, 45.2, 45.3, and 45.4.
3. Examine the sagittal section of the human skull, and locate the following features:

   nasal cavity
   - nostril (naris)
   - nasal septum
   - nasal conchae
   - meatuses

   paranasal sinuses
   - maxillary sinus
   - frontal sinus
   - ethmoidal sinus
   - sphenoidal sinus

4. Observe the larynx model, the thoracic organs model, and the human torso model. Locate the features listed in step 3. Also locate the following features:

   pharynx
   - nasopharynx
   - oropharynx
   - laryngopharynx

larynx (palpate your larynx)
- vocal cords
  - false vocal cords (vestibular folds)
  - true vocal cords (vocal folds)
- thyroid cartilage ("Adam's apple")
- cricoid cartilage
- cricothyroid ligament
- epiglottis
- epiglottic cartilage
- arytenoid cartilages
- corniculate cartilages
- cuneiform cartilages
- glottis

trachea (palpate your trachea)

bronchi
- primary (right and left main) bronchi
- secondary (lobar) bronchi
- tertiary (segmental) bronchi

bronchioles

lung
- hilum
- lobes
  - superior lobe
  - middle lobe (right only)
  - inferior lobe

visceral pleura

parietal pleura

pleural cavity

5. Complete Part A of Laboratory Report 45.

Demonstration

Observe the animal lung and the attached trachea. Identify the larynx, major laryngeal cartilages, trachea, and the incomplete cartilaginous rings of the trachea. Open the larynx and locate the vocal folds. Examine the visceral pleura on the surface of a lung. A bicycle pump could be used to demonstrate lung inflation. Section the lung and locate some bronchioles and alveoli. Squeeze a portion of a lung between your fingers. How do you describe the texture of the lung?

____

____

____

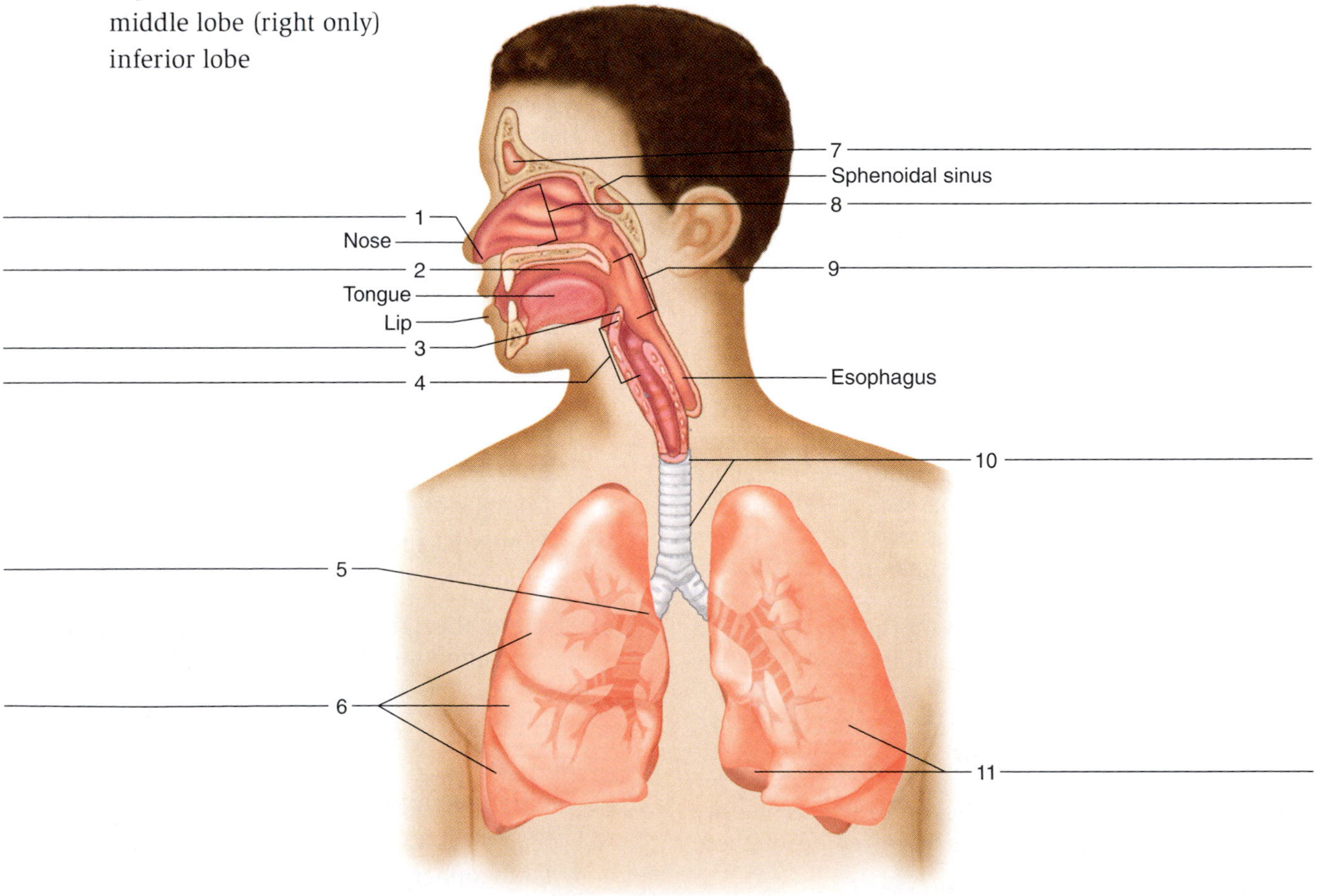

Figure 45.1 Label the major features of the respiratory system. 1

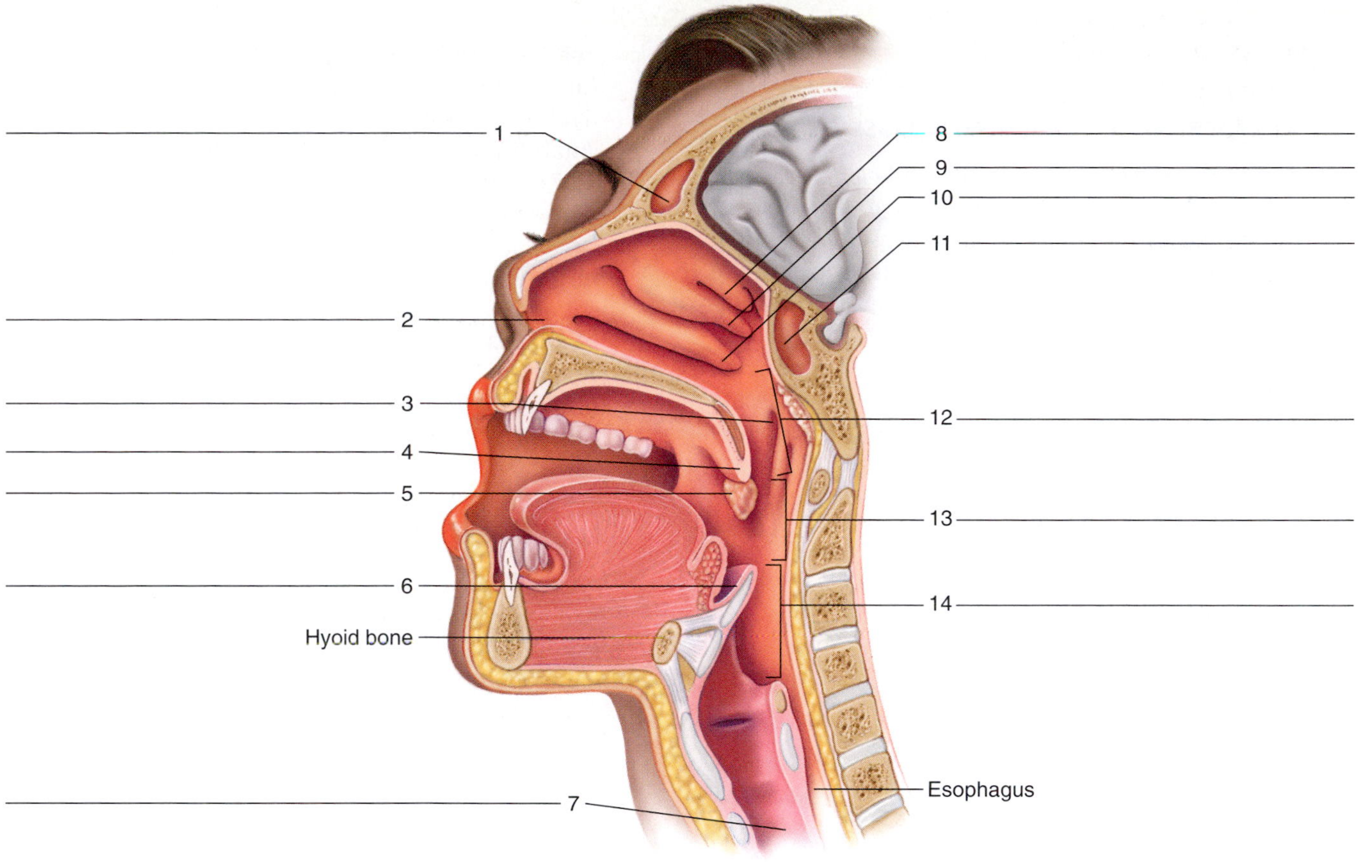

Figure 45.2 Label the features of this sagittal section of the upper respiratory tract. 1

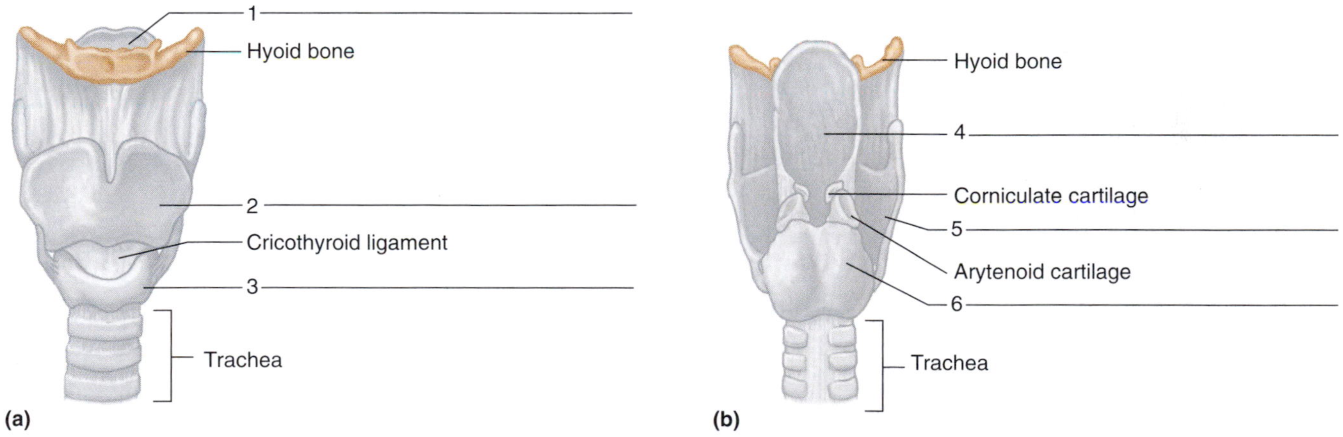

Figure 45.3 Label the major features of the larynx: (*a*) anterior view; (*b*) posterior view. 1

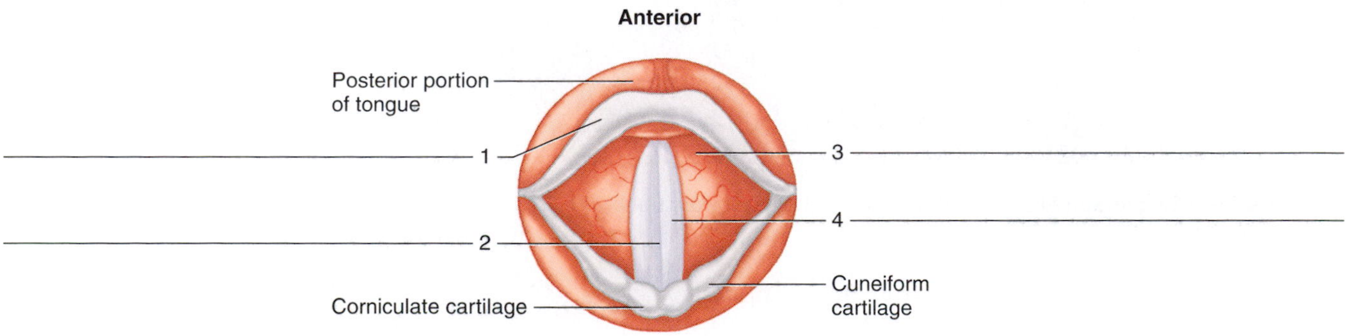

Figure 45.4 Label the features of the superior aspect of the larynx with the glottis closed. 1

EXPLORE

Procedure B—Respiratory Tissues

1. Obtain a prepared microscope slide of a trachea, and use low-power magnification to examine it. Notice the inner lining of ciliated pseudostratified columnar epithelium and the deep layer of hyaline cartilage, which represents a portion of an incomplete (C-shaped) tracheal ring (fig. 45.5).
2. Use high-power magnification to observe the cilia on the free surface of the epithelial lining. Locate the wineglass-shaped goblet cells, which secrete a protective mucus, in the epithelium (fig. 45.6).
3. Prepare a labeled sketch of a representative portion of the tracheal wall in Part B of the laboratory report.
4. Obtain a prepared microscope slide of human lung tissue. Examine it, using low-power magnification, and note the many open spaces of the air sacs (alveoli). Look for a bronchiole—a tube with a relatively thick wall and a wavy inner lining. Locate the smooth muscle tissue in the wall of this tube (fig. 45.7). You also may see a section of cartilage as part of the bronchiole wall.
5. Use high-power magnification to examine the alveoli (fig. 45.8). Their walls are composed of simple squamous epithelium. You also may see sections of blood vessels containing blood cells.
6. Prepare a labeled sketch of a representative portion of the lung in Part B of the laboratory report.
7. Complete Part C of the laboratory report.

Demonstration

Examine the prepared microscope slides of the lung tissue of a smoker and a person with emphysema, using low-power magnification. How does the smoker's lung tissue compare with that of the normal lung tissue that you examined previously?

____

____

____

____

____

How does the emphysema patient's lung tissue compare with the normal lung tissue?

____

____

____

____

____

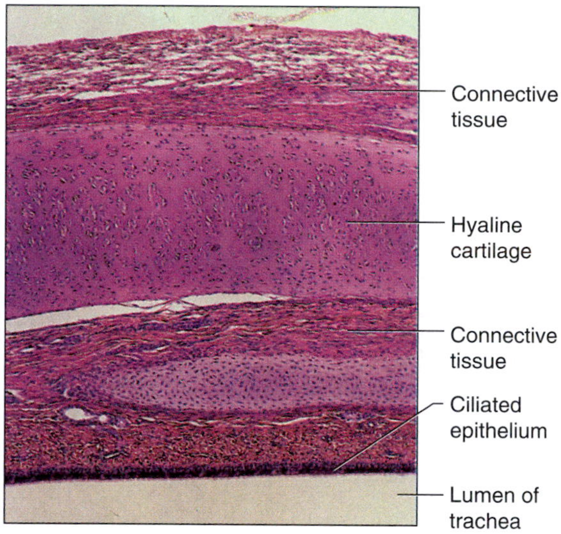

Figure 45.5 Micrograph of a section of the tracheal wall (63×).

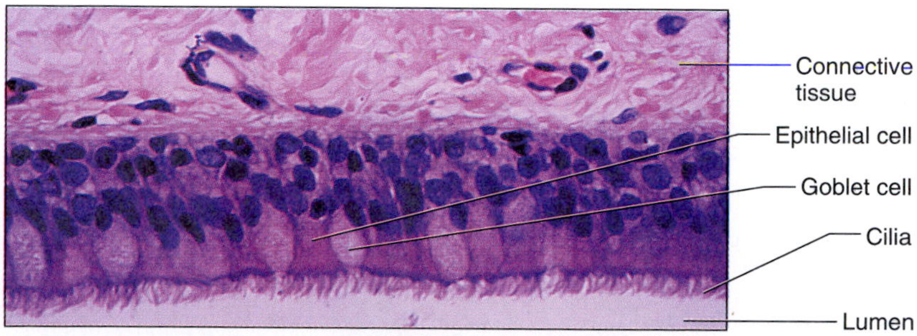

Figure 45.6 Micrograph of ciliated epithelium in the respiratory tract (275×).

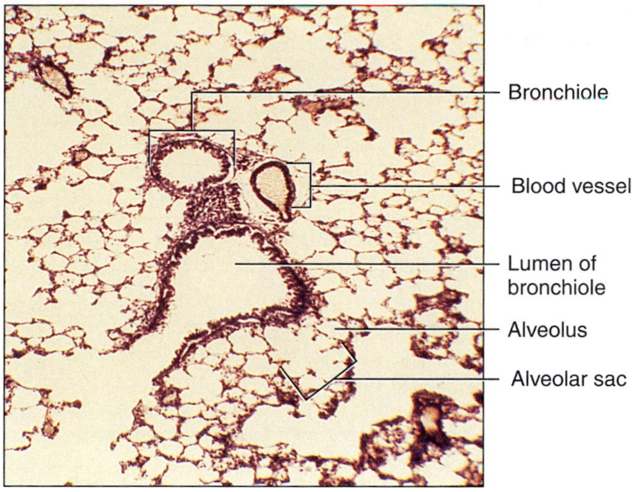

Figure 45.7 Micrograph of human lung tissue (35×).

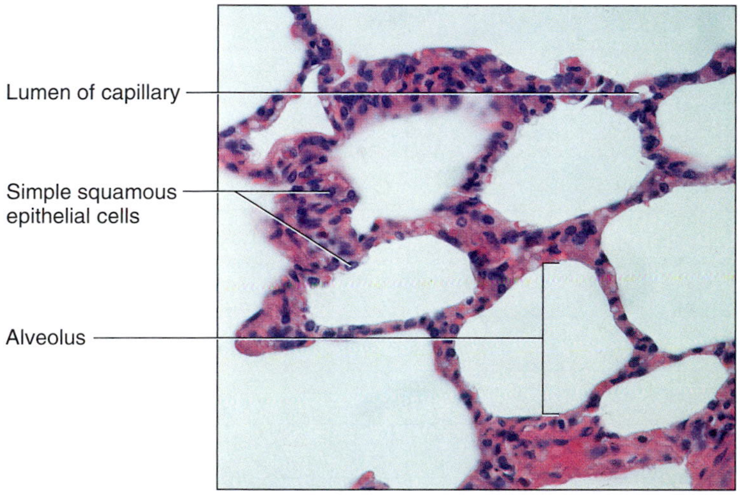

Figure 45.8 Micrograph of human lung tissue (250×).

Notes

Name ________________________________

Date ________________________________

Section ________________________________

The [A] corresponds to the indicated Learning Outcome(s) listed at the beginning of the laboratory exercise.

Laboratory Report 45

Respiratory Organs

Part A Assessments

Match the terms in column A with the descriptions in column B. Place the letter of your choice in the space provided. 1 2

Column A		Column B
a. Alveolus	_____	1. Potential space between visceral and parietal pleurae
b. Cricoid cartilage	_____	2. Most inferior portion of larynx
c. Epiglottis	_____	3. Serves as resonant chamber and reduces weight of skull
d. Glottis	_____	4. Microscopic air sac for gas exchange
e. Lung	_____	5. Consists of large lobes
f. Nasal concha	_____	6. Opening between vocal cords
g. Pharynx	_____	7. Fold of mucous membrane containing elastic fibers responsible for sounds
h. Pleural cavity	_____	8. Increases surface area of nasal mucous membrane
i. Sinus (paranasal sinus)	_____	9. Passageway for air and food
j. Vocal cord (true)	_____	10. Partially covers opening of larynx during swallowing

Part B Assessments

Sketch and label a portion of the tracheal wall and a portion of lung tissue. 3

Tracheal wall (______×)	Lung tissue (______×)

Part C Assessments

Complete the following:

1. What is the function of the mucus secreted by the goblet cells? 2 ______________________________

2. Describe the function of the cilia in the respiratory tubes. 2 ______________________________

3. How is breathing affected if the smooth muscle of the bronchial tree relaxes? 2 ______________________________

Critical Thinking Application

Why are the alveolar walls so thin? 2

____

Laboratory Exercise 46

Breathing and Respiratory Volumes

Materials Needed

Textbook
Spirometer, handheld (dry portable)
Alcohol swabs
Disposable mouthpieces
Nose clips
Meterstick

For Demonstration:
Lung function model

Safety

- Clean the spirometer with an alcohol swab before each use.
- Place a new disposable mouthpiece on the stem of the spirometer before each use.
- Dispose of the alcohol swabs and mouthpieces according to directions from your laboratory instructor.

Breathing involves the movement of air from outside the body through the bronchial tree and into the alveoli and the reversal of this air movement to allow gas (oxygen and carbon dioxide) exchange between air and blood. These movements are caused by changes in the size of the thoracic cavity that result from skeletal muscle contractions and from the elastic recoil of stretched tissues.

The volumes of air that move in and out of the lungs during various phases of breathing are called *respiratory air volumes* and *capacities*. These volumes can be measured by using an instrument called a *spirometer*. However, the values obtained vary with a person's age, sex, height, weight, stress, and physical fitness. Various respiratory capacities can be calculated by combining two or more of the respiratory volumes.

Purpose of the Exercise

To review the mechanisms of breathing and to measure or calculate certain respiratory air volumes and respiratory capacities.

LEARNING OUTCOMES

After completing this exercise, you should be able to

1. Differentiate between the mechanisms responsible for inspiration and expiration.
2. Match the respiratory air volumes and respiratory capacities with their definitions.
3. Measure respiratory air volumes using a spirometer.
4. Calculate respiratory capacities using the data obtained from respiratory air volumes.

EXPLORE

Procedure A—Breathing Mechanisms

1. Review the sections entitled "Inspiration" and "Expiration" in chapter 19 of the textbook.
2. Complete Part A of Laboratory Report 46.

Demonstration

Observe the mechanical lung function model (fig. 46.1). It consists of a heavy plastic bell jar with a rubber sheeting clamped over its wide open end. Its narrow upper opening is plugged with a rubber stopper through which a Y tube is passed. Small rubber balloons are fastened to the arms of the Y. What happens to the balloons when the rubber sheeting is pulled downward?

____

____

What happens when the sheeting is pushed upward?

____

____

How do you explain these changes?

____

____

____

What part of the respiratory system is represented by
the rubber sheeting? ______________________
the bell jar? ______________________
the Y tube? ______________________
the balloons? ______________________

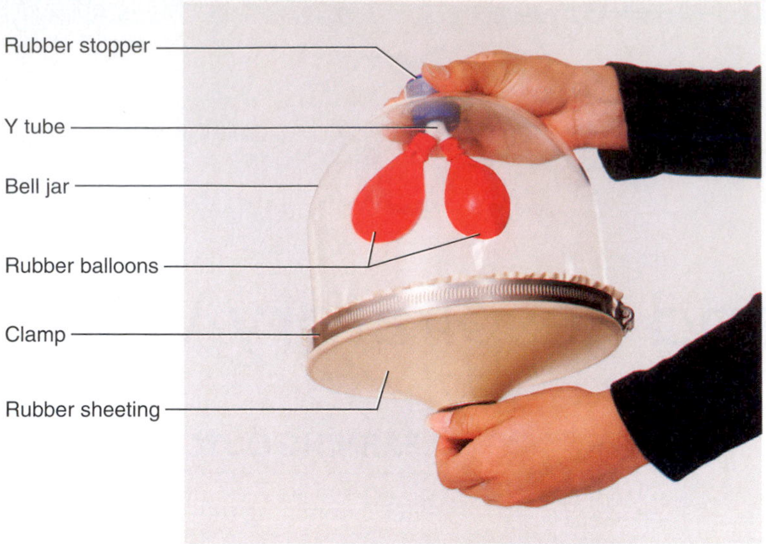

Figure 46.1 A lung function model.

> **WARNING:** If the subject begins to feel dizzy or light-headed while performing Procedure B, stop the exercise and breathe normally.

EXPLORE

Procedure B—Respiratory Volumes and Capacities

1. Review the section entitled "Respiratory Volumes and Capacities" in chapter 19 of the textbook.
2. Complete Part B of the laboratory report.
3. Get a handheld spirometer (fig. 46.2). Point the needle to zero by rotating the adjustable dial. Before using the instrument, clean it with an alcohol swab and place a new disposable mouthpiece over its stem. The instrument should be held with the dial upward, and air should be blown only into the disposable mouthpiece (fig. 46.3). If air tends to exit the nostrils during exhalation, use a nose clip or pinch your nose as a prevention when exhaling into the spirometer. Movement of the needle indicates the air volume that leaves the lungs. The exhalation should be slowly and forcefully performed. Too rapid an exhalation could result in erroneous data or damage to the spirometer.

Figure 46.2 A handheld spirometer can be used to measure respiratory air volumes.

Figure 46.3 Demonstration of a handheld spirometer. Air should only be blown slowly and forcefully into a disposable mouthpiece. Use a nose clip or pinch your nose when measuring expiratory reserve volume and vital capacity volume if air exits the nostrils.

4. *Tidal volume* (TV) (about 500 mL) is the volume of air that enters (or leaves) the lungs during a *respiratory cycle* (one inspiration plus the following expiration). *Resting tidal volume* is the volume of air that enters (or leaves) the lungs during normal, quiet breathing (fig. 46.4). To measure this volume, follow these steps:
   a. Sit quietly for a few moments.
   b. Position the spirometer dial so that the needle points to zero.
   c. Place the mouthpiece between your lips and exhale three ordinary expirations into it after inhaling through the nose each time. *Do not force air out of your lungs; exhale normally.*
   d. Divide the total value indicated by the needle by 3, and record this amount as your resting tidal volume on the table in Part C of the laboratory report.
5. *Expiratory reserve volume* (ERV) (about 1,100 mL) is the volume of air in addition to the tidal volume that leaves the lungs during forced expiration. To measure this volume, follow these steps:
   a. Breathe normally for a few moments. Set the needle to zero.
   b. At the end of an ordinary expiration, place the mouthpiece between your lips and exhale all of the air you can force from your lungs through the spirometer. Use a nose clip or pinch your nose to prevent air from exiting your nostrils.
   c. Record the results as your expiratory reserve volume in Part C of the laboratory report.
6. *Vital capacity* (VC) (about 4,600 mL) is the maximum volume of air that can be exhaled after taking the deepest breath possible. To measure this volume, follow these steps:
   a. Breathe normally for a few moments. Set the needle at zero.
   b. Breathe in and out deeply a couple of times, then take the deepest breath possible.
   c. Place the mouthpiece between your lips and exhale all the air out of your lungs, slowly and forcefully. Use a nose clip or pinch your nose to prevent air from exiting your nostrils.
   d. Record the value as your vital capacity in Part C of the laboratory report. Compare your result with that expected for a person of your sex, age, and height listed in tables 46.1 and 46.2. Use the meterstick to determine your height in centimeters if necessary or multiply your height in inches times 2.54 to calculate your height in centimeters. Considerable individual variations from the expected will be noted due to parameters other than sex, age, and height, which could include physical shape, health, medications, and others.

Critical Thinking Application

It can be noted from the data in tables 46.1 and 46.2 that vital capacities gradually decrease with age. Propose an explanation for this normal correlation.

____

____

____

____

7. *Inspiratory reserve volume* (IRV) (about 3,000 mL) is the volume of air in addition to the tidal volume that enters the lungs during forced inspiration. Calculate your inspiratory reserve volume by subtracting your tidal volume (TV) and your expiratory reserve volume (ERV) from your vital capacity (VC):

$$IRV = VC - (TV + ERV)$$

8. *Inspiratory capacity* (IC) (about 3,500 mL) is the maximum volume of air a person can inhale following

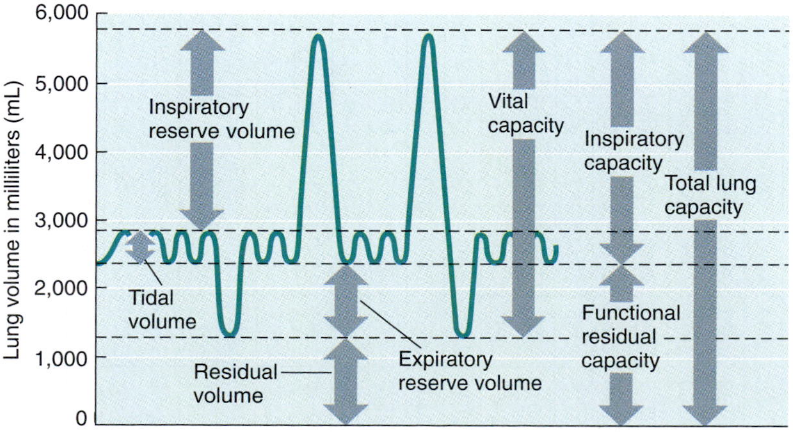

Figure 46.4 Graphic representation of respiratory volumes and capacities.

Table 46.1 Predicted Vital Capacities (in Milliliters) for Females

	Height in Centimeters																								
Age	146	148	150	152	154	156	158	160	162	164	166	168	170	172	174	176	178	180	182	184	186	188	190	192	194
16	2950	2990	3030	3070	3110	3150	3190	3230	3270	3310	3350	3390	3430	3470	3510	3550	3590	3630	3670	3715	3755	3800	3840	3880	3920
18	2920	2960	3000	3040	3080	3120	3160	3200	3240	3280	3320	3360	3400	3440	3480	3520	3560	3600	3640	3680	3720	3760	3800	3840	3880
20	2890	2930	2970	3010	3050	3090	3130	3170	3210	3250	3290	3330	3370	3410	3450	3490	3525	3565	3605	3645	3695	3720	3760	3800	3840
22	2860	2900	2940	2980	3020	3060	3095	3135	3175	3215	3255	3290	3330	3370	3410	3450	3490	3530	3570	3610	3650	3685	3725	3765	3800
24	2830	2870	2910	2950	2985	3025	3065	3100	3140	3180	3220	3260	3300	3335	3375	3415	3455	3490	3530	3570	3610	3650	3685	3725	3765
26	2800	2840	2880	2920	2960	3000	3035	3070	3110	3150	3190	3230	3265	3300	3340	3380	3420	3455	3495	3530	3570	3610	3650	3685	3725
28	2775	2810	2850	2890	2930	2965	3000	3040	3070	3115	3155	3190	3230	3270	3305	3345	3380	3420	3460	3495	3535	3570	3610	3650	3685
30	2745	2780	2820	2860	2895	2935	2970	3010	3045	3085	3120	3160	3195	3235	3270	3310	3345	3385	3420	3460	3495	3535	3570	3610	3645
32	2715	2750	2790	2825	2865	2900	2940	2975	3015	3050	3090	3125	3160	3200	3235	3275	3310	3350	3385	3425	3460	3495	3535	3570	3610
34	2685	2725	2760	2795	2835	2870	2910	2945	2980	3020	3055	3090	3130	3165	3200	3240	3275	3310	3350	3385	3425	3460	3495	3535	3570
36	2655	2695	2730	2765	2805	2840	2875	2910	2950	2985	3020	3060	3095	3130	3165	3205	3240	3275	3310	3350	3385	3420	3460	3495	3530
38	2630	2665	2700	2735	2770	2810	2845	2880	2915	2950	2990	3025	3060	3095	3130	3170	3205	3240	3275	3310	3350	3385	3420	3455	3490
40	2600	2635	2670	2705	2740	2775	2810	2850	2885	2920	2955	2990	3025	3060	3095	3135	3170	3205	3240	3275	3310	3345	3380	3420	3455
42	2570	2605	2640	2675	2710	2745	2780	2815	2850	2885	2920	2955	2990	3025	3060	3100	3135	3170	3205	3240	3275	3310	3345	3380	3415
44	2540	2575	2610	2645	2680	2715	2750	2785	2820	2855	2890	2925	2960	2995	3030	3060	3095	3130	3165	3200	3235	3270	3305	3340	3375
46	2510	2545	2580	2615	2650	2685	2715	2750	2785	2820	2855	2890	2925	2960	2995	3030	3060	3095	3130	3165	3200	3235	3270	3305	3340
48	2480	2515	2550	2585	2620	2650	2685	2715	2750	2785	2820	2855	2890	2925	2960	2995	3030	3060	3095	3130	3160	3195	3230	3265	3300
50	2455	2485	2520	2555	2590	2625	2655	2690	2720	2755	2785	2820	2855	2890	2925	2955	2990	3025	3060	3090	3125	3155	3190	3225	3260
52	2425	2455	2490	2525	2555	2590	2625	2655	2690	2720	2755	2790	2820	2855	2890	2925	2955	2990	3020	3055	3090	3125	3155	3190	3220
54	2395	2425	2460	2495	2530	2560	2590	2625	2655	2690	2720	2755	2790	2820	2855	2885	2920	2950	2985	3020	3050	3085	3115	3150	3180
56	2365	2400	2430	2460	2495	2525	2560	2590	2625	2655	2690	2720	2755	2790	2820	2855	2885	2920	2950	2980	3015	3045	3080	3110	3145
58	2335	2370	2400	2430	2460	2495	2525	2560	2590	2625	2655	2690	2720	2750	2785	2815	2850	2880	2920	2945	2975	3010	3040	3075	3105
60	2305	2340	2370	2400	2430	2460	2495	2525	2560	2590	2625	2655	2685	2720	2750	2780	2810	2845	2875	2915	2940	2970	3000	3035	3065
62	2280	2310	2340	2370	2405	2435	2465	2495	2525	2560	2590	2620	2655	2685	2715	2745	2775	2810	2840	2870	2900	2935	2965	2995	3025
64	2250	2280	2310	2340	2370	2400	2430	2465	2495	2525	2555	2585	2620	2650	2680	2710	2740	2770	2805	2835	2865	2895	2920	2955	2990
66	2220	2250	2280	2310	2340	2370	2400	2430	2460	2495	2525	2555	2585	2615	2645	2675	2705	2735	2765	2800	2825	2860	2890	2920	2950
68	2190	2220	2250	2280	2310	2340	2370	2400	2430	2460	2490	2520	2550	2580	2610	2640	2670	2700	2730	2760	2795	2820	2850	2880	2910
70	2160	2190	2220	2250	2280	2310	2340	2370	2400	2425	2455	2485	2515	2545	2575	2605	2635	2665	2695	2725	2755	2780	2810	2840	2870
72	2130	2160	2190	2220	2250	2280	2310	2335	2365	2395	2425	2455	2480	2510	2540	2570	2600	2630	2660	2685	2715	2745	2775	2805	2830
74	2100	2130	2160	2190	2220	2245	2275	2305	2335	2360	2390	2420	2450	2475	2505	2535	2565	2590	2620	2650	2680	2710	2740	2765	2795

From E. DeF. Baldwin and E. W. Richards, Jr., "Pulmonary Insufficiency 1, Physiologic Classification, Clinical Methods of Analysis, Standard Values in Normal Subjects" in *Medicine* 27:243, © by William & Wilkins. Used by permission.

Table 46.2 Predicted Vital Capacities (in Milliliters) for Males

	Height in Centimeters																								
Age	**146**	**148**	**150**	**152**	**154**	**156**	**158**	**160**	**162**	**164**	**166**	**168**	**170**	**172**	**174**	**176**	**178**	**180**	**182**	**184**	**186**	**188**	**190**	**192**	**194**
16	3765	3820	3870	3920	3975	4025	4075	4130	4180	4230	4285	4335	4385	4440	4490	4540	4590	4645	4695	4745	4800	4850	4900	4955	5005
18	3740	3790	3840	3890	3940	3995	4045	4095	4145	4200	4250	4300	4350	4405	4455	4505	4555	4610	4660	4710	4760	4815	4865	4915	4965
20	3710	3760	3810	3860	3910	3960	4015	4065	4115	4165	4215	4265	4320	4370	4420	4470	4520	4570	4625	4675	4725	4775	4825	4875	4930
22	3680	3730	3780	3830	3880	3930	3980	4030	4080	4135	4185	4235	4285	4335	4385	4435	4485	4535	4585	4635	4685	4735	4790	4840	4890
24	3635	3685	3735	3785	3835	3885	3935	3985	4035	4085	4135	4185	4235	4285	4330	4380	4430	4480	4530	4580	4630	4680	4730	4780	4830
26	3605	3655	3705	3755	3805	3855	3905	3955	4000	4050	4100	4150	4200	4250	4300	4350	4395	4445	4495	4545	4595	4645	4695	4740	4790
28	3575	3625	3675	3725	3775	3820	3870	3920	3970	4020	4070	4115	4165	4215	4265	4310	4360	4410	4460	4510	4555	4605	4655	4705	4755
30	3550	3595	3645	3695	3740	3790	3840	3890	3935	3985	4035	4080	4130	4180	4230	4275	4325	4375	4425	4470	4520	4570	4615	4665	4715
32	3520	3565	3615	3665	3710	3760	3810	3855	3905	3950	4000	4050	4095	4145	4195	4240	4290	4340	4385	4435	4485	4530	4580	4625	4675
34	3475	3525	3570	3620	3665	3715	3760	3810	3855	3905	3950	4000	4045	4095	4140	4190	4225	4285	4330	4380	4425	4475	4520	4570	4615
36	3445	3495	3540	3585	3635	3680	3730	3775	3825	3870	3920	3965	4010	4060	4105	4155	4200	4250	4295	4340	4390	4435	4485	4530	4580
38	3415	3465	3510	3555	3605	3650	3695	3745	3790	3840	3885	3930	3980	4025	4070	4120	4165	4210	4260	4305	4350	4400	4445	4495	4540
40	3385	3435	3480	3525	3575	3620	3665	3710	3760	3805	3850	3900	3945	3990	4035	4085	4130	4175	4220	4270	4315	4360	4410	4455	4500
42	3360	3405	3450	3495	3540	3590	3635	3680	3725	3770	3820	3865	3910	3955	4000	4050	4095	4140	4185	4230	4280	4325	4370	4415	4460
44	3315	3360	3405	3450	3495	3540	3585	3630	3675	3725	3770	3815	3860	3905	3950	3995	4040	4085	4130	4175	4220	4270	4315	4360	4405
46	3285	3330	3375	3420	3465	3510	3555	3600	3645	3690	3735	3780	3825	3870	3915	3960	4005	4050	4095	4140	4185	4230	4275	4320	4365
48	3255	3300	3345	3390	3435	3480	3525	3570	3615	3655	3700	3745	3790	3835	3880	3925	3970	4015	4060	4105	4150	4190	4235	4280	4325
50	3210	3255	3300	3345	3390	3430	3475	3520	3565	3610	3650	3695	3740	3785	3830	3870	3915	3960	4005	4050	4090	4135	4180	4225	4270
52	3185	3225	3270	3315	3355	3400	3445	3490	3530	3575	3620	3660	3705	3750	3795	3835	3880	3925	3970	4010	4055	4100	4140	4185	4230
54	3155	3195	3240	3285	3325	3370	3415	3455	3500	3540	3585	3630	3670	3715	3760	3800	3845	3890	3930	3975	4020	4060	4105	4145	4190
56	3125	3165	3210	3255	3295	3340	3380	3425	3465	3510	3550	3595	3640	3680	3725	3765	3810	3850	3895	3940	3980	4025	4065	4110	4150
58	3080	3125	3165	3210	3250	3290	3335	3375	3420	3460	3500	3545	3585	3630	3670	3715	3755	3800	3840	3880	3925	3965	4010	4050	4095
60	3050	3095	3135	3175	3220	3260	3300	3345	3385	3430	3470	3500	3555	3595	3635	3680	3720	3760	3805	3845	3885	3930	3970	4015	4055
62	3020	3060	3110	3150	3190	3230	3270	3310	3350	3390	3440	3480	3520	3560	3600	3640	3680	3730	3770	3810	3850	3890	3930	3970	4020
64	2990	3030	3080	3120	3160	3200	3240	3280	3320	3360	3400	3440	3490	3530	3570	3610	3650	3690	3730	3770	3810	3850	3900	3940	3980
66	2950	2990	3030	3070	3110	3150	3190	3230	3270	3310	3350	3390	3430	3470	3510	3550	3600	3640	3680	3720	3760	3800	3840	3880	3920
68	2920	2960	3000	3040	3080	3120	3160	3200	3240	3280	3320	3360	3400	3440	3480	3520	3560	3600	3640	3680	3720	3760	3800	3840	3880
70	2890	2930	2970	3010	3050	3090	3130	3170	3210	3250	3290	3330	3370	3410	3450	3480	3520	3560	3600	3640	3680	3720	3760	3800	3840
72	2860	2900	2940	2980	3020	3060	3100	3140	3180	3210	3250	3290	3330	3370	3410	3450	3490	3530	3570	3610	3650	3680	3720	3760	3800
74	2820	2860	2900	2930	2970	3010	3050	3090	3130	3170	3200	3240	3280	3320	3360	3400	3440	3470	3510	3550	3590	3630	3670	3710	3740

From E. DeF. Baldwin and E. W. Richards, Jr., "Pulmonary Insufficiency 1, Physiologic Classification, Clinical Methods of Analysis, Standard Values in Normal Subjects" in *Medicine* 27:243, © by William & Wilkins. Used by permission.

exhalation of the tidal volume. Calculate your inspiratory capacity by adding your tidal volume (TV) and your inspiratory reserve volume (IRV):

$$IC = TV + IRV$$

9. *Functional residual capacity* (FRC) (about 2,300 mL) is the volume of air that remains in the lungs following exhalation of the tidal volume. Calculate your functional residual capacity (FRC) by adding your expiratory reserve volume (ERV) and your residual volume (RV), which you can assume is 1,200 mL:

$$FRC = ERV + 1{,}200$$

10. *Residual volume* (RV) (about 1,200 mL) is the volume of air that always remains in the lungs after the most forceful expiration. Although it is part of *total lung capacity* (about 5,800 mL), it cannot be measured with a spirometer. The residual air allows gas exchange and the alveoli to remain open during the respiratory cycle.
11. Complete Part C of the laboratory report.

Learning Extension

Determine your *minute respiratory volume.* To do this, follow these steps:

1. Sit quietly for a while, and then to establish your breathing rate, count the number of times you breathe in 1 minute. This might be inaccurate because conscious awareness of breathing rate can alter the results. You might ask a laboratory partner to record your breathing rate at some time when you are not expecting it to be recorded. A normal breathing rate is about 12–15 breaths per minute.
2. Calculate your minute respiratory volume by multiplying your breathing rate by your tidal volume:

   ________ × ________ = ____________
   (breathing rate) (tidal volume) (minute respiratory volume)
3. This value indicates the total volume of air that moves into your respiratory passages during each minute of ordinary breathing.

Name ___________________

Date ___________________

Section ___________________

The [A] corresponds to the indicated Learning Outcome(s) listed at the beginning of the laboratory exercise.

Breathing and Respiratory Volumes

Laboratory Report 46

Part A Assessments

Complete the following statements:

1. Breathing also can be called ___________________.
2. The weight of air causes a force called ___________________ pressure.
3. The weight of air at sea level is sufficient to support a column of mercury within a tube __________ mm high.
4. If the pressure inside the lungs decreases, outside air is forced into the airways by ___________________. [1]
5. Nerve impulses are carried to the diaphragm by the ___________________ nerves. [1]
6. When the diaphragm contracts, the size of the thoracic cavity ___________________. [1]
7. The ribs are raised by contraction of the ___________________ muscles, which increases the size of the thoracic cavity. [1]
8. Only a thin film of lubricating serous fluid separates the parietal pleura from the ______________ of a lung.
9. A mixture of lipoproteins, called ___________________, acts to reduce the tendency of alveoli to collapse.
10. The force responsible for normal expiration comes from ____________ of tissues and from surface tension. [1]
11. Muscles that help to force out more than the normal volume of air by pulling the ribs downward and inward include the ___________________. [1]
12. The diaphragm can be forced to move higher than normal by contraction of the ______________ muscles. [1]

Part B Assessments

Match the air volumes in column A with the definitions in column B. Place the letter of your choice in the space provided. [2]

Column A		Column B
a. Expiratory reserve volume	_______	1. Volume in addition to tidal volume that leaves the lungs during forced expiration
b. Functional residual capacity	_______	2. Vital capacity plus residual volume
c. Inspiratory capacity	_______	3. Volume that remains in lungs after the most forceful expiration
d. Inspiratory reserve volume	_______	4. Volume that enters or leaves lungs during a respiratory cycle
e. Residual volume	_______	5. Volume in addition to tidal volume that enters lungs during forced inspiration
f. Tidal volume	_______	6. Maximum volume a person can exhale after taking the deepest possible breath
g. Total lung capacity	_______	7. Maximum volume a person can inhale following exhalation of the tidal volume
h. Vital capacity	_______	8. Volume of air remaining in the lungs following exhalation of the tidal volume

Part C Assessments

1. Test results for respiratory air volumes and capacities: 3 4

Respiratory Volume or Capacity	Expected Value* (approximate)	Test Results	Percent of Expected Value (test result/expected value × 100)
Tidal volume (resting) (TV)	500 mL		
Expiratory reserve volume (ERV)	1,100 mL		
Vital capacity (VC)	4,600 mL (enter yours from table 46.1 or 46.2)		
Inspiratory reserve volume (IRV)	3,000 mL		
Inspiratory capacity (IC)	3,500 mL		
Functional residual capacity (FRC)	2,300 mL		

**The values listed are most characteristic for a healthy, tall, young adult male. If your expected value for vital capacity is considerably different than 4,600 mL, your other values would vary accordingly.*

2. Complete the following:

   a. How do your test results compare with the expected values?

   ____

   ____

   ____

   b. How does your vital capacity compare with the average value for a person of your sex, age, and height?

   ____

   ____

   ____

   c. What measurement in addition to vital capacity is needed before you can calculate your total lung capacity? 4

   ____

3. If your experimental results are considerably different than the predicted vital capacities, propose reasons for the differences. As you write this paragraph, consider factors such as smoking, physical fitness, respiratory disorders, stress, and medications. (Your instructor might have you make some class correlations from class data.)

____

____

____

____

____

____

Laboratory Exercise 47

Control of Breathing

Materials Needed

Textbook
Clock with second hand
Paper bags, small

For Demonstration:
Flasks
Glass tubing
Rubber stoppers, two-hole
Calcium hydroxide solution (limewater)

For Learning Extension:
Pneumograph
Physiological recording apparatus

Several skeletal muscles contract during breathing. The principal muscles involved during inspiration are the diaphragm and external intercostals. The sternocleidomastoid, scalenes, and pectoralis minor are synergistic during forceful inhalation. During quiet respiration, there is minimal involvement of the expiratory muscles. During forced expiration, the principal muscles are the internal intercostals, but the rectus abdominis and the external oblique muscles can provide extra force.

Breathing is controlled from regions of the brainstem called the *respiratory areas,* which control both inspiration and expiration. These areas initiate and regulate nerve impulses that travel to various breathing muscles, causing rhythmic breathing movements and adjustments to the rate and depth of breathing to meet various cellular needs. The respiratory area includes the *medullary respiratory area,* which is composed of two bilateral groups of neurons in the medulla oblongata. They are called the *dorsal respiratory group* and the *ventral respiratory group.* Neurons in another part of the brainstem, the pons, compose the *pontine respiratory group* (formerly the pneumotaxic center). These neurons make connections with the medullary rhythmicity center, and they may contribute to the basic rhythm of breathing.

Various factors can influence the respiratory areas and thus affect the rate and depth of breathing. These factors include stretch of the lung tissues, emotional state, and the presence in the blood of certain chemicals, such as carbon dioxide, hydrogen ions, and oxygen. For example, the breathing rate increases as the blood concentration of carbon dioxide or hydrogen ions increases or as the concentration of oxygen decreases.

Purpose of the Exercise

To review the muscles and the mechanisms that control breathing and to investigate some of the factors that affect the rate and depth of breathing.

LEARNING OUTCOMES

After completing this exercise, you should be able to

1. Locate the breathing centers in the brainstem.
2. Describe the mechanisms that control and influence breathing.
3. Select the respiratory muscles involved in inspiration and forced expiration.
4. Test and record the effect of various factors on the rate and depth of breathing.

EXPLORE

Procedure A—Control of Breathing

1. Review the section entitled "Control of Breathing" in chapter 19 of the textbook.
2. Respiratory areas in the brainstem control the cycle of breathing. The dorsal respiratory group of neurons in the medulla oblongata receives input from the pontine respiratory group (pneumotaxic center), various regions of the brain, cranial nerves (vagus and glossopharyngeal), and chemosensitive areas of the medulla oblongata (fig. 47.1). Impulses from the dorsal respiratory group stimulate the muscles of inspiration and lead to the basic rhythm of breathing. This results in a normal respiratory rate of 12–15 breaths per minute. The ventral respiratory group of neurons will fire during inspiration

and expiration to stimulate the appropriate muscles of inspiration and expiration. The ventral respiratory group is more active during forceful breathing (fig. 47.2).

3. Complete Part A of Laboratory Report 47.

Demonstration

When a solution of calcium hydroxide is exposed to carbon dioxide, a chemical reaction occurs and a white precipitate of calcium carbonate is formed as indicated by the following reaction:

$$Ca(OH)_2 + CO_2 \longrightarrow CaCO_3 + H_2O$$

Thus, a clear water solution of calcium hydroxide (lime-water) can be used to detect the presence of carbon dioxide because the solution becomes cloudy if this gas is bubbled through it.

The laboratory instructor will demonstrate this test for carbon dioxide by drawing some air through limewater in an apparatus such as that shown in figure 47.3. Then the instructor will blow an equal volume of expired air through a similar apparatus. (*Note:* A new sterile mouthpiece should be used each time the apparatus is demonstrated.) Watch for the appearance of a precipitate that causes the limewater to become cloudy. Was there any carbon dioxide in the atmospheric air drawn through the limewater?

____

If so, how did the amount of carbon dioxide in the atmospheric air compare with the amount in the expired air?

____

____

____

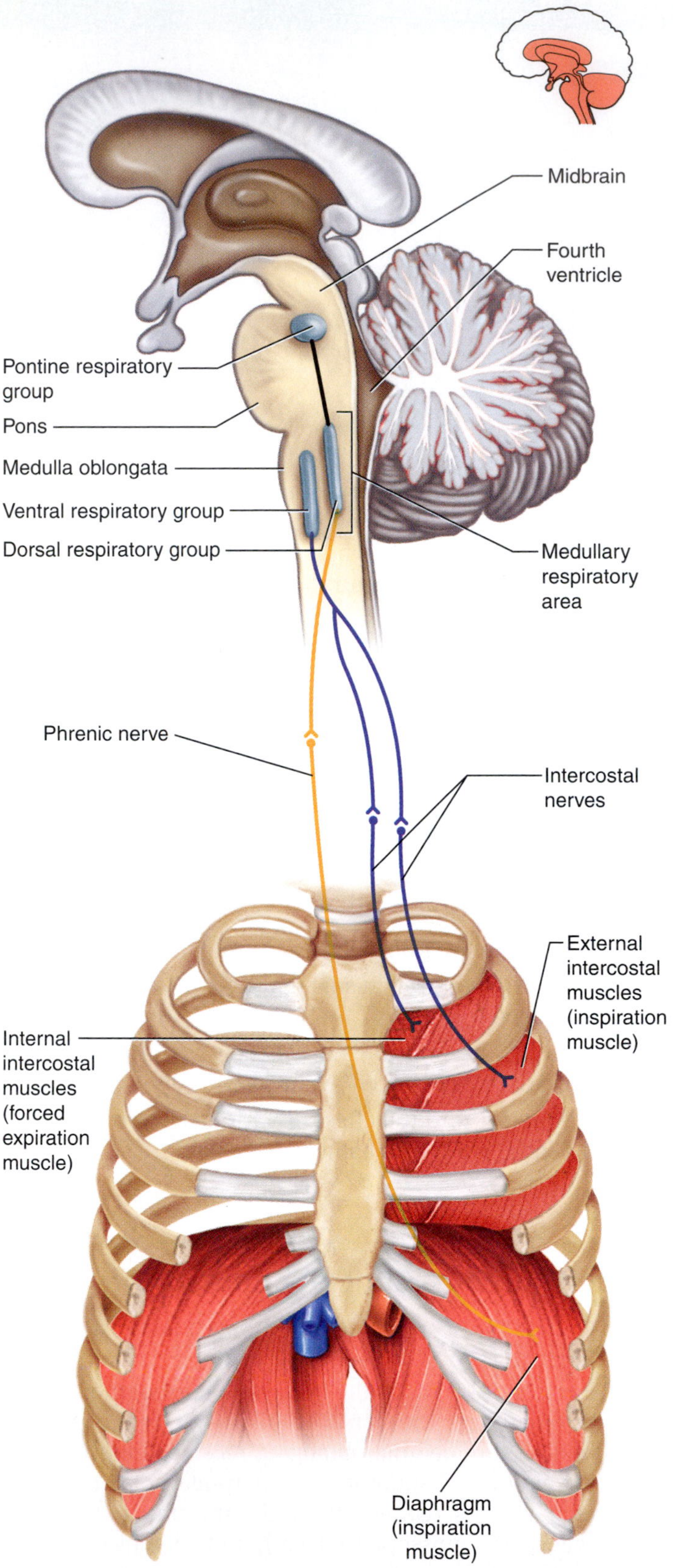

Figure 47.1 The respiratory areas are located in the pons and the medulla oblongata.

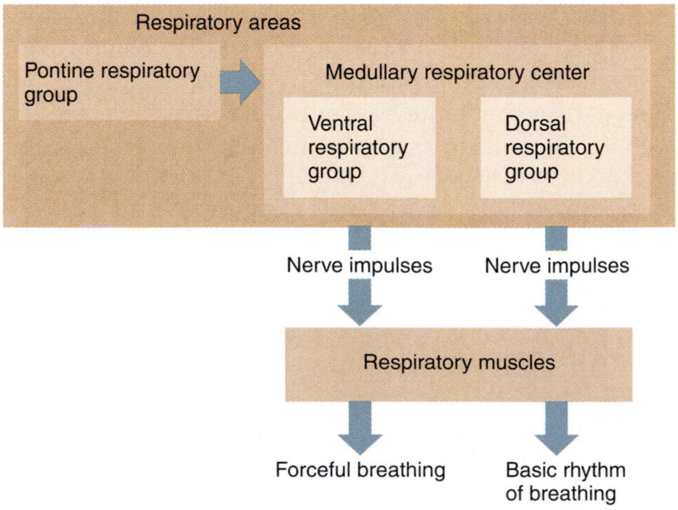

Figure 47.2 The medullary respiratory area and the pontine respiratory group control breathing.

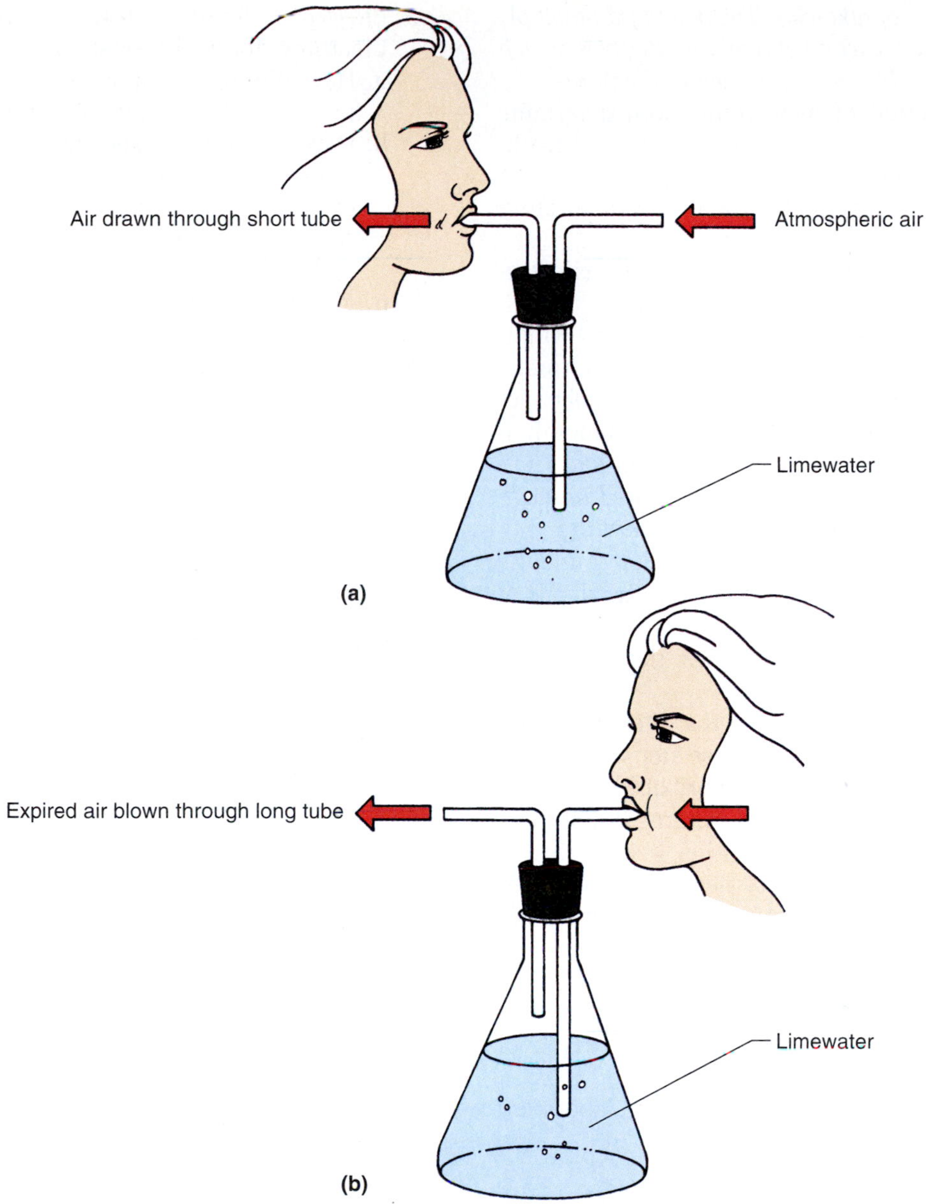

Figure 47.3 Apparatus used to demonstrate the presence of carbon dioxide in air: (*a*) atmospheric air is drawn through limewater; (*b*) expired air is blown through limewater.

EXPLORE

Procedure B—Factors Affecting Breathing

Perform each of the following tests, using your laboratory partner as a test subject.

1. *Normal breathing.* To determine the subject's normal breathing rate and depth, follow these steps:
   a. Have the subject sit quietly for a few minutes.
   b. After the rest period, ask the subject to count backwards mentally, beginning with five hundred.
   c. While the subject is distracted by counting, watch the subject's chest movements, and count the breaths taken in a minute. Use this value as the normal breathing rate (breaths per minute).
   d. Note the relative depth of the breathing movements.
   e. Record your observations in the table in Part B of the laboratory report.
2. *Effect of hyperventilation.* To test the effect of hyperventilation on breathing, follow these steps:
   a. Seat the subject and *guard to prevent the possibility of the subject falling over.*
   b. Have the subject breathe rapidly and deeply for a maximum of 1 minute. *If the subject begins to feel dizzy, the hyperventilation should be halted immediately to prevent the subject from fainting from*

complications of alkalosis. The increased blood pH causes vasoconstriction of cerebral arterioles, which decreases circulation and oxygen to the brain.

   c. After the period of hyperventilation, determine the subject's breathing rate, and judge the breathing depth as before.
   d. Record the results in Part B of the laboratory report.

3. *Effect of rebreathing air.* To test the effect of rebreathing air on breathing, follow these steps:
   a. Have the subject sit quietly (approximately 5 minutes) until the breathing rate returns to normal.
   b. Have the subject breathe deeply into a small paper bag that is held tightly over the nose and mouth. *If the subject begins to feel light-headed or like fainting, the rebreathing air should be halted immediately to prevent further acidosis and fainting.*
   c. After 2 minutes of rebreathing air, determine the subject's breathing rate, and judge the depth of breathing.
   d. Record the results in Part B of the laboratory report.
4. *Effect of breath holding.* To test the effect of breath holding on breathing, follow these steps:
   a. Have the subject sit quietly (approximately 5 minutes) until the breathing rate returns to normal.
   b. Have the subject hold his or her breath as long as possible. *If the subject begins to feel light-headed or like fainting, breath holding should be halted immediately to prevent further acidosis and fainting.*
   c. As the subject begins to breathe again, determine the rate of breathing, and judge the depth of breathing.
   d. Record the results in Part B of the laboratory report.
5. *Effect of exercise.* To test the effect of exercise on breathing, follow these steps:
   a. Have the subject sit quietly (approximately 5minutes) until breathing rate returns to normal.
   b. Have the subject exercise by moderately running in place for 3–5 minutes. *This exercise should be avoided by anyone with health risks.*
   c. After the exercise, determine the breathing rate, and judge the depth of breathing.
   d. Record the results in Part B of the laboratory report.
6. Complete Part B of the laboratory report.

Learning Extension

A *pneumograph* is a device that can be used together with some type of recording apparatus to record breathing movements. The laboratory instructor will demonstrate the use of this equipment to record various movements, such as those that accompany coughing, laughing, yawning, and speaking.

Devise an experiment to test the effect of some factor, such as hyperventilation, rebreathing air, or exercise, on the length of time a person can hold their breath. *After the laboratory instructor has approved your plan,* carry out the experiment, using the pneumograph and recording equipment. What conclusion can you draw from the results of your experiment?

____

____

____

____

____

Name ______________________________

Date ______________________________

Section ______________________________

The [A] corresponds to the indicated Learning Outcome(s) listed at the beginning of the laboratory exercise.

Control of Breathing

Laboratory Report 47

Part A Assessments

Complete the following statements:

1. The respiratory areas are widely scattered throughout the pons and ______________ of the brainstem. [1]
2. The ______________ respiratory group primarily stimulates the diaphragm. [1]
3. The ______________ respiratory group primarily stimulates the intercostals and abdominal muscles. [1]
4. Chemosensitive areas are stimulated by changes in the blood concentrations of carbon dioxide and ______________ ions. [2]
5. As the blood concentration of carbon dioxide increases, the breathing rate ______________. [2]
6. As a result of increased breathing, the blood concentration of carbon dioxide is ______________. [2]
7. The chemoreceptors that sense low blood oxygen concentration are located in the aortic bodies and the ______________ bodies. [2]
8. As a result of hyperventilation, breath-holding time is ______________. [2]
9. The principal muscles of forced expiration are the ______________. [3]
10. The principal muscles of inspiration are the ______________ and the external intercostal muscles. [3]

Part B Assessments

1. Record the results of your breathing tests in the table. [4]

Factor Tested	Breathing Rate (breaths/min)	Breathing Depth (+, + +, + + +)
Normal		
Hyperventilation		
Rebreathing air		
Breath holding		
Exercise		

2. Explain the reason for the changes in breathing that occurred in each of the following cases: 2

a. Hyperventilation

b. Rebreathing air

c. Breath holding

d. Exercise

3. Complete the following:

a. Why is it important to distract a person when you are determining the normal rate of breathing?

b. How can the depth of breathing be measured accurately?

Critical Thinking Application

Why is it dangerous for a swimmer to hyperventilate in order to hold the breath for a longer period of time? 2

Laboratory Exercise 48

Kidney Structure

Materials Needed

Textbook
Human torso model
Kidney model
Preserved pig (or sheep) kidney
Dissecting tray
Dissecting instruments
Long knife
Compound light microscope
Prepared microscope slides of the following:
- Kidney section
- Ureter

Safety

- Wear disposable gloves when working on the kidney dissection.
- Dispose of the kidney and gloves as directed by your laboratory instructor.
- Wash the dissecting tray and instruments as instructed.
- Wash your laboratory table.
- Wash your hands before leaving the laboratory.

The two kidneys are the primary organs of the urinary system. They are located in the abdominal cavity, against the posterior wall and behind the parietal peritoneum (retroperitoneal). Masses of adipose tissue associated with the kidneys hold them in place at a vertebral level between T12 and L3. The right kidney is slightly more inferior due to the large mass of the liver near its superior border. Ureters force urine by means of peristaltic waves into the urinary bladder, which temporarily stores urine. The urethra conveys urine to the outside of the body.

Each kidney contains over 1 million nephrons, which serve as the basic structural and functional units of the kidney. A glomerular capsule, proximal convoluted tubule, nephron loop, and distal convoluted tubule compose the microscopic, multicellular structure of a relatively long nephron tubule, which drains into a collecting duct. Approximately 80% of the nephrons are cortical nephrons with short nephron loops, while the remaining represent juxtamedullary nephrons, with long nephron loops extending deeper into the renal medulla. An elaborate network of blood vessels surrounds the entire nephron. Glomerular filtration, tubular reabsorption, and tubular secretion represent three processes resulting in urine as the final product.

A variety of functions occur in the kidneys. They remove metabolic wastes from the blood; help regulate blood volume, blood pressure, and pH of blood; control water and electrolyte concentrations; and secrete rennin and erythropoietin.

Purpose of the Exercise

To review the structure of the kidney, to dissect a kidney, and to observe the major structures of a nephron.

LEARNING OUTCOMES

After completing this exercise, you should be able to

1. Locate and identify the major structures of the urinary system.
2. Locate and identify the major structures of a kidney.
3. Identify and sketch the major structures of a nephron.
4. Trace the path of filtrate through a nephron.
5. Trace the path of blood and blood pressure changes through the renal blood vessels.

EXPLORE

Procedure A—Kidney Structure

1. Review the section entitled "Kidney Structure" in chapter 20 of the textbook.
2. As a review activity, label figures 48.1 and 48.2.
3. Complete Part A of Laboratory Report 48.

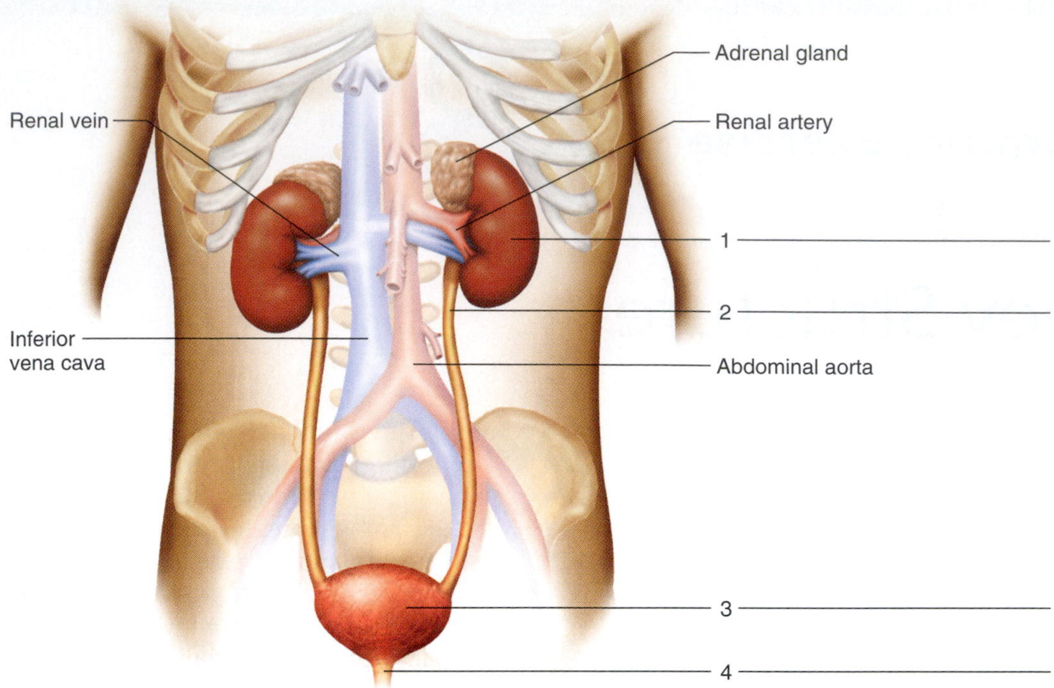

Figure 48.1 Label the major structures of the urinary system. 1

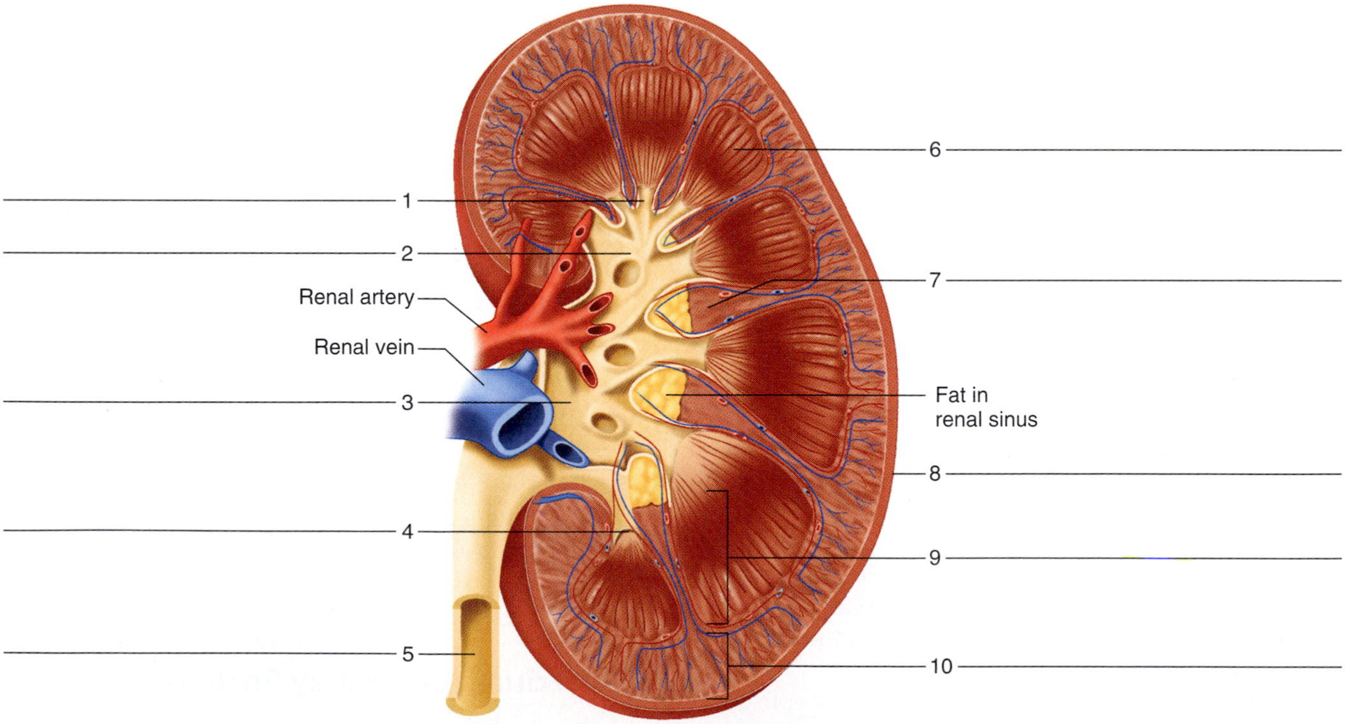

Figure 48.2 Label the major structures in the longitudinal section of a kidney. 2

4. Observe the human torso model and the kidney model. Locate the following:

   kidneys
   ureters
   urinary bladder
   urethra
   renal sinus
   renal pelvis
      major calyces
      minor calyces
   renal medulla
      renal pyramids
      renal papillae
   renal cortex
   renal columns
   nephrons
      cortical nephrons (80% of nephrons)
      juxtamedullary nephrons (20% of nephrons)

5. To observe the structure of a kidney, follow these steps:
   a. Obtain a pig or sheep kidney and rinse it with water to remove as much of the preserving fluid as possible.
   b. Carefully remove any adipose tissue from the surface of the specimen.
   c. Locate the following features:

      renal (fibrous) capsule
      hilum of kidney
      renal artery
      renal vein
      ureter

   d. Use a long knife to cut the kidney in half longitudinally along the frontal plane, beginning on the convex border.
   e. Rinse the interior of the kidney with water, and using figure 48.3 as a reference, locate the following:

      renal pelvis
         major calyces
         minor calyces
      renal cortex
      renal columns (extensions of renal cortical tissue between renal pyramids)
      renal medulla
         renal pyramids

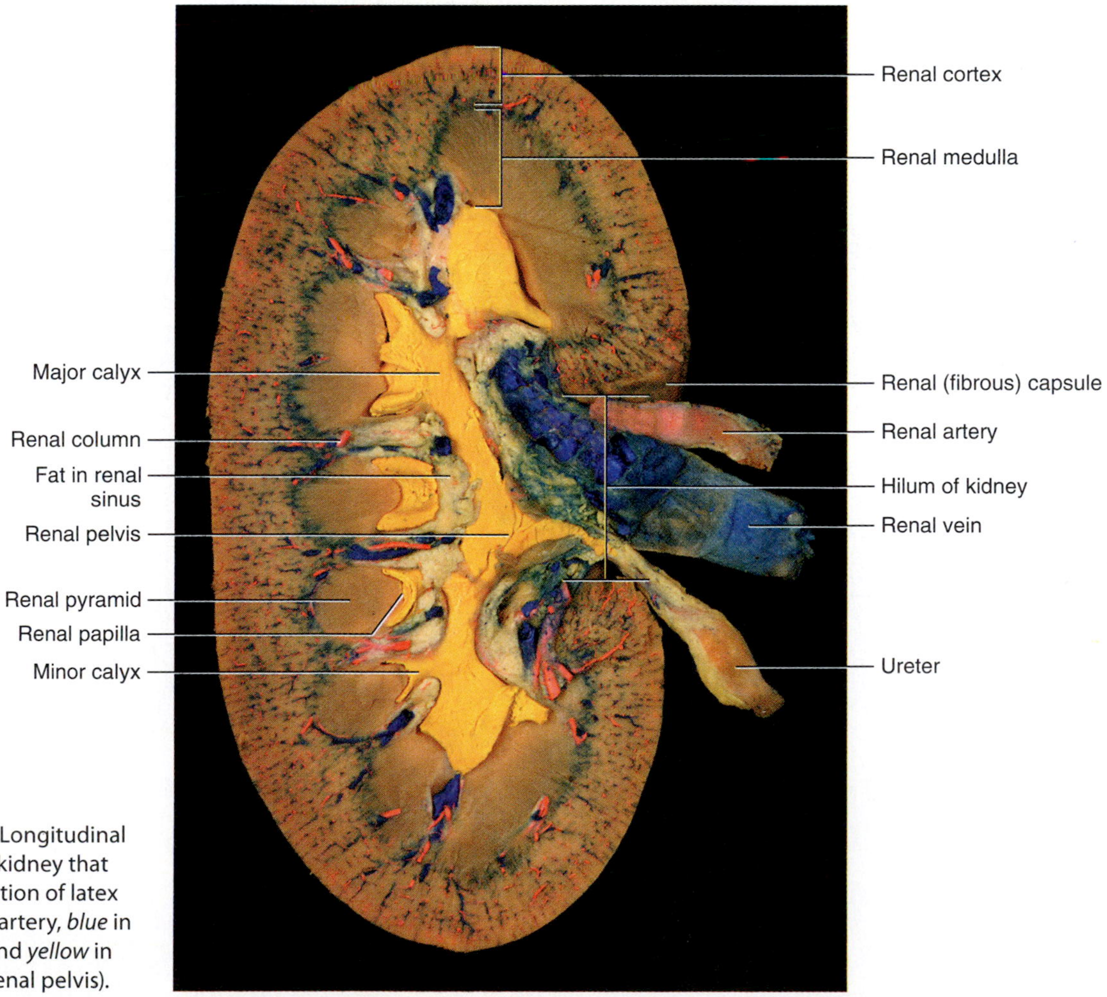

Figure 48.3 Longitudinal section of a pig kidney that has a triple injection of latex (*red* in the renal artery, *blue* in the renal vein, and *yellow* in the ureter and renal pelvis).

EXPLORE

Procedure B—The Renal Blood Vessels and Nephrons

1. Review the sections entitled "Renal Blood Vessels" and "Nephrons" in chapter 20 of the textbook.
2. As a review activity, label figure 48.4.
3. Complete Part B of the laboratory report.
4. Obtain a microscope slide of a kidney section, and examine it using low-power magnification. Locate the *renal capsule,* the *renal cortex* (which appears somewhat granular and may be more darkly stained than the other renal tissues,) and the *renal medulla* (fig. 48.5).
5. Examine the renal cortex, using high-power magnification. Locate a *renal corpuscle.* These structures appear as isolated circular areas. Identify the *glomerulus,* the capillary cluster inside the corpuscle, and the *glomerular capsule,* which appears as a clear area surrounding the glomerulus. Also note the numerous sections of renal tubules that occupy the spaces between renal corpuscles (fig. 48.5*a*).
6. Prepare a labeled sketch of a representative section of renal cortex in Part C of the laboratory report.
7. Examine the renal medulla, using high-power magnification. Identify longitudinal and cross sections of various collecting ducts. These ducts are lined with simple epithelial cells, which vary in shape from squamous to cuboidal (fig. 48.5*b*).
8. Prepare a labeled sketch of a representative section of renal medulla in Part C of the laboratory report.

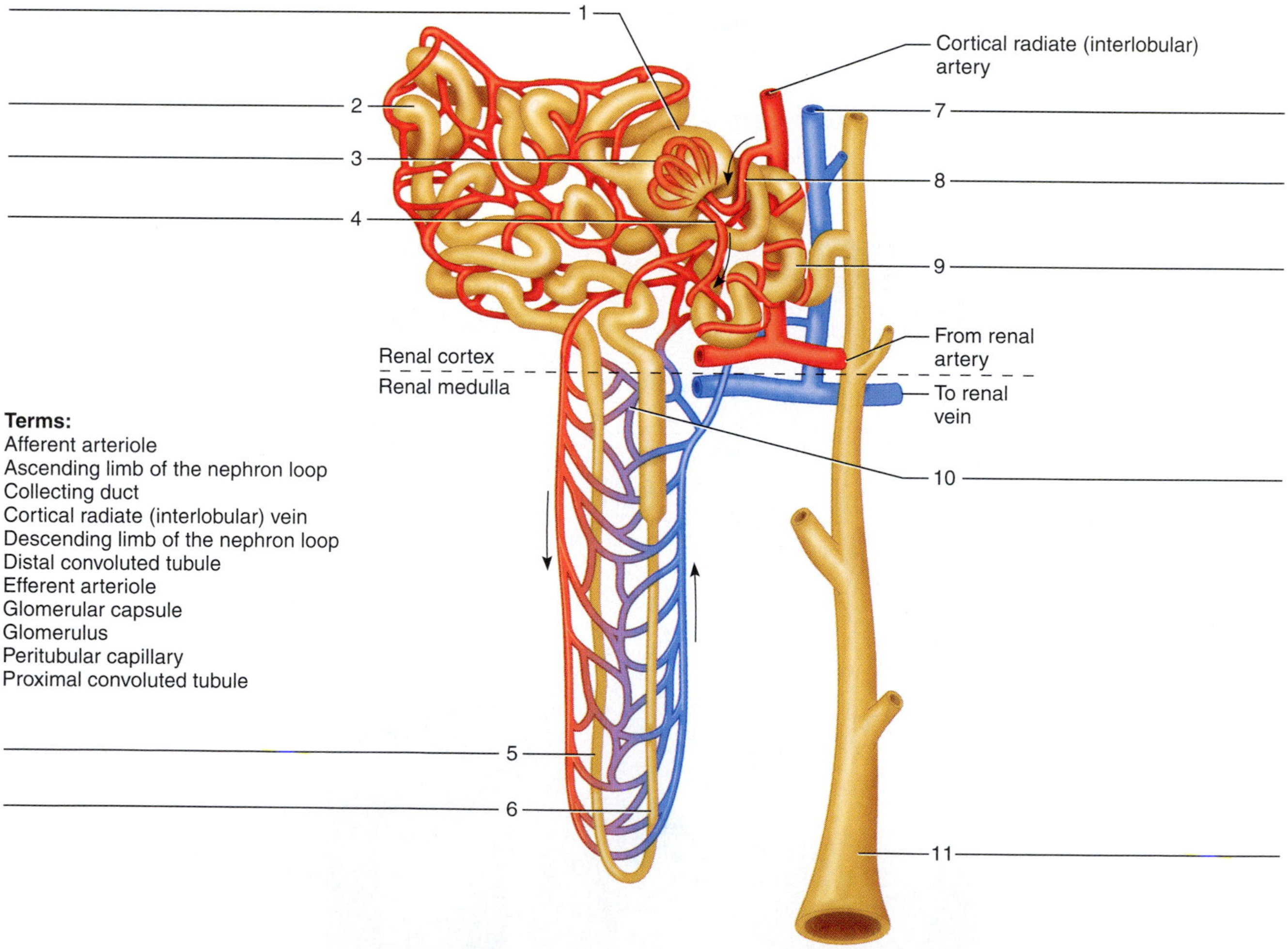

Terms:
Afferent arteriole
Ascending limb of the nephron loop
Collecting duct
Cortical radiate (interlobular) vein
Descending limb of the nephron loop
Distal convoluted tubule
Efferent arteriole
Glomerular capsule
Glomerulus
Peritubular capillary
Proximal convoluted tubule

Figure 48.4 Label the major structures of the nephron and the blood vessels associated with it, using the terms provided. 3

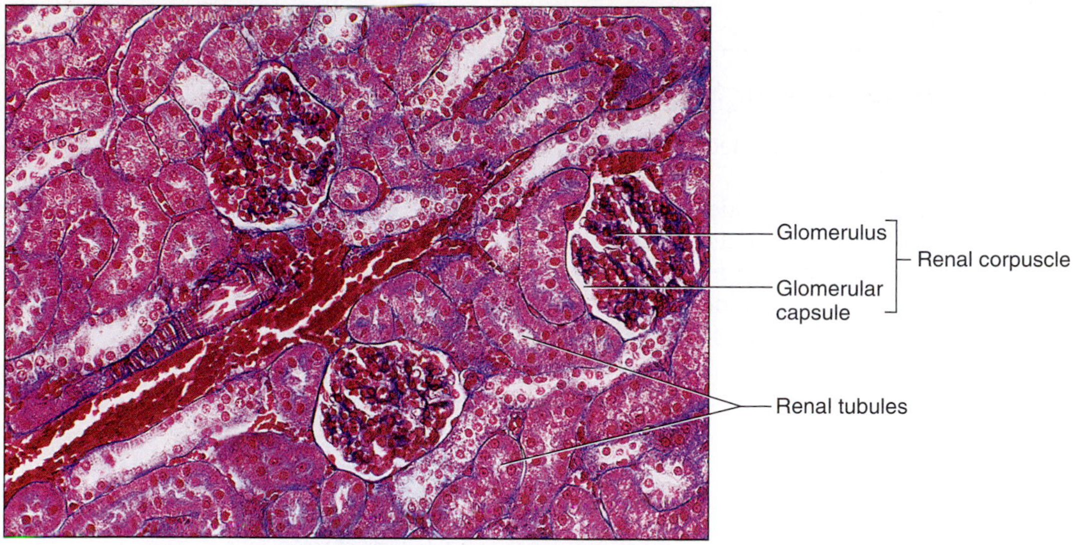

(a)

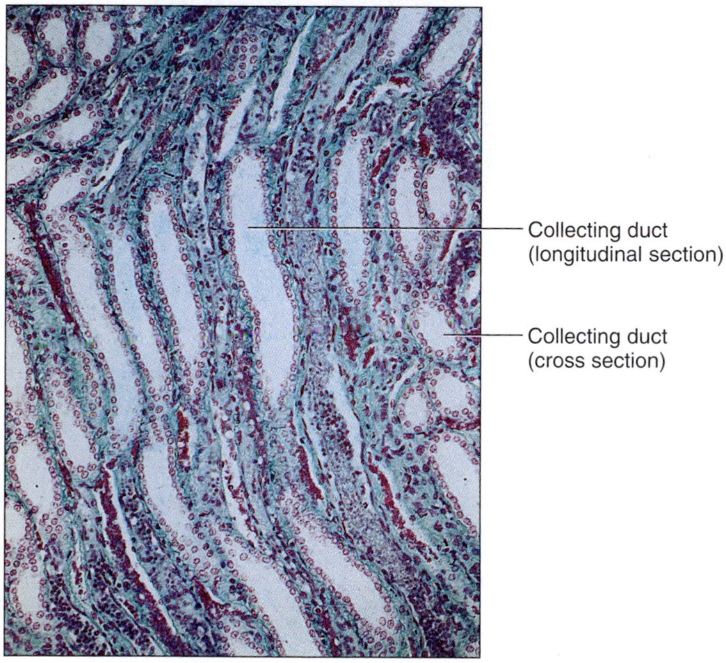

(b)

Figure 48.5 (*a*) Micrograph of a section of the renal cortex (220×). (*b*) Micrograph of a section of the renal medulla (80× micrograph enlarged to 200×).

EXPLORE

Procedure C—Ureter Structure

1. Review the section entitled "Ureters" in chapter 20 of the textbook.
2. Obtain a microscope slide of a cross section of a ureter, and examine it using low-power magnification. Locate the *mucous coat* layer next to the lumen. Examine the middle *muscular coat* composed of longitudinal and circular smooth muscle cells, responsible for the peristaltic waves propelling urine from the kidneys to the urinary bladder. The outer *fibrous coat,* composed of connective tissue, secures the ureter in the retroperitoneal position (fig. 48.6).
3. Examine the mucous coat using high-power magnification. The specialized tissue is transitional epithelium which allows changes in its thickness when unstretched and stretched.
4. Prepare a labeled sketch of a ureter in Part D of the laboratory report.

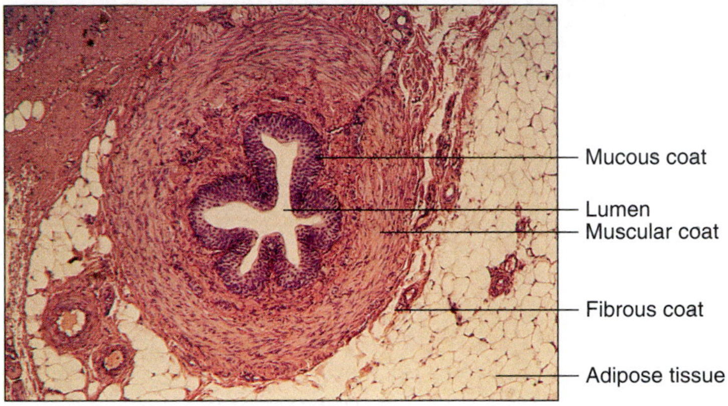

Figure 48.6 Cross section of a ureter (75×).

Name ______________________________

Date ______________________________

Section ______________________________

The [A] corresponds to the indicated Learning Outcome(s) listed at the beginning of the laboratory exercise.

Kidney Structure

Laboratory Report 48

Part A Assessments

Match the terms in column A with the descriptions in column B. Place the letter of your choice in the space provided. [2]

Column A		Column B
a. Calyces	_____	1. Shell around the renal medulla
b. Hilum of kidney	_____	2. Branches of renal pelvis to renal papillae
c. Nephron	_____	3. Conical mass of tissue within renal medulla
d. Renal column	_____	4. Projection with tiny openings into a minor calyx
e. Renal cortex	_____	5. Hollow chamber within kidney
f. Renal papilla	_____	6. Microscopic functional unit of kidney
g. Renal pelvis	_____	7. Cortical tissue between renal pyramids
h. Renal pyramid	_____	8. Superior funnel-shaped end of ureter inside the renal sinus
i. Renal sinus	_____	9. Medial depression for blood vessels and ureter to enter kidney chamber

Part B Assessments

Complete the following:

1. Distinguish between a renal corpuscle and a renal tubule. [3] ______________________________

____

____

2. Number the following structures to indicate their respective positions in relation to the nephron. Assign the number 1 to the structure nearest the glomerulus. [4]

_____ Ascending limb of nephron loop

_____ Collecting duct

_____ Descending limb of nephron loop

_____ Distal convoluted tubule

_____ Glomerular capsule

_____ Proximal convoluted tubule

_____ Renal papilla

3. Number the following structures to indicate their respective positions in the blood pathway within the kidney. Assign the number 1 to the vessel nearest the renal artery. 5

_______ Afferent arteriole
_______ Efferent arteriole
_______ Glomerulus
_______ Peritubular capillary
_______ Renal vein

Part C Assessments

Sketch a representative section of the renal cortex and the renal medulla. Label the glomerulus, glomerular capsule, and sections of renal tubules in the renal cortex and a longitudinal section and cross section of a collecting duct in the renal medulla. 3

Renal cortex (_____×)	Renal medulla (_____×)

Part D Assessments

Sketch a cross section of a ureter. Label the three layers and the lumen. Indicate the magnification used for your sketch.

Laboratory Exercise 49

Urinalysis

Materials Needed

Normal and abnormal simulated urine specimens are suggested as a substitute for collected urine.
Disposable urine-collecting container
Paper towel
Urinometer cylinder
Urinometer hydrometer
Laboratory thermometer
pH test paper
Reagent strips (individual or combination strips such as Chemstrip or Multistix) to test for the presence of the following:
- Glucose
- Protein
- Ketones
- Bilirubin
- Hemoglobin/occult blood

Compound light microscope
Microscope slide
Coverslip
Centrifuge
Centrifuge tube
Graduated cylinder, 10 mL
Medicine dropper
Sedi-stain

Safety

- Consider using simulated urine samples available from various laboratory supply houses.
- Wear disposable gloves when working with body fluids.
- Work only with your own urine sample.
- Use an appropriate disinfectant to wash the laboratory table before and after the procedures.
- Place glassware in a disinfectant when finished.
- Dispose of contaminated items as directed by your laboratory instructor.
- Wash your hands before leaving the laboratory.

Urine is the product of three processes of the nephrons within the kidneys: glomerular filtration, tubular reabsorption, and tubular secretion. As a result of these processes, various waste substances are removed from the blood and body fluid and electrolyte balance are maintained. Consequently, the composition of urine varies considerably because of differences in dietary intake and physical activity from day to day. The normal urinary output ranges from 1.0–1.8 liters per day. Typically, urine consists of 95% water and 5% solutes. The volume of urine produced by the kidneys varies with such factors as fluid intake, environmental temperature, relative humidity, respiratory rate, and body temperature.

An analysis of urine composition and volume often is used to evaluate the functions of the kidneys and other organs. This procedure, called *urinalysis,* is a clinical assessment and a diagnostic tool for certain pathological conditions and general overall health. A urinalysis and a complete blood analysis complement each other for an evaluation of certain diseases and general health.

A urinalysis involves three aspects: physical characteristics, chemical analysis, and a microscopic examination. Physical characteristics of urine that are noted include volume, color, transparency, and odor. The chemical analysis of solutes in urine addresses urea and other nitrogenous wastes; electrolytes; pigments; and possible glucose, protein, ketones, bilirubin, and hemoglobin. The specific gravity and pH of urine are greatly influenced by the components and amounts of solutes. An examination of microscopic solids, including cells, casts, and crystals, assists in the diagnosis of injury, various diseases, and urinary infections.

Purpose of the Exercise

To perform the observations and tests commonly used to analyze the characteristics and composition of urine.

LEARNING OUTCOMES

After completing this exercise, you should be able to

1. Evaluate the color, transparency, and specific gravity of a urine sample.
2. Measure the pH of a urine sample.
3. Test a urine sample for the presence of glucose, protein, ketones, bilirubin, and hemoglobin.
4. Perform a microscopic study of urine sediment.
5. Summarize the results of these observations and tests.

WARNING: While performing the following tests, you should wear disposable latex gloves so that skin contact with urine is avoided. Observe all safety procedures listed for this lab. (Normal and abnormal simulated urine specimens could be used instead of real urine for this lab.)

EXPLORE

Procedure—Urinalysis

1. Proceed to the restroom with a clean disposable container. The first small volume of urine should not be collected because it contains abnormally high levels of microorganisms from the urethra. Collect a midstream sample of about 50 mL of urine. The best collections are the first specimen in the morning or one taken 3 hours after a meal. Refrigerate samples if they are not used immediately.
2. Place a sample of urine in a clean, transparent container. Describe the *color* of the urine. Normal urine varies from light yellow to amber, depending on the presence of urochromes, end-product pigments produced during the decomposition of hemoglobin. Dark urine indicates a high concentration of pigments.

   Abnormal urine colors include yellow-brown or green, due to elevated concentrations of bile pigments, and red to dark brown, due to the presence of blood. Certain foods, such as beets or carrots, and various drug substances also may cause color changes in urine, but in such cases the colors have no clinical significance. Enter the results of this and the following tests in Part A of Laboratory Report 49.
3. Evaluate the *transparency* of the urine sample (judge whether the urine is clear, slightly cloudy, or very cloudy). Normal urine is clear enough to see through. You can read newsprint through slightly cloudy urine; you can no longer read newsprint through cloudy urine. Cloudy urine indicates the presence of various substances that may include mucus, bacteria, epithelial cells, fat droplets, or inorganic salts.
4. Determine the *specific gravity* of the urine sample, which indicates the solute concentration. Specific gravity is the ratio of the weight of something to the weight of an equal volume of pure water. For example, mercury (at 15°C) weighs 13.6 times as much as an equal volume of water; thus, it has a specific gravity of 13.6. Although urine is mostly water, it has substances dissolved in it and is slightly heavier than an equal volume of water. Thus, urine has a specific gravity of more than 1.000. Actually, the specific gravity of normal urine varies from 1.003 to 1.035. If the specific gravity is too low, the urine contains few solutes and represents dilute urine, a likely result of excessive fluid intake or the use of diuretics. A specific gravity above the normal range represents a higher concentration of solutes, likely from a limited fluid intake. Concentrated urine over an extended time increases the risk of the formation of kidney stones.

   To determine the specific gravity of a urine sample, follow these steps:
   - **a.** Pour enough urine into a clean urinometer cylinder to fill it about three-fourths full. Any foam that appears should be removed with a paper towel.
   - **b.** Use a laboratory thermometer to measure the temperature of the urine.
   - **c.** Gently place the urinometer hydrometer into the urine, and *make sure that the float is not touching the sides or the bottom of the cylinder* (fig. 49.1).
   - **d.** Position your eye at the level of the urine surface. Determine which line on the stem of the hydrometer intersects the lowest level of the concave surface (meniscus) of the urine.
   - **e.** Liquids tend to contract and become denser as they are cooled, or to expand and become less dense as they are heated, so it may be necessary to make a temperature correction to obtain an accurate specific gravity measurement. To do this, add 0.001 to the hydrometer reading for each 3 degrees of urine temperature above 25°C or subtract 0.001 for each 3 degrees below 25°C. Enter this calculated value in the table of the laboratory report as the test result.
5. Reagent strips can be used to perform a variety of urine tests. In each case, directions for using the strips are found on the strip container. *Be sure to read them.*

   To perform each test, follow these steps:
   - **a.** Obtain a urine sample and the proper reagent strip.
   - **b.** Read the directions on the strip container.
   - **c.** Dip the strip in the urine sample.
   - **d.** Remove the strip at an angle and let it touch the inside rim of the urine container to remove any excess liquid.

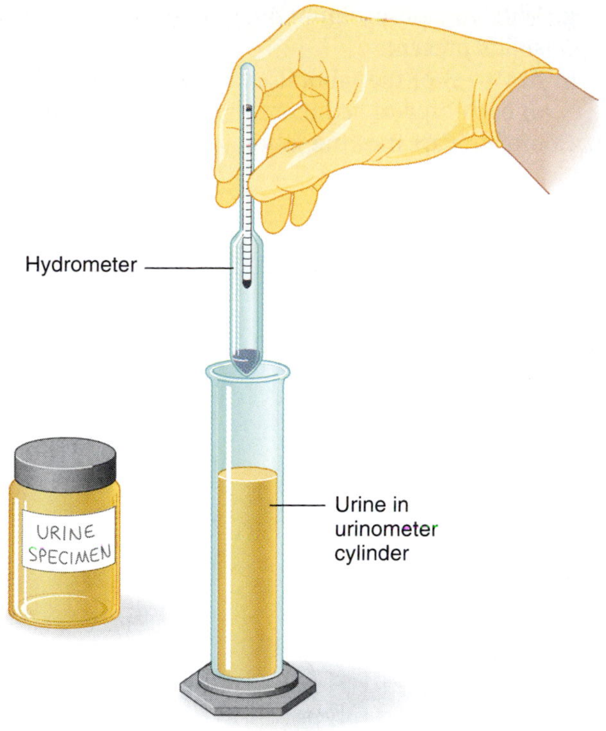

Figure 49.1 Float the hydrometer in the urine, making sure that it does not touch the sides or the bottom of the cylinder.

e. Wait for the length of time indicated by the directions on the container before you compare the color of the test strip with the standard color scale on the side of the container. The value or amount represented by the matching color should be used as the test result and recorded in Part A of the laboratory report.

> ### Alternative Procedure
>
> If combination reagent test strips (Multistix or Chemstrip) are being used, locate the appropriate color chart for each test being evaluated. Wait the designated time for each color reaction before the comparison is made to the standard color scale.

6. Perform the *pH test.* The pH of normal urine varies from 4.6 to 8.0, but most commonly, it is near 6.0 (slightly acidic). The pH of urine may decrease as a result of a diet high in protein, or it may increase with a vegetarian diet. Significant daily variations within the broad normal range are results of concentrations of excesses from variable diets.
7. Perform the *glucose test.* Normally, there is no glucose in urine. However, glucose may appear in the urine temporarily following a meal high in carbohydrates. Glucose also may appear in the urine as a result of uncontrolled diabetes mellitus.
8. Perform the *protein test.* Normally, proteins of large molecular size are not present in urine. However, those of small molecular sizes, such as albumins, may appear in trace amounts, particularly following strenuous exercise. Increased amounts of proteins also may appear as a result of kidney diseases in which the glomeruli are damaged or as a result of high blood pressure.
9. Perform the *ketone test.* Ketones are products of fat metabolism. Usually they are not present in urine. However, they may appear in the urine if the diet fails to provide adequate carbohydrate, as in the case of prolonged fasting or starvation, or as a result of insulin deficiency (diabetes mellitus).
10. Perform the *bilirubin test.* Bilirubin, which results from hemoglobin decomposition in the liver, normally is absent in urine. It may appear, however, as a result of liver disorders that cause obstructions of the biliary tract. Urochrome, a normal yellow component of urine, is a result of additional breakdown of bilirubin.
11. Perform the *hemoglobin/occult blood test.* Hemoglobin occurs in the red blood cells, and because such cells normally do not pass into the renal tubules, hemoglobin is not found in normal urine. Its presence in urine usually indicates a disease process, a transfusion reaction, an injury to the urinary organs, or menstrual blood.
12. Complete Part A of the laboratory report.
13. A urinalysis usually includes a study of urine sediment—the microscopic solids present in a urine sample. This sediment normally includes mucus; certain crystals; and a variety of cells, such as the epithelial cells that line the urinary tubes and an occasional white blood cell. Other types of solids, such as casts or red blood cells, may indicate a disease or injury if they are present in excess. (Casts are cylindrical masses of cells or other substances that form in the renal tubules and are flushed out by the flow of urine.)

    To observe urine sediment, follow these steps:

    a. Thoroughly stir or shake a urine sample to suspend the sediment, which tends to settle to the bottom of the container.
    b. Pour 10 mL of urine into a clean centrifuge tube and centrifuge it for 5 minutes at slow speed (1,500 rpm). Be sure to balance the centrifuge with an even number of tubes filled to the same levels.
    c. Carefully decant 9 mL (leave 1 mL) of the liquid from the sediment in the bottom of the centrifuge tube, as directed by your laboratory instructor. Resuspend the 1 mL of sediment.
    d. Use a medicine dropper to remove some of the sediment and place it on a clean microscope slide.

e. Add a drop of Sedi-stain to the sample, and add a coverslip.
f. Examine the sediment with low-power (reduce the light when using low power) and high-power magnifications.
g. With the aid of figure 49.2, identify the types of solids present.
h. In Part B of the laboratory report, make a sketch of each type of sediment that you observed.

14. Complete Part B of the laboratory report.

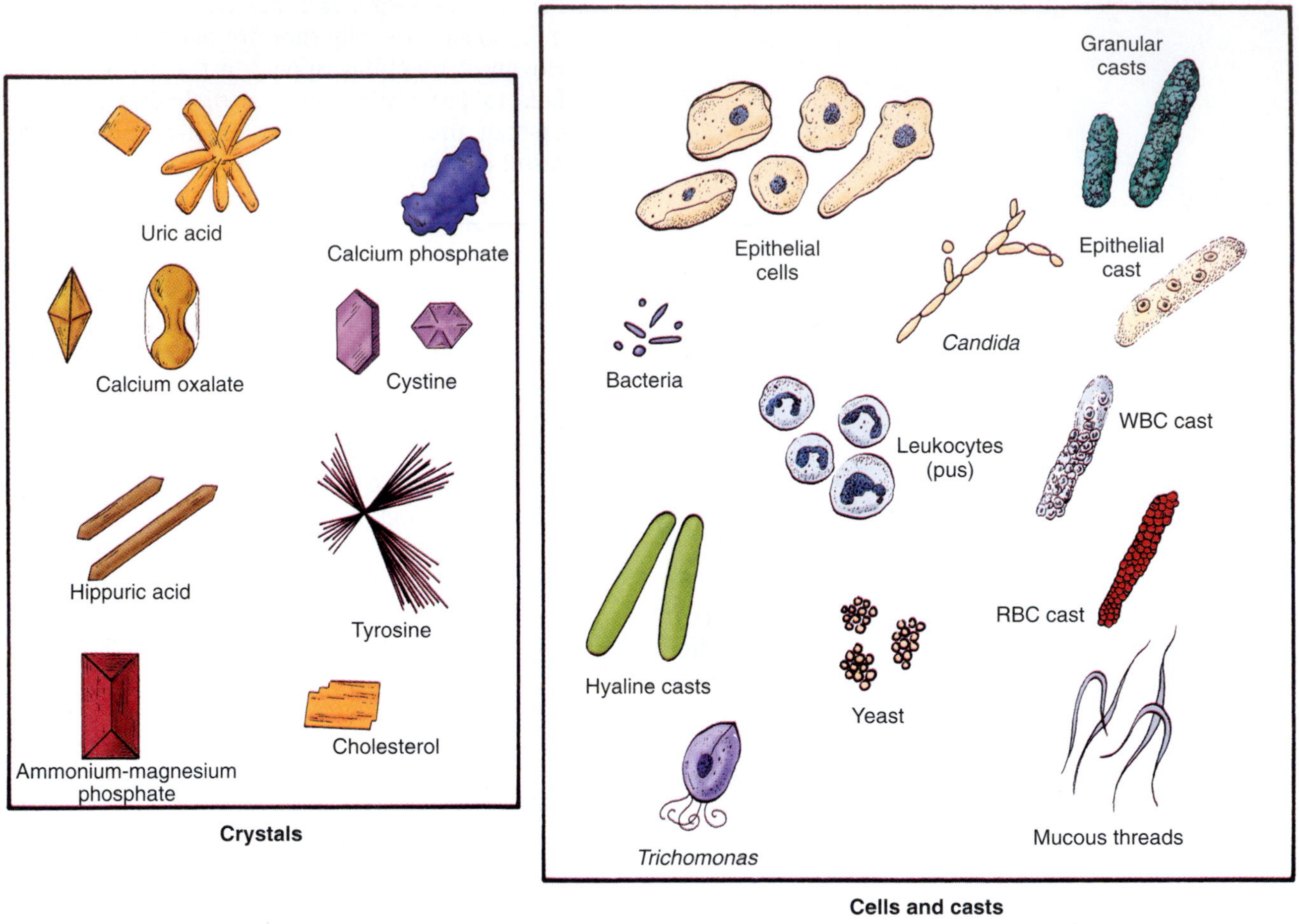

Figure 49.2 Types of urine sediment. Healthy individuals lack many of these sediments and possess only occasional to trace amounts of others. (*Note:* Shades of white to purple sediments are most characteristic when using Sedi-stain.)

Name ___________________________

Date ___________________________

Section ___________________________

The A corresponds to the indicated Learning Outcome(s) listed at the beginning of the laboratory exercise.

Laboratory Report 49

Urinalysis

Part A Assessments

1. Enter your observations, test results, and evaluations in the following table: 1 2 3

Urine Characteristics	Observations and Test Results	Normal Values	Evaluations
Color		Light yellow to amber	
Transparency		Clear	
Specific gravity (corrected for temperature)		1.003–1.035	
pH		4.6–8.0	
Glucose		0 (negative)	
Protein		0 to trace	
Ketones		0	
Bilirubin		0	
Hemoglobin/occult blood		0	
(Other)			
(Other)			

2. Summarize the results of the urinalysis. 5 ___________________________

____

____

____

____

____

Critical Thinking Application

Why do you think it is important to refrigerate a urine sample if an analysis cannot be performed immediately after collecting it?

Part B Assessments

1. Make a sketch for each type of sediment you observed. Label any from those shown in figure 49.2. 4

2. Summarize the results of the urine sediment study. 5

Laboratory Exercise 50

Male Reproductive System

Materials Needed

Textbook
Human torso model
Model of the male reproductive system
Anatomical chart of the male reproductive system
Compound light microscope
Prepared microscope slides of the following:
- Testis section
- Epididymis, cross section
- Penis, cross section

The organs of the male reproductive system are specialized to produce and maintain the male sex cells, to transport these cells together with supporting fluids to the female reproductive tract, and to produce and secrete male sex hormones.

These organs include the testes, in which sperm cells and male sex hormones are produced, and sets of internal and external accessory organs. The internal organs include various tubes and glands, whereas the external structures are the scrotum and the penis.

Purpose of the Exercise

To review the structure and functions of the male reproductive organs and to examine some of these organs.

LEARNING OUTCOMES

After completing this exercise, you should be able to

1. Locate and identify the organs of the male reproductive system.
2. Describe the functions of these organs.
3. Sketch and label the major features from microscopic sections of the testis, epididymis, and penis.

EXPLORE

Procedure A—Male Reproductive Organs

1. Review the sections entitled "Testes," "Male Internal Accessory Organs," and "Male External Reproductive Organs" in chapter 22 of the textbook.
2. As a review activity, label figures 50.1 and 50.2.
3. Observe the human torso model, the model of the male reproductive system, and the anatomical chart of the male reproductive system. Locate the following features:

testes
- seminiferous tubules
- interstitial cells (Leydig cells)

inguinal canal
spermatic cord
epididymis
ductus deferens (vas deferens)
ejaculatory duct
seminal vesicles
prostate gland
bulbourethral glands
scrotum
penis
- corpora cavernosa
- corpus spongiosum
- tunica albuginea
- glans penis
- external urethral orifice
- prepuce (foreskin)

4. Complete Part A of Laboratory Report 50.

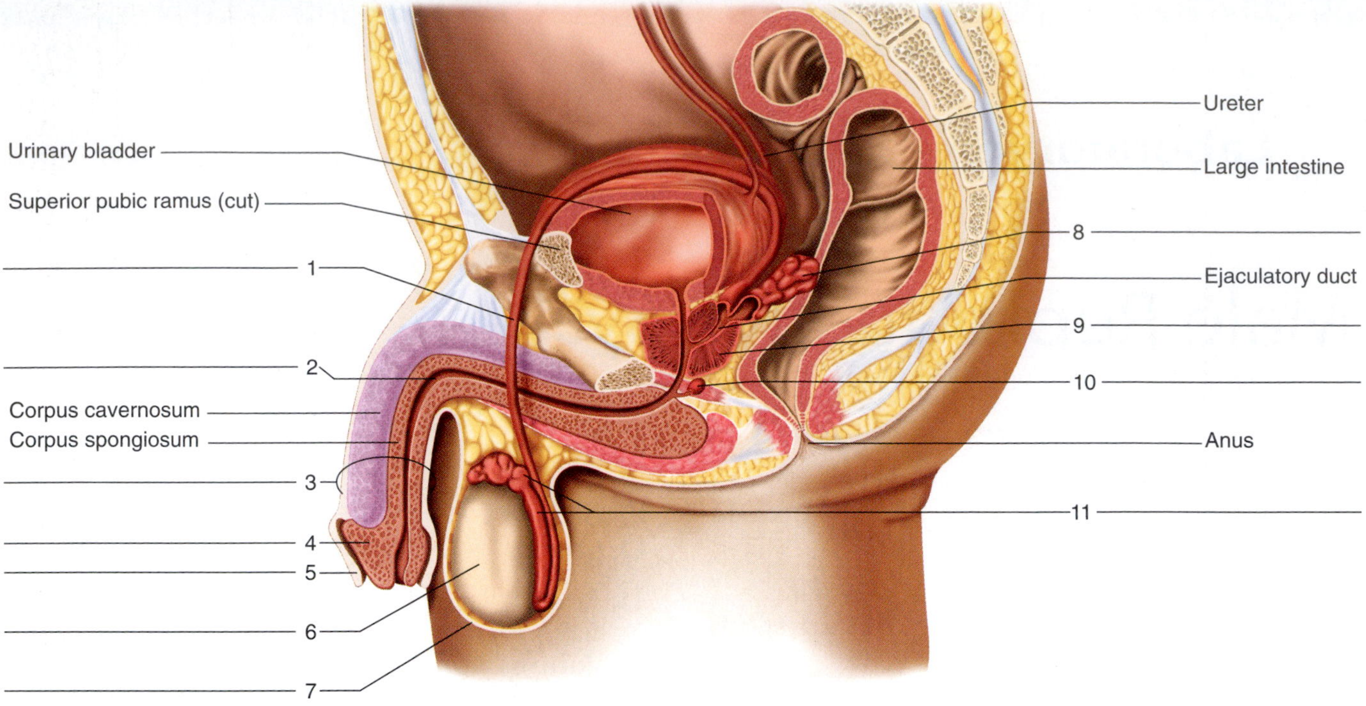

Figure 50.1 Label the major structures of the male reproductive system in this sagittal view. 1

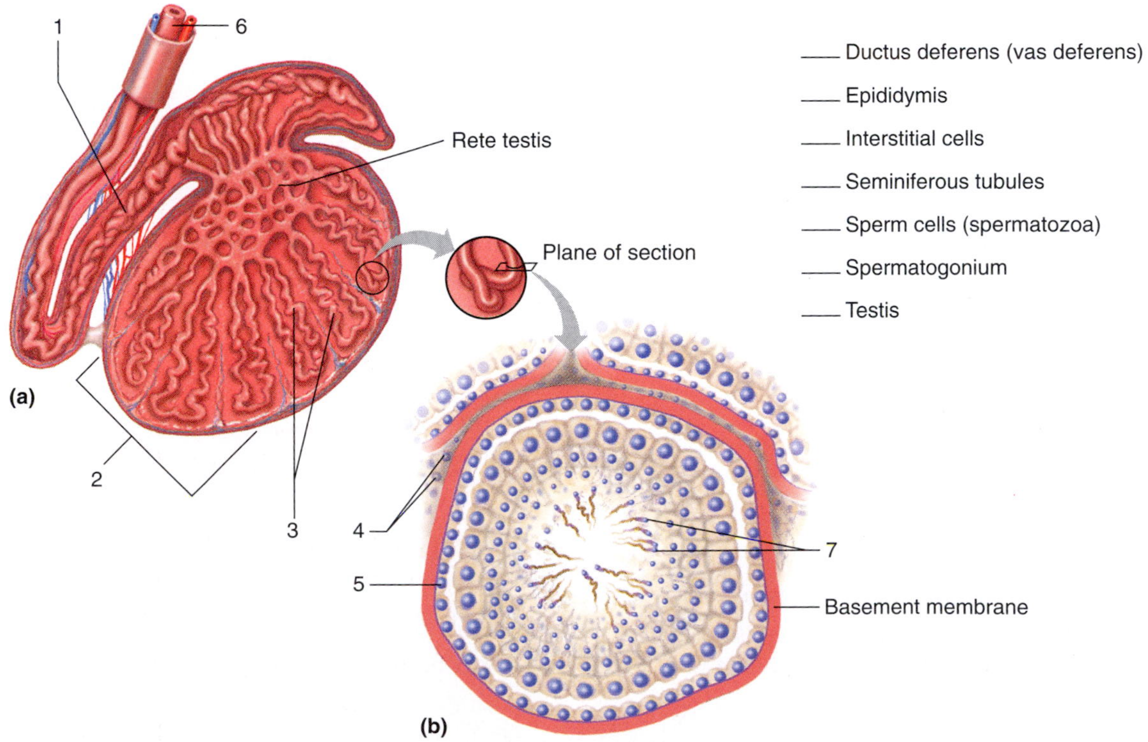

Figure 50.2 Label the diagram of (*a*) the sagittal section of a testis and (*b*) a cross section of a seminiferous tubule by placing the correct numbers in the spaces provided. 1

EXPLORE

Procedure B—Microscopic Anatomy

1. Obtain a microscope slide of human testis section and examine it, using low-power magnification (fig. 50.3). Locate the thick *fibrous capsule* (tunica albuginea) on the surface and the numerous sections of *seminiferous tubules* inside.
2. Focus on some of the seminiferous tubules, using high-power magnification (fig. 50.4). Locate the *basement membrane* and the layer of *spermatogonia* just beneath the basement membrane. Identify some *sustentacular cells* (supporting cells; Sertoli cells), which have pale, oval-shaped nuclei, and some *spermatogenic cells,* which have smaller, round nuclei. Spermatogonia give rise to spermatogenic cells that are in various stages of spermatogenesis as they are forced toward the lumen. Near the lumen of the tube, find some darkly stained, elongated heads of developing sperm cells. In the spaces between adjacent seminiferous tubules, locate some isolated *interstitial cells* (Leydig cells) of the endocrine system. Interstitial cells produce the hormone testosterone, transported by the blood.
3. Prepare a labeled sketch of a representative section of the testis in Part B of the laboratory report.
4. Obtain a microscope slide of a cross section of *epididymis* (fig. 50.5). Examine its wall, using high-power magnification. Note the elongated, *pseudostratified columnar epithelial cells* that comprise most of the inner lining. These cells have nonmotile stereocilia (microvilli) on their free surfaces. Also note the thin layer of smooth muscle and connective tissue surrounding the tube.

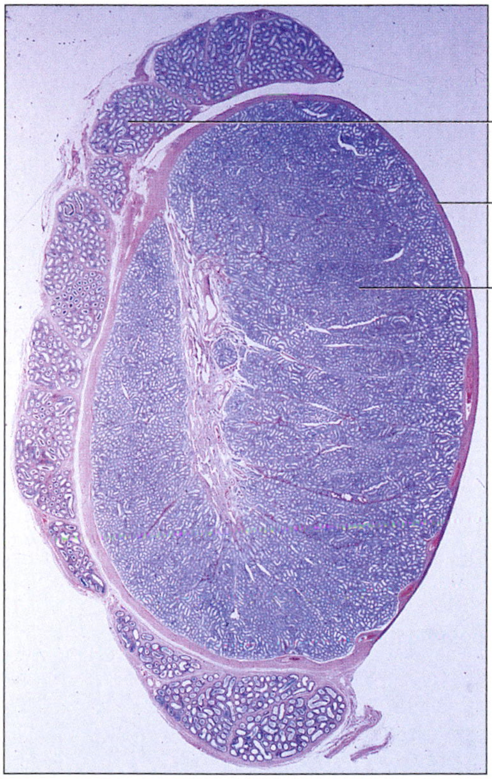

Figure 50.3 Micrograph of a human testis (1.7×).

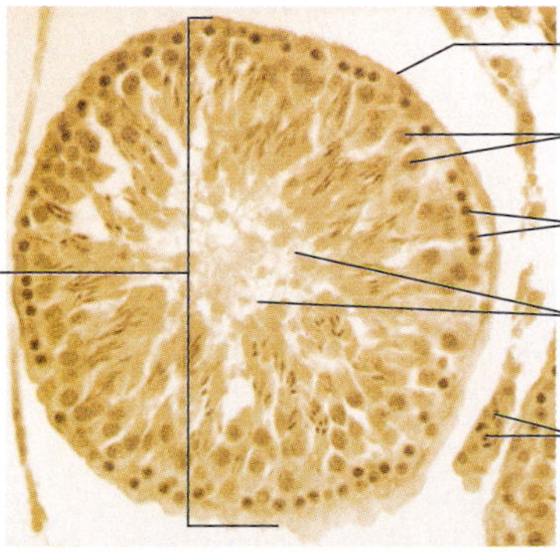

Figure 50.4 Micrograph of a seminiferous tubule (250×).

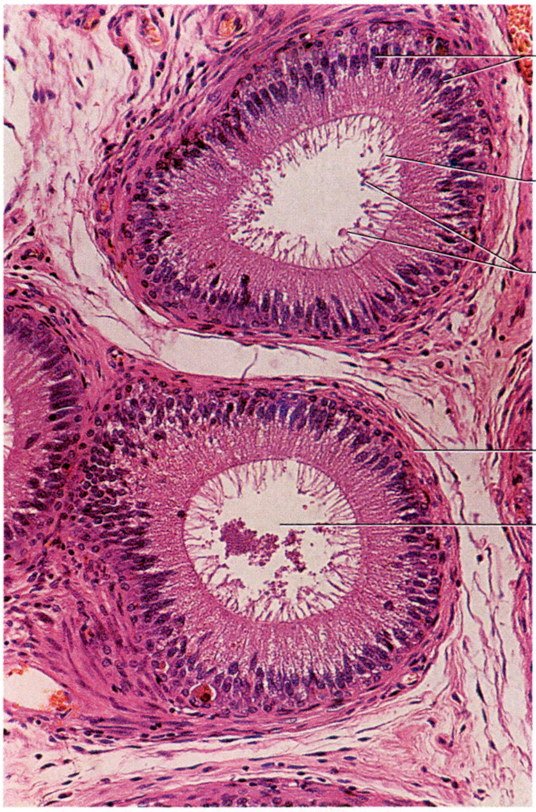

Figure 50.5 Micrograph of a cross section of a human epididymis (200×).

5. Prepare a labeled sketch of the epididymis wall in Part B of the laboratory report.
6. Obtain a microscope slide of a *penis* cross section, and examine it with low-power magnification (fig. 50.6). Identify the following features:

   corpora cavernosa
   corpus spongiosum
   tunica albuginea
   urethra
   skin

7. Prepare a labeled sketch of a penis cross section in Part B of the laboratory report.
8. Complete Part B of the laboratory report.

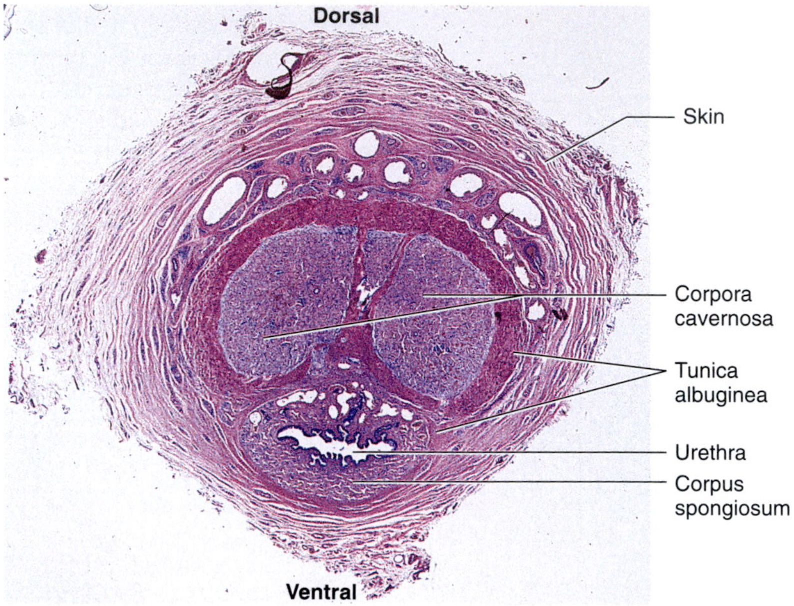

Figure 50.6 Micrograph of a cross section of the body of the penis (5×).

Name ______________________________

Date ______________________________

Section ______________________________

The [A] corresponds to the indicated Learning Outcome(s) listed at the beginning of the laboratory exercise.

Male Reproductive System

Laboratory Report 50

Part A Assessments

Complete the following statements:

1. Each testis is suspended by a (an) ______________________ within the scrotum. [1]
2. As the testes descend, they pass through the ______________________ of the abdominal wall into the scrotum. [1]
3. Connective tissue subdivides a testis into many ______________________, which contain seminiferous tubules. [1]
4. The ______________________ is a highly coiled tube on the surface of the testis. [1]
5. The ______________________ cells of the epithelium that line the seminiferous tubule give rise to sperm cells. [2]
6. ______________________ is the process by which sperm cells are formed. [2]
7. The number of chromosomes normally present in a human sperm cell is ______________________. [1]
8. Sperm cells undergo maturation while they are stored in the ______________________. [2]
9. The secretion of the seminal vesicles is rich in the monosaccharide called ______________________. [2]
10. Prostate gland secretion helps neutralize seminal fluid because the prostate secretion is ______________________. [2]
11. The secretion of the ______________________ glands lubricates the end of the penis in preparation for sexual intercourse. [2]
12. The sensitive, cone-shaped end of the penis is called the ______________________. [1]
13. ______________________ is the movement of sperm cells and various glandular secretions into the urethra. [2]
14. ______________________ is the process by which semen is forced out through the urethra. [2]

Part B Assessments

1. Sketch and label a representative section of the testis. [3]

2. Sketch and label a region of the epididymis. 3

3. Sketch and label a penis cross section. 3

4. Briefly describe the function of each of the following: 2

a. scrotum

____

____

b. spermatogenic cell

____

____

c. interstitial cell (Leydig cell)

____

____

d. epididymis

____

____

e. corpora cavernosa and corpus spongiosum

____

____

Laboratory Exercise 51

Female Reproductive System

Materials Needed

Textbook
Human torso model
Model of the female reproductive system
Anatomical chart of the female reproductive system
Compound light microscope
Prepared microscope slides of the following:
- Ovary section with maturing follicles
- Uterine tube, cross section
- Uterine wall section

For Demonstration:

Prepared microscope slides of the following:
- Uterine wall, early proliferative phase
- Uterine wall, secretory phase
- Uterine wall, early menstrual phase

The organs of the female reproductive system are specialized to produce and maintain the female sex cells, to transport these cells to the site of fertilization, to provide a favorable environment for a developing offspring, to move the offspring to the outside, and to produce female sex hormones.

These organs include the ovaries, which produce the egg cells and female sex hormones, and sets of internal and external accessory structures. The internal accessory structures include the uterine tubes, uterus, and vagina. The external structures are the labia majora, labia minora, clitoris, and vestibular glands.

Purpose of the Exercise

To review the structure and functions of the female reproductive organs and to examine some of their features.

LEARNING OUTCOMES

After completing this exercise, you should be able to

1. Locate and identify the organs of the female reproductive system.
2. Describe the functions of these organs.
3. Sketch and label the major features from microscopic sections of the ovary, uterine tube, and uterine wall.

EXPLORE

Procedure A—Female Reproductive Organs

1. Review the sections entitled, "Ovaries," "Female Internal Accessory Organs," "Female External Reproductive Organs," and "Mammary Glands" in chapter 22 of the textbook.
2. As a review activity, label figures 51.1, 51.2, 51.3, 51.4, and 51.5.
3. Observe the human torso model, the model of the female reproductive system, and the anatomical chart of the female reproductive system. Locate the following features:

ovaries
- medulla
- cortex

ligaments
- broad ligament
- suspensory ligament
- ovarian ligament
- round ligament

uterine tubes (oviducts; fallopian tubes)
- infundibulum
- fimbriae

uterus
- fundus
- body
- cervix
- cervical orifice
- uterine wall
  - endometrium
  - myometrium
  - perimetrium (serous membrane)

rectouterine pouch

vagina
- fornices
- vaginal orifice
- hymen
- mucosal layer

vulva (external accessory organs)
- mons pubis
- labia majora
- labia minora
- vestibule
- vestibular glands
- clitoris

breasts
- nipple
- areola
- alveolar glands (compose milk-producing parts of mammary glands)
- lactiferous ducts
- adipose tissue

4. Complete Part A of Laboratory Report 51.

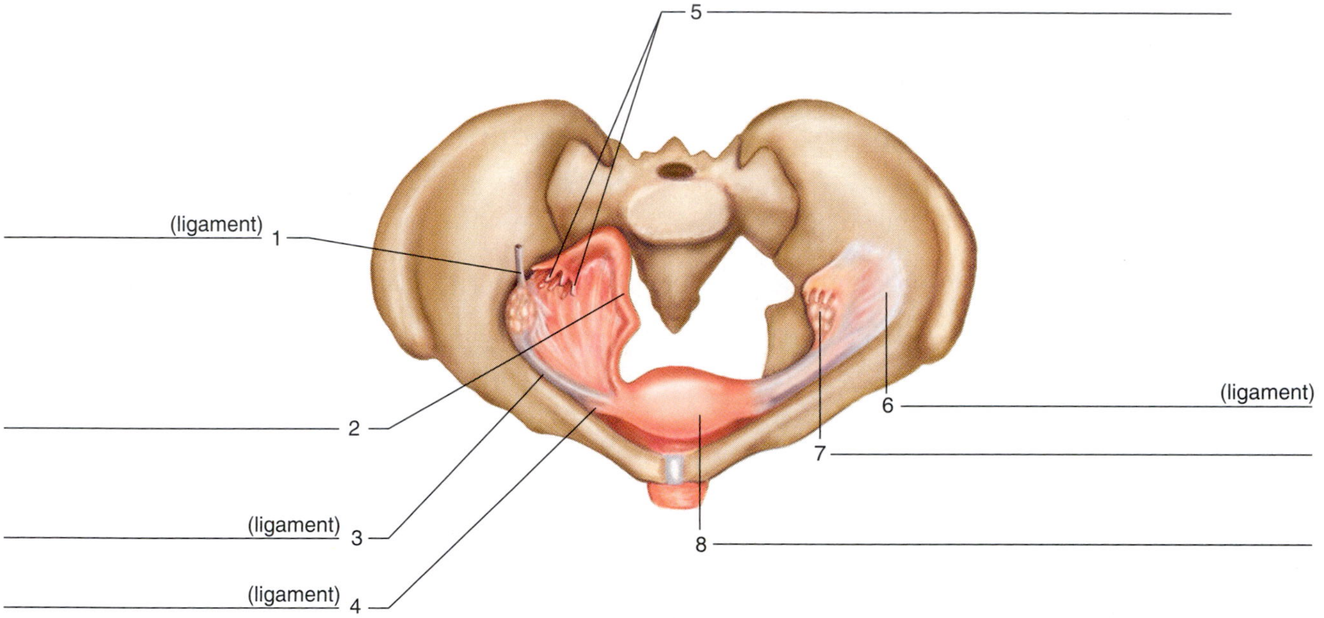

Figure 51.1 Label the ligaments and other structures associated with the female internal reproductive organs.

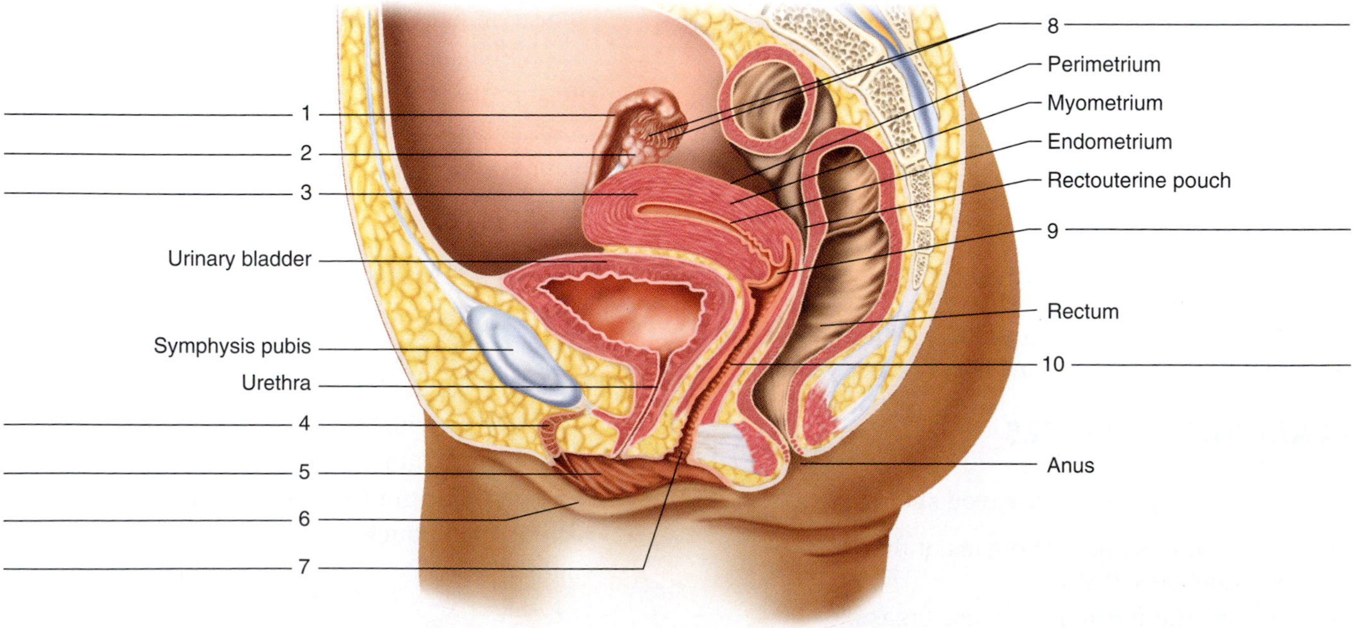

Figure 51.2 Label the structures of the female reproductive system in this sagittal view.

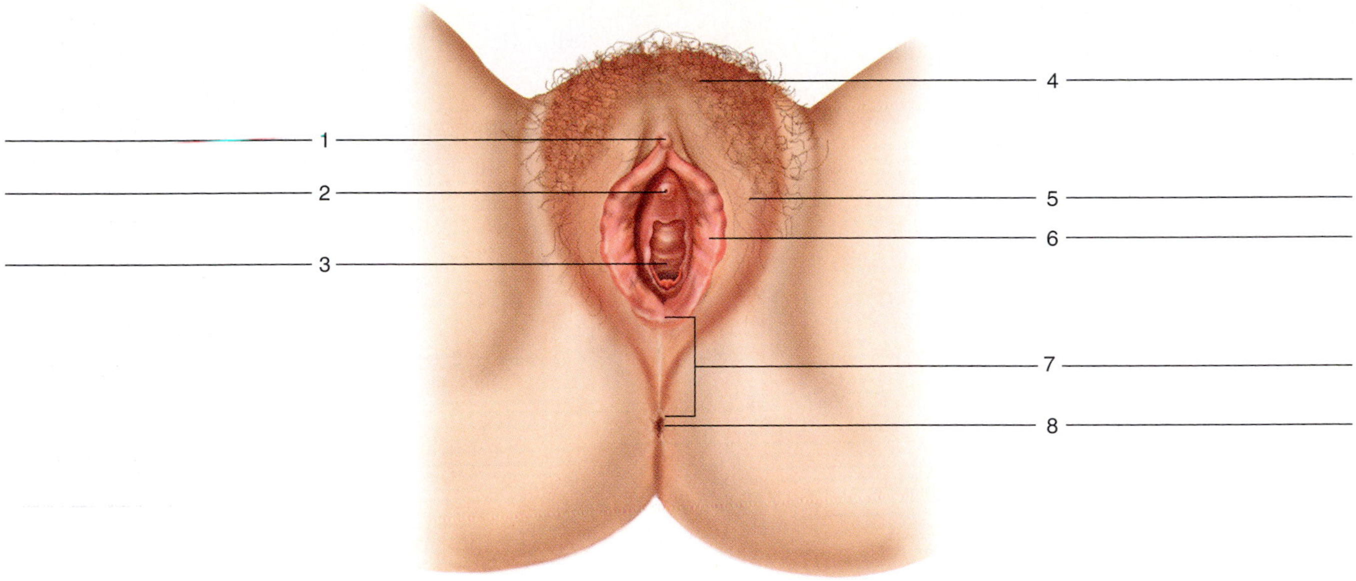

Figure 51.3 Label the female external reproductive organs and associated structures.

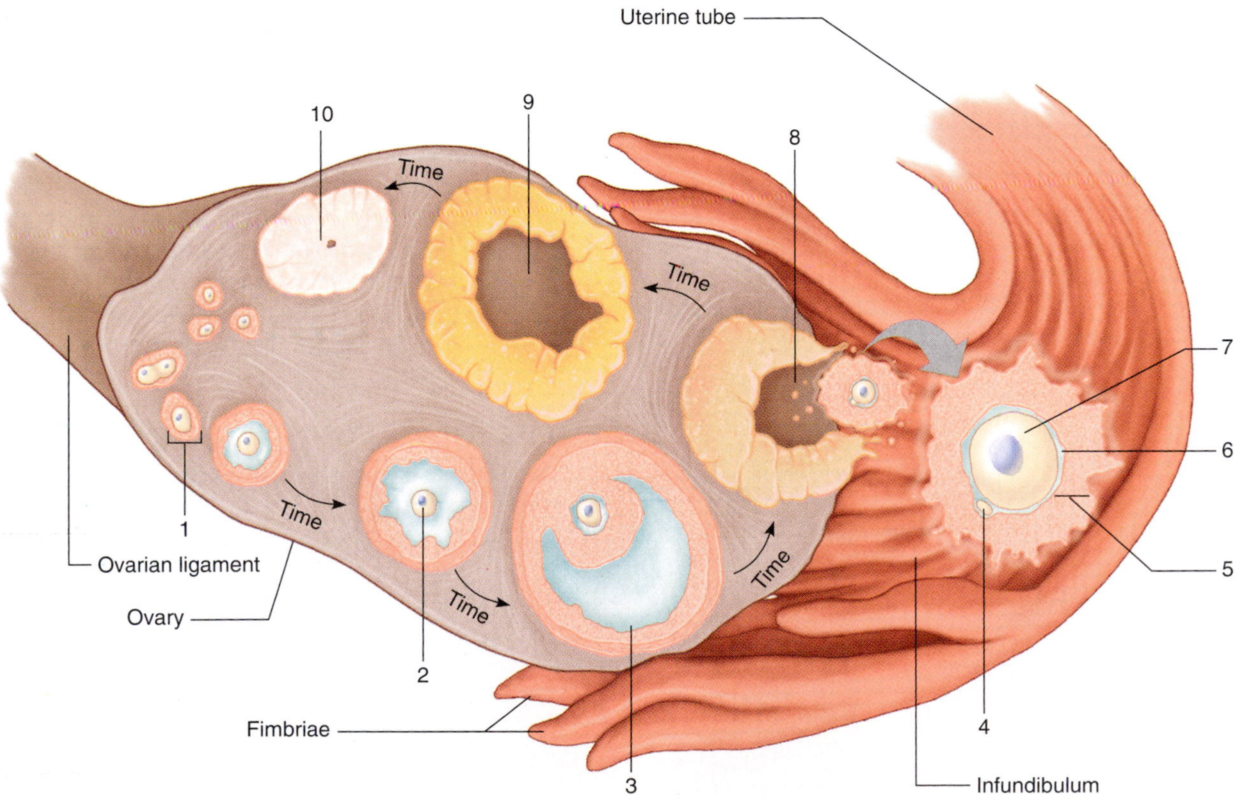

____ Corona radiata

____ Corpus albicans

____ Corpus luteum

____ First polar body

____ Follicular fluid

____ Ovulation site

____ Primary follicle

____ Primary oocyte

____ Secondary oocyte

____ Zona pellucida

Figure 51.4 Label this ovary by placing the correct numbers in the spaces provided.

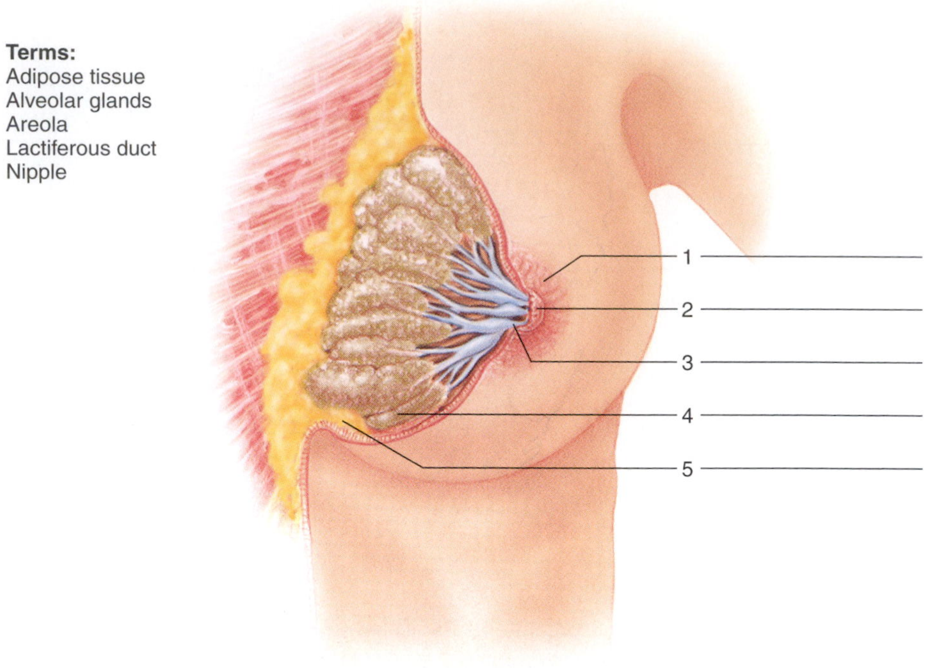

Figure 51.5 Label the structures of the breast (anterior view), using the terms provided. 1

EXPLORE

Procedure B—Microscopic Anatomy

1. Obtain a microscope slide of an ovary section with maturing follicles, and examine it with low-power magnification (fig. 51.6). Locate the outer layer, or *cortex,* composed of densely packed cells, and the inner layer, or *medulla,* which largely consists of loose connective tissue.
2. Focus on the cortex of the ovary, using high-power magnification (fig. 51.7). Note the thin layer of small cuboidal cells on the free surface. These cells comprise the *germinal epithelium.* Also locate some *primordial follicles* just beneath the germinal epithelium. Each follicle consists of a single, relatively large *primary oocyte* with a prominent nucleus and a covering of *follicular cells.*
3. Prepare a labeled sketch of the ovarian cortex in Part B of the laboratory report.

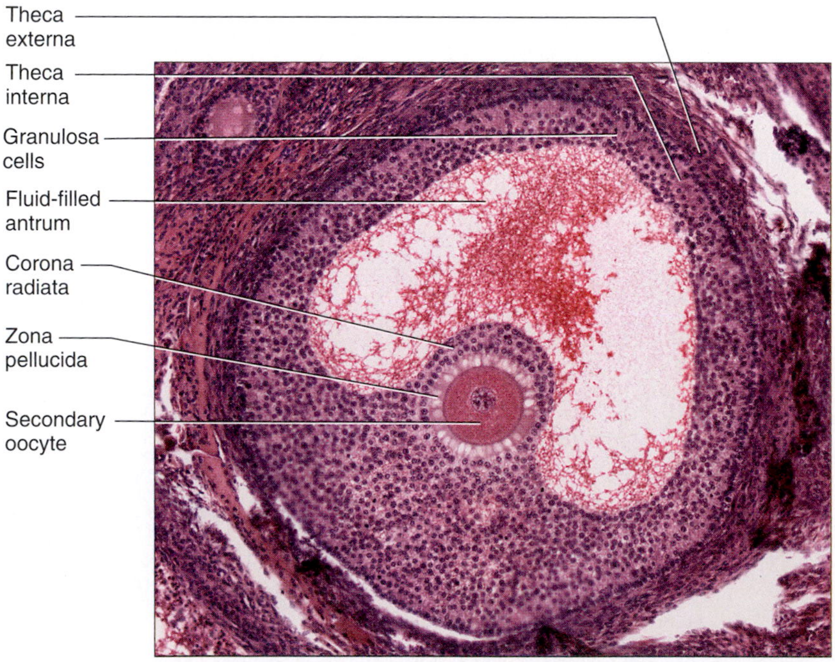

Figure 51.6 Micrograph of the ovary (250×).

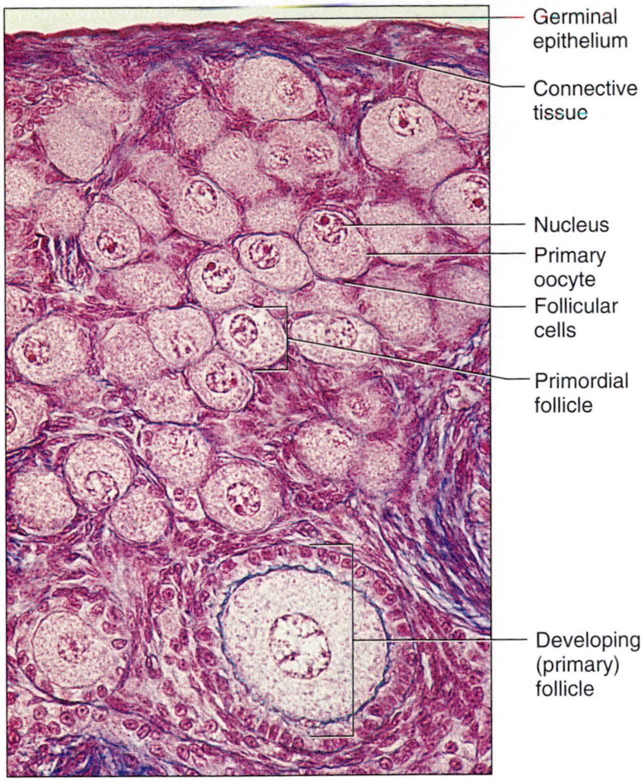

Figure 51.7 Micrograph of the ovarian cortex (100× micrograph enlarged to 200×).

4. Use low-power magnification to search the ovarian cortex for maturing follicles in various stages of development. Prepare three labeled sketches in Part B of the laboratory report to illustrate the changes that occur in a follicle as it matures.
5. Obtain a microscope slide of a cross section of a uterine tube. Examine it, using low-power magnification (fig. 51.8). The shape of the lumen is very irregular.
6. Focus on the inner lining of the uterine tube, using high-power magnification. The lining is composed of *simple columnar epithelium,* and some of the epithelial cells are ciliated on their free surfaces.
7. Prepare a labeled sketch of a representative region of the wall of the uterine tube in Part B of the laboratory report.
8. Obtain a microscope slide of the uterine wall section (fig. 51.9). Examine it, using low-power magnification, and locate the following:

endometrium (inner mucosal layer)
myometrium (middle, thick muscular layer)
perimetrium (outer serosal layer)

9. Prepare a labeled sketch of a representative section of the uterine wall in Part B of the laboratory report.
10. Complete Part B of the laboratory report.

Demonstration

Observe the slides in the demonstration microscopes. Each slide contains a section of uterine mucosa taken during a different phase in the reproductive cycle. In the *early proliferative phase,* note the simple columnar epithelium on the free surface of the mucosa and the many sections of tubular uterine glands in the tissues beneath the epithelium. In the *secretory phase,* the endometrium is thicker and the uterine glands appear more extensive and they are coiled. In the *early menstrual phase,* the endometrium is thinner because its surface layer has been lost. Also, the uterine glands are less apparent, and the spaces between the glands contain many leukocytes. What is the significance of these changes?

____

____

____

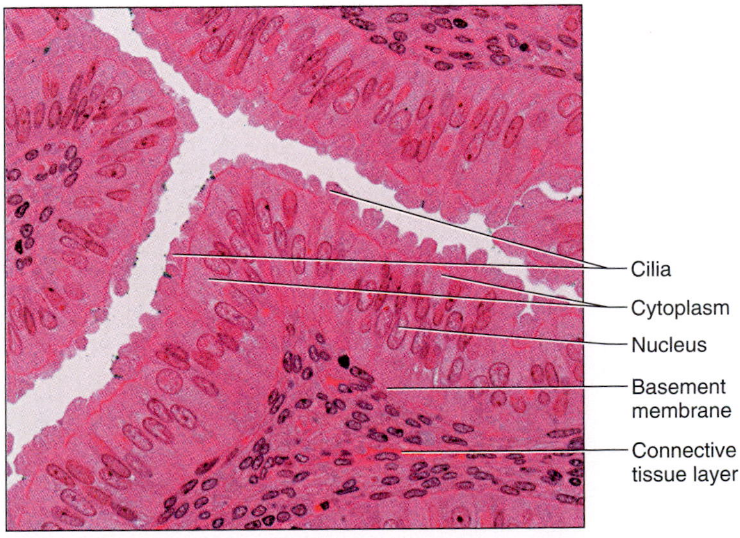

Figure 51.8 Micrograph of a cross section of the uterine tube (800×).

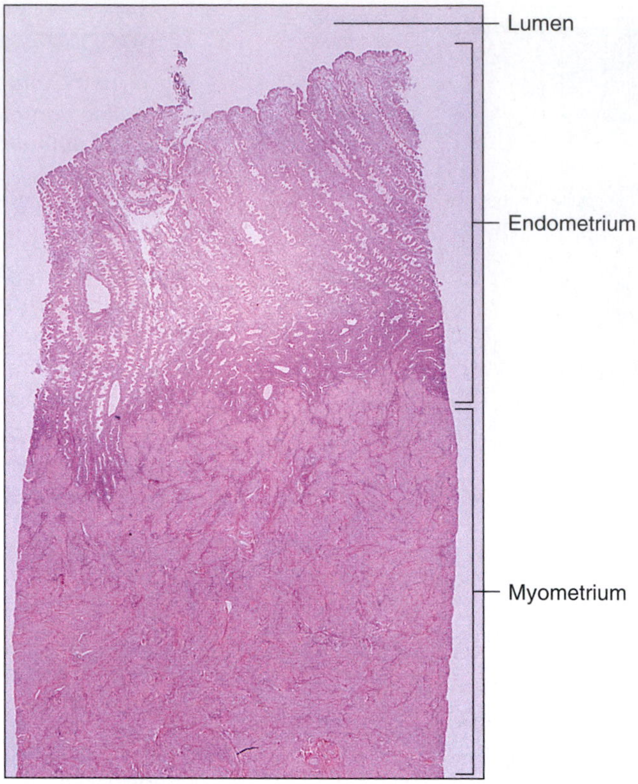

Figure 51.9 Micrograph of a section of the uterine wall (10× micrograph enlarged to 35×).

Name ____________________

Date ____________________

Section ____________________

The A corresponds to the indicated Learning Outcome(s) listed at the beginning of the laboratory exercise.

Female Reproductive System

Laboratory Report 51

Part A Assessments

Complete the following statements:

1. The ovaries are located in the lateral wall of the __________ cavity. 1
2. The largest of the ovarian attachments is called the ______________ ligament. 1
3. The ovarian cortex appears granular because of the presence of ______________________. 1
4. The meiosis of egg formation is called ______________________. 2
5. A primary oocyte is closely surrounded by flattened epithelial cells called ______________________ cells. 1
6. When a primary oocyte divides, a secondary oocyte and a(an) ______________________ are produced. 2
7. Primordial follicles are stimulated to develop into primary follicles by the hormone called ______________. 2
8. ______________________ is the process by which a secondary oocyte is released from the ovary.
9. Uterine tubes are also called ______________________. 1
10. The ______________________ is the funnel-shaped expansion at the end of a uterine tube. 1
11. A portion of the uterus called the ______________ extends downward into the upper portion of the vagina. 1
12. The inner mucosal lining of the uterus is called the ______________________. 1
13. The myometrium is largely composed of ______________________ tissue. 1
14. The vaginal orifice is partially closed by a thin membrane called the ______________________. 1
15. The group of external accessory organs that surround the openings of the urethra and vagina comprise the ______________________ . 1
16. The rounded mass of fatty tissue overlying the symphysis pubis of the female is called the ______________. 1
17. The female organ that corresponds to the male penis is the ______________________. 1
18. The ______________________ of the female corresponds to the bulbourethral glands of the male. 1

Part B Assessments

1. Sketch and label a representative region of the ovarian cortex. 3

2. Sketch and label a series of three changes to illustrate follicular maturation. 3

3. Sketch and label a representative section of the wall of a uterine tube. 3

4. Sketch and label a representative section of uterine wall. 3

5. Complete the following:
   a. Describe the fate of a mature follicle. 2 ______________________________
   ____
   ____
   b. Describe the function of the cilia in the lining of the uterine tube. 2 ______________________________
   ____
   ____
   c. Briefly describe the changes that occur in the uterine lining during a reproductive cycle. 2 ____________
   ____
   ____

Laboratory Exercise 52

Fertilization and Early Development

Materials Needed

Textbook
Sea urchin egg suspension*
Sea urchin sperm suspension*
Compound light microscope
Depression microscope slide
Coverslip
Medicine droppers
Prepared microscope slide of the following:
  Sea urchin embryos (early and late cleavage)
Models of human embryos

For Learning Extension:

Vaseline
Toothpick

For Demonstration:

Preserved mammalian embryos

*See the Instructor's Manual for a source of materials.

Fertilization is the process by which the nuclei of an egg and a sperm cell come together and combine their chromosomes, forming a single cell called a zygote.

Ordinarily, before fertilization can occur in a human female, an egg cell must be released from the ovary and must be carried into a uterine tube. Also, semen containing sperm cells must be deposited in the vagina; some of these sperm cells must travel through the uterus and into the uterine tubes. Although many sperm cells may reach an egg cell, only one will participate in the fertilization of the egg.

Shortly after fertilization, the zygote undergoes division (mitosis) to form two cells. These two cells become four, they in turn divide into eight, and so forth. The resulting mass of cells continues to grow and undergoes developmental changes and growth that give rise to an offspring.

Purpose of the Exercise

To review the process of fertilization, to observe sea urchin eggs being fertilized, and to examine embryos in early stages of development.

LEARNING OUTCOMES

After completing this exercise, you should be able to

1. Describe the process and structures of fertilization.
2. Describe and distinguish the early developmental stages of a human.
3. Sketch the early developmental stages of a sea urchin.

EXPLORE

Procedure A—Fertilization

1. Review the sections entitled "Fertilization" and "Period of Cleavage" in chapter 23 of the textbook.
2. As a review activity, label figure 52.1.
3. Complete Part A of Laboratory Report 52.
4. Although it is difficult to observe fertilization in animals since the process occurs internally, it is possible to view forms of external fertilization. For example, egg and sperm cells can be collected from sea urchins, and the process of fertilization can be observed microscopically. To make this observation, follow these steps:
   a. Place a drop of sea urchin egg-cell suspension in the chamber of a depression slide, and add a coverslip.
   b. Examine the egg cells, using low-power magnification.
   c. Focus on a single egg cell with high-power magnification, and sketch the cell in Part B of the laboratory report.
   d. Remove the coverslip and add a drop of sea urchin sperm-cell suspension to the depression slide. Replace the coverslip, and observe the sperm cells with high-power magnification as they cluster around the egg cells. This attraction is stimulated by gamete secretions.
   e. Observe the egg cells with low-power magnification once again, and watch for the appearance of *fertilization membranes*. Such a membrane forms as soon as an egg cell is penetrated by a sperm cell; it looks like a clear halo surrounding the egg cell.
   f. Focus on a single fertilized egg cell, and sketch it in Part B of the laboratory report.

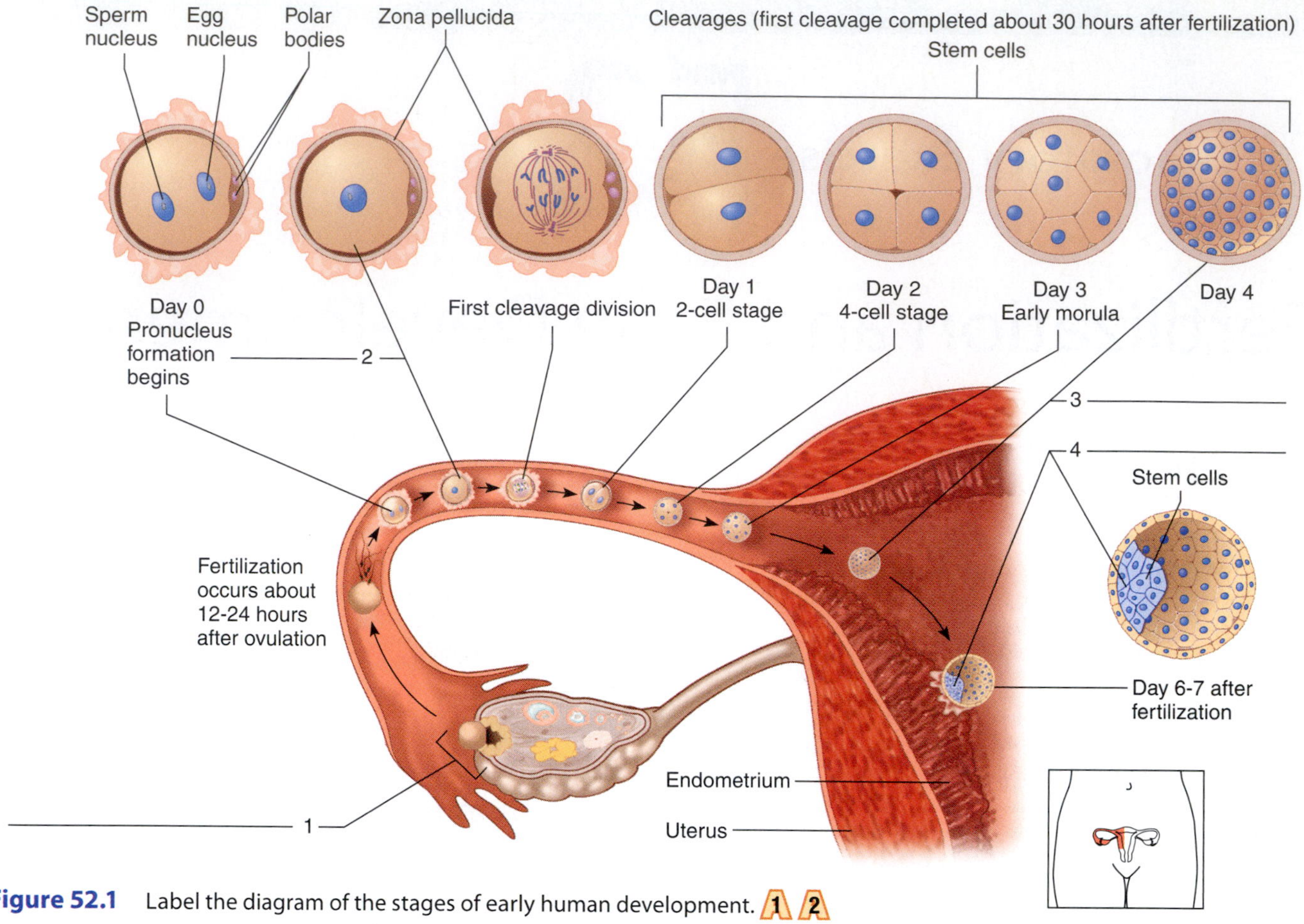

Figure 52.1 Label the diagram of the stages of early human development. 1 2

> **Learning Extension**
>
> Use a toothpick to draw a thin line of Vaseline around the chamber of the depression slide containing the fertilized sea urchin egg cells. Place a coverslip over the chamber, and gently press it into the Vaseline to seal the chamber and prevent the liquid inside from evaporating. Keep the slide in a cool place so that the temperature never exceeds 22°C (72°F). Using low-power magnification, examine the slide every 30 minutes, and look for the appearance of two-, four-, and eight-cell stages of developing sea urchin embryos.

EXPLORE

Procedure B—Sea Urchin Early Development

1. Obtain a prepared microscope slide of developing sea urchin embryos. This slide contains embryos in various stages of cleavage. Search the slide, using low-power magnification, and locate embryos in two-, four-, and eight-cell and morula stages. Observe that cleavage results in an increase of cell numbers; however, the cells get progressively smaller.
2. Prepare a sketch of each stage in Part C of the laboratory report.

EXPLORE

Procedure C—Human Early Development

1. Review the sections entitled "Period of Cleavage" and "Embryonic Stage" in chapter 23 of the textbook.
2. As a review activity, study and label figures 52.2 and 52.3.
3. Complete Part D of the laboratory report.
4. Observe the models of human embryos, and identify the following features:

blastocyst

- inner cell mass
- trophoblast
- blastocyst cavity (blastocoel)

primary germ layers
- ectoderm
- endoderm
- mesoderm

chorion
chorionic villi
amnion
amniotic fluid
umbilical cord
- umbilical arteries (2)
- umbilical vein (1)

yolk sac
allantois
placenta

Demonstration

Observe the preserved mammalian embryos that are on display. In addition to observing the developing external body structures, identify such features as the chorion, chorionic villi, amnion, yolk sac, umbilical cord, and placenta. What special features provide clues to the types of mammals these embryos represent?

____

____

____

____

____

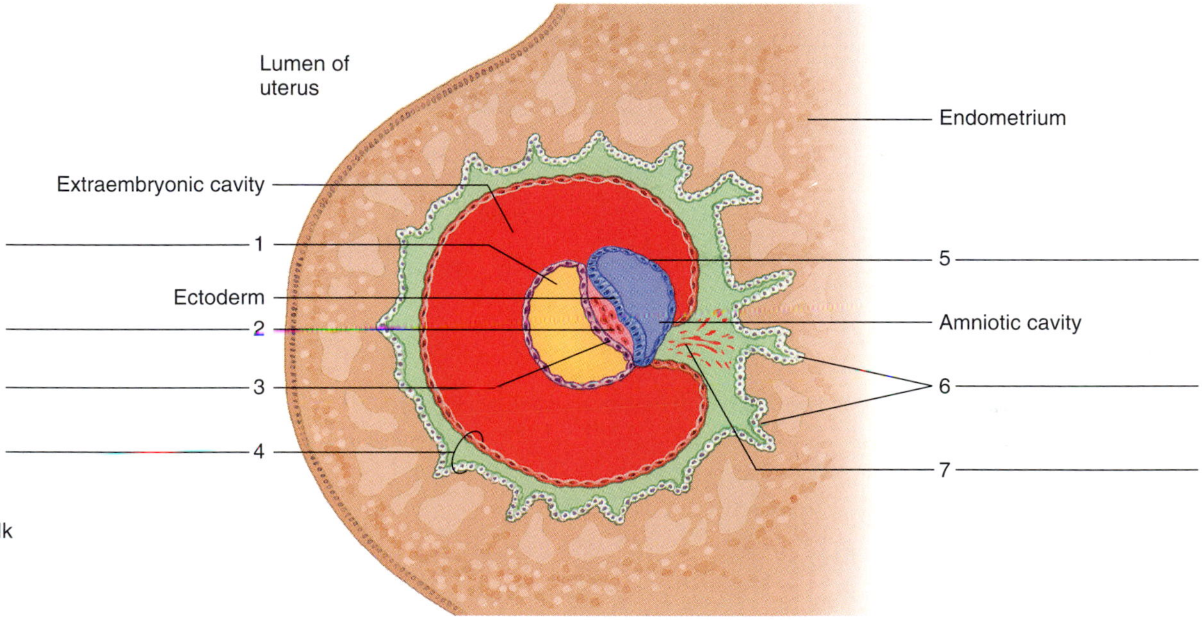

Terms:
Amnion
Chorion
Chorionic villi
Connecting stalk
Endoderm
Mesoderm
Yolk sac

Figure 52.2 Label the major features of the early embryo and the structures associated with it, using the terms provided. 2

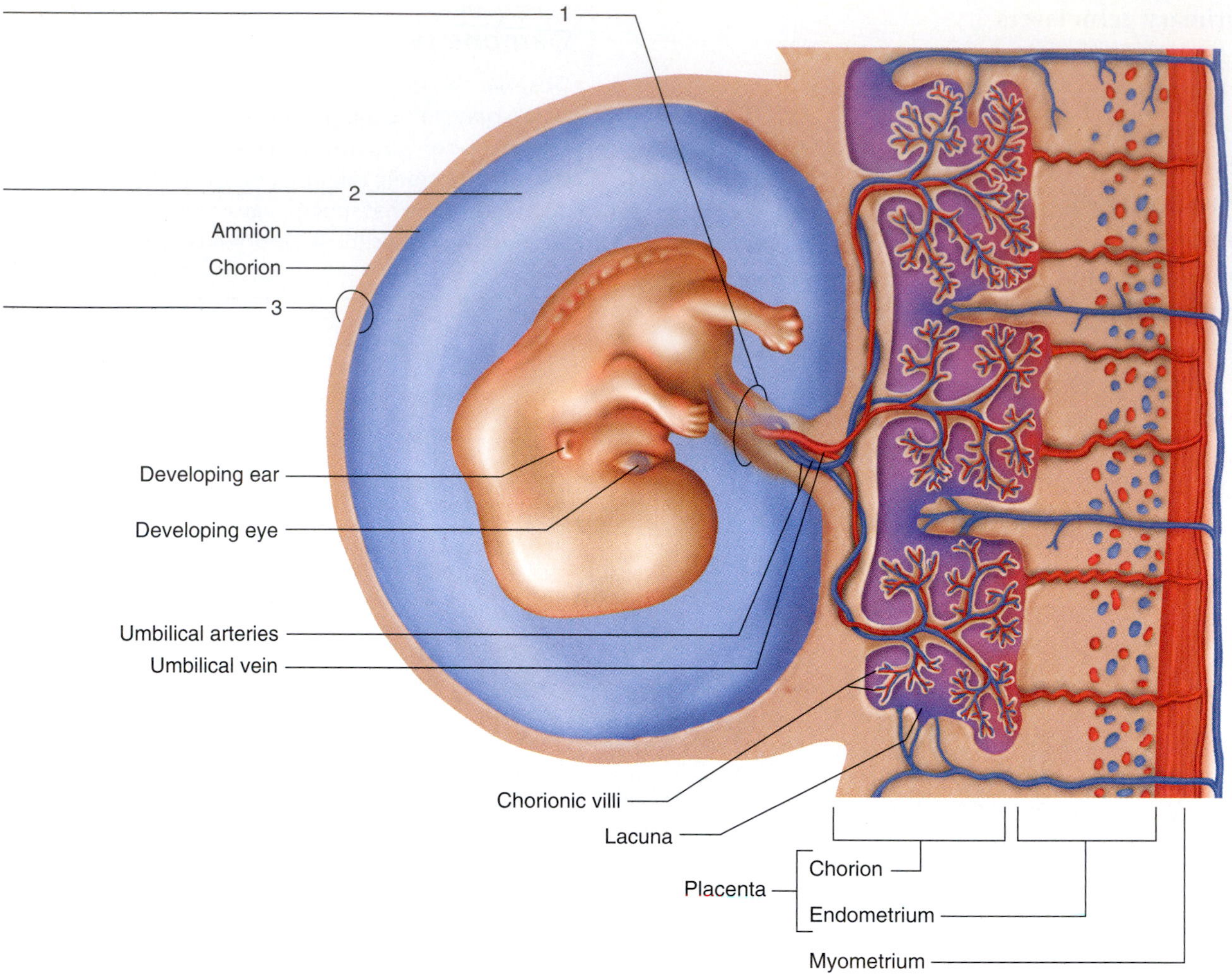

Figure 52.3 Label the structures associated with this seven-week-old embryo. 2

Name ____

Date ____

Section ____

The [A] corresponds to the indicated Learning Outcome(s) listed at the beginning of the laboratory exercise.

Fertilization and Early Development

Laboratory Report 52

Part A Assessments

Complete the following statements:

1. The zona pellucida surrounds the cell membrane of a(an) ________________________. 1
2. Enzymes secreted by the ________________________ of a sperm cell help it to penetrate the zona pellucida. 1
3. The cell resulting from fertilization is called a(an) ________________________. 1
4. The cell resulting from fertilization divides by the process of ________________________. 1
5. ________________________ is the phase of development during which cellular divisions result in smaller and smaller cells. 2
6. ________________________ on the inner lining of the uterine tube aid in moving a developing embryo. 1
7. The trip from the site of fertilization to the uterine cavity takes about ________________________ days. 1
8. The hollow ball of cells formed early in development is called a(an) ________________________. 2
9. A human offspring is called a(an) ____________________ until the end of the eighth week of development. 2
10. After the eighth week, a developing human is called a(an) ______________________ until the time of birth. 2

Part B Assessments

Prepare sketches of the following: 3

Single sea urchin egg (___×)	Fertilized sea urchin egg (___×)

Part C Assessments

Prepare sketches of the following: 3

Two-cell sea urchin embryo (____×)	Four-cell sea urchin embryo (____×)
Eight-cell sea urchin embryo (____×)	Morula sea urchin embryo (____×)

Part D Assessments

Match the terms in column A with the descriptions in column B. Place the letter of your choice in the space provided. 2

Column A

a. Amniotic fluid
b. Blastocyst
c. Chorionic villi
d. Ectoderm
e. Endoderm
f. Mesoderm
g. Morula
h. Placenta
i. Trophoblast
j. Umbilical cord

Column B

_______ **1.** Germ layer that gives rise to muscle and bone tissues
_______ **2.** Composed of parts of chorion and uterine wall
_______ **3.** Protects the embryo from being jarred
_______ **4.** Hollow ball of cells
_______ **5.** Cells forming wall of blastocyst
_______ **6.** Solid ball of about sixteen cells
_______ **7.** Inner germ layer of embryonic disc
_______ **8.** Slender extensions that grow out from the trophoblast
_______ **9.** Contains two arteries and one vein
_______ **10.** Outer germ layer of embryonic disc

Laboratory Exercise 53

Genetics

Materials Needed

Textbook
Pennies (or other coins)
Dice
PTC paper
Astigmatism chart
Ichikawa's or Ishihara's color plates for colorblindness test

Many examples of human genetics in this laboratory exercise are basic, external features to observe. Some traits are based upon simple Mendelian genetics. As our knowledge of genetics continues to develop, traits such as tongue roller and free earlobe may no longer be considered examples of simple Mendelian genetics. We may need to continue to abandon some simple Mendelian models. New evidence may involve polygenic inheritance, effects of other genes, or environmental factors as more appropriate explanations. Therefore, the analysis of family genetics is not appropriate nor is it the purpose of this laboratory exercise.

Genetics is the study of the inheritance of characteristics. The genes that transmit this information are coded in segments of DNA in chromosomes. Homologous chromosomes possess the same gene at the same *locus.* These genes may exist in variant forms, called *alleles.* If a person possesses two identical alleles, the condition is *homozygous.* If a person possesses two different alleles, the condition is *heterozygous.* The particular combination of these gene variants (alleles) represents the person's *genotype;* the appearance of the individual that develops as a result of the way the genes are expressed represents the person's *phenotype.*

If one allele determines the phenotype by masking the expression of the other allele in a heterozygous individual, the allele is termed *dominant.* The allele whose expression is masked is termed *recessive.* If the heterozygous condition determines an intermediate phenotype, the inheritance represents *incomplete dominance.* However, different alleles are *codominant* if both are expressed in the heterozygous condition. Some characteristics inherited on the sex chromosomes result in phenotype frequencies that might be more prevalent in males or females. Such characteristics are called sex-linked (X-linked or Y-linked) characteristics.

As a result of meiosis during the formation of eggs and sperm, a mother and father each transmit an equal number of chromosomes (the haploid number 23) to form the zygote (diploid number 46). An offspring will receive one allele from each parent. These gametes combine randomly in the formation of each offspring. Hence, the *laws of probability* can be used to predict possible genotypes and phenotypes of offspring. A genetic tool called a *Punnett square* simulates all possible combinations (probabilities) that can occur in offspring genotypes and resulting phenotypes.

Purpose of the Exercise

To observe some selected human traits, to use pennies and dice to demonstrate laws of probability, and to solve some genetic problems using a Punnett square.

LEARNING OUTCOMES

After completing this exercise, you should be able to

1. Examine and record twelve genotypes and phenotypes of selected human traits.
2. Demonstrate the laws of probability using tossed pennies and dice and interpret the results.
3. Predict genotypes and phenotypes of complete dominance, codominance, and sex-linked problems using Punnett squares.

EXPLORE

Procedure A—Human Genotypes and Phenotypes

A complete set of genetic instructions in one human cell constitutes one's *genome.* The human genome contains about 2.9 billion base pairs representing approximately 20,500 protein-encoding genes. These instructions represent our genotypes and are expressed as phenotypes

Table 53.1 Examples of Some Common Human Phenotypes

Dominant Traits and Genotypes	Recessive Traits and Genotypes
Tongue roller (*R*__)	Nonroller (*rr*)
Freckles (*F*__)	No freckles (*ff*)
Widow's peak (*W*__)	Straight hairline (*ww*)
Dimples (*D*__)	No dimples (*dd*)
Free earlobe (*E*__)	Attached earlobe (*ee*)
Normal skin coloration (*M*__)	Albinism (*mm*)
Astigmatism (*A*__)	Normal vision (*aa*)
Natural curly hair (*C*__)	Natural straight hair (*cc*)
PTC taster (*T*__)	Nontaster (*tt*)
Blood type A (I^A__), B (I^B__), or AB ($I^A I^B$)	Blood type O (*ii*)
Normal color vision ($X^C X^C$), ($X^C X^c$), or ($X^C Y$)	Red-green colorblindness ($X^c X^c$) or ($X^c Y$)

sometimes clearly observable on our bodies. Some of these traits are listed in table 53.1 and are discernible in figure 53.1.

A dominant trait might be homozygous or heterozygous, so only one capital letter is used along with a blank for the possible second dominant or recessive allele. For a recessive trait, two lowercase letters represent the homozygous recessive genotype for that characteristic. Dominant does not always correlate with the predominance of the allele in the gene pool; dominant means one allele will determine the appearance of the phenotype.

1. **Tongue roller/nonroller:** The dominant allele (*R*) determines the person's ability to roll the tongue into a U-shaped trough. The homozygous recessive condition (*rr*) prevents this tongue rolling (fig. 53.1). Record your results in the table in Part A of Laboratory Report 53.
2. **Freckles/no freckles:** The dominant allele (*F*) determines the appearance of freckles. The homozygous recessive condition (*ff*) does not produce freckles (fig. 53.1). Record your results in the table in Part A of the laboratory report.
3. **Widow's peak/straight hairline:** The dominant allele (*W*) determines the appearance of a hairline above the forehead that has a distinct downward point in the center, called a widow's peak. The homozygous recessive condition (*ww*) produces a straight hairline (fig. 53.1). A receding hairline would prevent this phenotype determination. Record your results in the table in Part A of the laboratory report.
4. **Dimples/no dimples:** The dominant allele (*D*) determines the appearance of a distinct dimple in one or both cheeks upon smiling. The homozygous recessive condition (*dd*) results in the absence of dimples (fig. 53.1). Record your results in the table in Part A of the laboratory report.
5. **Free earlobe/attached earlobe:** The dominant allele (*E*) codes for the appearance of an inferior earlobe that hangs freely below the attachment to the head. The homozygous recessive condition (*ee*) determines the earlobe attaching directly to the head at its inferior border (fig. 53.1). Record your results in the table in Part A of the laboratory report.
6. **Normal skin coloration/albinism:** The dominant allele (*M*) determines the production of some melanin, producing normal skin coloration. The homozygous recessive condition (*mm*) determines albinism due to the inability to produce or use the enzyme tyrosinase in pigment cells. An albino does not produce melanin in the skin, hair, or the middle tunic (choroid coat, ciliary body, and iris) of the eye. The absence of melanin in the middle tunic allows the pupil to appear slightly red to nearly black. Remember that the pupil is an opening in the iris filled with transparent aqueous humor. An albino human has pale white skin, flax-white hair, and a pale blue iris. Record your results in the table in Part A of the laboratory report.
7. **Astigmatism/normal vision:** The dominant allele (*A*) results in an abnormal curvature to the cornea or the lens. As a consequence, some portions of the image projected on the retina are sharply focused, and other portions are blurred. The homozygous recessive condition (*aa*) generates normal cornea and lens shapes and normal vision. Use the astigmatism chart and directions to assess this possible defect described in Laboratory Exercise 35. Other eye defects, such as nearsightedness (myopia) and farsightedness (hyperopia), are different genetic

Figure 53.1 Representative genetic traits comparing dominant and recessive phenotypes: (*a*) tongue roller; (*b*) nonroller; (*c*) freckles; (*d*) no freckles; (*e*) widow's peak; (*f*) straight hairline; (*g*) dimples; (*h*) no dimples; (*i*) free earlobe; (*j*) attached earlobe.

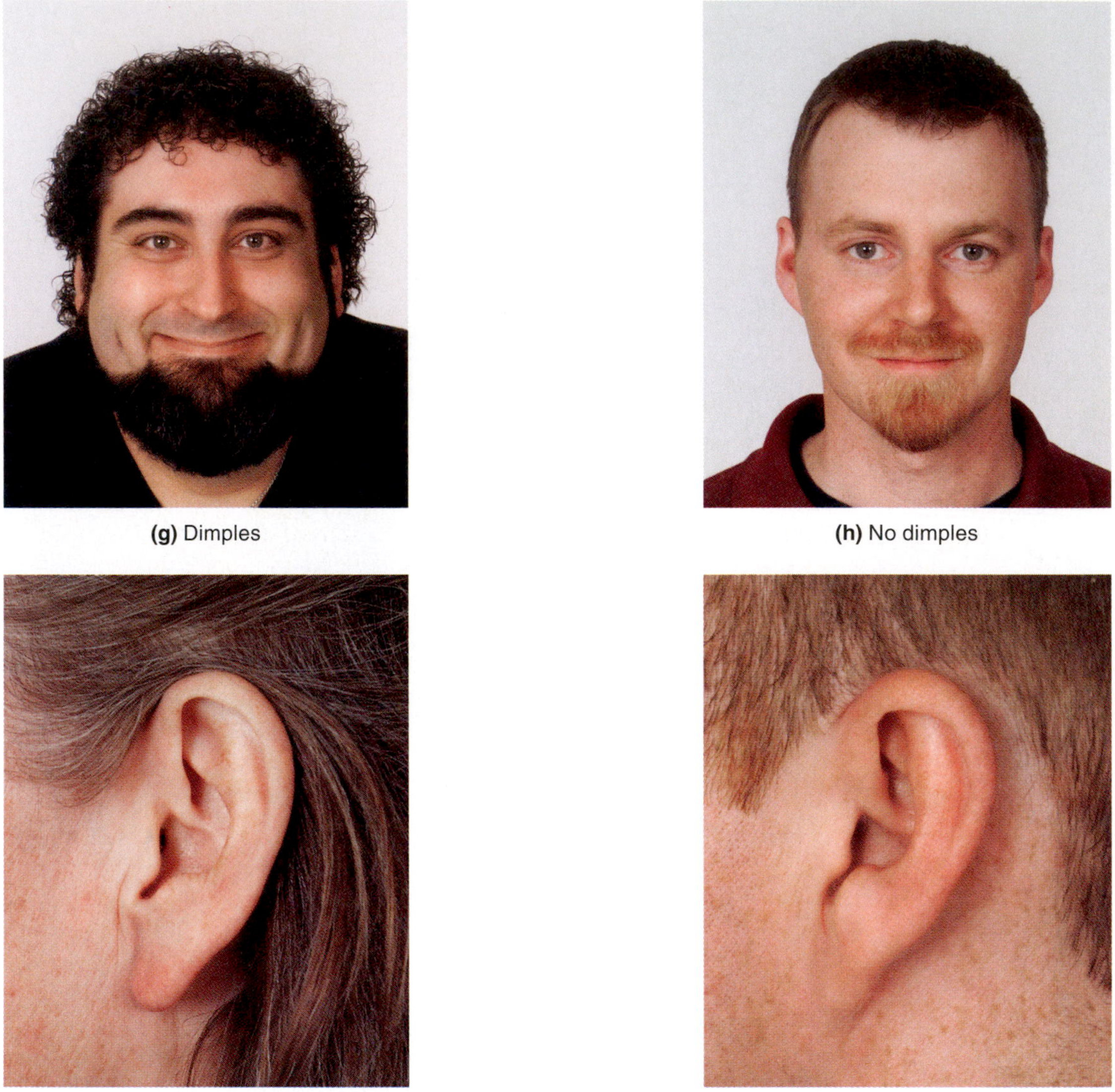

Dominant Traits | Recessive Traits

(g) Dimples | (h) No dimples

(i) Free earlobe | (j) Attached earlobe

Figure 53.1 *Continued.*

traits due to genes at other locations. Record your results in the table in Part A of the laboratory report.

8. **Curly hair/straight hair:** The dominant allele (*C*) determines the appearance of curly hair. Curly hair is somewhat flattened in cross section, as the hair follicle of a similar shape served as a mold for the root of the hair during its formation. The homozygous recessive condition (*cc*) produces straight hair. Straight hair is nearly round in cross section from being molded into this shape in the hair follicle. In some populations (Caucasians) the heterozygous condition (*Cc*) expresses the intermediate wavy hair phenotype (incomplete dominance). This trait determination assumes no permanents or hair straightening procedures have been performed. Such hair alterations do not change the hair follicle shape, and future hair growth results in original genetic hair conditions. Record your unaltered hair appearance in the table in Part A of the laboratory report.
9. **PTC taster/nontaster:** The dominant allele (*T*) determines the ability to experience a bitter sensation when PTC paper is placed on the tongue. About 70% of people possess this dominant gene. The homozygous recessive condition (*tt*) makes a person unable to notice the substance. Place a piece of PTC (phenylthiocarbamide) paper on the upper tongue surface and chew it slightly to see if you notice a bitter sensation from this harmless

chemical. The nontaster of the PTC paper does not detect any taste at all from this substance. Record your results in the table in Part A of the laboratory report.

10. **Blood type A, B, or AB/blood type O:** There are three alleles (I^A, I^B, and i) in the human population affecting RBC membrane structure. These alleles are located on a single pair of homologous chromosomes, so a person could possess either two of the three alleles or two of the same allele. All of the possible combinations of these alleles of genotypes and the resulting phenotypes are depicted in table 53.2. The expression of the blood type AB is a result of both codominant alleles located in the same individual. Possibly you have already determined your blood type in Laboratory Exercise 37 or have it recorded on a blood donor card. (If simulated blood-typing kits were used for Laboratory Exercise 37, those results would not be valid for your genetic factors.) Record your results in the table in Part A of the laboratory report.
11. **Sex determination:** A person with sex chromosomes XX displays a female phenotype. A person with sex chromosomes XY displays a male phenotype. Record your results in the table in Part A of the laboratory report.
12. **Normal color vision/red-green colorblindness:** This condition is a sex-linked (X-linked) characteristic. The alleles for color vision are also on the X chromosome, but they are absent on the Y chromosome. As a result, a female might possess both alleles (C and c), one on each of the X chromosomes. The dominant allele (C) determines normal color vision; the homozygous recessive condition (cc) results in red-green colorblindness. However, a male would possess only one of the two alleles for color vision because there is only a single X chromosome in a male. Hence a male with even a single recessive gene for colorblindness possesses the defect. Note all the possible genotypes and phenotypes for this condition (table 53.1). Review the color vision test in Laboratory Exercise 35 using the color plates in figure 35.4 and Ichikawa's or Ishihara's book. Record your results in the table in Part A of the laboratory report.
13. Complete Part A of Laboratory Report 53.

Table 53.2 **Genotypes and Phenotypes (Blood Types)**

Genotypes	Phenotypes (Blood Types)
I^AI^A or I^Ai	A
I^BI^B or I^Bi	B
I^AI^B (codominant)	AB
ii	O

EXPLORE

Procedure B—Laws of Probability

The laws of probability provide a mathematical way to determine the likelihood of events occurring by chance. This prediction is often expressed as a ratio of the number of results from experimental events to the number of results considered possible. For example, when tossing a coin there is an equal chance of the results displaying heads or tails. Hence the probability is one-half of obtaining either a heads or a tails (there are two possibilities for each toss). When all of the probabilities of all possible outcomes are considered for the result, they will always add up to a 1. To predict the probability of two or more events occurring in succession, multiply the probabilities of each individual event. For example, the probability of tossing a die and displaying a 4 two times in a row is $1/6 \times 1/6 = 1/36$ (there are six possibilities for each toss). Each toss in a sequence is an *independent event* (chance has no memory). The same laws apply when parents have multiple children (each fertilization is an independent event). Perform the following experiments to demonstrate the laws of probability:

1. Use a single penny (or other coin) and toss it 20 times. Predict the number of heads and tails that would occur from the 20 tosses. Record your prediction and the actual results observed in Part B of the laboratory report.
2. Use a single die (*pl.* dice) and toss it 24 times. Predict the number of times a number below 3 (numbers 1 and 2) would occur from the 24 tosses. Record your prediction and the actual results in Part B of the laboratory report.
3. Use two pennies and toss them simultaneously 32 times. Predict the number of times two heads, a heads and a tails, and two tails occur. Record your prediction and the actual results in Part B of the laboratory report.
4. Use a pair of dice and toss them simultaneously 32 times. Predict the number of times for both dice coming up with odd numbers, one die an odd and the other an even number, and both dice coming up with even numbers. Record your prediction and the actual results in Part B of the laboratory report.
5. Obtain class totals for all of the coins and dice tossed by adding your individual results to a class tally location as on the blackboard.
6. A Punnett square can be used for a visual representation to demonstrate the probable results for two pennies tossed simultaneously. For the purpose of a genetic comparison, an h (heads) will represent one "allele" on the coin; a t (tails) will represent a different "allele" on the coin.

Coin #1 Possibilities

		h	*t*	
Coin #2 Possibilities	*h*	*hh*	*ht*	Possible Combinations of Two Tossed Pennies (in Boxes)
	t	*ht*	*tt*	

7. Complete Part B of the laboratory report.

EXPLORE

Procedure C—Genetic Problems

1. A Punnett square can be constructed to demonstrate a visual display of the predicted offspring from parents with known genotypes. Recall that in complete dominance, a dominant allele is expressed in the phenotype, as it can mask the other recessive allele on the homologous chromosome pair. Recall also that during meiosis, the homologous chromosomes with their alleles separate (Mendel's Law of Segregation) into different gametes. An example of such a cross might be a homozygous dominant mother for dimples (*DD*) has offspring with a father homozygous recessive (*dd*) for the same trait. The results of such a cross, according to the laws of probability, would be represented by the following Punnett square:

Female Gametes

		D	*D*	
Male Gametes	*d*	*Dd*	*Dd*	Possible Genotypes of Offspring (in Boxes)
	d	*Dd*	*Dd*	

Results: Genotypes: 100% *Dd* (all heterozygous)

Phenotypes: 100% dimples

In another example, assume that both parents are heterozygous (*Dd*) for dimples. The results of such a cross, according to the laws of probability, would be represented by the following Punnett square:

Female Gametes

		D	*d*	
Male Gametes	*D*	*DD*	*Dd*	Possible Genotypes of Offspring
	d	*Dd*	*dd*	

Results: Genotypes: 25% *DD* (homozygous dominant); 50% Dd (heterozygous); 25% *dd* (homozygous recessive) (1:2:1 genotypic ratio)

Phenotypes: 75% dimples; 25% no dimples (3:1 phenotypic ratio)

2. Work the genetic problems 1 and 2 in Part C of the laboratory report.
3. The ABO blood type inheritance represents an example of codominance. Review table 53.2 for the genotypes and phenotypes for the expression of this trait. A similar technique to that used previously can be used to predict the offspring of parents of known genotypes. In this example, assume the genotype of the mother is $I^A I^B$, and the father is *ii*. The results of such a cross would be represented by the following Punnett square:

Female Gametes

		I^A	I^B	
Male Gametes	*i*	$I^A i$	$I^B i$	Possible Genotypes of Offspring
	i	$I^A i$	$I^B i$	

Results: Genotypes: 50% I^Ai (heterozygous for A); 50% I^Bi (heterozygous for B) (1:1 genotypic ratio)

Phenotypes: 50% blood type A; 50% blood type B (1:1 phenotypic ratio)

Note: In this particular cross, all of the children would have blood types unlike either parent.

4. Work the genetic problems 3 and 4 in Part C of the laboratory report.
5. Review the inheritance of red-green colorblindness, an X-linked characteristic, in table 53.1. A similar technique to that used to identify complete dominance can be used to predict the offspring of parents of known genotypes. In this example, assume the genotype of the mother is heterozygous $X^C X^c$ (normal color vision, but a carrier for the colorblindness defect), and the father is X^C Y (normal color vision; no allele on the Y chromosome). The results of such a cross would be represented by the following Punnett square:

Female Gametes

		X^C	X^c	
Male Gametes	X^C	X^CX^C	X^CX^c	Possible Genotypes of Offspring
	Y	X^CY	X^cY	

Results: Genotypes: 25% X^CX^C; 25% X^CX^c; 25% X^CY; 25% X^cY (1:1:1:1 genotypic ratio)

Phenotypes for sex determination: 50% females; 50% males (1:1 phenotypic ratio)

Phenotypes for color vision: Females 100% normal color vision (however, 50% are heterozygous carriers for colorblindness)

Phenotypes for color vision: Males 50% normal; 50% with red-green colorblindness (1:1 phenotypic ratio)

Note: In X-linked inheritance the colorblind males received the recessive gene from their mothers.

6. Complete Part C of the laboratory report.

Name ______________________

Date ______________________

Section ______________________

The [A] corresponds to the indicated Learning Outcome(s) listed at the beginning of the laboratory exercise.

Laboratory Report 53

Genetics

Part A Assessments

1. Enter your test results for genotypes and phenotypes in the table. Circle your particular phenotype and genotype for each of the twelve traits. [A1]

Trait	Dominant Phenotype	Genotype	Recessive Phenotype	Genotype
Tongue movement	Roller	*R*__	Nonroller	*rr*
Freckles	Freckles	*F*__	No freckles	*ff*
Hairline	Widow's peak	*W*__	Straight	*ww*
Dimples	Dimples	*D*__	No dimples	*dd*
Earlobe	Free	*E*__	Attached	*ee*
Skin coloration	Normal (some melanin)	*M*__	Albinism	*mm*
Vision	Astigmatism	*A*__	Normal	*aa*
Hair shape*	Curly	*C*__	Straight	*cc*
Taste	PTC taster	*T*__	Nontaster for PTC	*tt*
Blood type	A, B, or AB	I^A__; I^B__; or I^AI^B	O	*ii*
Sex		XX or XY		
Color vision	Normal	X^CX__ or X^CY	Red-green colorblindness	X^cX^c or X^cY

In some populations (Caucasians) the heterozygous condition (Cc) results in the appearance of wavy hair, which actually represents an example of incomplete dominance for this trait.

2. Analyze all the genotypes that you circled for the dominant phenotypes. If it is feasible to observe your biological parents and siblings for any of these traits, are you able to determine if any of your dominant genotypes are homozygous dominant or heterozygous? ________ If so, which ones? Explain the rationale for your response.

____

____

____

____

____

Part B Assessments

1. Single penny tossed 20 times and counting heads and tails: 2

   Probability (prediction): ___/20 heads ___/20 tails

   (Note: Traditionally, probabilities are converted to the lowest fractional representation.)

   Actual results: ____ heads ____ tails

   Class totals: ____ heads _____ tails

2. Single die tossed 24 times and counting the number of times a number below 3 occurs: 2

   Probability: ____/24 number below 3 (numbers 1 and 2)

   Actual results: ____ number below 3

   Class totals: ____ number below 3 ____ total tosses by class members

3. Two pennies tossed simultaneously 32 times and counting the number of two heads, a heads and a tails, and two tails: 2

   Probability: ____/32 of two heads ____/32 of a heads and a tails ____/32 of two tails

   Actual results: ____two heads ____heads and tails ____two tails

   Class totals: ____two heads ____heads and tails ____two tails

4. Two dice tossed simultaneously 32 times and counting the number of two odd numbers, an odd and an even number, and two even numbers: 2

   Probability: ____/32 of two odd numbers ____/32 of an odd and an even number ____/32 of two even numbers

   Actual results: ____ two odd numbers ____ an odd and an even number ____ two even numbers

   Class totals: ____ two odd numbers ____ an odd and an even number ____ two even numbers

5. Use the example of the two dice tossed 32 times and construct a Punnett square to represent the possible combinations that could be used to determine the probability (prediction) of odd and even numbers for the resulting tosses. Your construction should be similar to the Punnett square for the two coins tossed that is depicted in Procedure B of the laboratory exercise. 2

6. Complete the following:

   a. Are the class totals closer to the predicted probabilities than your results? ________ Explain your response.

   ____

   ____

   b. Does the first toss of the penny or the first toss of the die have any influence on the next toss? __________ Explain your response.

   ____

c. Assume a family has two boys or two girls. They wish to have one more child, but hope for the child to be of the opposite sex from the two they already have. What is the probability that the third child will be of the opposite sex? _____ Explain your response.

____

d. What is the probability (prediction) that a couple without children will eventually have four children, all girls?

________________ Explain your response.

____

____

Part C Assessments

For each of the genetic problems, (a) determine the parents' genotypes, (b) determine the possible gametes for each parent, (c) construct a Punnett square, and (d) record the resulting genotypes and phenotypes as ratios from the cross. Problems 1 and 2 involve examples of complete dominance; problems 3 and 4 are examples of codominance; problem 5 is an example of sex-linked (X-linked) inheritance.

1. Determine the results from a cross of a mother who is heterozygous (*Rr*) for tongue rolling with a father who is homozygous recessive (*rr*). 3

2. Determine the results from a cross of a mother and a father who are both heterozygous for freckles. 3

3. Determine the results from a mother who is heterozygous for blood type B and a father who is homozygous dominant for blood type A. 3

4. Determine the results from a mother who is heterozygous for blood type A and a father who is heterozygous for blood type B.

5. Colorblindness is an example of X-linked inheritance. Hemophilia is another example of X-linked inheritance, also from a recessive allele (*h*). The dominant allele (*H*) determines whether the person possesses normal blood clotting. A person with hemophilia has a permanent tendency for hemorrhaging due to a deficiency of one of the clotting factors (VIII—antihemophilic factor). Determine the offspring from a cross of a mother who is a carrier (heterozygous) for the disease and a father with normal blood coagulation.

Critical Thinking Application

Assume that the genes for hairline and earlobes are on different pairs of homologous chromosomes. Determine the genotypes and phenotypes of the offspring from a cross if both parents are heterozygous for both traits. (1) First determine the genotypes for each parent. (2) Determine the gametes, but remember each gamete has one allele for each trait (gametes are haploid). (3) Construct a Punnett square with 16 boxes that has four different gametes from each parent along the top and the left edges. (This is an application to demonstrate Mendel's Law of Independent Assortment.) (4) List the results of genotypes and phenotypes as ratios.

Laboratory Exercise 54

Cat Dissection: Musculature

Materials Needed

Textbook
Preserved cat (double injection)
Dissecting tray
Dissecting instruments
Large plastic bag
Identification tag
Disposable gloves
Bone shears
Human torso model
Human upper and lower limb models

For Demonstration:

Cat skeleton

Safety

- Wear disposable gloves when working on the cat dissection.
- Dispose of tissue remnants and gloves as instructed.
- Wash the dissecting tray and instruments as instructed.
- Wash your laboratory table.
- Wash your hands before leaving the laboratory.

Although the aim of this exercise is to become more familiar with the human musculature, human cadavers are not always available for dissection. Instead, preserved cats often are used for dissection because they are relatively small and can be purchased from biological suppliers. Also, as mammals, cats have many features in common with humans, including similar skeletal muscles (with similar names).

On the other hand, cats make use of four limbs for support, whereas humans use only two limbs. Because the musculature of each type of organism is adapted to provide for its special needs, comparisons of the muscles of cats and humans may not be precise.

As you continue your dissection of various systems of the cat, many anatomical similarities between cats and humans will be observed. These fundamental similarities are called homologous structures. Although homologous structures have a similar structure and embryological origin, the functions are sometimes different.

Purpose of the Exercise

To observe the musculature of the cat and to compare it with that of the human.

LEARNING OUTCOMES

After completing this exercise, you should be able to

1. Name and locate the major skeletal muscles of the cat.
2. Name the origins, insertions, and actions of the muscles designated by the laboratory instructor.
3. Name and locate the corresponding muscles of the human.

Muscle Dissection Techniques

Dissect: to expose the entire length of the muscle from origin to insertion. Most of the procedures are accomplished with blunt probes used to separate various connective tissues that hold adjacent structures together. It does not mean to remove or to cut into the muscles or other organs.

Transect: to cut through the muscle near its midpoint. The cut is perpendicular to the muscle fibers.

Reflect: to lift a transected muscle aside to expose deeper muscles or other organs.

EXPLORE

Procedure A—Skin Removal

1. Cats usually are preserved with a solution that prevents microorganisms from causing the tissues to decompose. However, because fumes from this preserving fluid may be annoying and may irritate

skin, be sure to work in a well-ventilated room and wear disposable gloves to protect your hands.

2. Obtain a preserved cat, a dissecting tray, a set of dissecting instruments, a large plastic bag, and an identification tag.
3. If the cat is in a storage bag, dispose of any excess preserving fluid in the bag, as directed by the laboratory instructor. However, you may want to save some of the preserving fluid to keep the cat moist until you have completed your work later in the year.
4. If the cat is too wet with preserving fluid, blot the cat dry with paper towels.
5. Remove the skin from the cat. To do this, follow these steps:
   a. Place the cat in the dissecting tray with its ventral surface down.
   b. Use a sharp scalpel to make a short, shallow incision through the skin in the dorsal midline at the base of the cat's neck.
   c. Insert the pointed blade of a dissecting scissors through the opening in the skin and cut along the dorsal midline, following the tips of the vertebral spinous processes to the base of the tail (fig. 54.1).

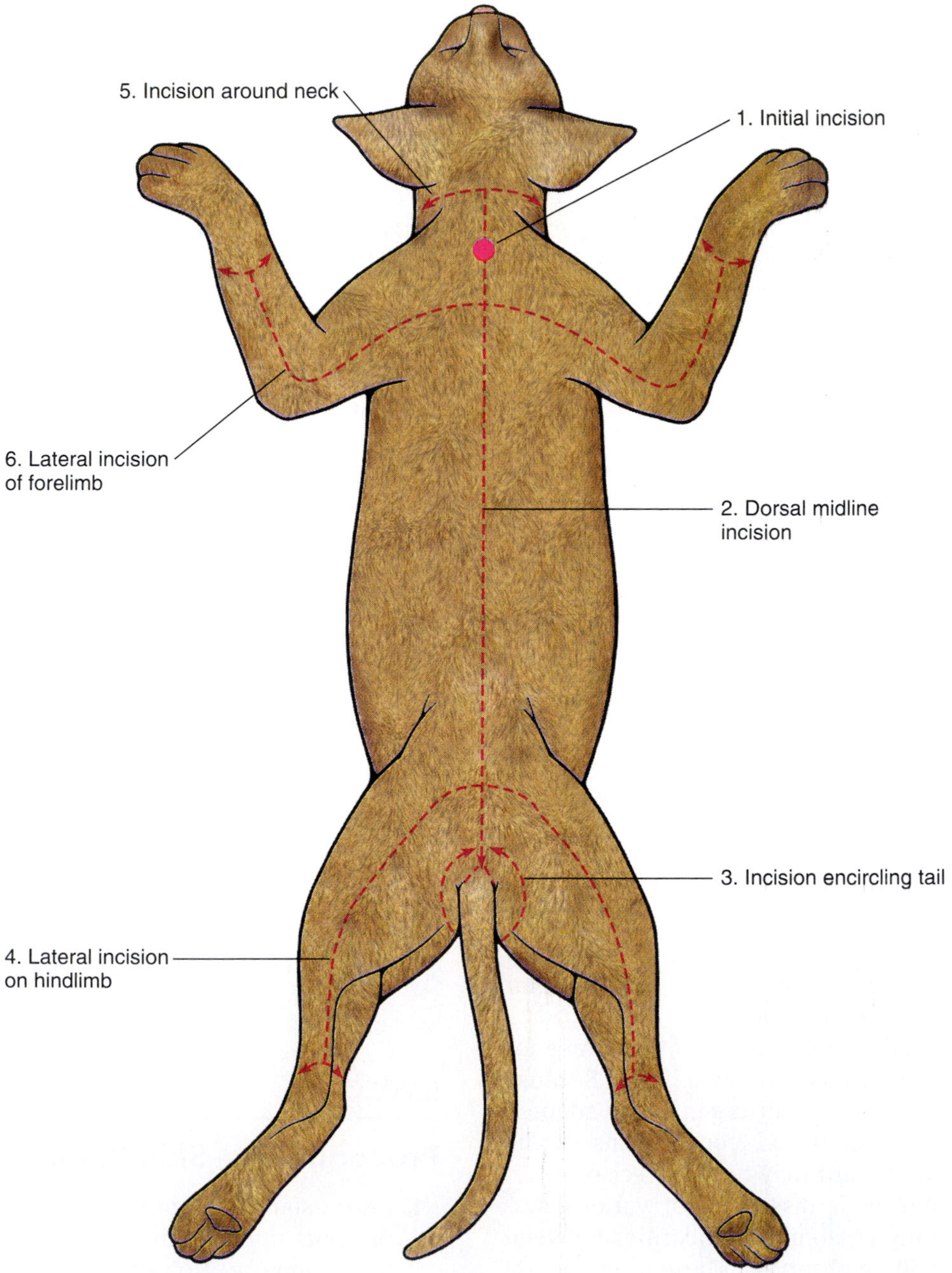

Figure 54.1 Incisions to be made for removing the skin of the cat.

d. Use the scissors to make an incision encircling the region of the tail, anus, and genital organs. Do not remove the skin from this area.
e. Use bone shears or a bone saw to sever the tail at its base and discard it (optional procedure).
f. From the incision at the base of the tail, cut along the lateral surface of each hindlimb to the paw, and encircle the ankle. Also clip off the claws from each paw to prevent them from tearing the plastic storage bag and scratching your skin.
g. Make an incision from the initial cut at the base of the neck to encircle the neck.
h. From the incision around the neck, cut along the lateral surface of each forelimb to the paw and encircle the wrist (fig. 54.1).
i. Grasp the skin on either side of the dorsal midline incision, and carefully pull the skin laterally away from the body. At the same time, use your fingers or a blunt probe to help remove the skin from the underlying muscles by separating the loose connective tissue (superficial fascia). As you pull the skin away, you may note a thin sheet of muscle tissue attached to it. This muscle is called the *cutaneous maximus,* and it functions to move the cat's skin. Humans lack the cutaneous maximus, but a similar sheet of muscle (platysma) is present in the neck of a human.
j. As you remove the skin, work toward the ventral surface, then work toward the head, and finally work toward the tail. Pull the skin over each limb as if you were removing a glove.
k. If the cat is a female, note the mammary glands on the ventral surface of the thorax and abdomen (fig. 54.2). These glands will adhere to the skin and can be removed from it and discarded.
l. After the skin has been pulled away, carefully remove as much of the remaining connective tissue as possible to expose the underlying skeletal muscles. The muscles should appear light brown and fibrous.

6. After skinning the cat, follow these steps:
   a. Discard the tissues you have removed as directed by the laboratory instructor.
   b. Wrap the skin around the cat to help keep its body moist, and place it in a plastic storage bag.
   c. Write your name in pencil on an identification tag, and tie the tag to the storage bag so that you can identify your specimen.
7. Observe the recommended safety procedures for the conclusion of a laboratory session.

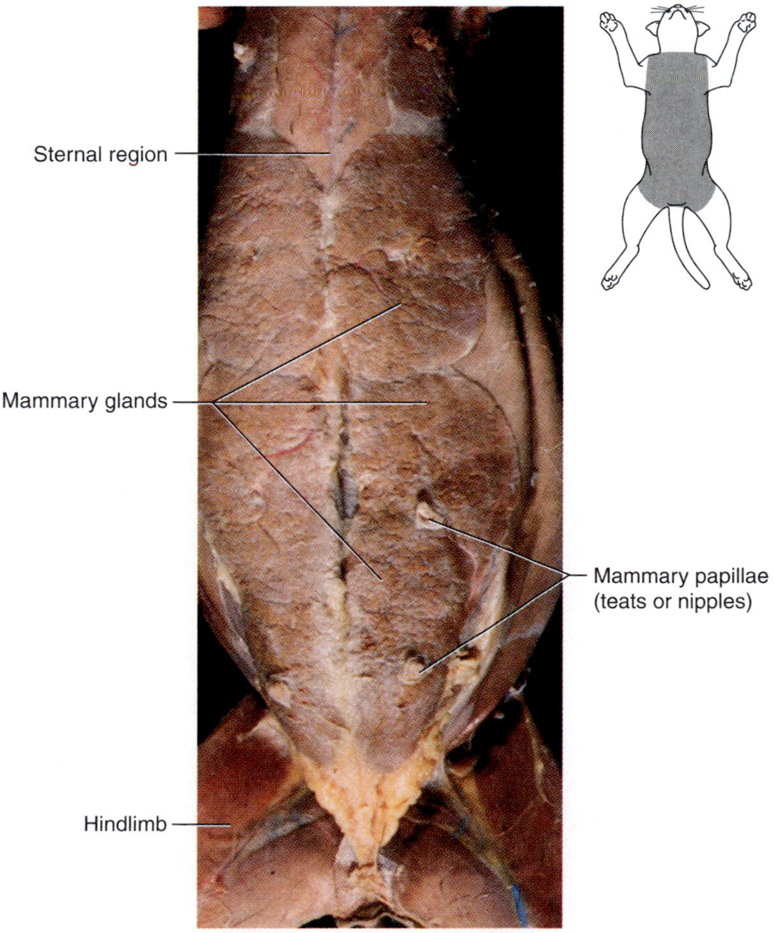

Figure 54.2 Mammary glands on the ventral surface of the cat's thorax and abdomen.

EXPLORE

Procedure B—Skeletal Muscle Dissection

1. The purpose of a skeletal muscle dissection is to separate the individual muscles from any surrounding tissues and thus expose the muscles for observation. To do a muscle dissection, follow these steps:
   a. Use the appropriate figure as a guide and locate the muscle to be dissected in the specimen.
   b. Use a blunt probe or a finger to separate the muscle from the surrounding connective tissue along its natural borders. The muscle should separate smoothly. If the border appears ragged, you probably have torn the muscle fibers.
2. If it is necessary to cut through (transect) a superficial muscle to observe a deeper one, use scissors to transect the muscle about halfway between its origin and insertion. Then, lift aside (reflect) the cut ends, leaving their attachments intact.
3. The following procedures will instruct you to dissect some of the larger and more easily identified muscles of the cat. In each case, the procedure will include the names of the muscles to be dissected, figures illustrating their locations, and tables listing the origins, insertions, and actions of the muscles. (More detailed dissection instructions can be obtained by consulting an additional guide or atlas for cat dissection.) As you dissect each muscle, study the figures and tables in chapter 9 of the textbook and identify any corresponding (homologous) muscles of the human body. Also locate these homologous muscles in the human torso model or models of the human upper and lower limbs.

Demonstration

Refer to the cat skeleton to help you to locate origins and insertions of the cat's muscles. Reference to the cat skeleton will also help you visualize the actions of these muscles, if you remember that a muscle's insertion is pulled toward its origin.

EXPLORE

Procedure C—Muscles of the Head and Neck

1. Place the cat in the dissecting tray with its ventral surface up.
2. Remove the skin and underlying connective tissues from one side of the neck, forward to the chin, and up to the ear.
3. Study figures 54.3 and 54.4, and then locate and dissect the following muscles:

sternomastoid
mylohyoid
sternohyoid
masseter
digastric

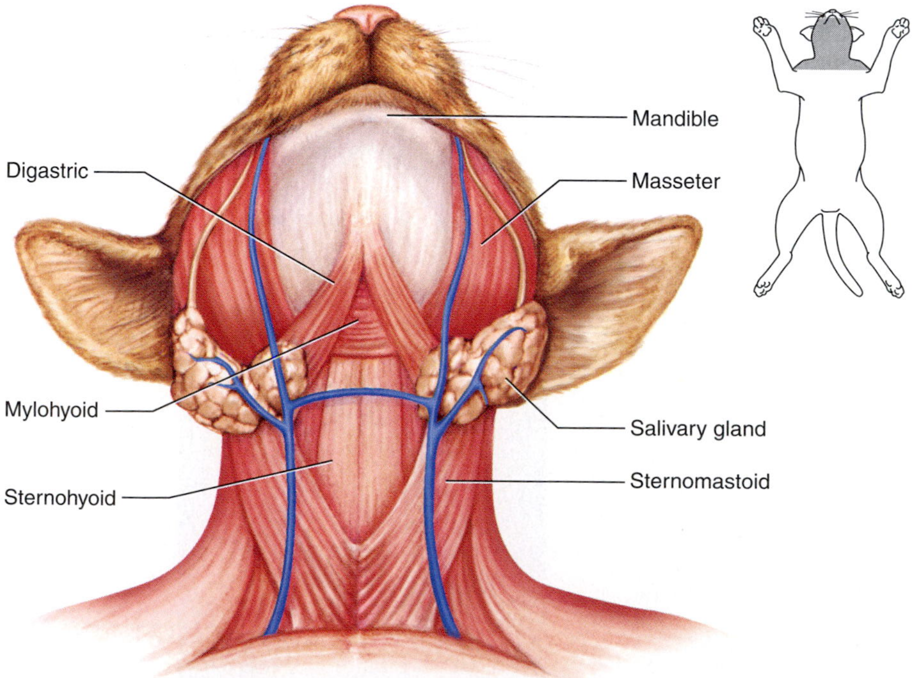

Figure 54.3 Muscles on the ventral surface of the cat's head and neck.

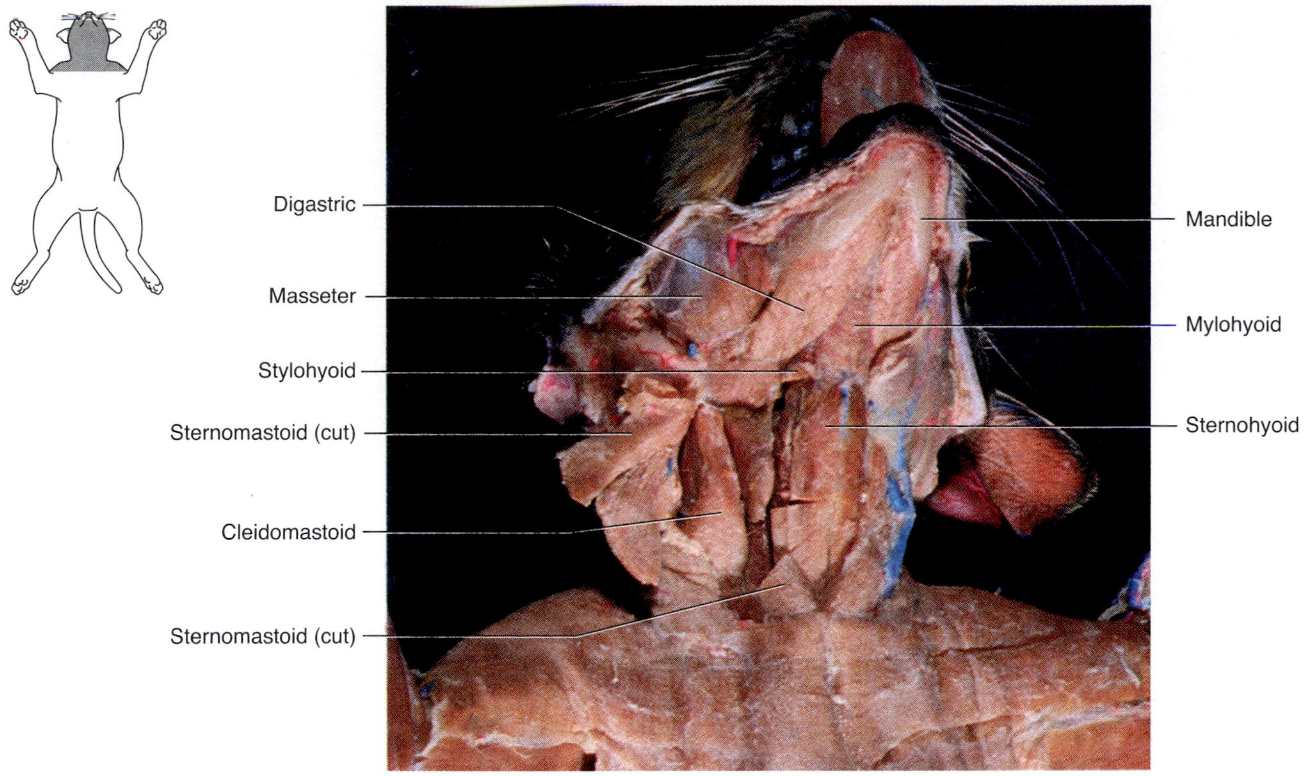

Figure 54.4 Deep muscles on the right ventral surface of the cat's neck.

4. To find the deep muscles of the cat's neck and chin, carefully remove the sternomastoid, sternohyoid, and mylohyoid muscles from the right side.
5. Study figure 54.5, and locate the following muscles:

   hyoglossus
   cleidomastoid
   styloglossus
   thyrohyoid
   geniohyoid
   sternothyroid

6. See table 54.1 for the origins, insertions, and actions of these head and neck muscles.
7. Complete Part A of Laboratory Report 54.

Table 54.1 Muscles of the Head and Neck

Muscle	Origin	Insertion	Action
Sternomastoid	Manubrium of sternum	Mastoid process of temporal bone	Turns and depresses head
Sternohyoid	Costal cartilage	Hyoid bone	Depresses hyoid bone
Digastric	Mastoid process and occipital bone	Ventral border of mandible	Depresses mandible
Mylohyoid	Inner surface of mandible	Hyoid bone	Raises floor of mouth
Masseter	Zygomatic arch	Coronoid fossa of mandible	Elevates mandible
Hyoglossus	Hyoid bone	Tongue	Pulls tongue back
Styloglossus	Styloid process	Tip of tongue	Pulls tongue back
Geniohyoid	Inner surface of mandible	Hyoid bone	Pulls hyoid bone forward
Cleidomastoid	Clavicle	Mastoid process	Turns head to side
Thyrohyoid	Thyroid cartilage	Hyoid bone	Raises larynx
Sternothyroid	Sternum	Thyroid cartilage	Pulls larynx back

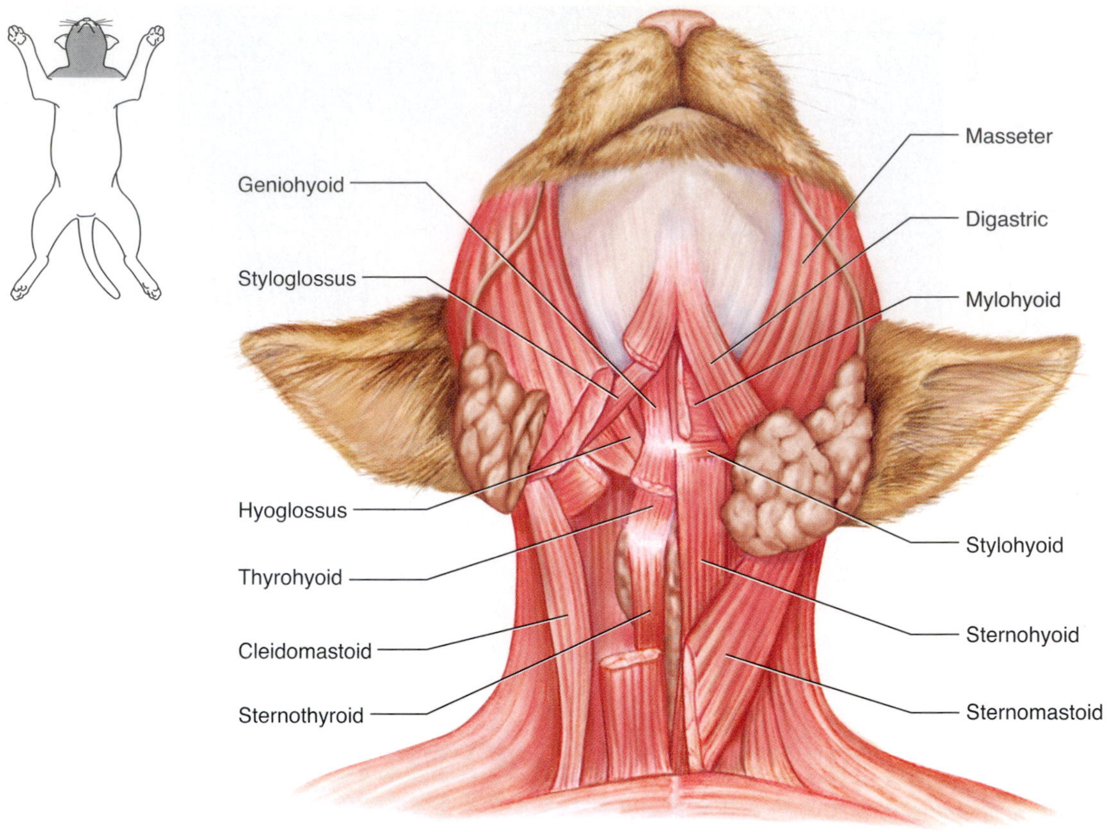

Figure 54.5 Deep muscles of the cat's neck and chin regions.

EXPLORE

Procedure D—Muscles of the Thorax

1. Place the cat in the dissecting tray with its ventral surface up.
2. Remove any remaining fat and connective tissue to expose the muscles in the walls of the thorax and abdomen.
3. Study figures 54.6 and 54.7. Transect the superficial pectoantebrachialis to expose the entire pectoralis major. Locate and dissect the following pectoral muscles:

   pectoantebrachialis
   pectoralis minor
   pectoralis major
   xiphihumeralis

4. To find the deep thoracic muscles, transect the pectoralis minor, pectoralis major, and pectoantebrachialis muscles, and reflect their cut edges to the sides. Remove the underlying fat and connective tissues to expose the following muscles, as shown in figure 54.8:

   scalenus
   levator scapulae
   serratus ventralis
   transversus costarum

5. See table 54.2 for the origins, insertions, and actions of these muscles.

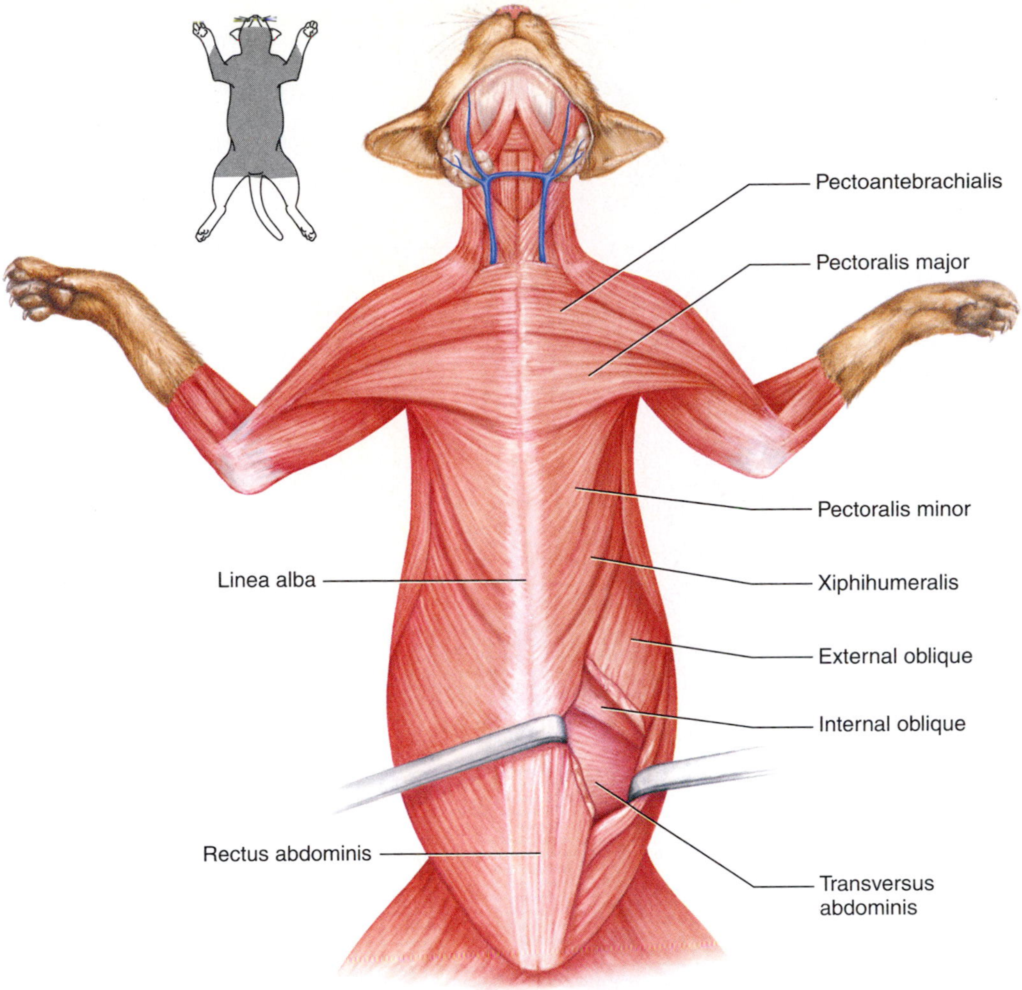

Figure 54.6 Muscles of the cat's thorax and abdomen.

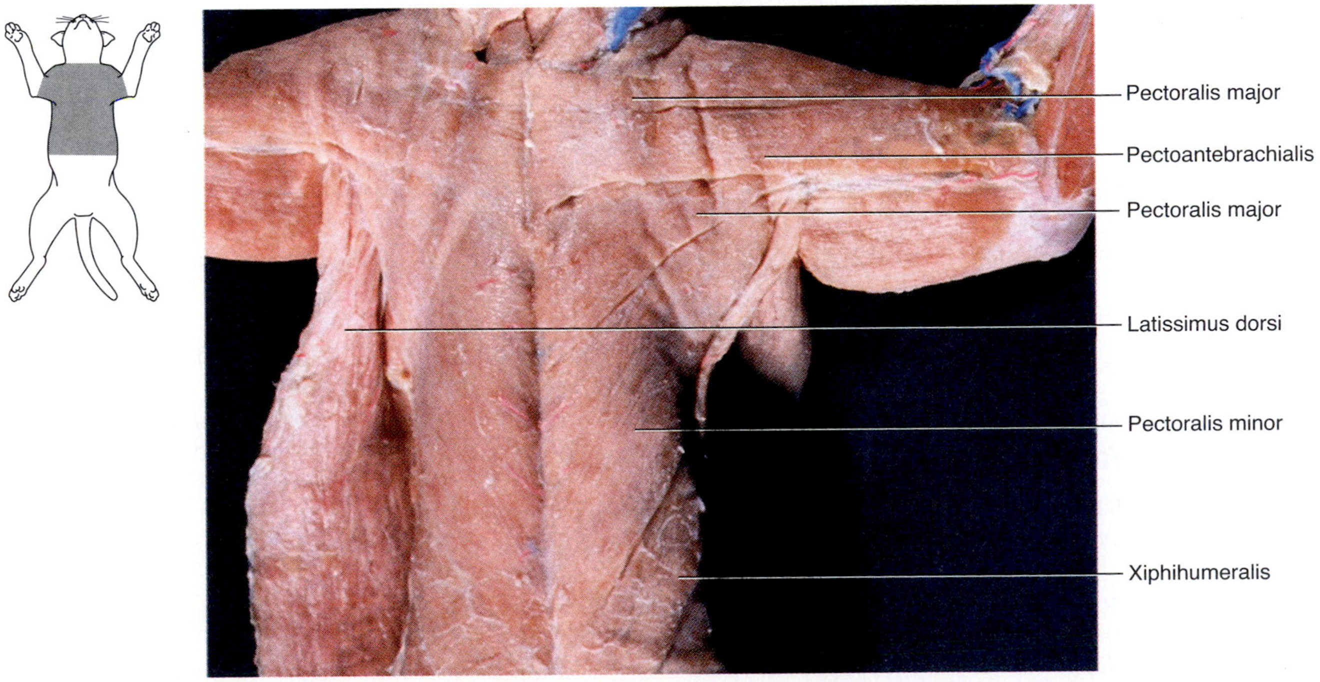

Figure 54.7 Superficial muscles of the cat's thorax.

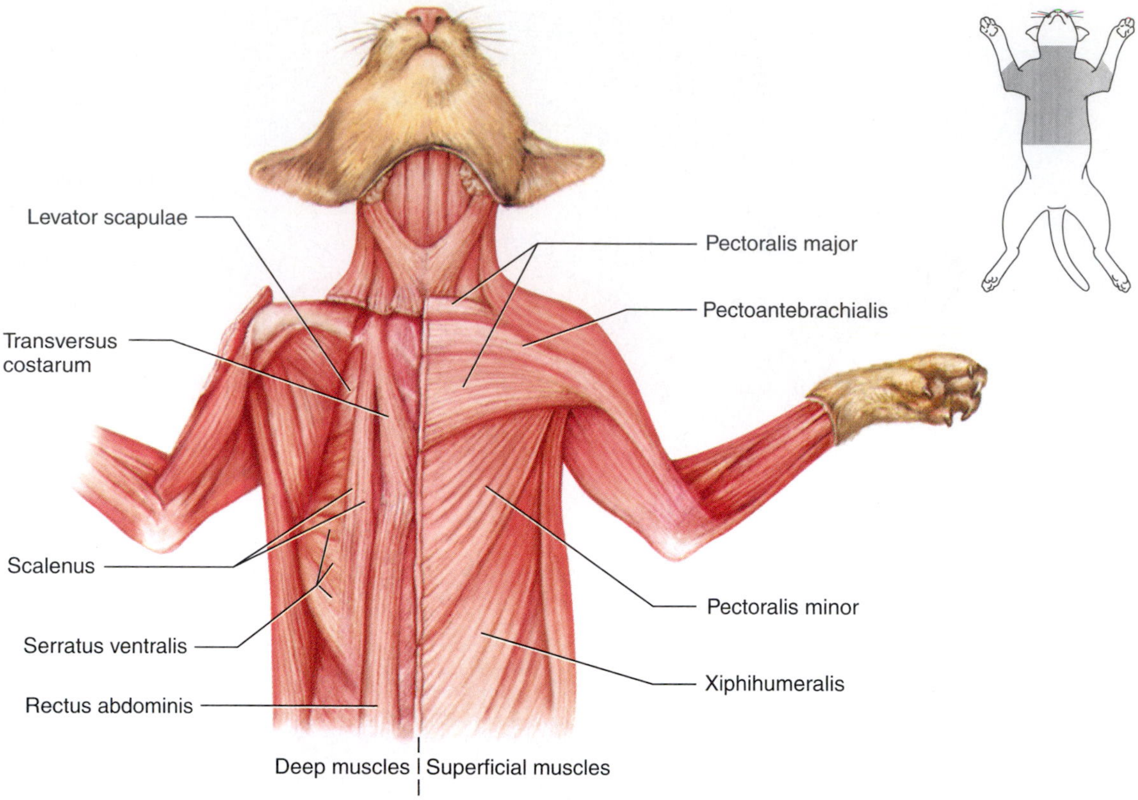

Figure 54.8 Superficial and deep muscles of the cat's thorax.

Table 54.2 Muscles of the Thorax

Muscle	Origin	Insertion	Action
Pectoantebrachialis	Manubrium of sternum	Fascia of forelimb	Adducts forelimb
Pectoralis major	Manubrium of sternum and costal cartilages	Greater tubercle of humerus	Adducts and rotates forelimb
Pectoralis minor	Sternum	Proximal end of humerus	Adducts and rotates forelimb
Xiphihumeralis	Xiphoid process of sternum	Proximal end of humerus	Adducts forelimb
Scalenus	Ribs	Cervical vertebrae	Flexes neck
Serratus ventralis	Ribs	Scapula	Pulls scapula ventrally and posteriorly
Levator scapulae	Cervical vertebrae	Scapula	Pulls scapula ventrally and anteriorly
Transversus costarum	Sternum	First rib	Pulls sternum toward head

EXPLORE

Procedure E—Muscles of the Abdominal Wall

1. Study figures 54.6 and 54.9
2. Locate the *external oblique* muscle in the abdominal wall.
3. Make a shallow, longitudinal incision through the external oblique. Reflect the cut edge and expose the *internal oblique* muscle beneath (fig. 54.6). The fibers of the internal oblique run at a right angle to those of the external oblique.
4. Make a longitudinal incision through the internal oblique. Reflect the cut edge and expose the *transversus abdominis* muscle.
5. Expose the *rectus abdominis* muscle on one side of the midventral line. This muscle lies beneath an aponeurosis.
6. See table 54.3 for the origins, insertions, and actions of these muscles.
7. Complete Part B of the laboratory report.

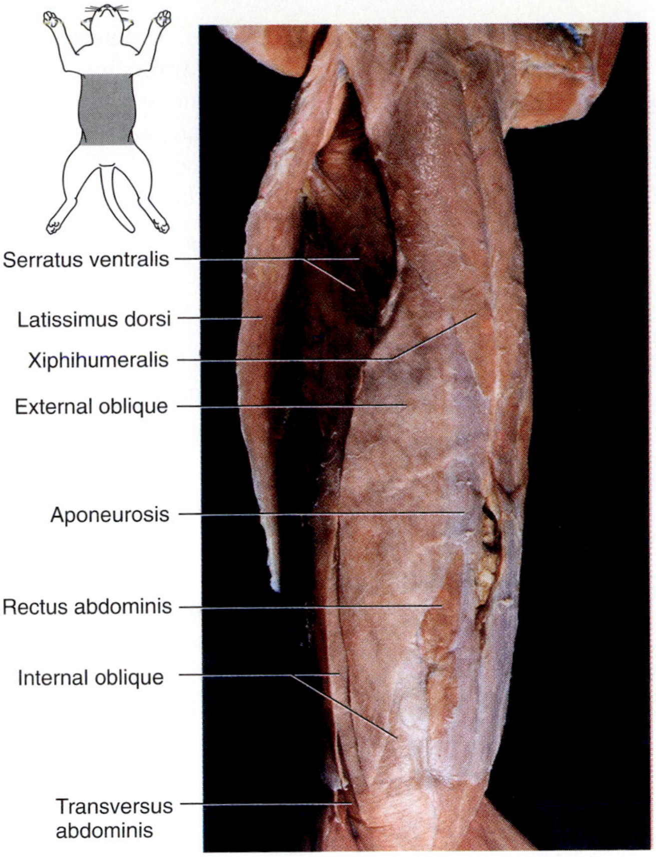

Figure 54.9 Muscles of the cat's thorax and abdomen.

Table 54.3 Muscles of the Abdominal Wall

Muscle	Origin	Insertion	Action
External oblique	Ribs and fascia of back	Linea alba	Compresses abdominal wall
Internal oblique	Fascia of back	Linea alba	Compresses abdominal wall
Transversus abdominis	Lower ribs	Linea alba	Compresses abdominal wall
Rectus abdominis	Symphysis pubis	Sternum and costal cartilages	Compresses abdominal wall and flexes trunk

EXPLORE

Procedure F—Muscles of the Shoulder and Back

1. Place the cat in the dissecting tray with its dorsal surface up.
2. Remove any remaining fat and connective tissue to expose the muscles of the shoulder and back.
3. Study figures 54.10 and 54.11, and then locate and dissect the following superficial muscles:

   clavotrapezius
   acromiotrapezius
   spinotrapezius
   clavodeltoid
   acromiodeltoid
   spinodeltoid
   latissimus dorsi

   Note: The clavodeltoid (clavobrachialis) appears as a continuation of the clavotrapezius.

4. Using scissors, transect the latissimus dorsi and the group of trapezius muscles. Reflect their cut edges and remove any underlying fat and connective tissue. Study figures 54.10 and 54.12, and then locate and dissect the following deep muscles of the shoulder and back:

   supraspinatus
   infraspinatus
   teres major
   rhomboid
   rhomboid capitis
   splenius

5. See table 54.4 for the origins, insertions, and actions of these muscles in the shoulder and back.
6. Review the locations, origins, insertions, and actions designated by the laboratory instructor without the aid of the figures and tables. 1 2
7. Complete Part C of the laboratory report.

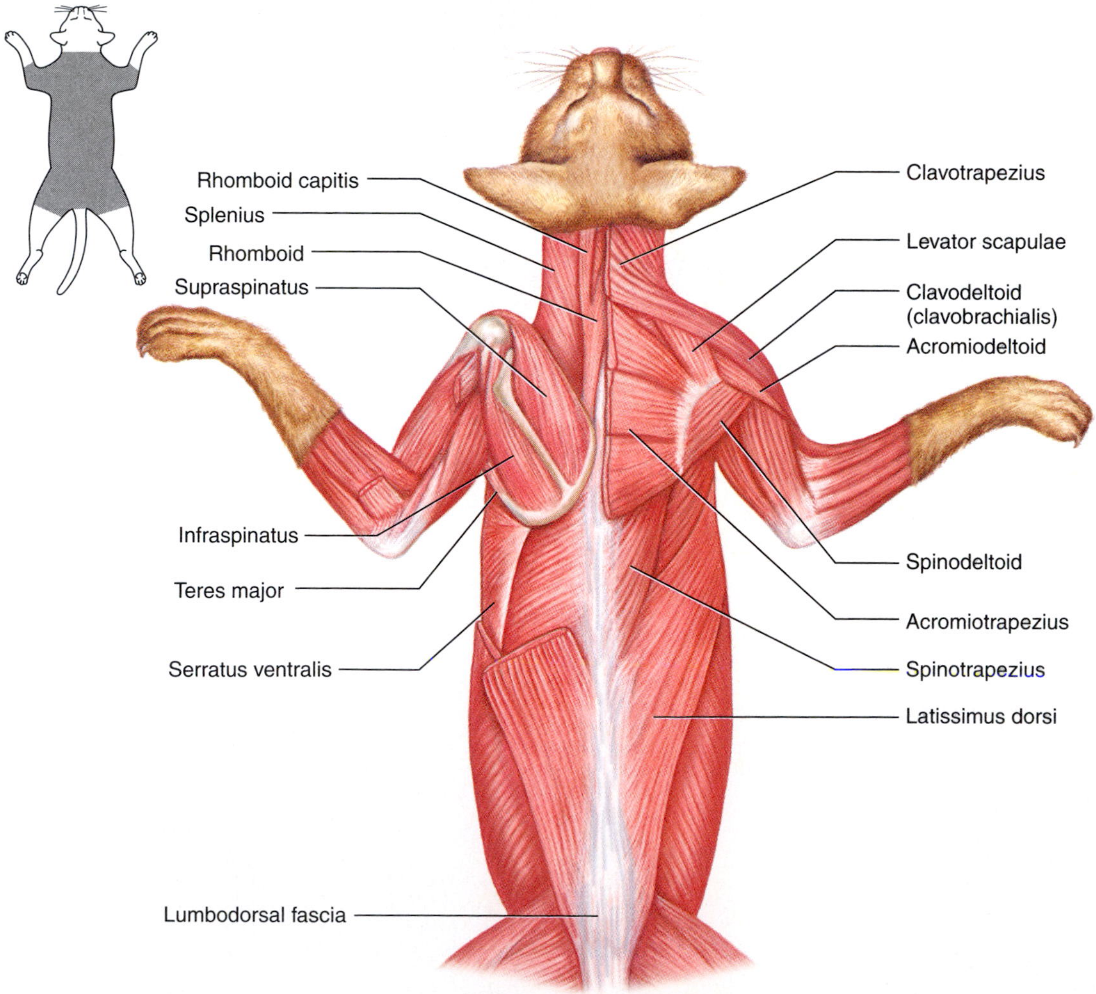

Figure 54.10 Superficial muscles of the cat's shoulder and back (*right side*); deep muscles of the shoulder and back (*left side*).

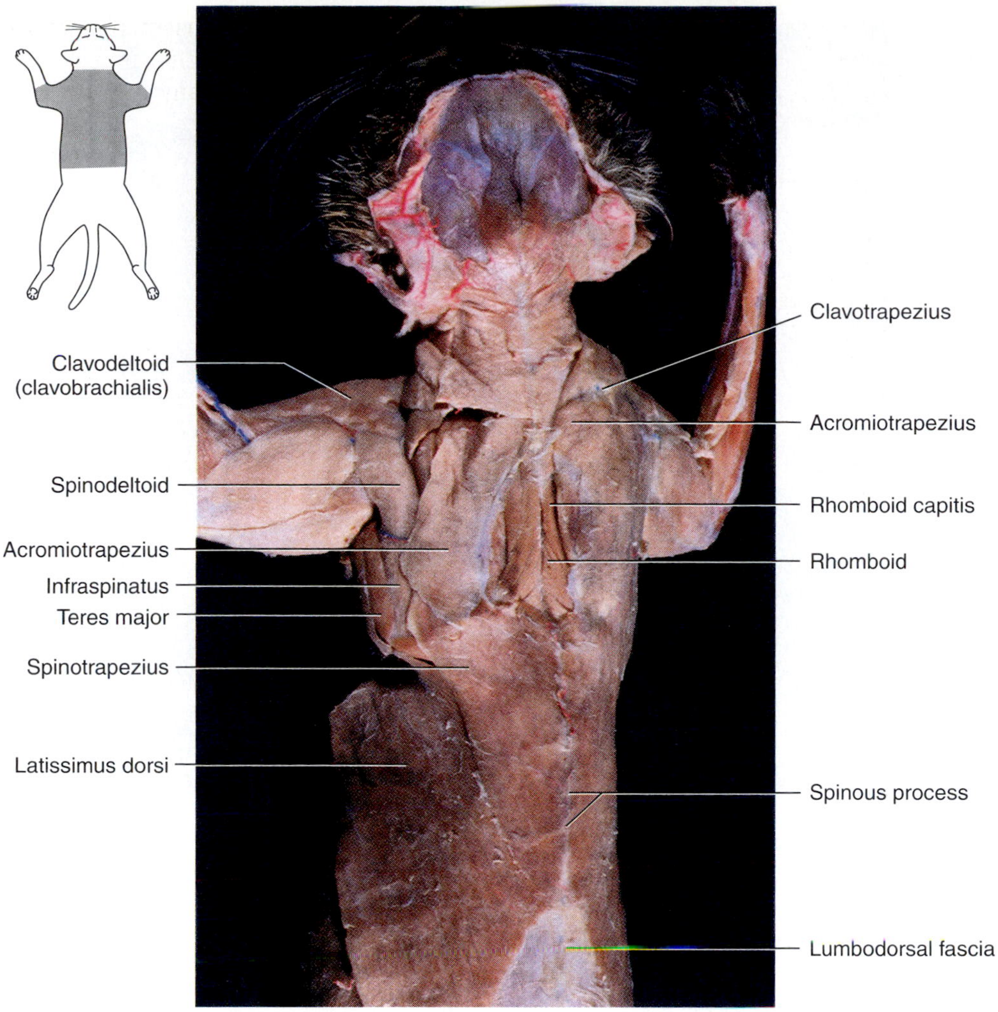

Figure 54.11 Muscles of the cat's shoulder and back.

Table 54.4 Muscles of the Shoulder and Back

Muscle	Origin	Insertion	Action
Clavotrapezius	Occipital bone of skull and spines of cervical vertebrae	Clavicle	Pulls clavicle upward
Acromiotrapezius	Spines of cervical and thoracic vertebrae	Spine and acromion process of scapula	Pulls scapula upward
Spinotrapezius	Spines of thoracic vertebrae	Spine of scapula	Pulls scapula upward and back
Clavodeltoid	Clavicle	Ulna near semilunar notch	Flexes forelimb
Acromiodeltoid	Acromion process of scapula	Proximal end of humerus	Flexes and rotates forelimb
Spinodeltoid	Spine of scapula	Proximal end of humerus	Flexes and rotates forelimb
Latissimus dorsi	Thoracic and lumbar vertebrae	Proximal end of humerus	Pulls forelimb upward and back
Supraspinatus	Fossa above spine of scapula	Greater tubercle of humerus	Extends forelimb
Infraspinatus	Fossa below spine of scapula	Greater tubercle of humerus	Rotates forelimb
Teres major	Posterior border of scapula	Proximal end of humerus	Rotates forelimb
Rhomboid	Spines of cervical and thoracic vertebrae	Medial border of scapula	Pulls scapula back
Rhomboid capitis	Occipital bone of skull	Medial border of scapula	Pulls scapula forward
Splenius	Fascia of neck	Occipital bone	Turns and raises head

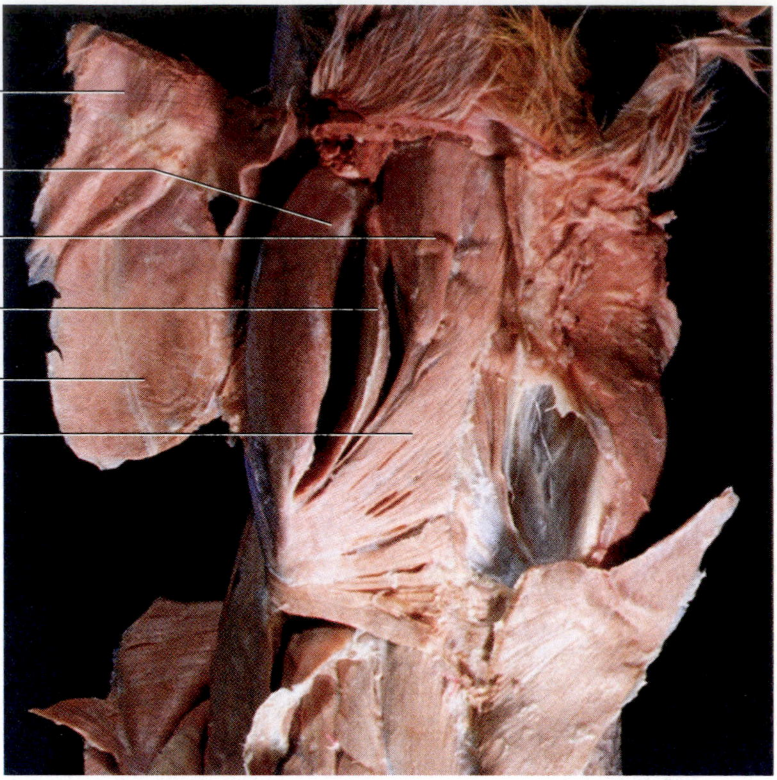

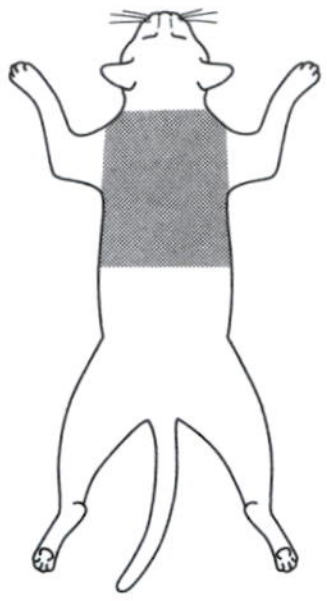

Figure 54.12 Deep muscles of the cat's back.

EXPLORE

Procedure G—Muscles of the Forelimb

1. Place the cat in the dissecting tray with its ventral surface up.
2. Remove any remaining fat and connective tissue from a forelimb to expose the muscles.
3. Study figure 54.13*a*, and then locate and dissect the following muscles from the medial surface of the upper forelimb:

   epitrochlearis
   triceps brachii

4. Transect the pectoral muscles (pectoantebrachialis, pectoralis major, and pectoralis minor) and reflect their cut edges.
5. Study figures 54.13*b* and 54.14, and then locate and dissect the following muscles:

   coracobrachialis
   biceps brachii
   long head of the triceps brachii
   medial head of the triceps brachii

6. Study figures 54.15*a* and 54.16, which show the lateral surface of the upper forelimb.
7. Locate the lateral head of the triceps muscle, transect it, and reflect its cut edges.
8. Study figure 54.15*b*. Locate and dissect the following deeper muscles:

   anconeus
   brachialis

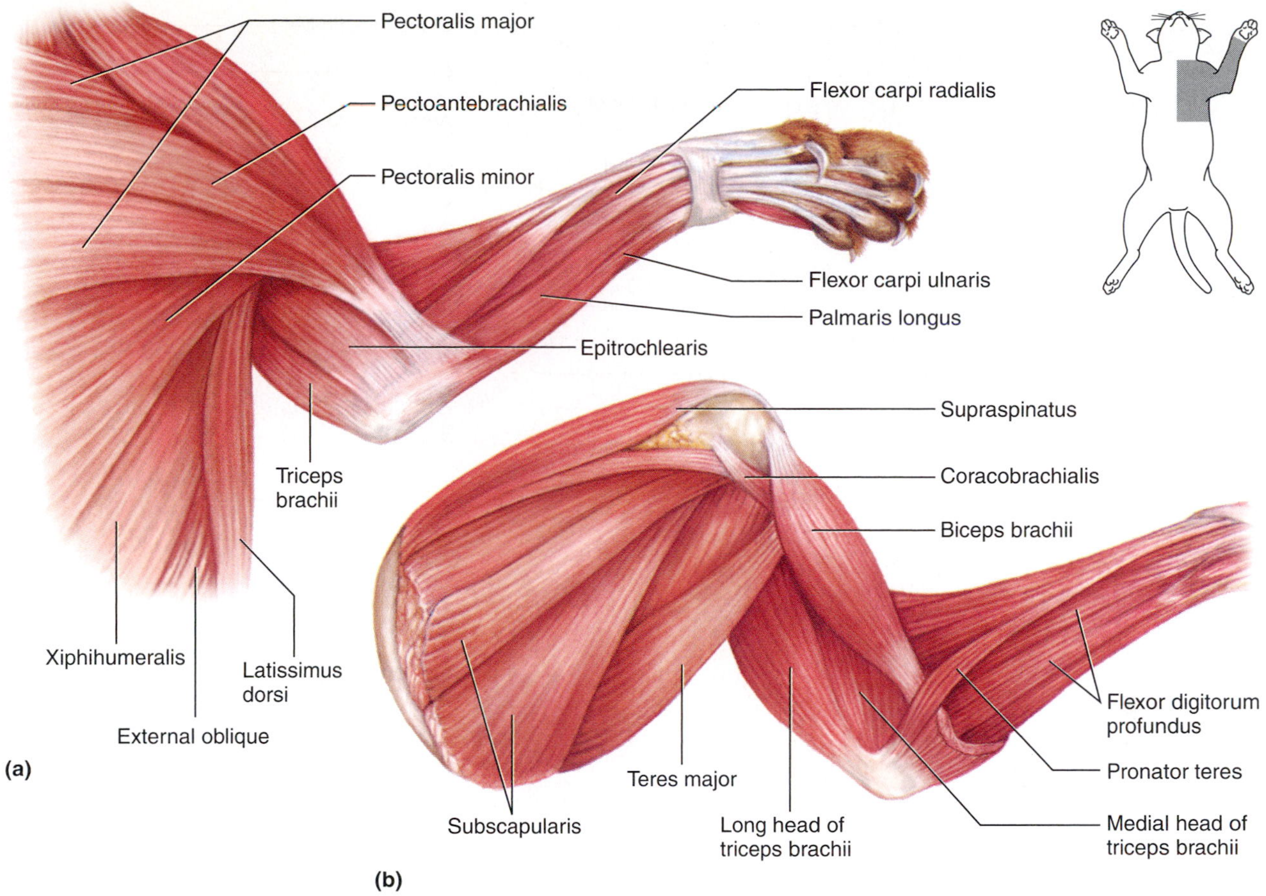

Figure 54.13 (*a*) Superficial muscles of the cat's left medial shoulder and forelimb. (*b*) Deep muscles of the left medial shoulder and forelimb.

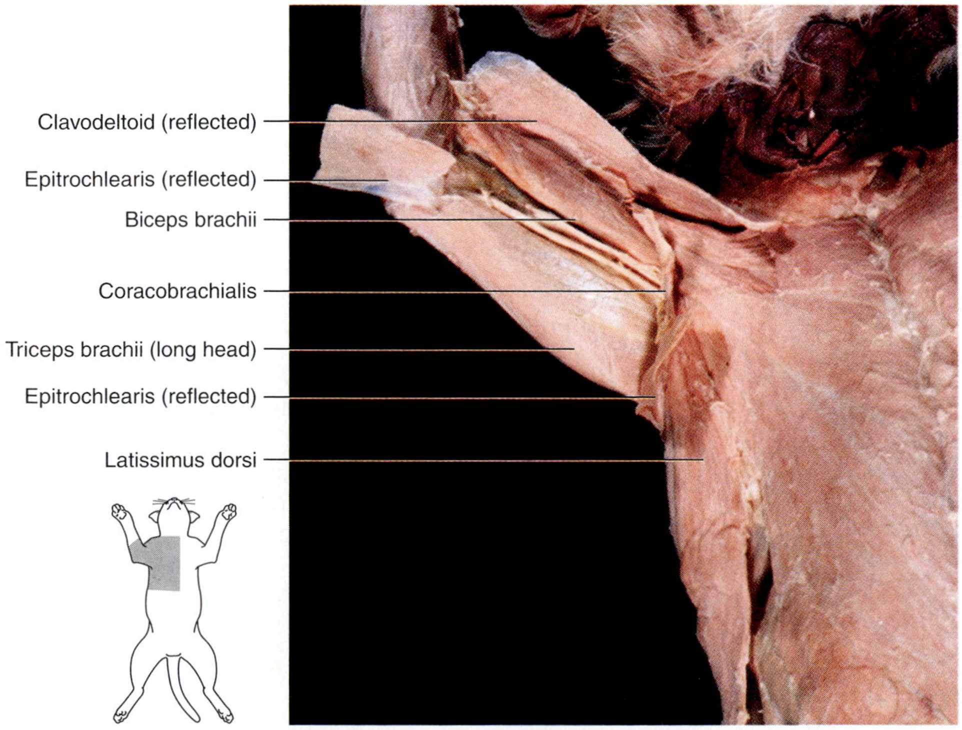

Figure 54.14 Deep muscles of the cat's right medial forelimb.

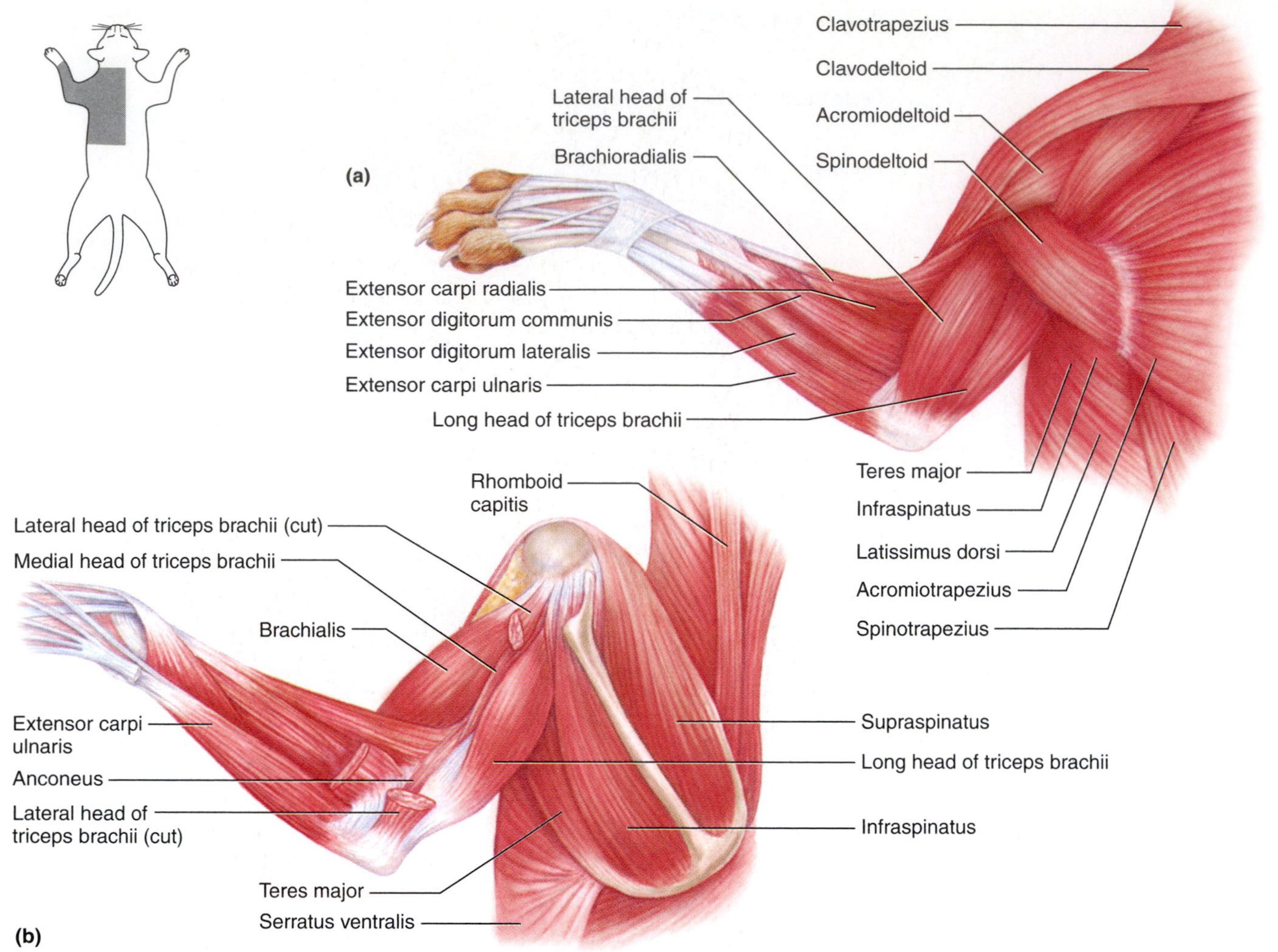

Figure 54.15 (*a*) Superficial muscles of the cat's left lateral shoulder and forelimb. (*b*) Deep muscles of the left lateral shoulder and forelimb.

9. Separate the muscles in the lower forelimb into an *extensor group* (fig. 54.15*a*) and a *flexor group* (fig. 54.13*a*). Generally, the extensors are located on the posterior surface of the lower forelimb, and the flexors are located on the anterior surface. A thick, tough layer of fascia (antebrachial fascia) surrounds the forelimb. Remove this fascia to expose and dissect the following muscles:

   brachioradialis
   extensor carpi radialis
   extensor digitorum communis
   extensor digitorum lateralis
   extensor carpi ulnaris
   flexor carpi ulnaris
   palmaris longus
   flexor carpi radialis
   pronator teres

10. Transect the flexor carpi ulnaris, palmaris longus, and flexor carpi radialis to expose the flexor digitorum profundus.
11. See table 54.5 for the origins, insertions, and actions of these muscles of the forelimb.
12. Complete Part D of the laboratory report.

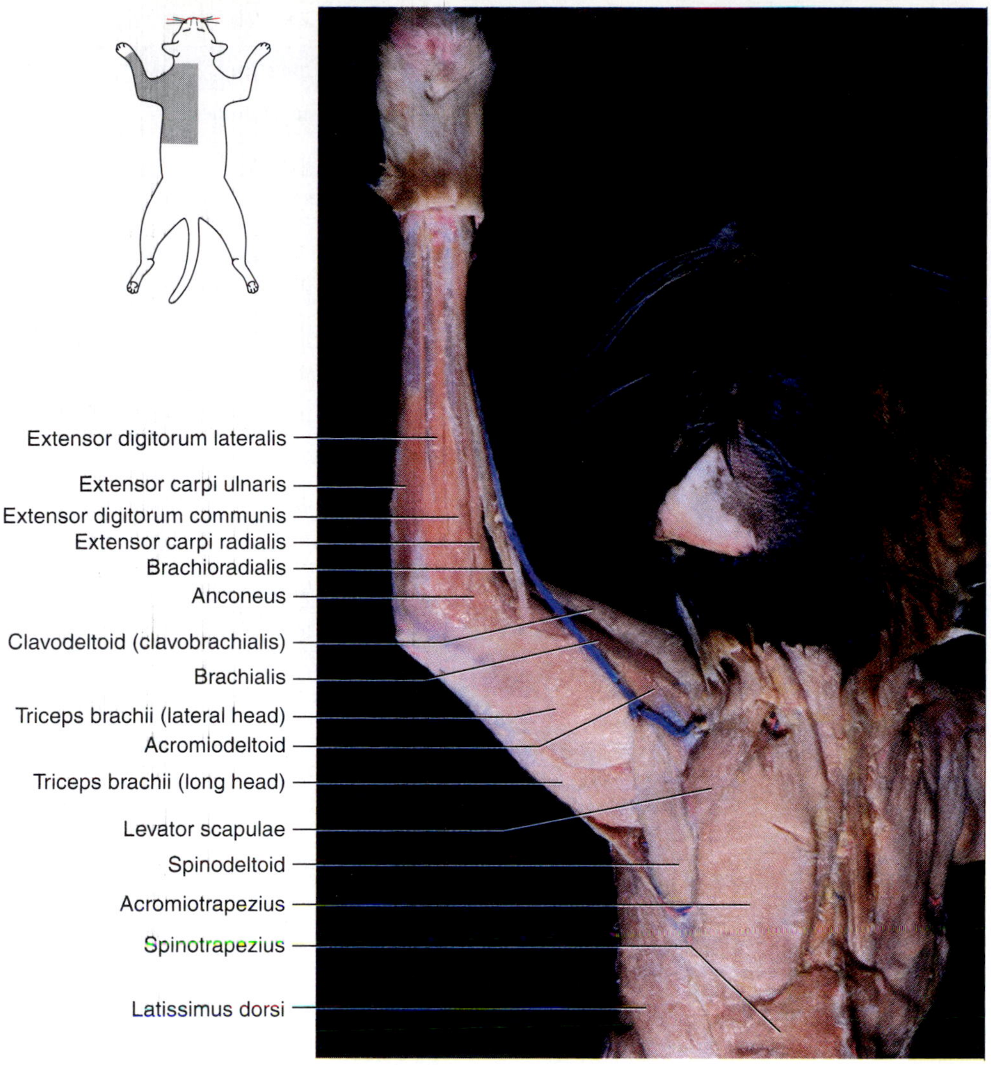

Figure 54.16 Superficial muscles of the cat's left lateral shoulder and forelimb, dorsal view.

Table 54.5 Muscles of the Forelimb

Muscle	Origin	Insertion	Action
Coracobrachialis	Coracoid process of scapula	Proximal end of humerus	Adducts forelimb
Epitrochlearis	Fascia of latissimus dorsi	Olecranon process of ulna	Extends forelimb
Biceps brachii	Border of glenoid cavity of scapula	Radial tuberosity of radius	Flexes forelimb
Triceps brachii			
Lateral head	Deltoid tuberosity of humerus	Olecranon process of ulna	Extends forelimb
Long head	Border of glenoid cavity of scapula	Olecranon process of ulna	Extends forelimb
Medial head	Shaft of humerus	Olecranon process of ulna	Extends forelimb
Anconeus	Lateral epicondyle of humerus	Olecranon process of ulna	Extends forelimb
Brachialis	Lateral surface of humerus	Proximal end of ulna	Flexes forelimb
Brachioradialis	Shaft of humerus	Distal end of radius	Supinates lower forelimb
Extensor carpi radialis	Lateral surface of humerus	Second and third metacarpals	Extends wrist
Extensor digitorum communis	Lateral surface of humerus	Digits two to five	Extends digits
Extensor digitorum lateralis	Lateral surface of humerus	Digits two to five	Extends digits
Extensor carpi ulnaris	Lateral epicondyle of humerus and ulna	Fifth metacarpal	Extends wrist
Flexor carpi ulnaris	Medial epicondyle of humerus	Carpals	Flexes wrist
Palmaris longus	Medial epicondyle of humerus	Digits	Flexes digits
Flexor carpi radialis	Medial epicondyle of humerus	Second and third metacarpals	Flexes metacarpals
Pronator teres	Medial epicondyle of humerus	Radius	Pronates lower forelimb
Flexor digitorum profundus	Radius, ulna, and medial epicondyle of humerus	Distal phalanges	Flexes digits

EXPLORE

Procedure H—Muscles of the Hip and Hindlimb

1. Place the cat in the dissecting tray with its ventral surface up.
2. Remove any remaining fat and connective tissue from the hip and hindlimb to expose the muscles.
3. Study figures 54.17*a* and 54.18, and then locate and dissect the following muscles from the medial surface of the thigh:

   sartorius
   gracilis

4. Using scissors, transect the sartorius and gracilis, and reflect their cut edges to observe the deeper muscles of the thigh.
5. Study figures 54.17*b* and 54.19, and then locate and dissect the following muscles:

   rectus femoris
   vastus medialis
   adductor longus
   adductor femoris
   semimembranosus
   semitendinosus
   tensor fasciae latae

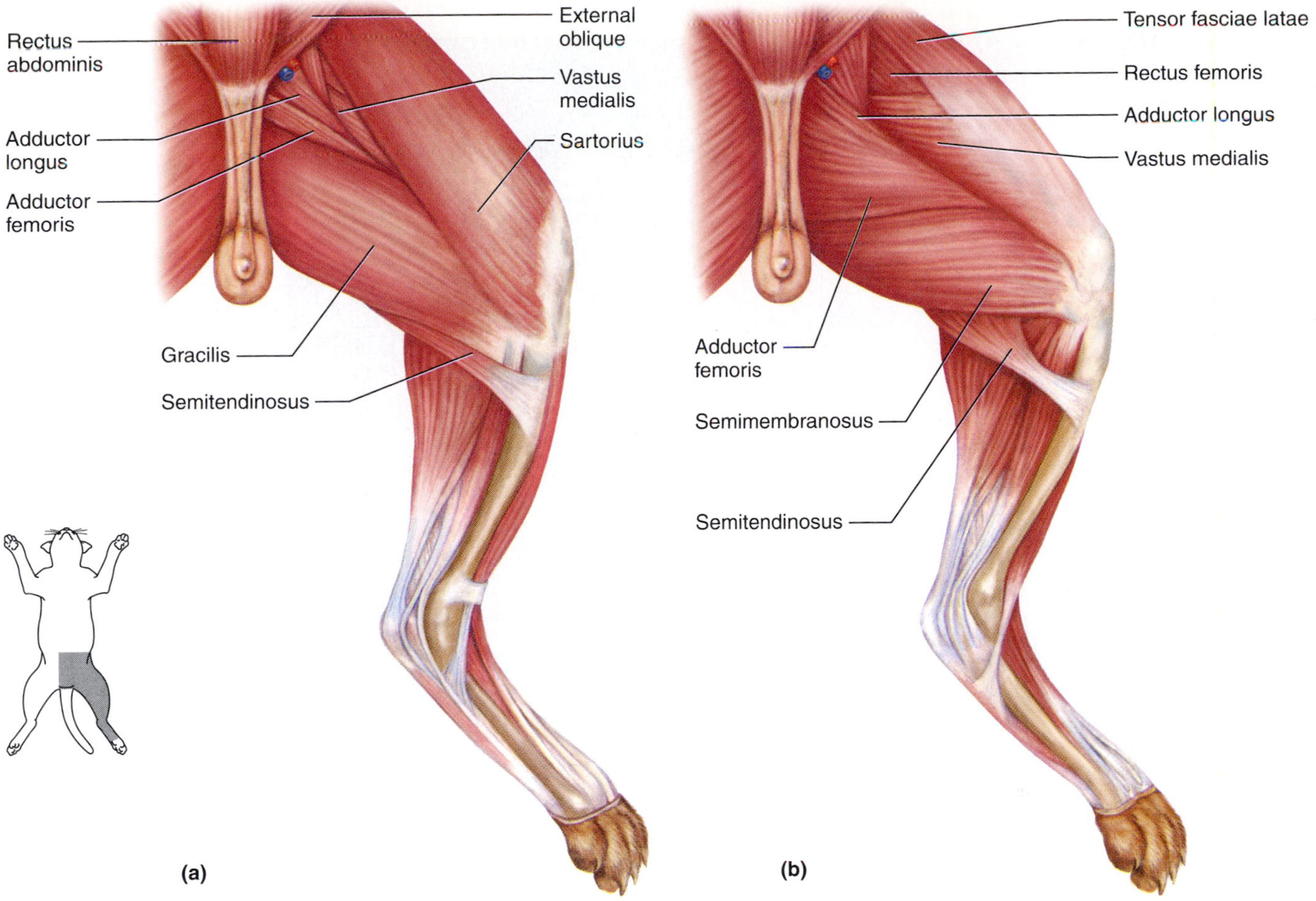

Figure 54.17 (*a*) Superficial muscles of the cat's left medial hindlimb. (*b*) Deep muscles of the left medial hindlimb.

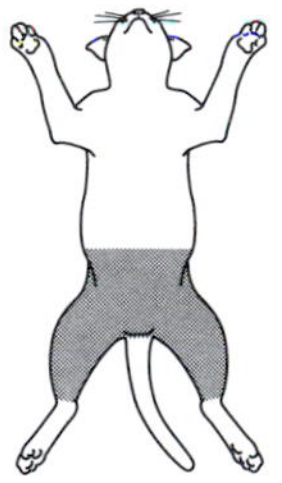

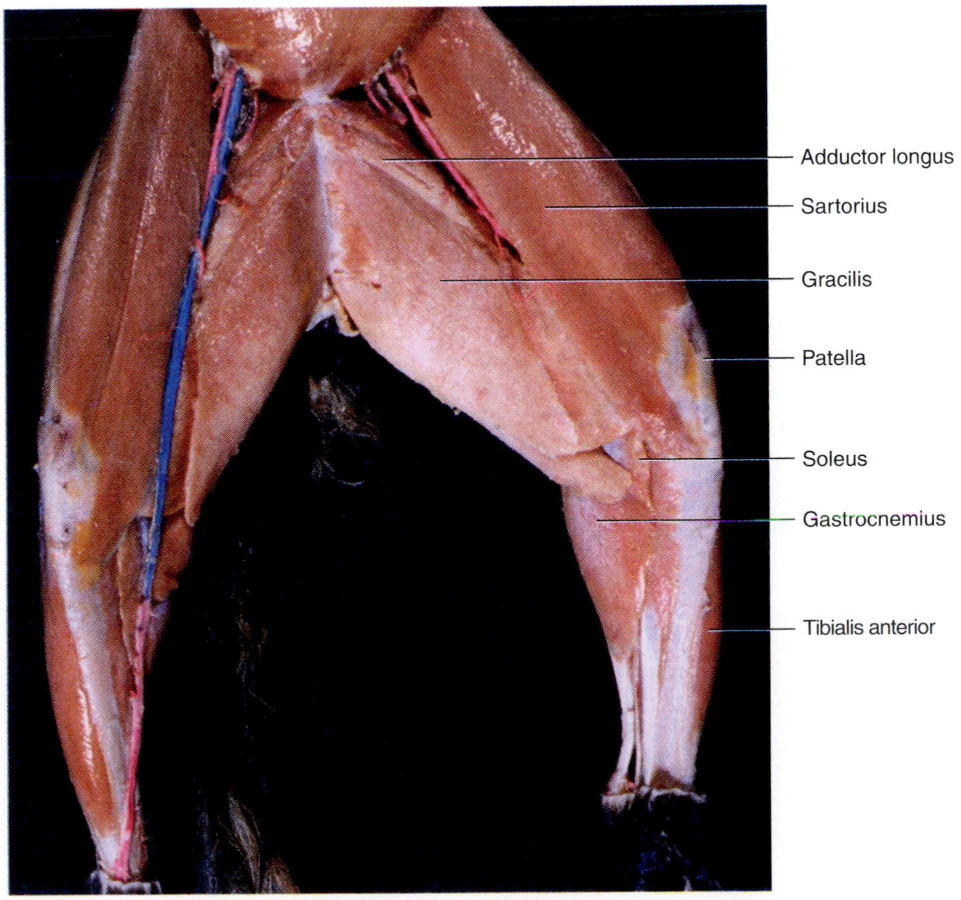

Figure 54.18 Superficial muscles of the cat's hindlimb, medial view.

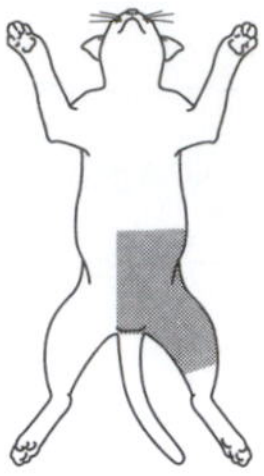

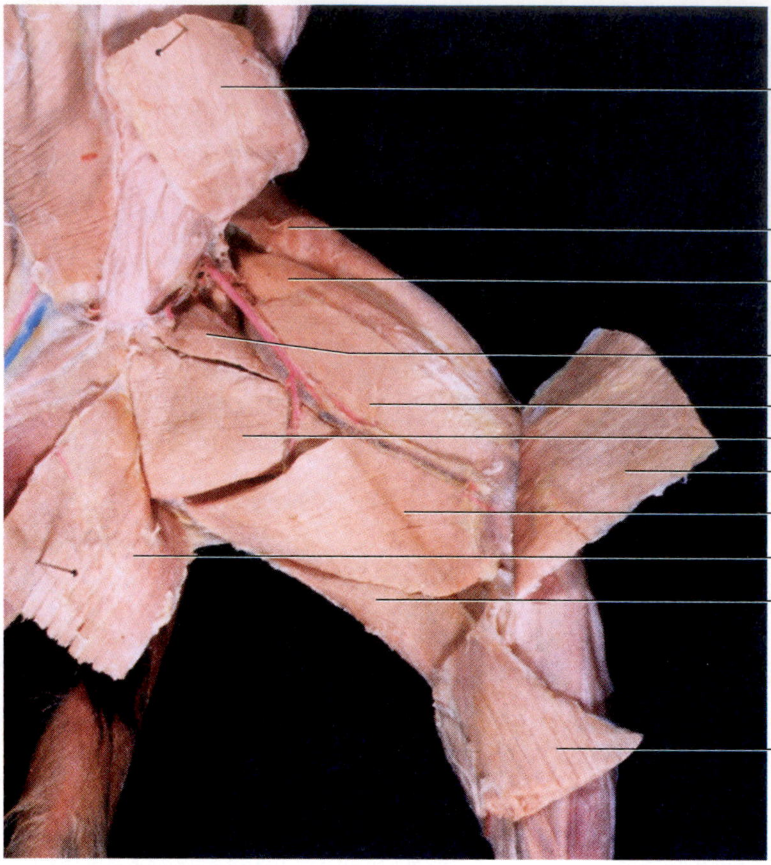

Figure 54.19 Deep muscles of the cat's left hindlimb, medial view.

6. Transect the tensor fasciae latae and rectus femoris muscles and turn their ends aside. Locate the *vastus intermedius* and *vastus lateralis* muscles beneath (fig. 54.20). (*Note:* In some specimens, the vastus intermedius, vastus lateralis, and vastus medialis muscles are closely united by connective tissue and are difficult to separate.)
7. Study figures 54.21*a* and 54.22, and then locate and dissect the following muscles from the lateral surface of the hip and thigh:

   biceps femoris
   caudofemoralis
   gluteus maximus

8. Using scissors, transect the tensor fasciae latae and biceps femoris, and reflect their cut edges to observe the deeper muscles of the thigh.
9. Locate and dissect the following muscles:

   tenuissimus
   gluteus medius

   (*Note:* The tenuissimus muscle and the sciatic nerve may adhere tightly to the dorsal surface of the biceps femoris muscle. Take care to avoid cutting these structures.)

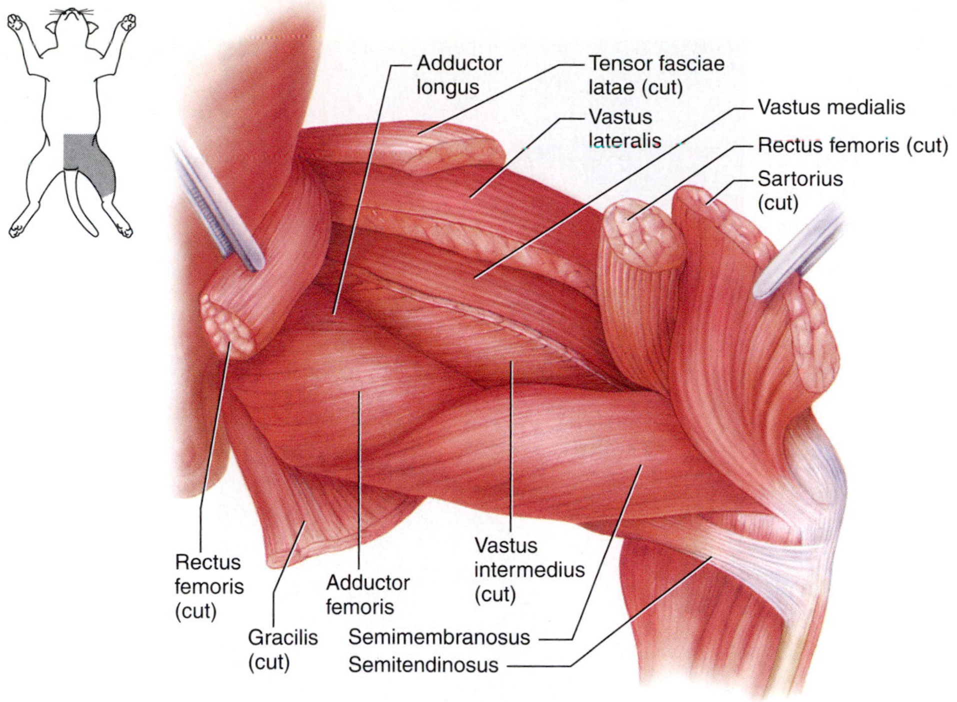

Figure 54.20 Other deep muscles of the cat's left thigh, medial view.

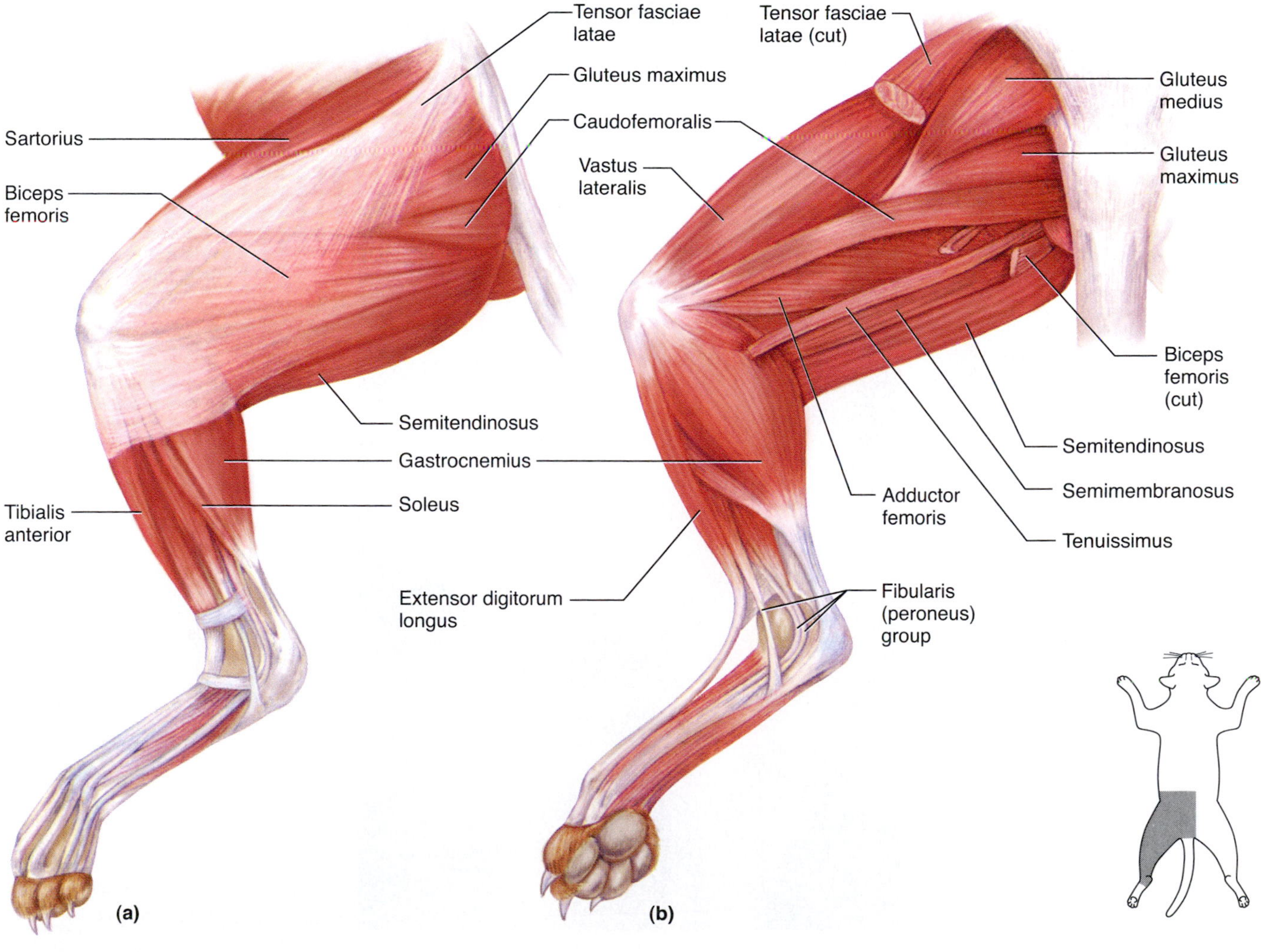

Figure 54.21 (*a*) Superficial muscles of the cat's left lateral hip and hindlimb. (*b*) Deep muscles of the left lateral hip and hindlimb.

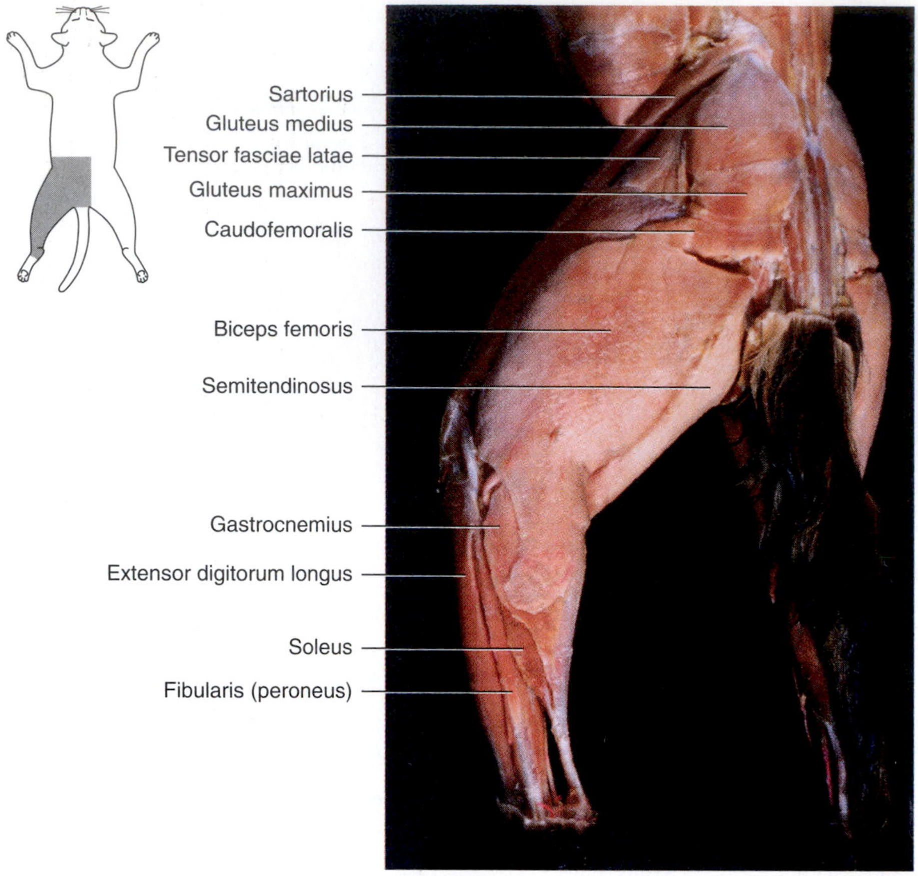

Figure 54.22 Muscles of the cat's left hip and hindlimb, lateral view.

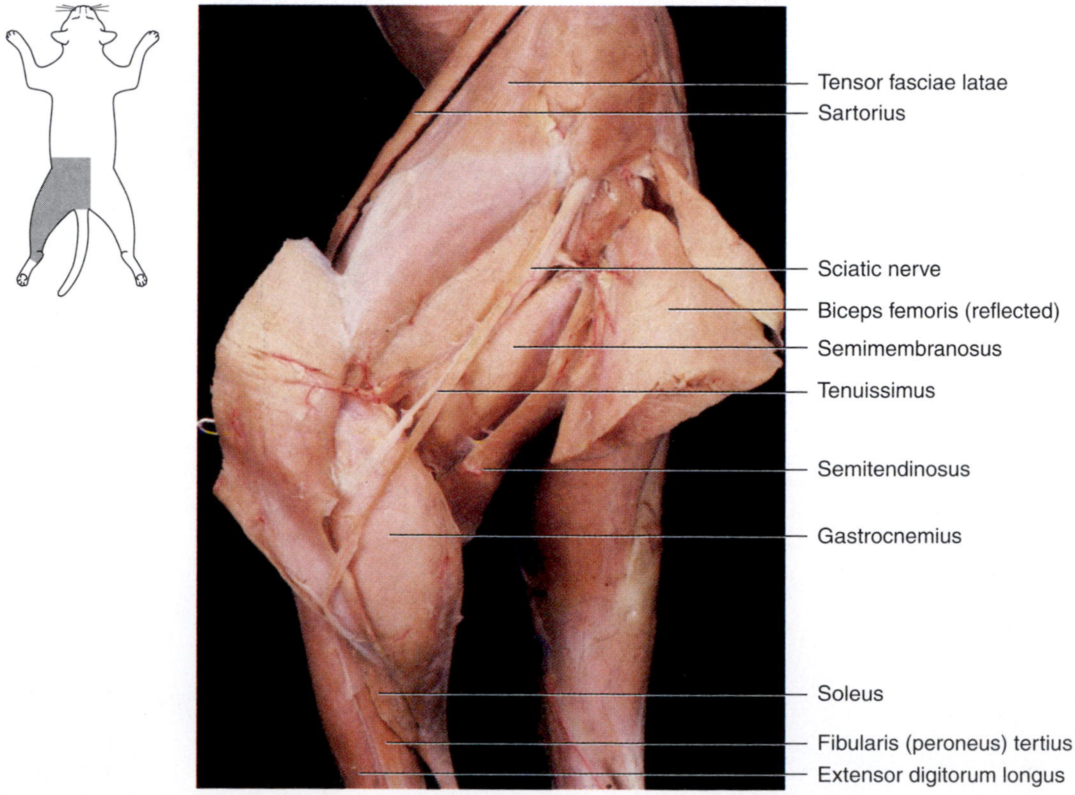

Figure 54.23 Deep muscles of the cat's left hindlimb, lateral view.

10. On the lateral surface of the lower hindlimb (figs. 54.21 and 54.23), locate and dissect the following muscles:

gastrocnemius
soleus
tibialis anterior
fibularis (peroneus) group

11. See table 54.6 for the origins, insertions, and actions of these muscles of the hip and hindlimb.
12. Review the locations, origins, insertions, and actions designated by the laboratory instructor without the aid of the figures and tables. 1 2
13. Complete Part E of the laboratory report.

Table 54.6 Muscles of the Hip and Hindlimb

Muscle	Origin	Insertion	Action
Sartorius	Crest of ilium	Fascia of knee and proximal end of tibia	Rotates and extends hindlimb
Gracilis	Symphysis pubis and ischium	Proximal end of tibia	Adducts hindlimb
Quadriceps femoris			
Vastus lateralis	Shaft of femur and greater trochanter	Patella	Extends hindlimb
Vastus intermedius	Shaft of femur	Patella	Extends hindlimb
Rectus femoris	Ilium	Patella	Extends hindlimb
Vastus medialis	Shaft of femur	Patella	Extends hindlimb
Adductor longus	Pubis	Proximal end of femur	Adducts hindlimb
Adductor femoris	Pubis and ischium	Shaft of femur	Adducts hindlimb
Semimembranosus	Ischial tuberosity	Medial epicondyle of femur	Extends thigh
Tensor fasciae latae	Ilium	Fascia of thigh	Extends thigh
Biceps femoris	Ischial tuberosity	Patella and tibia	Abducts thigh and flexes lower hindlimb
Tenuissimus	Second caudal vertebra	Tibia and fascia of biceps femoris	Abducts thigh and flexes lower hindlimb
Semitendinosus	Ischial tuberosity	Crest of tibia	Flexes lower hindlimb
Caudofemoralis	Caudal vertebrae	Patella	Abducts thigh and extends lower hindlimb
Gluteus maximus	Sacral and caudal vertebrae	Greater trochanter of femur	Abducts thigh
Gluteus medius	Ilium, and sacral and caudal vertebrae	Greater trochanter of femur	Abducts thigh
Gastrocnemius	Lateral and medial epicondyles of femur	Calcaneus	Extends foot
Soleus	Proximal end of fibula	Calcaneus	Extends foot
Tibialis anterior	Proximal end of tibia and fibula	First metatarsal	Flexes foot
Fibularis (peroneus) group	Shaft of fibula	Metatarsals	Flexes foot

Notes

Name ______________________________

Date ______________________________

Section ______________________________

The [A] corresponds to the indicated Learning Outcome(s) listed at the beginning of the laboratory exercise.

Cat Dissection: Musculature

Laboratory Report 54

Part A Assessments

Complete the following statements:

1. The ____________________ muscle of the human is homologous to the sternomastoid muscle of the cat. [3]
2. The ____________________ muscle elevates the mandible in the human and in the cat. [2] [3]
3. Two muscles of the cat that are inserted on the hyoid bone are the ____________________ and the ____________________. [1] [2]

Part B Assessments

Complete the following:

1. Name two muscles that are found in the thoracic wall of the cat but are absent in the human. [1] [3]

   ____________________ ____________________

2. Name two muscles that are found in the thoracic wall of the cat and the human. [3]

   ____________________ ____________________

3. Name four muscles that are found in the abdominal wall of the cat and the human. [3]

   ____________________ ____________________

   ____________________ ____________________

Part C Assessments

Complete the following:

1. Name three muscles of the cat that together correspond to the trapezius muscle in the human. [1] [3]

   ____

   ____

   ____

2. Name three muscles of the cat that together correspond to the deltoid muscle in the human. [1] [3]

   ____

   ____

   ____

3. Name the muscle in the cat and in the human that occupies the fossa above the spine of the scapula. [1] [3]

   ____

4. Name the muscle in the cat and in the human that occupies the fossa below the spine of the scapula. 1 3

____

5. Name two muscles found in the cat and in the human that can rotate the forelimb. 2 3

_______________ _______________

Part D Assessments

Complete the following:

1. Name two muscles found in the cat and in the human that can flex the forelimb. 2 3

_______________ _______________

2. Name a muscle found in the cat but absent in the human that can extend the forelimb. 2 3

____

3. Name a muscle that has three heads, can extend the forelimb, and is found in the cat and in the human. 2 3

____

Part E Assessments

Observe the muscles on the human torso model and the upper and lower limb models. Each of the muscles in figure 54.24 is found both in the cat and in the human. Identify each of the numbered muscles in the figure by placing its name in the space next to its number.

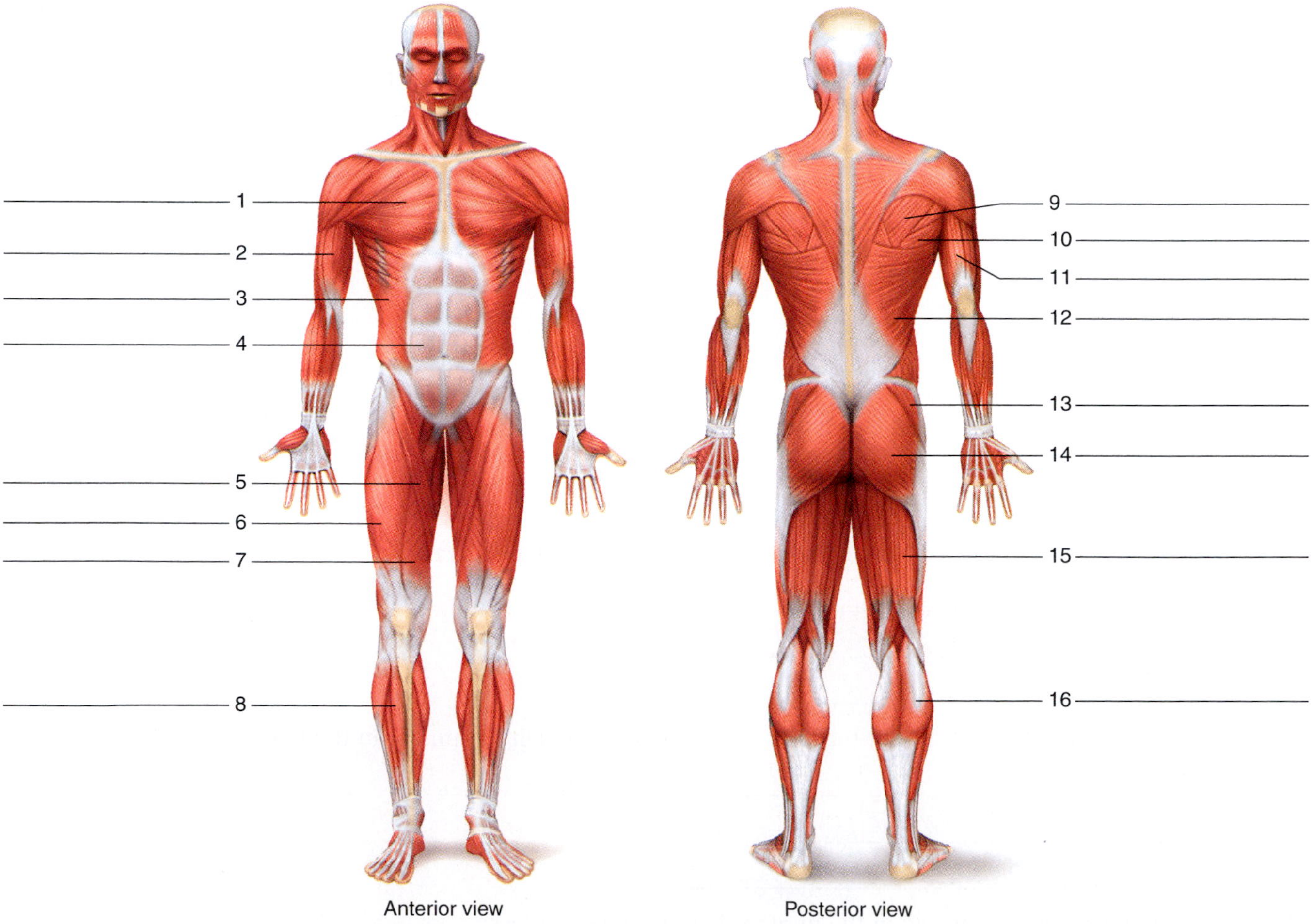

Figure 54.24 Identify the numbered muscles that occur in both the cat and the human. 3

Laboratory Exercise 55

Cat Dissection: Cardiovascular System

Materials Needed

Preserved cat
Dissecting tray
Dissecting instruments
Disposable gloves
Human torso model

Safety

- Wear disposable gloves when working on the cat dissection.
- Dispose of tissue remnants and gloves as instructed.
- Wash the dissecting tray and instruments as instructed.
- Wash your laboratory table.
- Wash your hands before leaving the laboratory.

In this laboratory exercise, you will dissect the major organs of the cardiovascular system of the cat. As before, while you are examining the organs of the cat, compare them with the corresponding organs of the human torso model.

If the cardiovascular system of the cat has been injected, the arteries will be filled with red latex, and the veins will be filled with blue latex. This will make it easier for you to trace the vessels as you dissect them.

Purpose of the Exercise

To examine the major organs of the cardiovascular system of the cat and to compare them with the corresponding organs of the human torso model.

LEARNING OUTCOMES

After completing this exercise, you should be able to

1. Locate and identify the major organs of the cardiovascular system of the cat.
2. Compare the features of the cardiovascular system of the cat with those of the human.
3. Identify the corresponding organs in the human torso model.

EXPLORE

Procedure A—The Arterial System

1. Place the preserved cat in the dissecting tray with its ventral surface up.
2. Open the thoracic cavity, and expose its contents. To do this, follow these steps:
   - **a.** Make a longitudinal incision passing anteriorly from the diaphragm along one side of the sternum. Continue the incision through the neck muscles to the mandible. Try to avoid damaging the internal organs as you cut.
   - **b.** Make a lateral cut on each side along the anterior surface of the diaphragm, and cut the diaphragm loose from the thoracic wall.
   - **c.** Spread the sides of the thoracic wall outward, and use a scalpel to make a longitudinal cut along each side of the inner wall of the rib cage to weaken the ribs. Continue to spread the thoracic wall laterally to break the ribs so that the flaps of the wall will remain open (figs. 55.1 and 55.2).
3. Note the location of the heart and the large blood vessels associated with it. Slit the outer *parietal pericardium* that surrounds the heart by cutting with scissors along the midventral line. Note how this membrane is connected to the *visceral pericardium* that is attached to the surface of the heart. Locate the *pericardial cavity,* the space between the two layers of the pericardium.
4. Examine the heart and closely associated blood vessels (figs. 55.1, 55.3, and 55.4). The arrangement of blood vessels coming off the aortic arch is different in the cat than in the human body. Locate the following:

   right atrium
   left atrium
   right ventricle
   left ventricle
   pulmonary trunk
   aorta
   coronary arteries

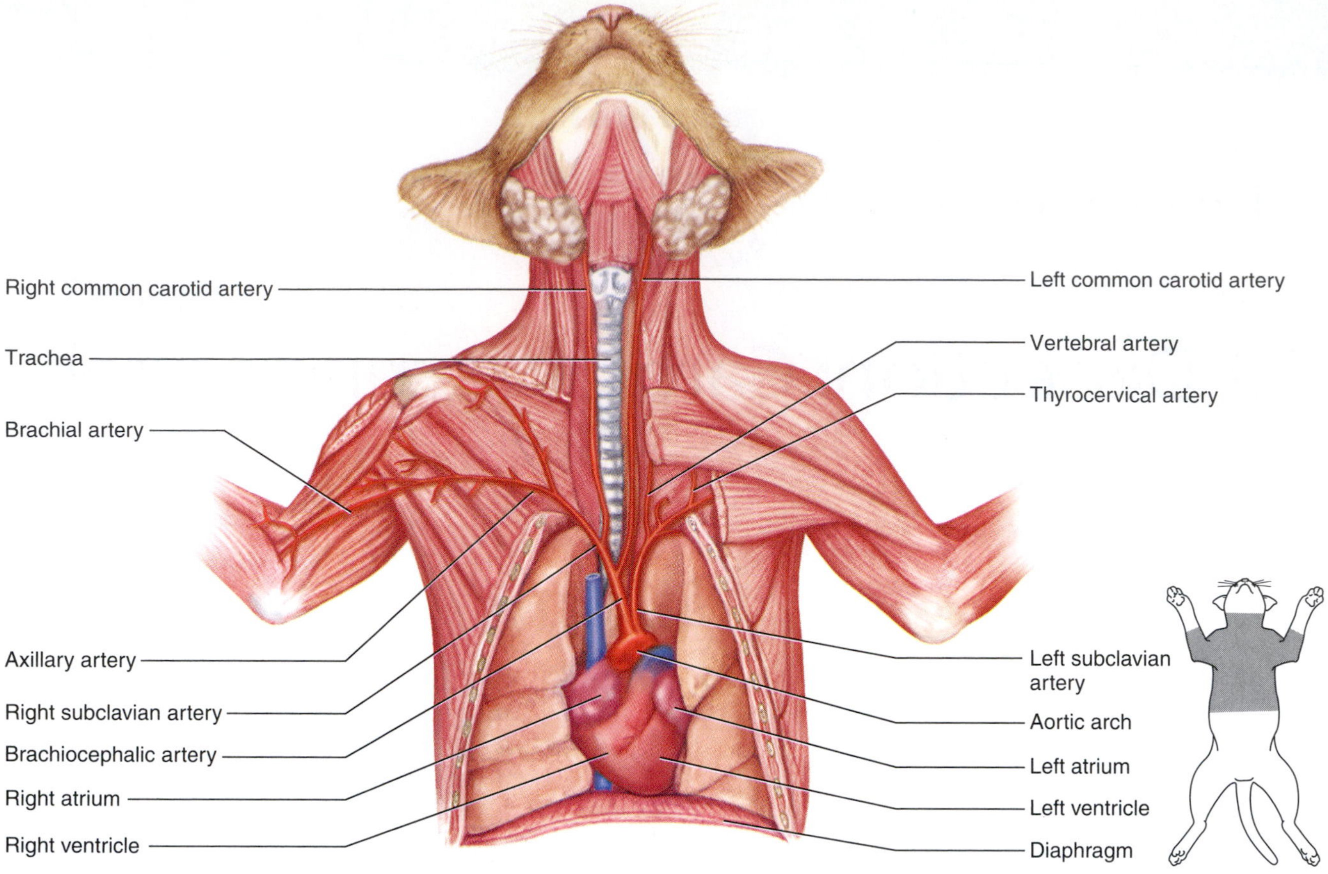

Figure 55.1 Arteries of the cat's thorax, neck, and forelimb.

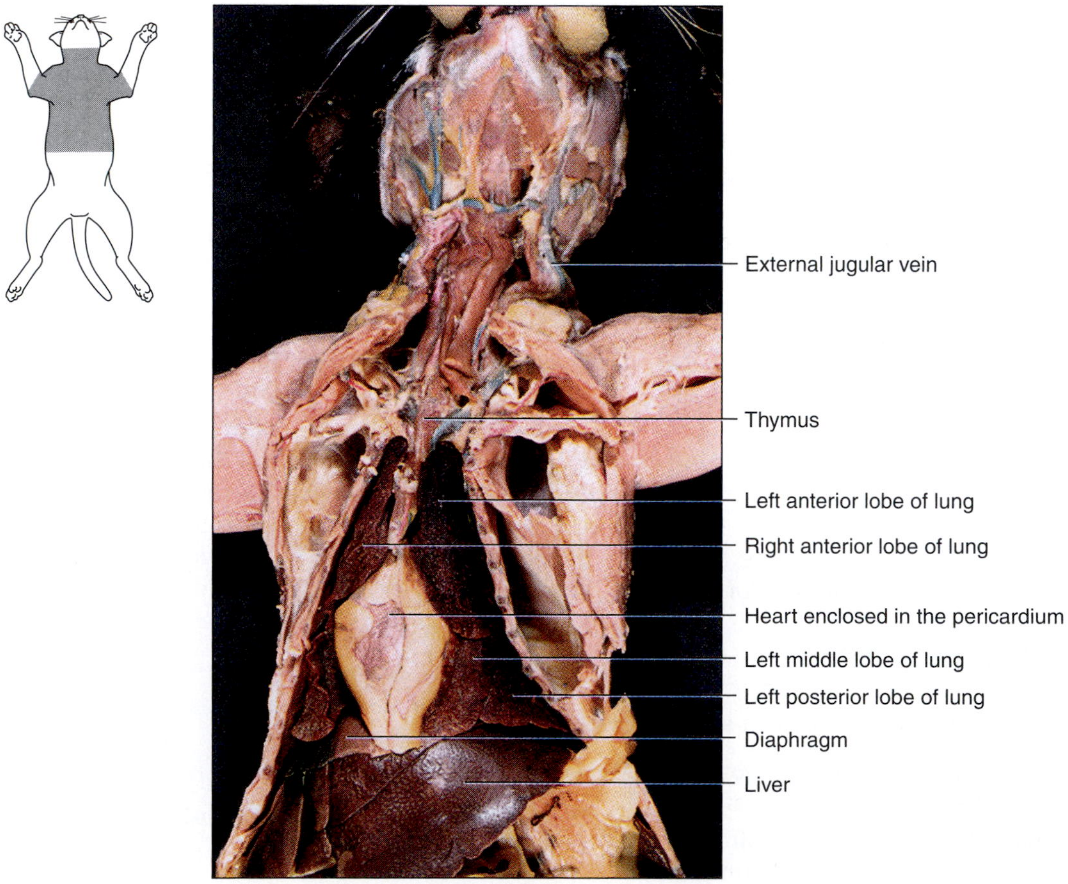

Figure 55.2 Features of the cat's neck and thoracic cavity.

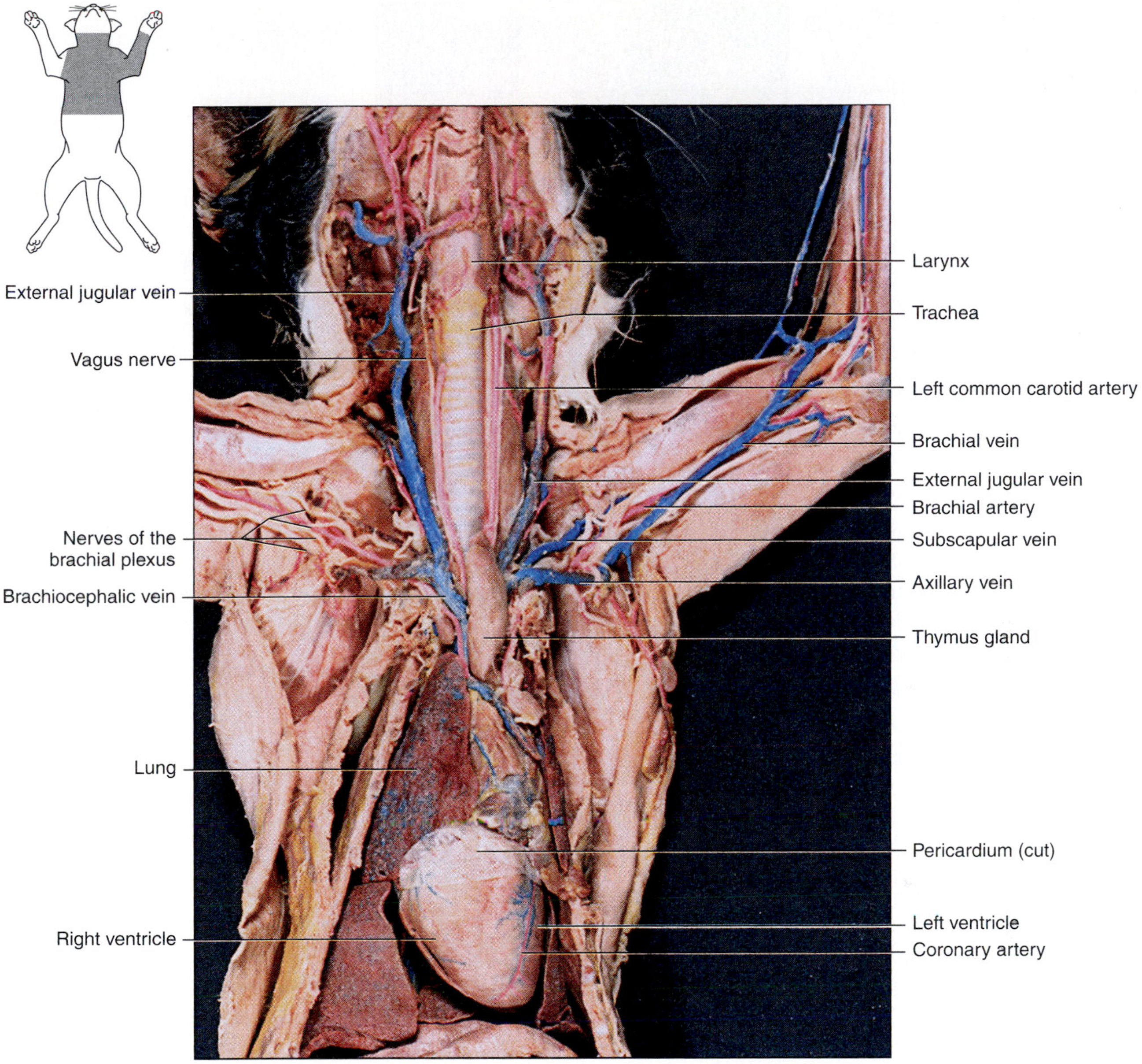

Figure 55.3 Blood vessels of the cat's upper forelimb and neck.

5. Use a scalpel to open the heart chambers by making a cut along the frontal plane from its apex to its base. Remove any remaining latex from the chambers. Examine the valves between the chambers, and note the relative thicknesses of the chamber walls (fig. 55.4). Do not remove the heart as it is needed for future examination of the relationship between major blood vessels.
6. Using figures 55.1, 55.3, and 55.4 as guides, locate and dissect the following arteries of the thorax and neck:

aortic arch
brachiocephalic artery (on right only; first branch of the aortic arch)
right subclavian artery
left subclavian artery
right common carotid artery
left common carotid artery

7. Trace the right subclavian artery into the forelimb, and locate the following arteries:

vertebral artery
thyrocervical artery
subscapular artery
axillary artery
brachial artery

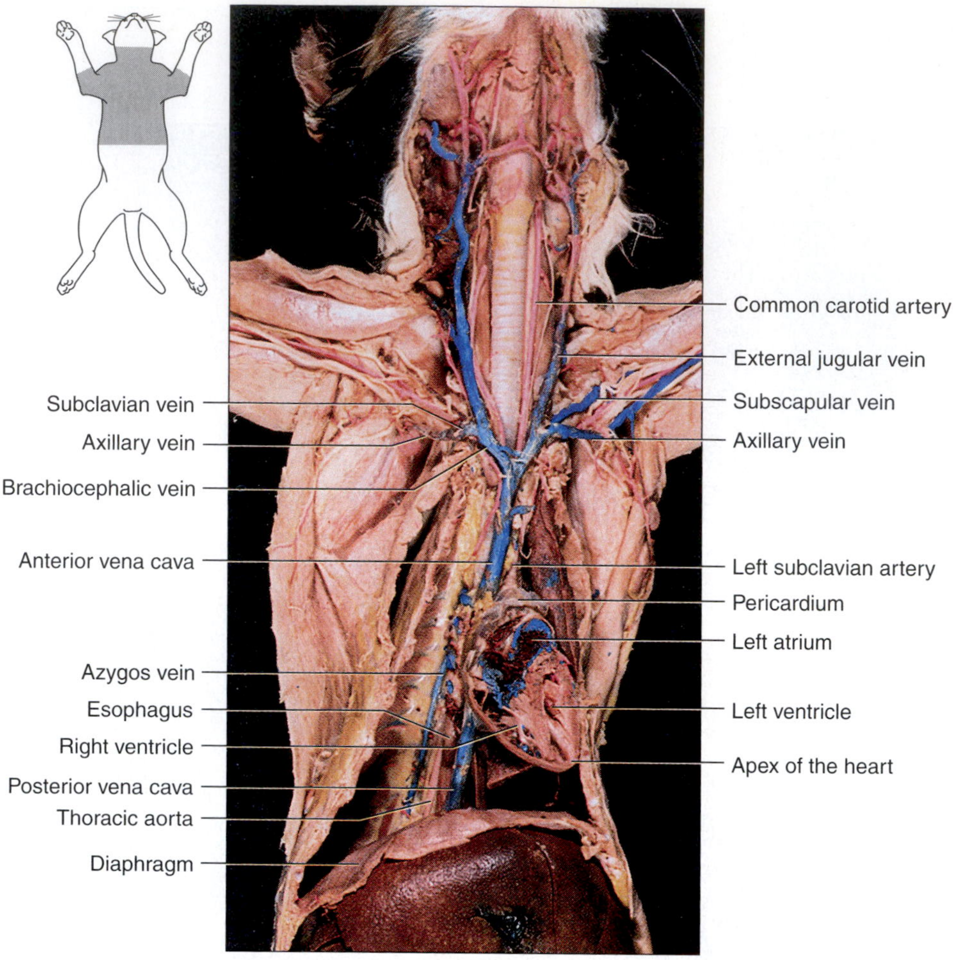

Figure 55.4 Blood vessels of the cat's neck and thorax.

8. Open the abdominal cavity. To do this, follow these steps:
   a. Use scissors to make an incision through the body wall along the midline from the symphysis pubis to the diaphragm.
   b. Make a lateral incision through the body wall along either side of the inferior border of the diaphragm and along the bases of the thighs.
   c. Reflect the flaps created in the body wall as you would open a book, and expose the contents of the abdominal cavity.
   d. Note the *parietal peritoneum* that forms the inner lining of the abdominal wall. Also note the *greater omentum,* a structure composed of a double layer of peritoneum that hangs from the border of the stomach and covers the lower abdominal organs like a fatty apron (fig. 55.5).
9. As you expose and dissect blood vessels, try not to destroy other visceral organs needed for future studies.

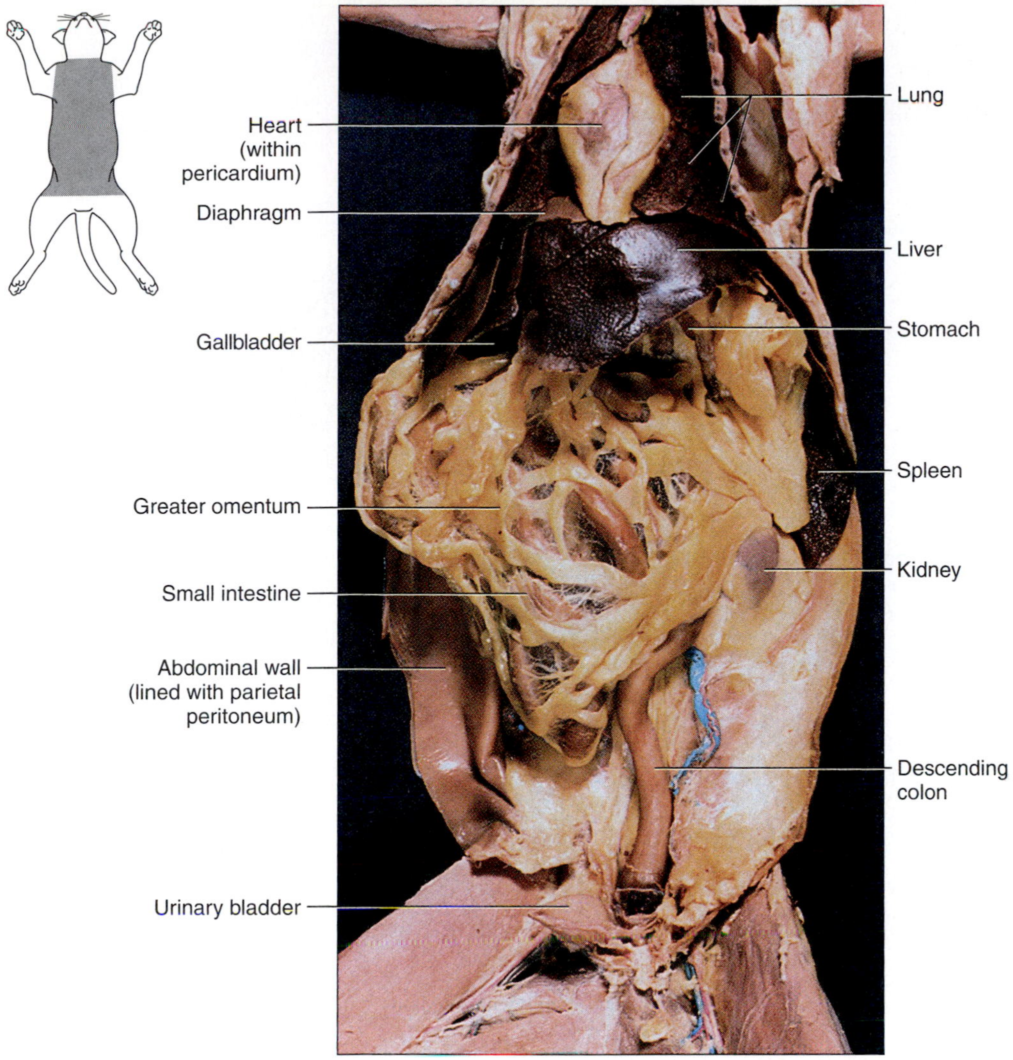

Figure 55.5 Organs of the cat's thoracic and abdominal cavities.

Using figures 55.6, 55.7, and 55.8 as guides, locate and dissect the following arteries of the abdomen:

abdominal aorta (unpaired)
celiac artery (unpaired)
hepatic artery (unpaired)
gastric artery (unpaired)
splenic artery (unpaired)
anterior mesenteric artery (corresponds to superior mesenteric artery)
renal arteries (paired)
adrenolumbar arteries (paired)
gonadal arteries (paired)
posterior mesenteric artery (corresponds to inferior mesenteric artery)
iliolumbar arteries (paired)
external iliac arteries (paired)
internal iliac arteries (paired)
caudal artery (unpaired)

10. Trace the external iliac artery into the right hindlimb (fig. 55.9), and locate the following:

femoral artery
deep femoral artery

11. Review the locations of the heart structures and arteries of the cat without the aid of the figures. 1
12. Examine the human torso model along with figures 55.1, 55.4, 55.6, and 55.8. Identify the arteries of the cat that correspond with those of the human. 3
13. Complete Part A of Laboratory Report 55.

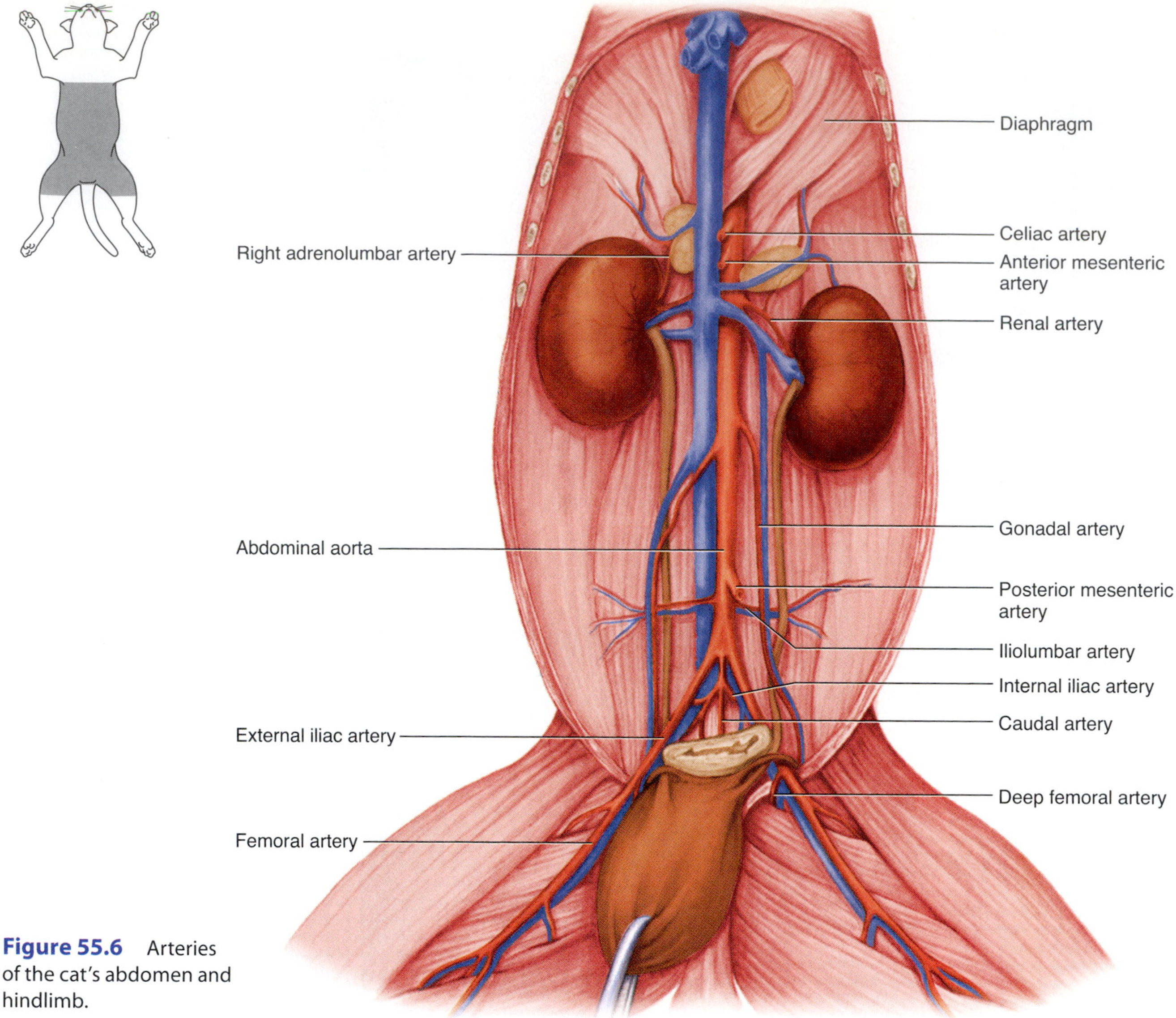

Figure 55.6 Arteries of the cat's abdomen and hindlimb.

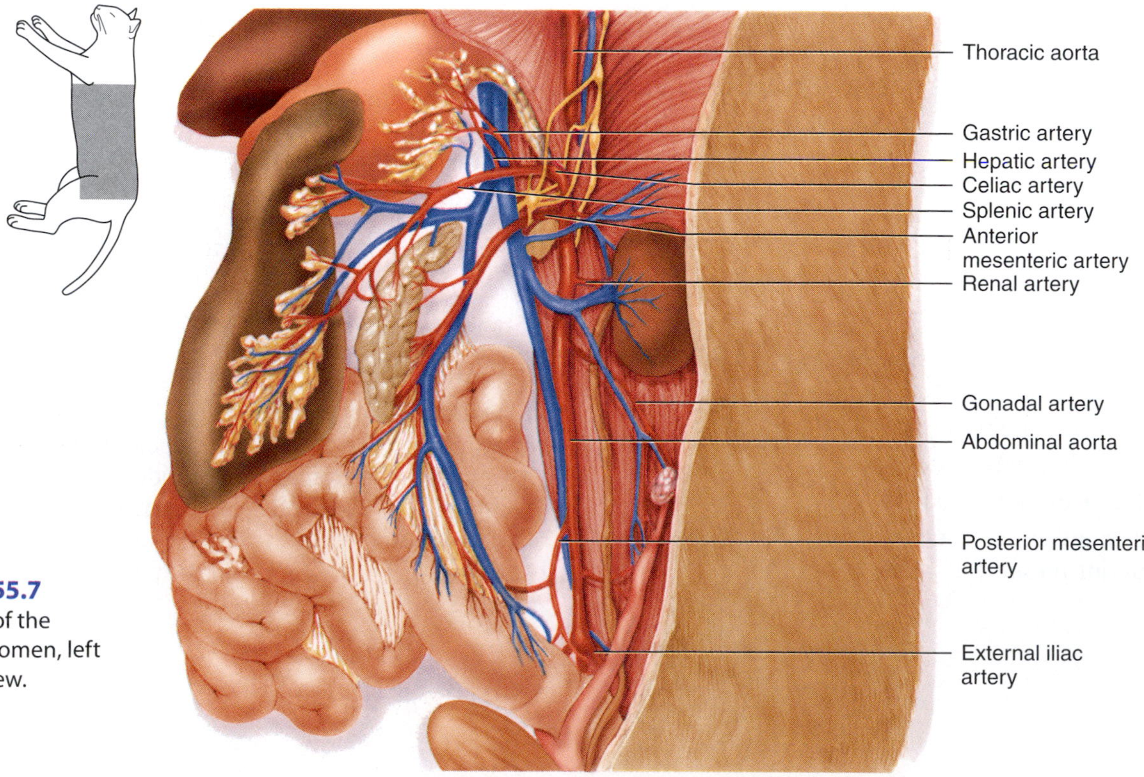

Figure 55.7 Arteries of the cat's abdomen, left lateral view.

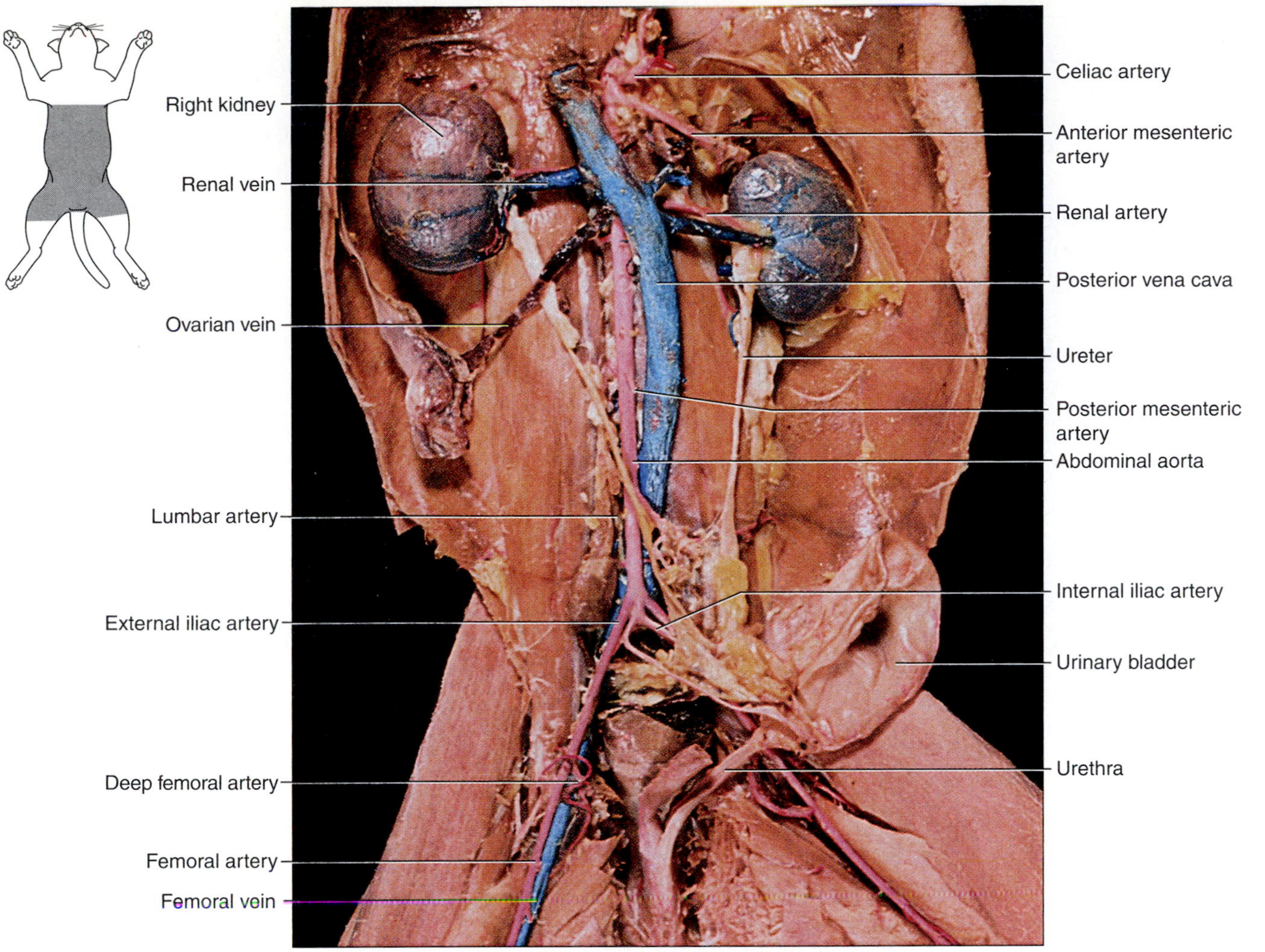

Figure 55.8 Blood vessels of the cat's abdominal cavity, with the digestive organs removed.

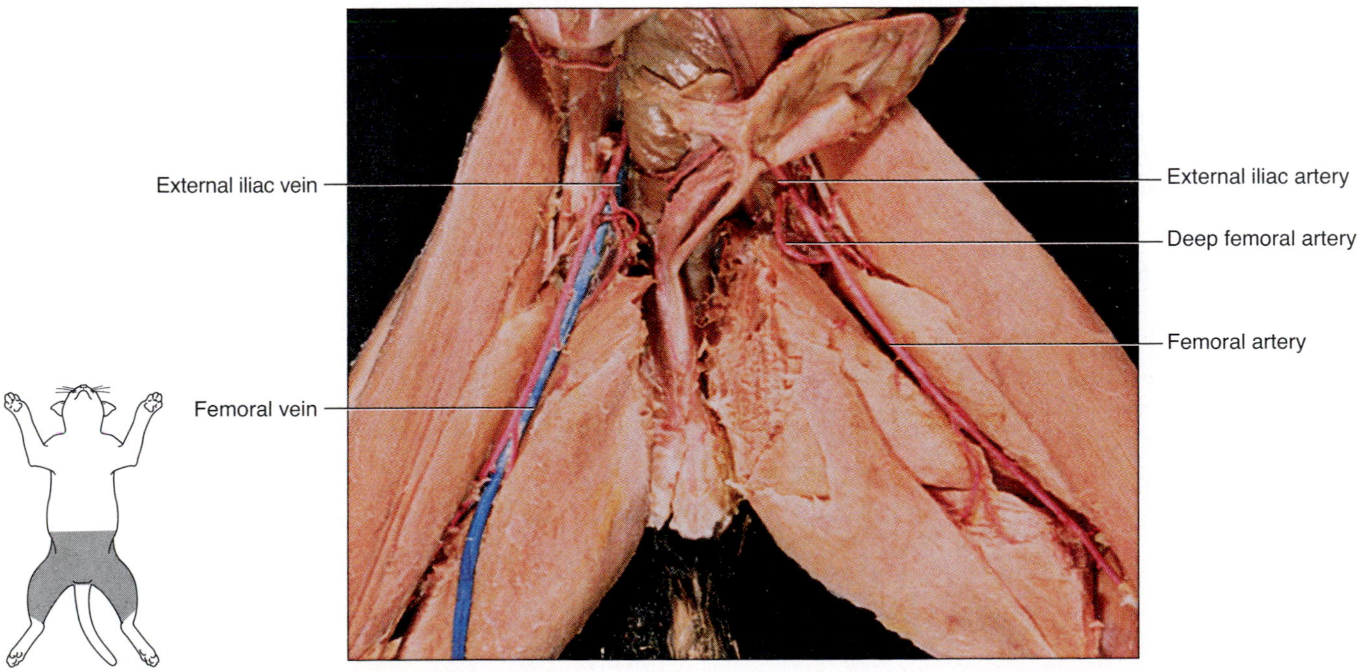

Figure 55.9 Blood vessels of the cat's upper hindlimb.

EXPLORE

Procedure B—The Venous System

1. Examine the heart again, and locate the following veins:

   anterior vena cava (corresponds to superior vena cava)

   posterior vena cava (corresponds to inferior vena cava)

   pulmonary veins

2. Using figures 55.3, 55.4, and 55.10 as guides, locate and dissect the following veins in the thorax and neck:

   right brachiocephalic vein
   left brachiocephalic vein
   right subclavian vein
   left subclavian vein
   internal jugular vein
   external jugular vein

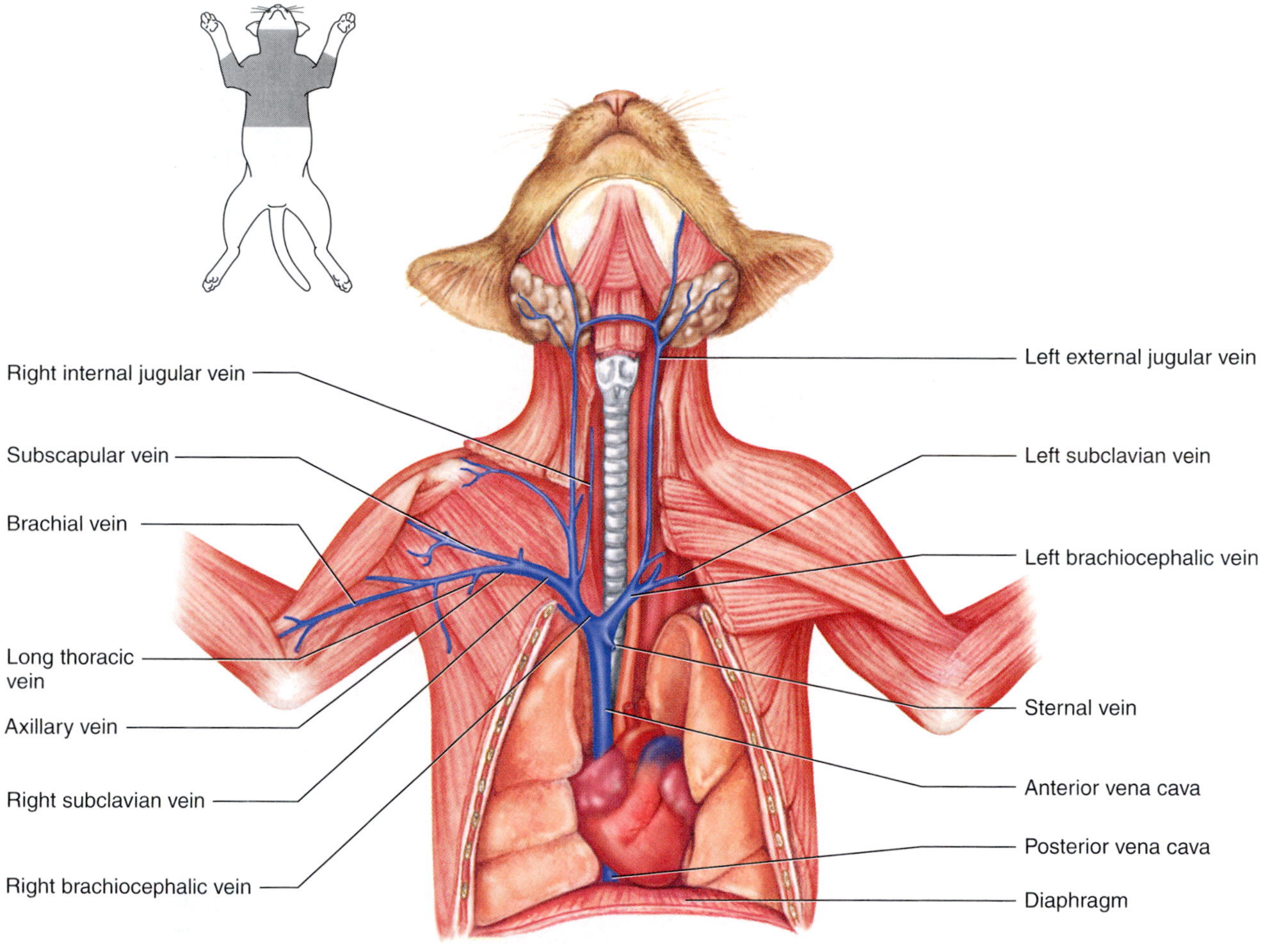

Figure 55.10 Veins of the cat's thorax, neck, and forelimb.

3. Trace the right subclavian vein into the forelimb, and locate the following veins:

axillary vein
subscapular vein
brachial vein

4. Using figures 55.8, 55.11, and 55.12 as guides, locate and dissect the following veins in the abdomen:

posterior vena cava
adrenolumbar vein
anterior mesenteric vein (corresponds to superior mesenteric vein)
posterior mesenteric vein (corresponds to inferior mesenteric vein)
hepatic portal vein
splenic vein
renal vein
gonadal vein
iliolumbar vein
common iliac vein
internal iliac vein
external iliac vein

5. Trace the external iliac vein into the hindlimb (fig. 55.9), and locate the following veins:

femoral vein
deep femoral vein

6. Review the locations of the veins of the cat without the aid of the figures. 1
7. Examine the human torso model along with figures 55.4, 55.8, 55.10, and 55.11. Identify the veins of the cat that correspond with those of the human. 3
8. Complete Part B of the laboratory report.

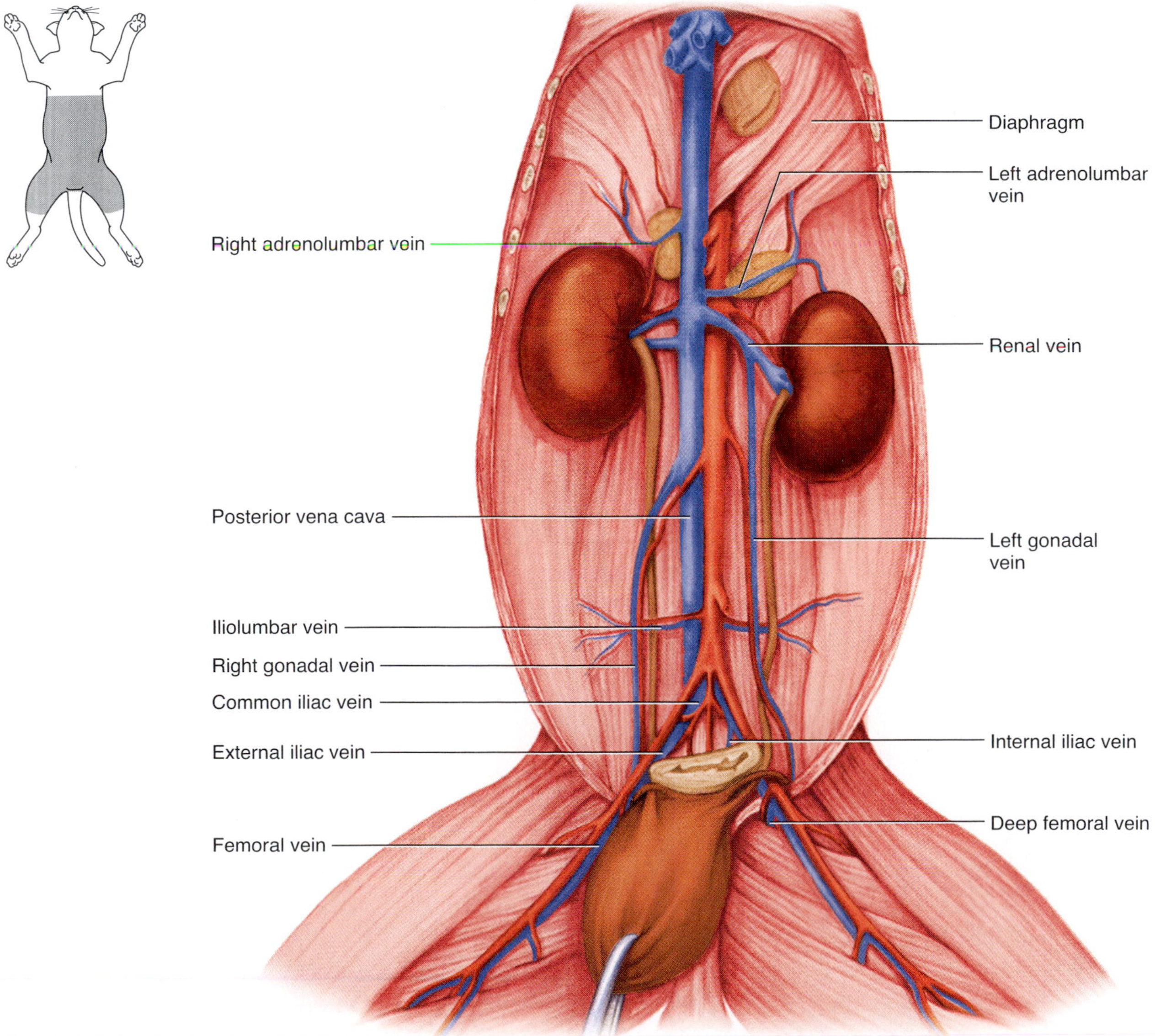

Figure 55.11 Veins of the cat's abdomen and hindlimb.

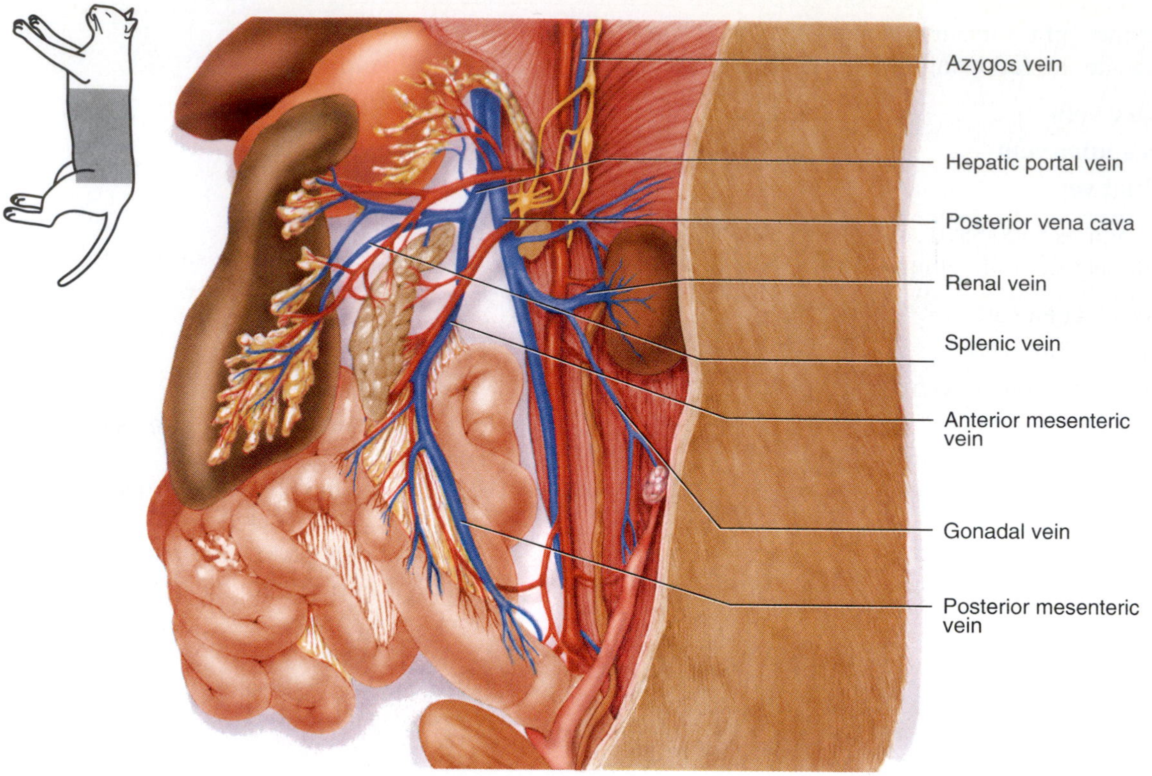

Figure 55.12 Veins of the cat's abdomen and hindlimb.

Name ______________________________

Date ______________________________

Section ______________________________

The [A] corresponds to the indicated Learning Outcome(s) listed at the beginning of the laboratory exercise.

Cat Dissection: Cardiovascular System

Laboratory Report 55

Part A Assessments

Complete the following:

1. Describe the position and attachments of the parietal pericardium of the heart of the cat. [1] ______________________________

2. Describe the relative thicknesses of the walls of the heart chambers of the cat. [1] ______________________________

3. Explain how the wall thicknesses are related to the functions of the chambers. [1] ______________________________

4. Compare the origins of the common carotid arteries of the cat with those of the human. [2] ______________________________

5. Compare the origins of the iliac arteries of the cat with those of the human. [2] ______________________________

Part B Assessments

Complete the following:

1. Compare the origins of the brachiocephalic veins of the cat with those of the human. 2 ______________________

2. Compare the relative sizes of the external and internal jugular veins of the cat with those of the human. 2

3. List twelve veins that cats and humans have in common. 2 3 ______________________

Laboratory Exercise 56

Cat Dissection: Digestive System

Materials Needed

Preserved cat
Dissecting tray
Dissecting instruments
Bone cutter
Disposable gloves
Hand lens
Human torso model

Safety

- Wear disposable gloves when working on the cat dissection.
- Dispose of tissue remnants and gloves as instructed.
- Wash the dissecting tray and instruments as instructed.
- Wash your laboratory table.
- Wash your hands before leaving the laboratory.

In this laboratory exercise, you will dissect the major digestive organs of the cat. As you observe these organs, compare them with those of the human by observing the parts of the human torso model; however, keep in mind that the cat is a carnivore (flesh-eating mammal) and that the organs of its digestive system are adapted to capturing, holding, eating, and digesting the bodies of other animals. In contrast, humans are omnivores (eat both plant and animal substances). Because humans are adapted to eating and digesting a greater variety of foods, comparisons of the cat and the human digestive organs may not be precise.

Purpose of the Exercise

To examine the major digestive organs of the cat and to compare these organs with those of the human.

LEARNING OUTCOMES

After completing this exercise, you should be able to

1. Locate and identify the major digestive organs of the cat.
2. Compare the digestive system of the cat with that of the human.
3. Identify the corresponding organs in the human torso model.

EXPLORE

Procedure—Digestive System Dissection

1. Place the preserved cat in the dissecting tray on its left side.
2. Locate the major salivary glands on one side of the head (fig. 56.1). To do this, follow these steps:
   a. Clear away any remaining fascia and other connective tissue from the region below the ear and near the joint of the mandible.
   b. Identify the *parotid gland,* a relatively large mass of glandular tissue just below the ear. Note the parotid duct (Stensen's duct) that passes over the surface of the masseter muscle and opens into the mouth.
   c. Look for the *submandibular gland* just below the parotid gland, near the angle of the jaw.
   d. Locate the *sublingual gland* that is adjacent and medial to the submandibular gland.
3. Open the oral cavity. To do this, follow these steps:
   a. Use scissors to cut through the soft tissues at the angle of the mouth.
   b. When you reach the bone of the jaw, use a bone cutter to cut through the bone, thus freeing the mandible.

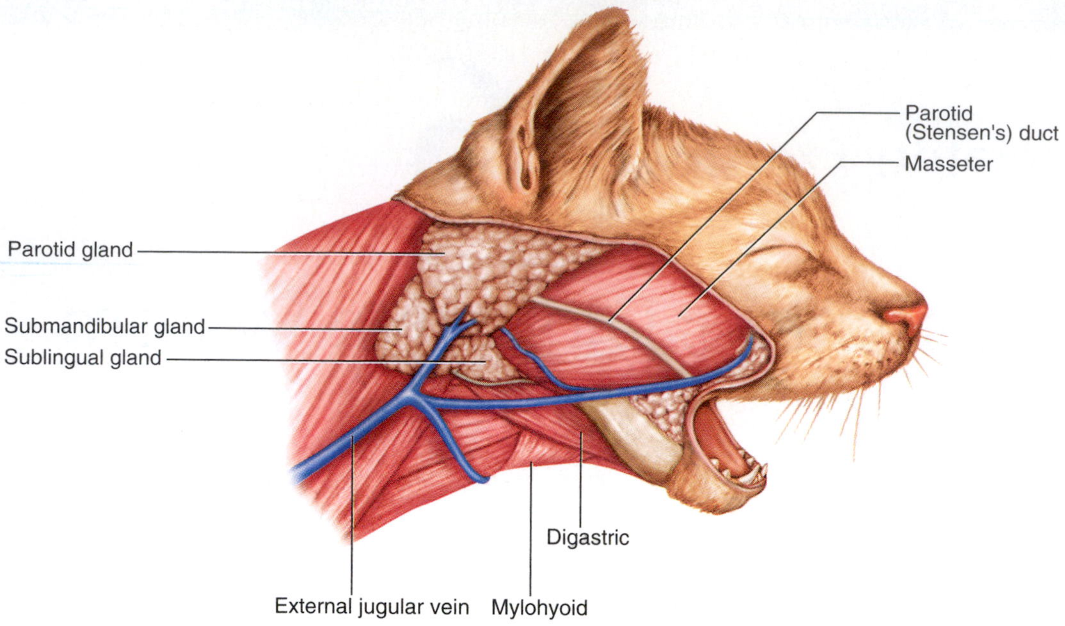

Figure 56.1 Major salivary glands of the cat.

c. Open the mouth wide, and locate the following features:

cheek
lip
vestibule
palate
hard palate
soft palate
palatine tonsils (small, rounded masses of glandular tissue in the lateral wall of the soft palate)
tongue
frenulum
papillae (examine with a hand lens)

4. Examine the teeth of the upper jaw. The adult cat has 6 incisors, 2 canines, 8 premolars, and 2 molars. The lower jaw teeth are similar, except 4 premolars are present.
5. Complete Part A of Laboratory Report 56.
6. Examine organs in the abdominal cavity with the cat positioned with its ventral surface up. Review the normal position and function of the greater omentum.
7. Examine the *liver,* which is located just beneath the diaphragm and is attached to the central portion of the diaphragm by the *falciform ligament.* Also, locate the *spleen,* which is posterior to the liver on the left side (fig. 56.2). The liver has five lobes—a right medial, right lateral (which is subdivided into two parts by a deep cleft), left medial, left lateral, and caudate lobe. The caudate lobe is the smallest lobe and is located in the median line, where it projects into the curvature of the stomach. The caudate lobe is covered by a sheet of mesentery, the *lesser omentum,* that connects the liver to the stomach (figs. 56.3 and 56.4). Find the greenish *gallbladder* on the inferior surface of the liver on the right side. Also note the *cystic duct,* by which the gallbladder is attached to the *bile duct,* and the *hepatic duct,* which originates in the liver and attaches to the cystic duct. Trace the bile duct to its connection with the duodenum.
8. Locate the *stomach* in the left side of the abdominal cavity. At its anterior end, note the union with the *esophagus,* which passes through the diaphragm. Identify the *cardia, fundus, body,* and *pyloric part* (listed from entrance to exit) of the stomach. Use scissors to make an incision along the convex border of the stomach from the cardia to the pyloric part. The lining of the stomach has numerous *gastric folds* (*rugae*). Examine the *pyloric sphincter,* which creates a constriction between the stomach and small intestine.
9. Locate the *pancreas.* It appears as a grayish, two-lobed elongated mass of glandular tissue. One lobe lies dorsal to the stomach and extends across to the duodenum. The other lobe is enclosed by the *mesentery,* which supports the duodenum (figs. 56.3 and 56.4).
10. Trace the *small intestine,* beginning at the pyloric sphincter. The first portion, the *duodenum,* travels posteriorly for several centimeters. Then it loops back around a lobe of the pancreas. The proximal half of the remaining portion of the small intestine

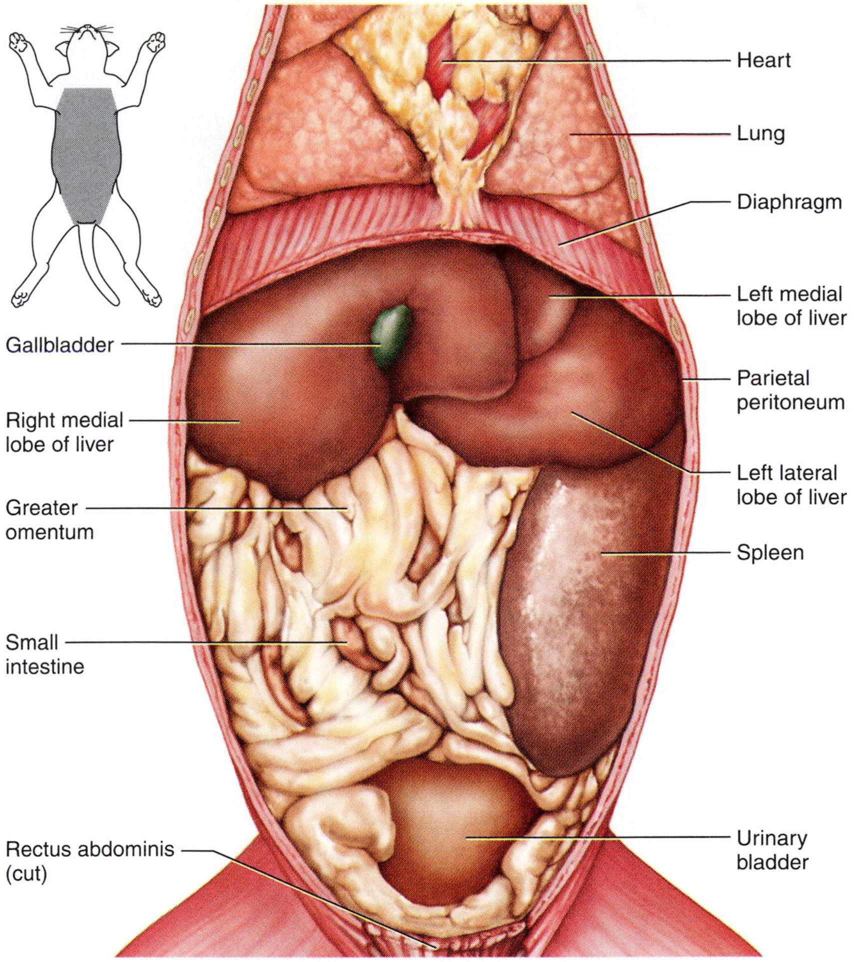

Figure 56.2 Ventral view of abdominal organs of the cat.

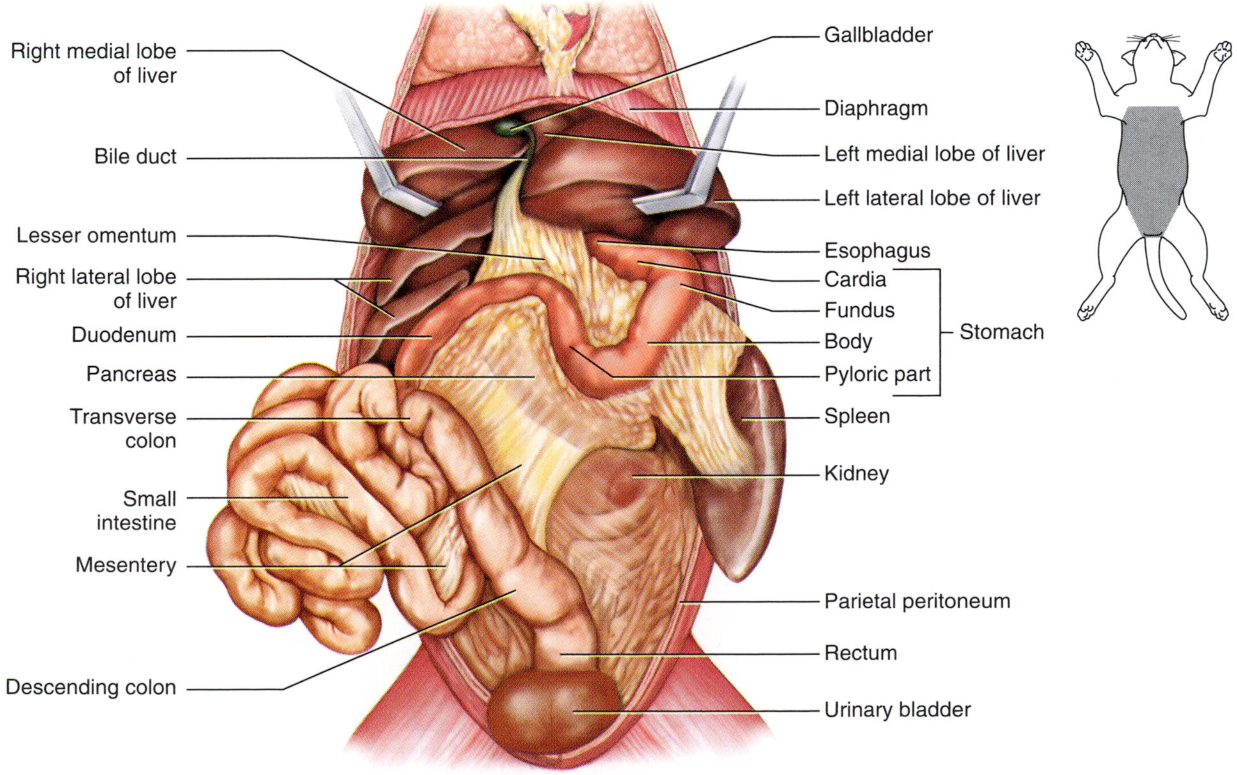

Figure 56.3 Ventral view of the cat's abdominal organs with intestines reflected to the right side.

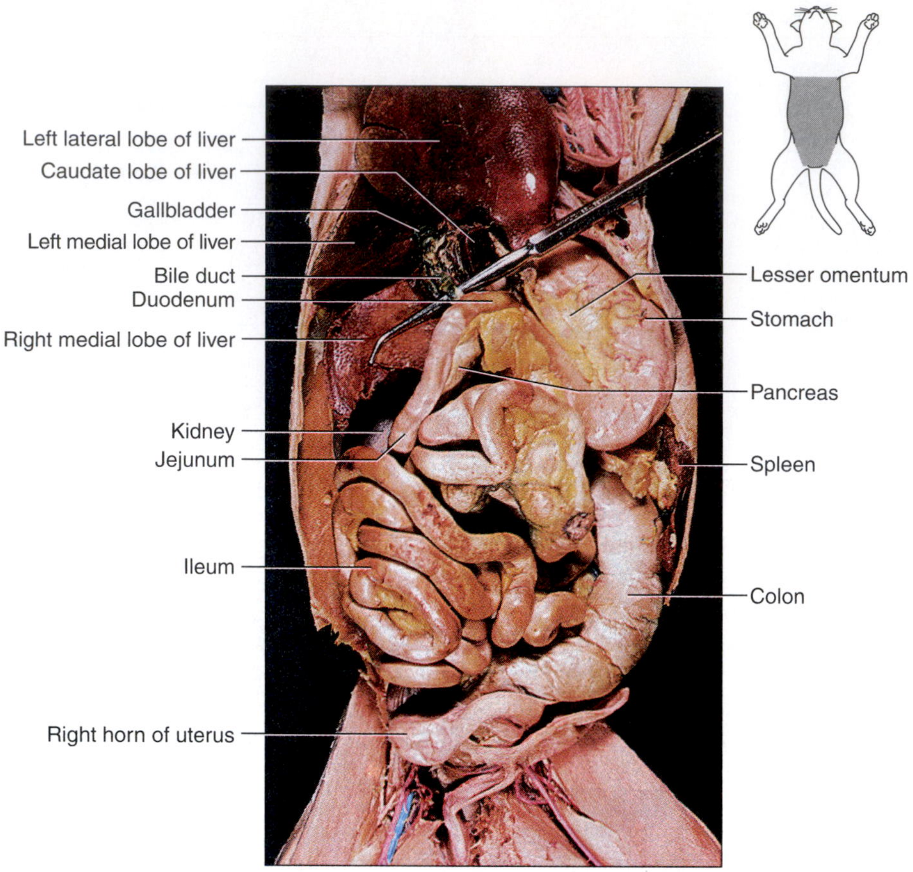

Figure 56.4 Abdominal organs of the cat with the greater omentum removed. The left liver lobes have been reflected to the right side.

is the *jejunum,* and the distal half is the *ileum.* Open the duodenum and note the velvety appearance of the villi. Note how the mesentery supports the small intestine from the dorsal body wall. The small intestine terminates on the right side, where it joins the large intestine.

11. Locate the *large intestine,* and identify the *cecum, ascending colon, transverse colon,* and *descending colon.* Also locate the *rectum,* which extends through the pelvic cavity to the *anus.* Make an incision at the junction between the ileum and cecum, and look for the *ileocecal sphincter.*
12. Review the locations of the digestive organs without the aid of the figures. 1
13. Examine the human torso model along with figures 56.1, 56.2, 56.3, and 56.4. For the human digestive organs, identify corresponding digestive organs of the cat. 3
14. Complete Part B of the laboratory report.

Name ___

Date ___

Section ___

The [A] corresponds to the indicated Learning Outcome(s) listed at the beginning of the laboratory exercise.

Cat Dissection: Digestive System

56 Laboratory Report

Part A Assessments

Complete the following:

1. Compare the locations of the major salivary glands of the human with those of the cat. [2] ___

2. Compare the types and numbers of teeth present in the cat's upper and lower jaws with those of the human. [2] ___

3. In what ways do the cat's teeth seem to be adapted to a special diet? [1] ___

4. What part of the human soft palate is lacking in the cat? [2] ___

5. What do you think is the function of the transverse ridges (rugae) in the hard palate of the cat? [1] ___

6. How do the papillae on the surface of the human tongue compare with those of the cat? [2] ___

Part B Assessments

Complete the following:

1. Describe how the peritoneum and mesenteries are associated with the organs in the abdominal cavity. 1

2. Describe the inner lining of the stomach. 1

3. Compare the structure of the human liver with that of the cat. 2

4. Compare the structure and location of the human pancreas with those of the cat. 2

5. What feature of the human cecum is lacking in the cat? 2

Laboratory Exercise 57

Cat Dissection: Respiratory System

Materials Needed

Preserved cat
Dissecting tray
Dissecting instruments
Disposable gloves
Human torso model

Safety

- Wear disposable gloves when working on the cat dissection.
- Dispose of tissue remnants and gloves as instructed.
- Wash the dissecting tray and instruments as instructed.
- Wash your laboratory table.
- Wash your hands before leaving the laboratory.

In this laboratory exercise, you will dissect the major respiratory organs of a cat. As you observe these structures in the cat, compare them with those of the human torso model.

Purpose of the Exercise

To examine the major respiratory organs of the cat and to compare these organs with those of the human.

LEARNING OUTCOMES

After completing this exercise, you should be able to

1. Locate and identify the major respiratory organs of the cat.
2. Compare the respiratory system of the human with that of the cat.
3. Identify the corresponding organs in the human torso model.

EXPLORE

Procedure—Respiratory System Dissection

1. Place the preserved cat in a dissecting tray with its ventral surface up.
2. Examine the *nostrils* (*external nares*) and *nasal septum.*
3. Open the oral cavity wide to expose the *hard palate* and the *soft palate* (fig. 57.1). Cut through the tissues of the soft palate, and observe the *nasopharynx* above it. Locate the small openings of the *auditory tubes* in the lateral walls of the nasopharynx. Insert a probe into an opening of an auditory tube.
4. Pull the tongue forward, and locate the *epiglottis* near its base. Also identify the *glottis,* which is the opening between the *vocal folds* (*vocal cords*) into the larynx. Locate the *esophagus,* which is dorsal to the larynx (figs. 57.1 and 57.2).
5. Open the thoracic cavity, and expose its contents.
6. Dissect the *trachea* in the neck, and expose the *larynx* at the anterior end near the base of the tongue. Note the *tracheal rings,* and locate the lobes of the *thyroid gland* on each side of the trachea, just posterior to the larynx. Also note the *thymus gland,* a rather diffuse mass of glandular tissue extending along the ventral surface of the trachea into the thorax (figs. 57.1 and 57.2).
7. Examine the larynx, and identify the *thyroid cartilage* and *cricoid cartilage* in its wall. Make a longitudinal incision through the ventral wall of the larynx, and locate the vocal folds (vocal cords) inside.
8. Trace the trachea posteriorly to where it divides into *primary bronchi,* which pass into the lungs.
9. Examine the *lungs,* each of which is subdivided into three lobes—an *anterior* (*cranial*), a *middle,* and a *posterior (caudal) lobe.* The right lung of the cat has a small fourth *accessory* (*mediastinal; cardiac*) lobe associated with the right posterior lobe. Notice the

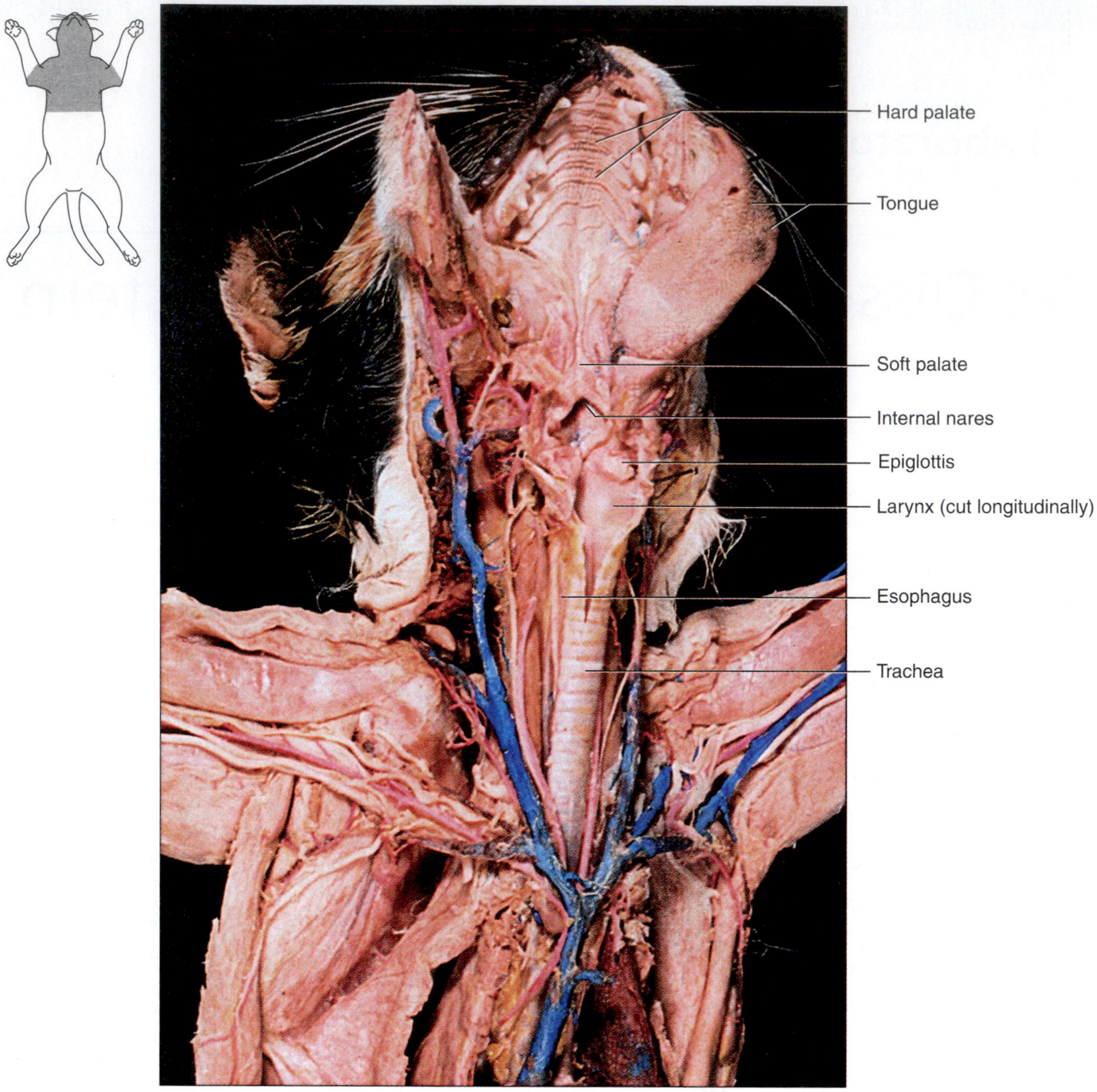

Figure 57.1 Features of the cat's neck and oral cavity, ventral view.

thin membrane, the *visceral pleura,* on the surface of each lung. Also notice the *parietal pleura,* which forms the inner lining of the thoracic wall, and locate the spaces of the *pleural cavities* (fig. 57.2).

10. Make an incision through a lobe of a lung, and examine its interior. Note the branches of the smaller air passages.
11. Examine the *diaphragm,* and note that its central portion is tendinous.
12. Locate the *phrenic nerve.* This nerve appears as a white thread passing along the side of the heart to the diaphragm.
13. Examine the organs located in the *mediastinum.* Note that the mediastinum separates the right and left lungs and pleural cavities within the thorax.
14. Review the locations of the respiratory organs without the aid of the figures. 1
15. Examine the human torso model along with figures 57.1 and 57.2. For each human respiratory organ, identify the corresponding respiratory organ of the cat. 3
16. Complete Laboratory Report 57.

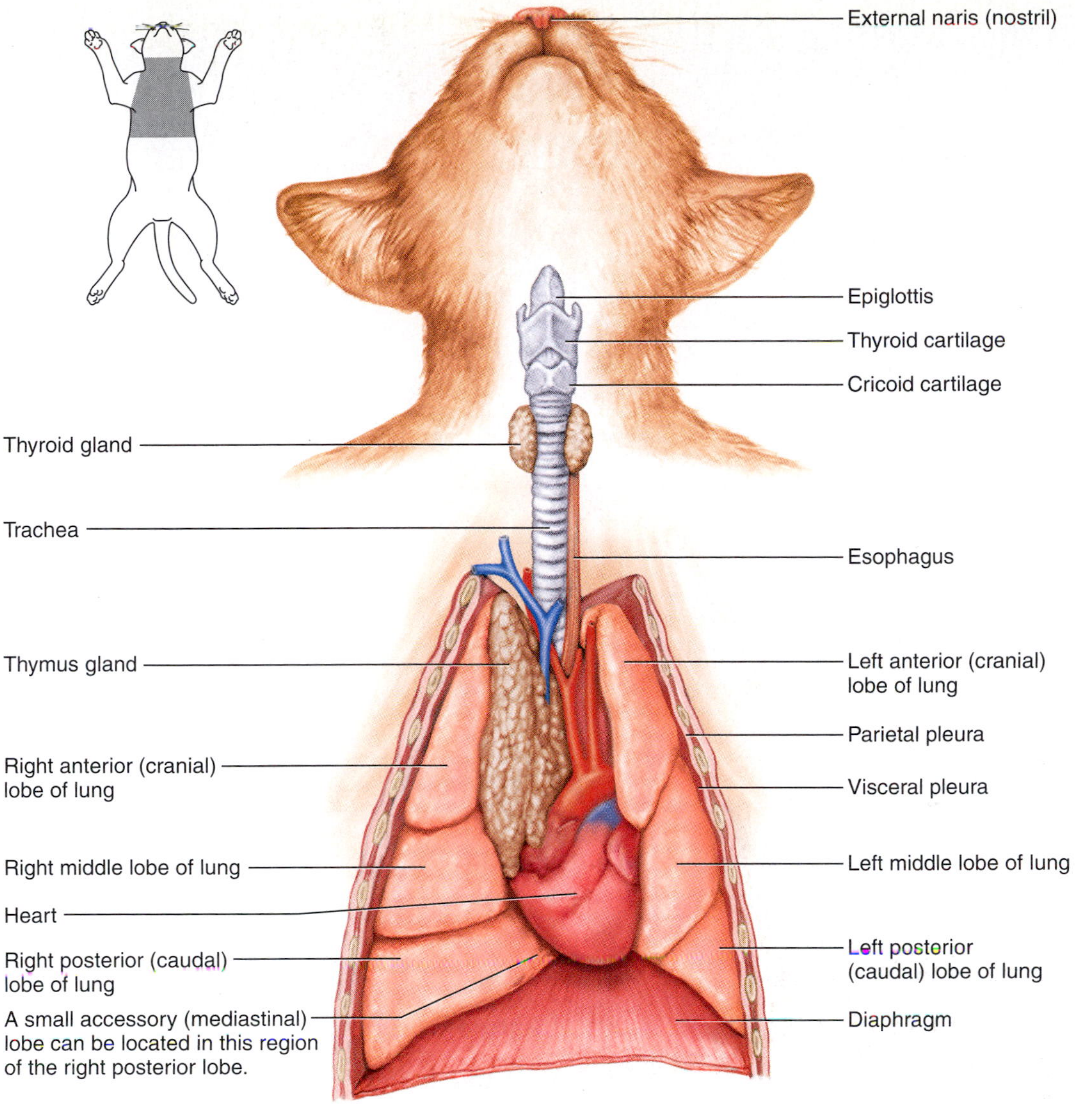

Figure 57.2 Ventral view of the cat's thoracic cavity.

Notes

Name ______________________

Date ______________________

Section ______________________

The A corresponds to the indicated Learning Outcome(s) listed at the beginning of the laboratory exercise.

Laboratory Report 57

Cat Dissection: Respiratory System

Complete the following:

1. Why do the auditory tubes open into the nasopharynx? 1 ______________________

2. Distinguish between the glottis and the epiglottis. 1 ______________________

3. Are the tracheal rings of the cat complete or incomplete? 1 ______________________ How does this feature compare with that of the human? 2 ______________________

4. How does the structure of the primary bronchi compare with that of the trachea? 1 ______________________

5. Compare the number of lobes in the human lungs with the number of lobes in the cat. 2 ______________________

6. Describe the attachments of the diaphragm. 1 ______________________

7. What major structures are located within the mediastinum of the cat? 1 ______________________

   How does this compare to the human mediastinum? 2 ______________________

Notes

Laboratory Exercise 58

Cat Dissection: Urinary System

Materials Needed

Preserved cat
Dissecting tray
Dissecting instruments
Disposable gloves
Human torso model

Safety

- Wear disposable gloves when working on the cat dissection.
- Dispose of tissue remnants and gloves as instructed.
- Wash the dissecting tray and instruments as instructed.
- Wash your laboratory table.
- Wash your hands before leaving the laboratory.

In this laboratory exercise, you will dissect the urinary organs of the cat. As you observe these structures, compare them with the corresponding human organs by observing the parts of the human torso model.

Purpose of the Exercise

To examine the urinary organs of the cat and to compare them with those of the human.

LEARNING OUTCOMES

After completing this exercise, you should be able to

1. Locate and identify the urinary organs of the cat.
2. Compare the urinary organs of the cat with those of the human.
3. Identify the corresponding organs in the human torso model.

EXPLORE

Procedure—Urinary System Dissection

1. Place the preserved cat in a dissecting tray with its ventral surface up.
2. Open the abdominal cavity, and remove the liver, stomach, and spleen (fig. 58.1).
3. Push the intestines to one side, and locate the *kidneys* in the dorsal abdominal wall on either side of the vertebral column. The kidneys are located dorsal to the *parietal peritoneum* (retroperitoneal).
4. Carefully remove the parietal peritoneum and the adipose tissue surrounding the kidneys. Locate the following, using figures 58.1 and 58.2 as guides:

   ureters
   renal arteries
   renal veins

5. Locate the *adrenal glands*, which lie medially and above the kidneys and are surrounded by connective tissues. The adrenal glands are attached to blood vessels and are separated by the abdominal aorta and posterior vena cava.
6. Expose the ureters by cleaning away the connective tissues along their lengths. They enter the *urinary bladder* on the dorsal surface.
7. Examine the urinary bladder, attached to the abdominal wall by folds of peritoneum that form ligaments. Locate the *medial ligament* on the ventral surface of the bladder that connects it with the linea alba and the *two lateral ligaments* on the dorsal surface.
8. Remove the connective tissue from around the urinary bladder. Use a sharp scalpel to open the bladder, and examine its interior. Locate the openings of the ureters and the urethra on the inside.
9. Expose the *urethra* at the posterior of the urinary bladder (figs. 58.1 and 58.3).

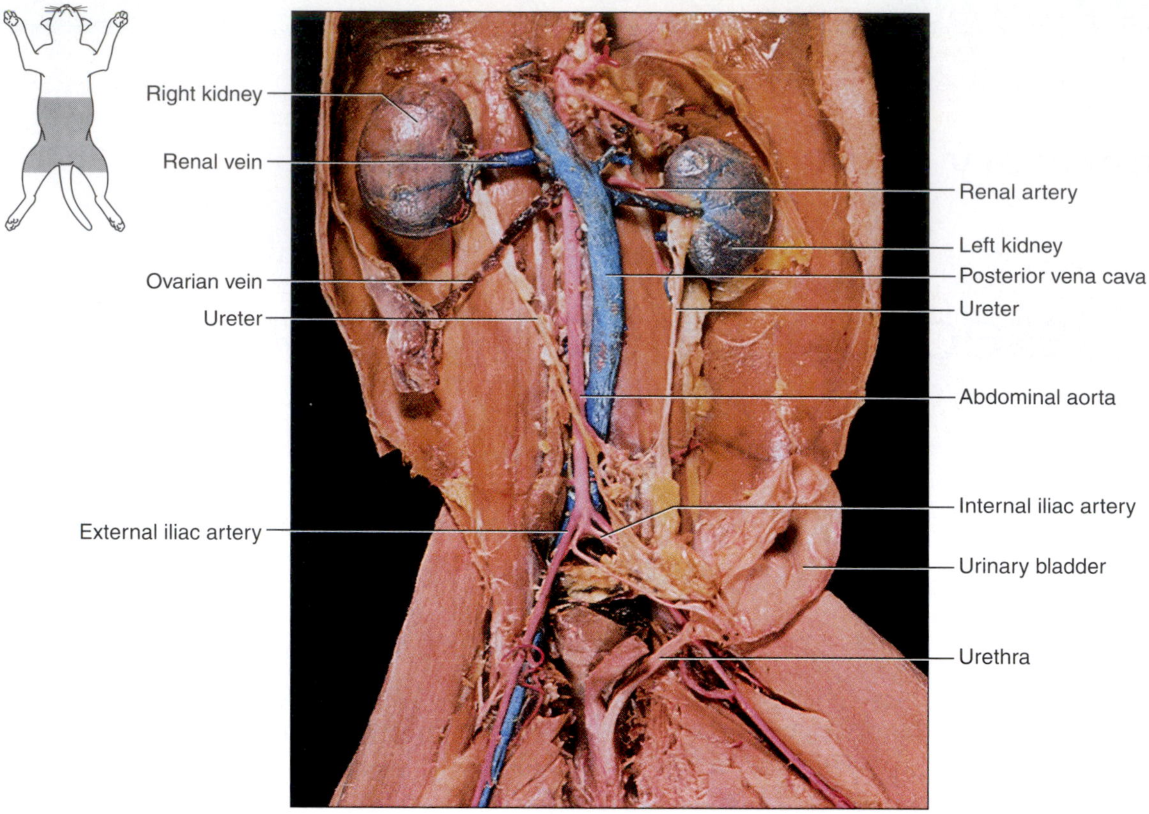

Figure 58.1 Urinary system of the female cat's abdominal cavity, with the digestive organs removed.

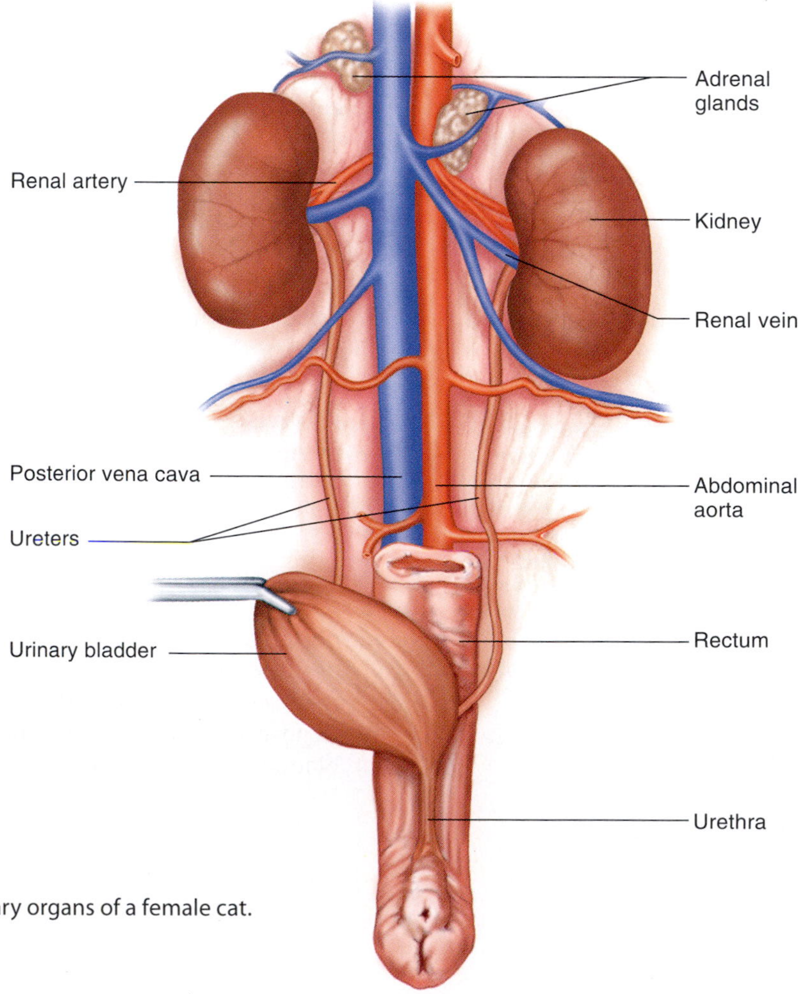

Figure 58.2 The urinary organs of a female cat.

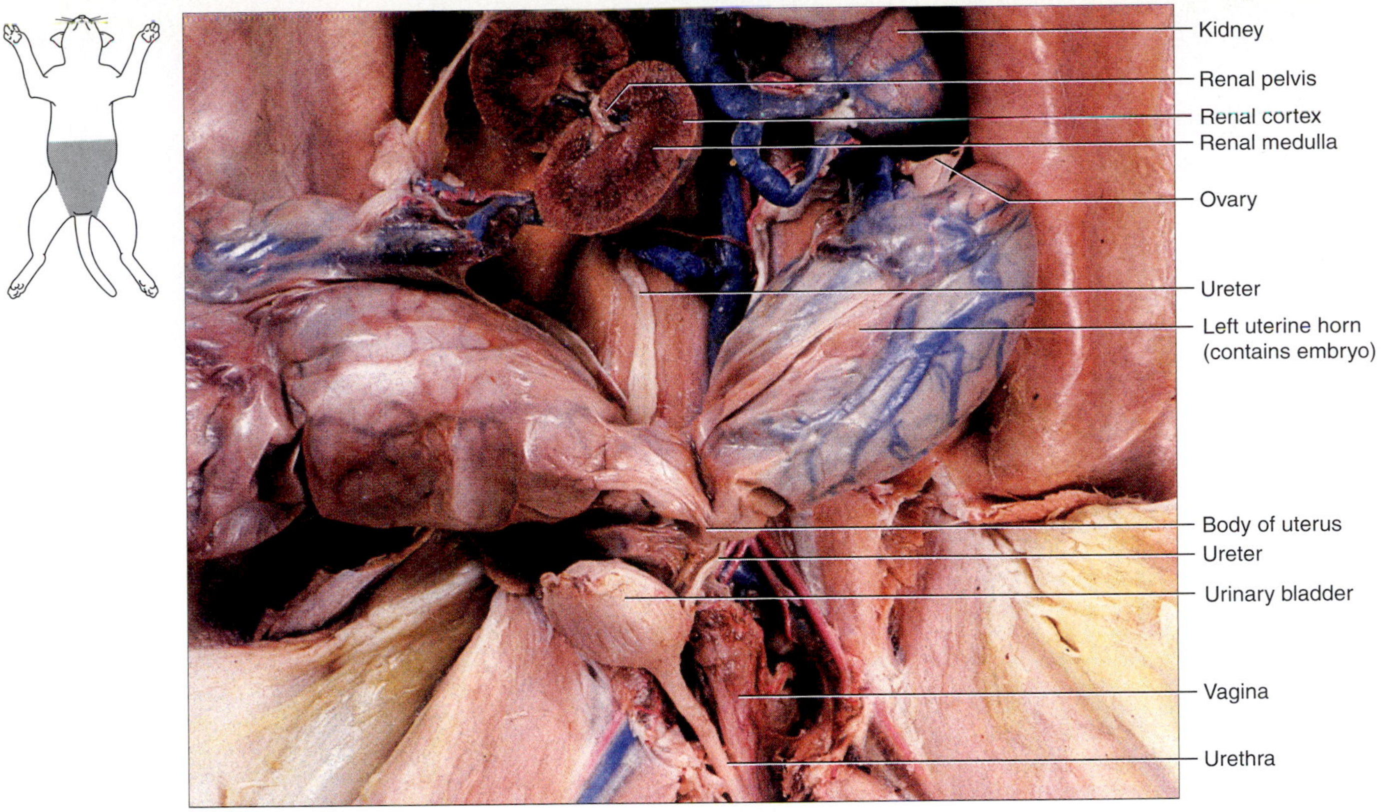

Figure 58.3 Female urinary and reproductive systems of a pregnant cat.

10. Remove one kidney, and section it longitudinally (figs. 58.3 and 58.4). Identify the following features:

    fibrous capsule
    renal cortex (superficial lighter region)
    renal medulla (deeper darker region)
    renal pyramid (single in cat)
    renal papilla (single in cat)
    renal sinus
    renal pelvis
    hilum of kidney

11. Review the locations of the urinary organs without the aid of the figures. 1
12. Discard the organs and tissues that were removed from the cat, as directed by the laboratory instructor.
13. Examine the human torso model along with figures 58.1, 58.2, 58.3, and 58.4. Identify urinary organs of the cat that correspond with those of the human. 3
14. Complete Laboratory Report 58.

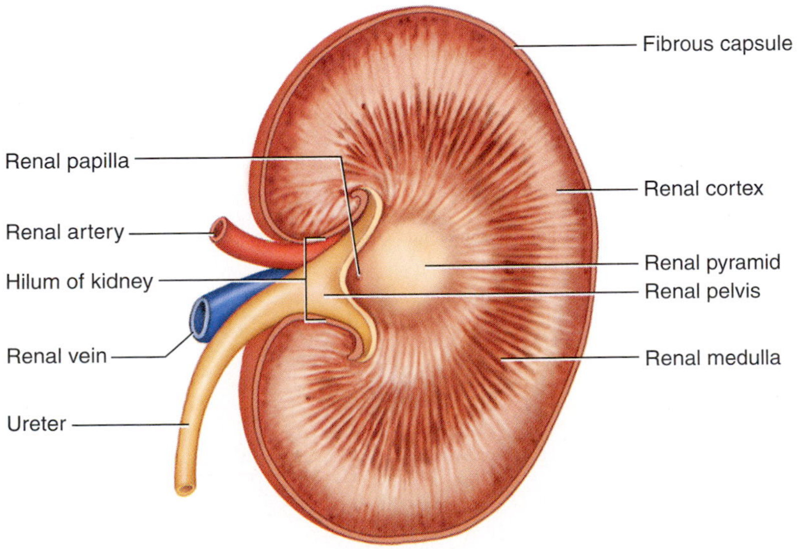

Figure 58.4 Longitudinal section of a cat's kidney.

Notes

Name ____________________

Date ____________________

Section ____________________

The [A] corresponds to the indicated Learning Outcome(s) listed at the beginning of the laboratory exercise.

Laboratory Report 58

Cat Dissection: Urinary System

Complete the following:

1. Compare the positions of the kidneys in the cat with those in the human. [2] ____________________

2. Compare the locations of the adrenal glands in the cat with those in the human. [2] ____________________

3. What structures of the cat urinary system are retroperitoneal? [1] ____________________

4. Describe the wall of the urinary bladder of the cat. [1] ____________________

5. Compare the renal pyramids and renal papillae of the human kidney with those of the cat. [2] ____________________

Notes

Laboratory Exercise 59

Cat Dissection: Reproductive Systems

Materials Needed

Preserved cat
Dissecting tray
Dissecting instruments
Bone cutters
Disposable gloves
Models of human reproductive systems

Safety

- Wear disposable gloves when working on the cat dissection.
- Dispose of tissue remnants and gloves as instructed.
- Wash the dissecting tray and instruments as instructed.
- Wash your laboratory table.
- Wash your hands before leaving the laboratory.

In this laboratory exercise, you will dissect the reproductive system of the cat. If you have a female cat, begin with Procedure A. If you have a male cat, begin with Procedure B. After completing the dissection, exchange cats with someone who has dissected one of the opposite sex, and examine its reproductive organs.

As you observe the cat reproductive organs, compare them with the corresponding human organs by examining the models of the human reproductive systems.

Purpose of the Exercise

To examine the reproductive organs of the cat and to compare them with the corresponding organs of the human.

LEARNING OUTCOMES

After completing this exercise, you should be able to

1. Locate and identify the reproductive organs of a cat.
2. Compare the reproductive organs of the cat with those of the human.
3. Identify the corresponding organs in models of the human reproductive systems.

EXPLORE

Procedure A—Female Reproductive System

1. Place the preserved female cat in a dissecting tray with its ventral surface up, and open its abdominal cavity.
2. Locate the small oval *ovaries* just posterior to the kidneys (figs. 59.1 and 59.2).
3. Examine an ovary, suspended from the dorsal body wall by a fold of peritoneum called the *mesovarium*. Another attachment, the *ovarian ligament*, connects the ovary to the tubular *uterine horn* (fig. 59.1).
4. Near the anterior end of the ovary, locate the funnel-shaped *infundibulum*, which is at the end of the *uterine tube* (*oviduct*). Note the tiny projections, or *fimbriae*, that create a fringe around the edge of the infundibulum. Trace the uterine tube around the ovary to its connection with the uterine horn near the attachment of the ovarian ligament.
5. Examine the uterine horn. It is suspended from the body wall by a fold of peritoneum, the *broad ligament*, that is continuous with the mesovarium of the ovary. Also note the fibrous *round ligament*, which extends from the uterine horn laterally and posteriorly to the body wall (figs. 59.1 and 59.2).
6. To observe the remaining organs of the reproductive system more easily, remove the left hindlimb by severing it near the hip with bone cutters. Then use the bone cutters to remove the left anterior portion of the pelvic girdle. Also remove the necessary muscles and connective tissue to expose the structures shown in figure 59.3.
7. Trace the uterine horns posteriorly. They unite to form the *uterine body*, which is located between the urethra and rectum. This Y-shaped uterus allows ample space for several offspring to develop at the same time in the uterine horns (fig. 59.2). The uterine body is continuous with the *vagina*, which leads to the outside.
8. Trace the *urethra* from the urinary bladder posteriorly. The urethra and the vagina open into a

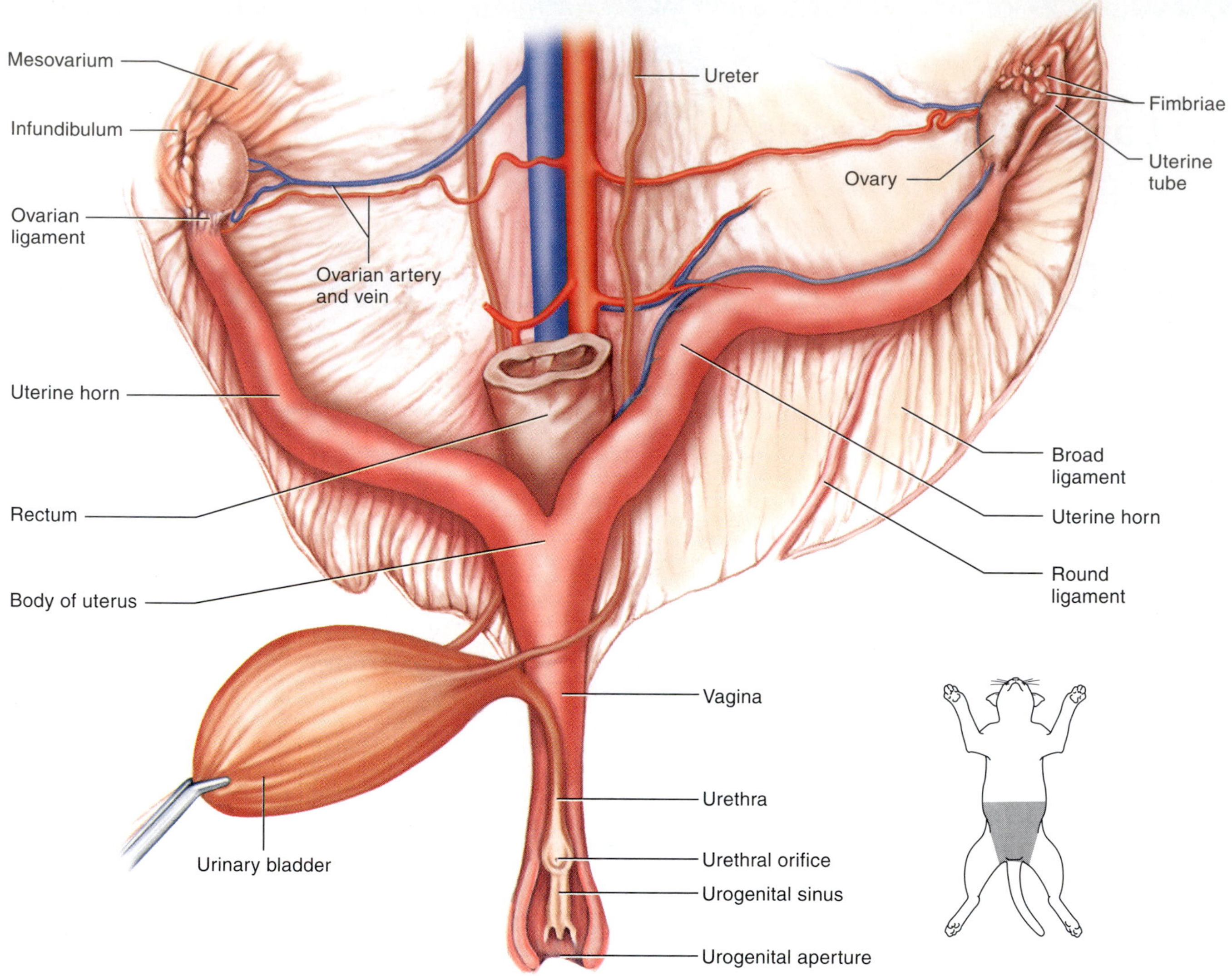

Figure 59.1 Ventral view of the female cat's reproductive system.

common chamber, called the *urogenital sinus.* The opening of this chamber, ventral to the anus, is called the *urogenital aperture.* Locate the *labia majora,* which are folds of skin on either side of the urogenital aperture.

9. Use scissors to open the vagina along its lateral wall, beginning at the urogenital aperture and continuing to the body of the uterus. Note the *urethral orifice* in the ventral wall of the urogenital sinus, and locate the small, rounded *cervix* of the uterus, which projects into the vagina at its deep end. The *clitoris* is located in the ventral wall of the urogenital sinus near its opening to the outside.
10. Review the locations of the reproductive organs without the aid of the figures. 1
11. Examine the models of the human reproductive systems along with figures 59.1, 59.2, and 59.3. Identify the reproductive organs of the cat that correspond with those of the human. 3
12. Complete Part A of Laboratory Report 59.

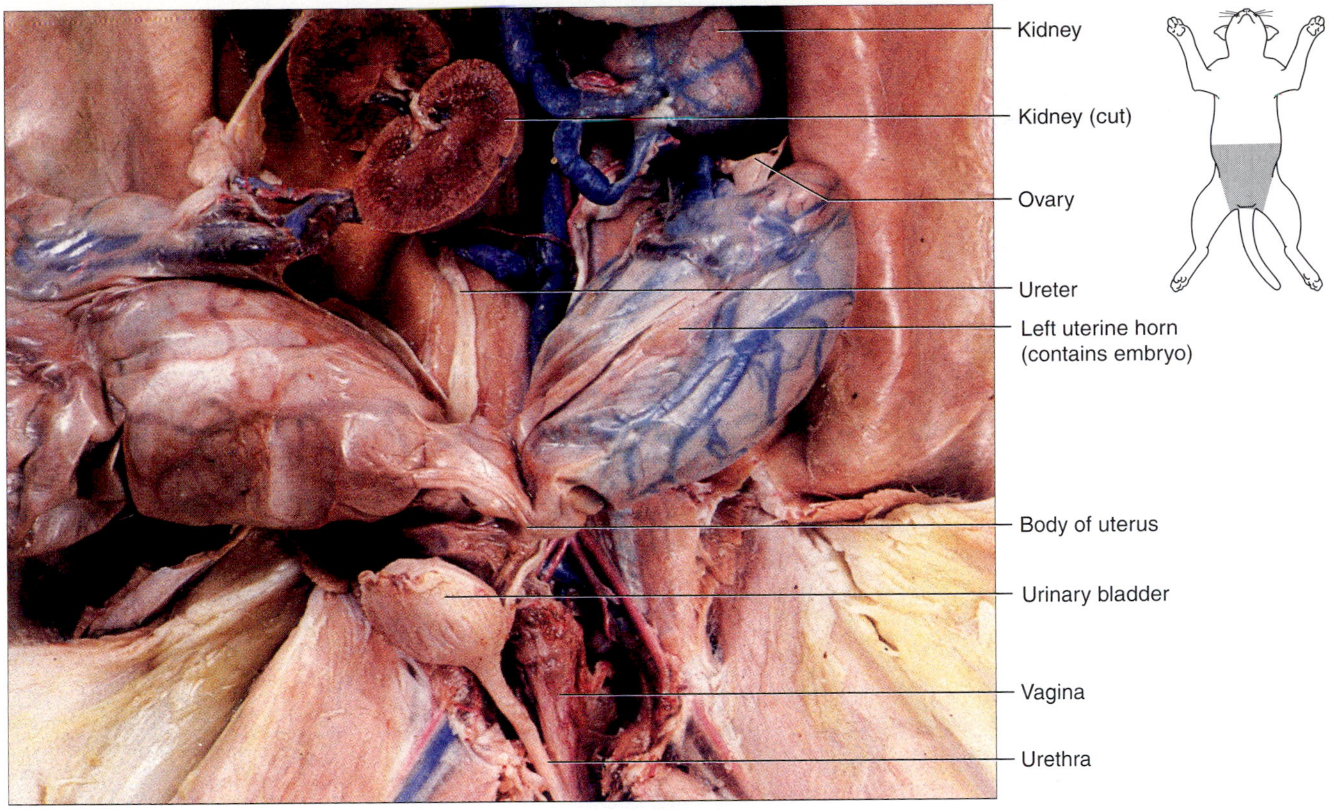

Figure 59.2 Female urinary and reproductive systems of a pregnant cat.

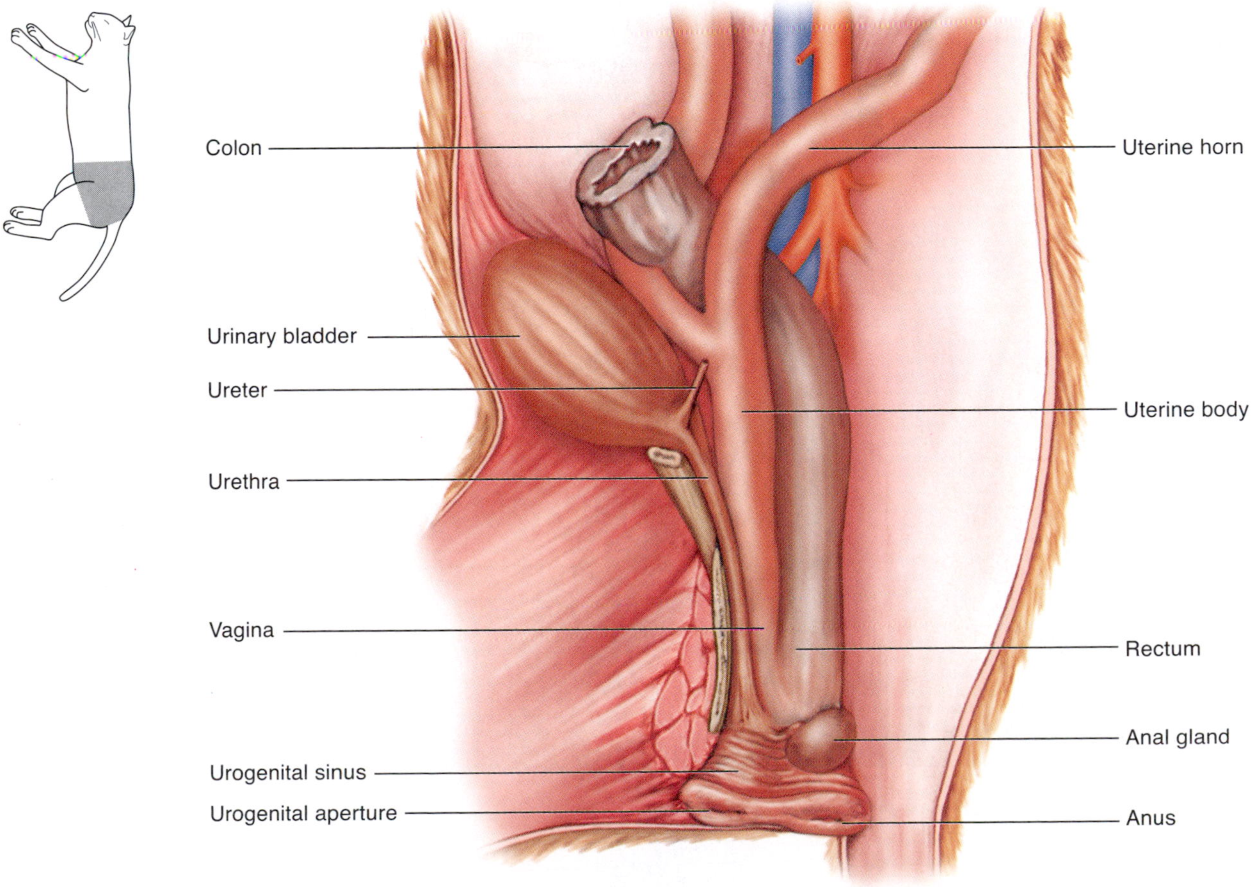

Figure 59.3 Lateral view of the female cat's reproductive system.

EXPLORE

Procedure B—Male Reproductive System

1. Place the preserved male cat in a dissecting tray with its ventral surface up.
2. Locate the *scrotum* between the hindlimbs. Make an incision along the midline of the scrotum, and expose the *testes* inside (fig. 59.4). The testes are separated by a septum, and each testis is enclosed in a sheath of connective tissue.
3. Remove the sheath surrounding the testes, and locate the convoluted *epididymis* on the dorsal surface of each testis (fig. 59.5).
4. Locate the *spermatic cord* on the right side, leading away from the testis. This spermatic cord contains the *ductus (vas) deferens,* which is continuous with the epididymis, as well as with the nerves and blood vessels that supply the testis on that side. Trace the spermatic cord to the body wall, where its contents pass through the *inguinal canal* and enter the pelvic cavity (fig. 59.5).
5. Locate the *penis,* and identify the *prepuce,* which forms a sheath around the penis. Make an incision through the skin of the prepuce, and expose the *glans penis.* The glans penis has minute spines on its surface.
6. To observe the remaining structures of the reproductive system more easily, remove the left hindlimb by severing it near the hip with bone cutters. Then use the bone cutters to remove the left anterior portion of the pelvic girdle. Also remove the necessary muscles and connective tissues to expose the organs shown in figure 59.6.
7. Trace the ductus deferens from the inguinal canal to the penis. Note that the ductus deferens loops over the ureter within the pelvic cavity and passes

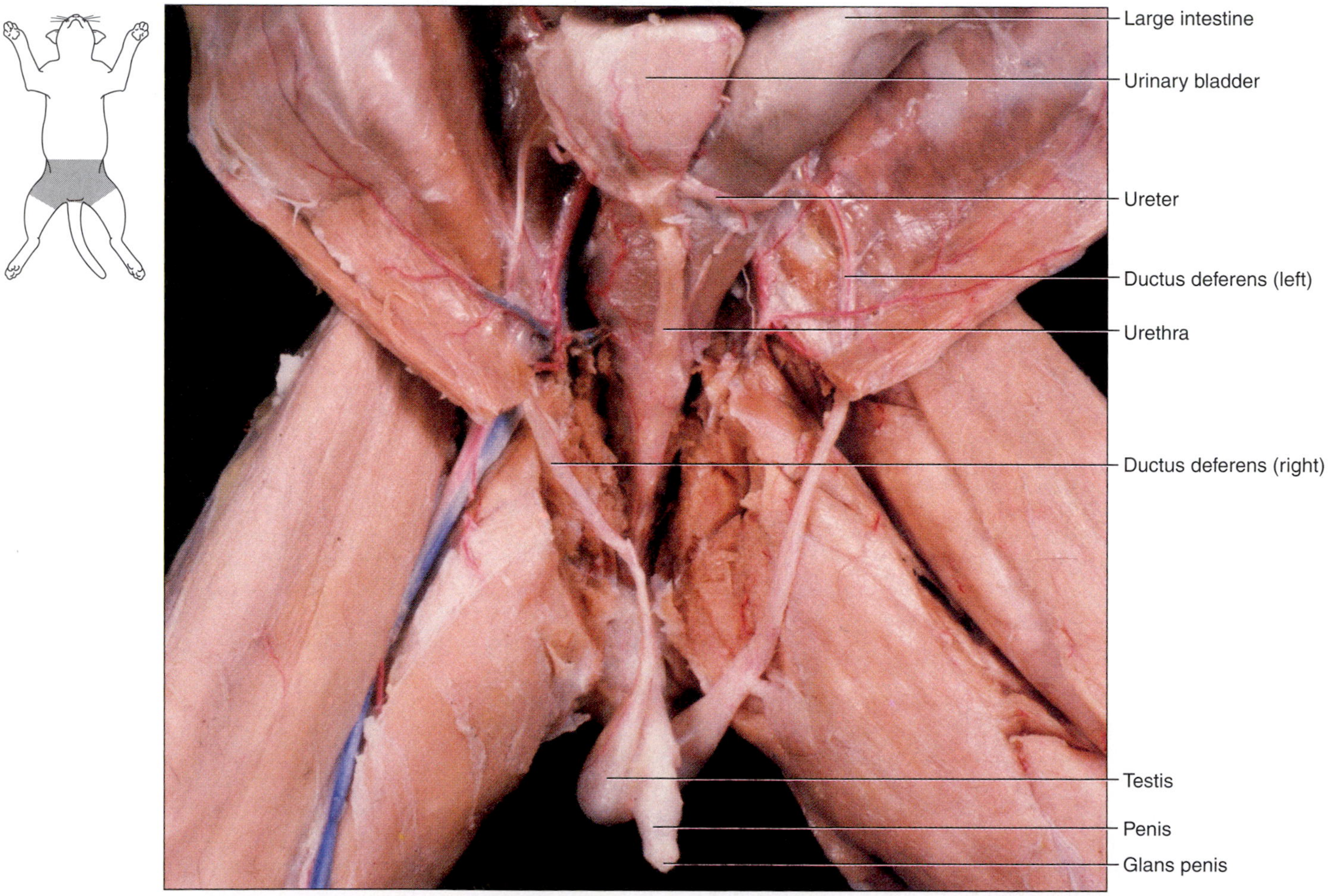

Figure 59.4 Male reproductive system of the cat.

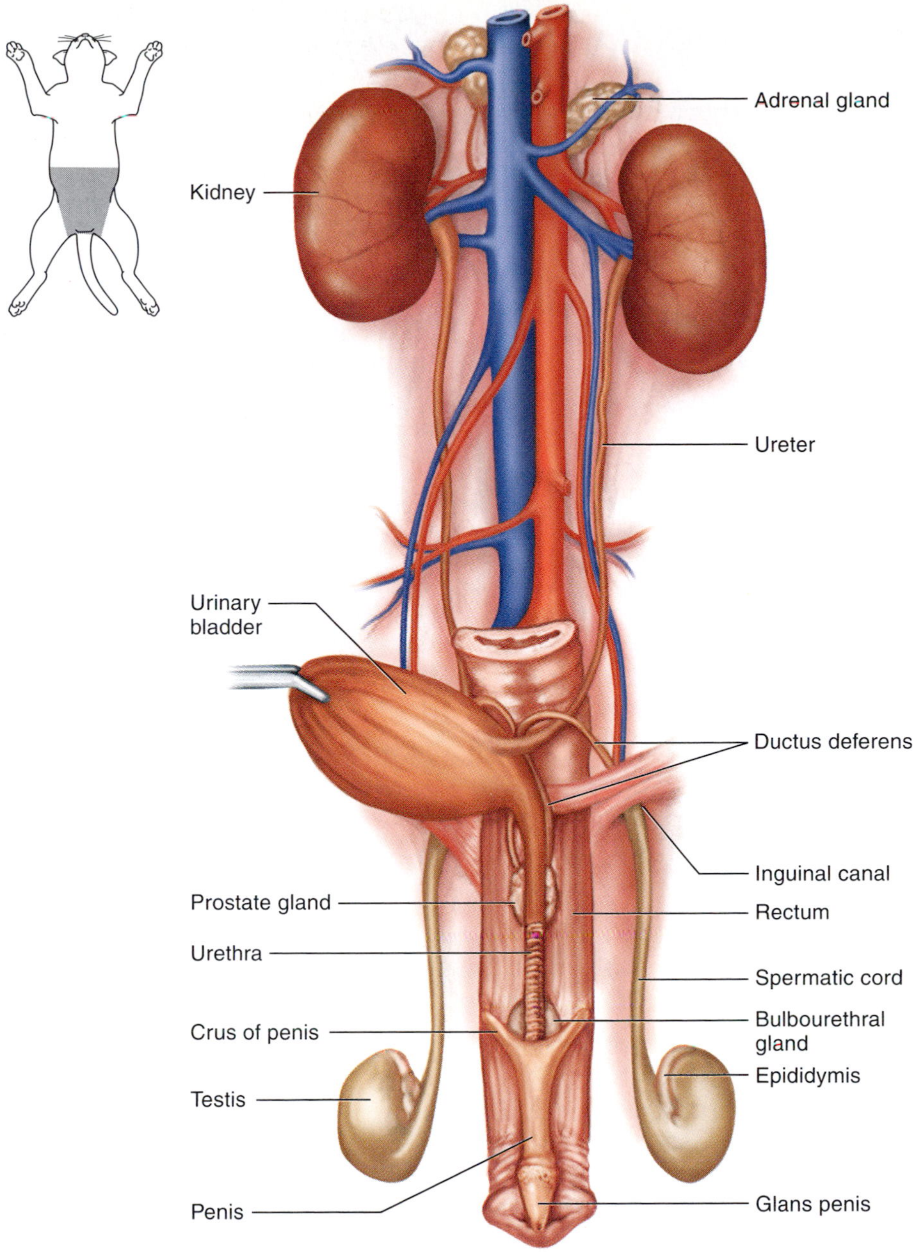

Figure 59.5 Ventral view of the male cat's reproductive system.

downward behind the urinary bladder to join the urethra. Locate the *prostate gland,* which appears as an enlargement at the junction of the ductus deferens and urethra.

8. Trace the urethra to the penis. Locate the *bulbourethral glands,* which form small swellings on either side at the proximal end of the penis.
9. Use a sharp scalpel to cut a transverse section of the penis. Identify the *corpora cavernosa,* each of which contains many blood spaces surrounded by a sheath of connective tissue.
10. Review the locations of the reproductive organs without the aid of the figures. 1
11. Examine the models of the human reproductive systems along with figures 59.4, 59.5, and 59.6. Identify the organs of the cat that correspond with those of the human. 3
12. Complete Part B of the laboratory report.

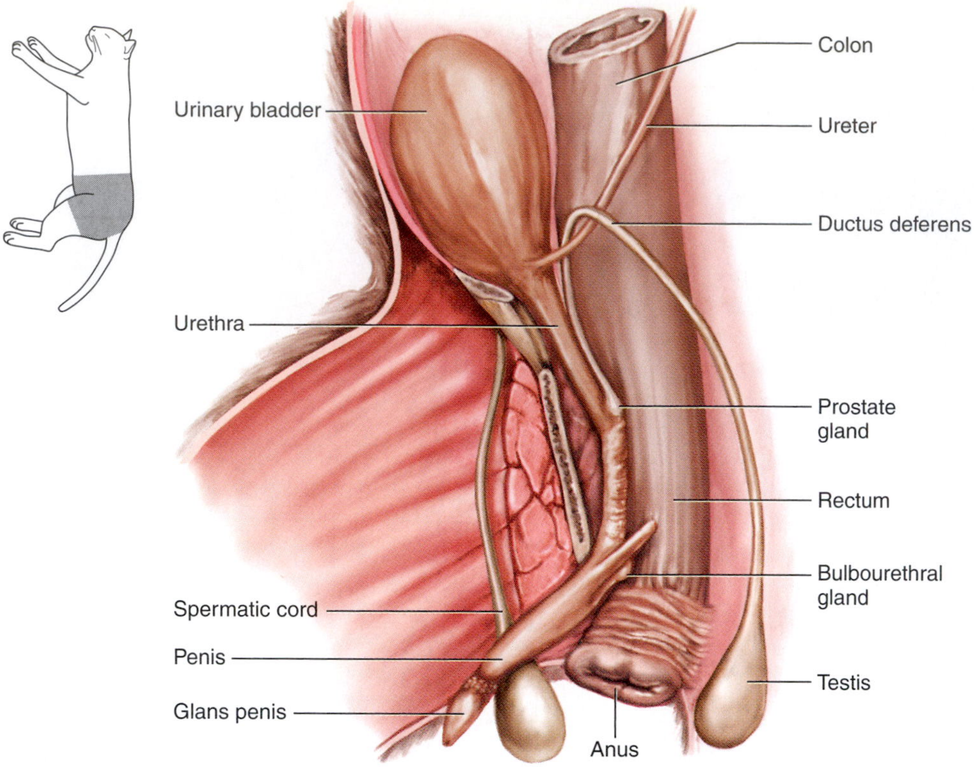

Figure 59.6 Lateral view of the male cat's reproductive system.

Name ______________________________

Date ______________________________

Section ______________________________

The [A] corresponds to the indicated Learning Outcome(s) listed at the beginning of the laboratory exercise.

Cat Dissection: Reproductive Systems

Laboratory Report 59

Part A Assessments

Complete the following:

1. Compare the relative lengths and paths of the uterine tubes (oviducts) of the cat and the human. [2] ______________________________

2. How do the shape and structure of the uterus of the cat compare with that of the human? [2] ______________________________

3. Assess the function of the uterine horns of the cat. [1] ______________________________

4. Compare the relationship of the urethra and the vagina in the cat and in the human. [2] ______________________________

Part B Assessments

Complete the following:

1. Compare the glans penis in the cat and in the human. 2 ____________________

2. How do the location and the relative size of the prostate gland of the cat compare with that of the human? 2

3. What glands associated with the human ductus deferens are missing in the cat? 2 ____________________

4. Compare the prepuce in the cat and in the human. 2 ____________________

Appendix 1

Preparation of Solutions

Amylase solution, 0.5%

Place 0.5 g of bacterial amylase in a graduated cylinder or volumetric flask. Add distilled water to the 100 mL level. Stir until dissolved. The amylase should be free of sugar for best results; a low-maltose solution of amylase yields good results. (Store amylase powder in a freezer until mixing this solution.)

Benedict's solution

Prepared solution is available from various suppliers.

Caffeine, 0.2%

Place 0.2 g of caffeine in a graduated cylinder or volumetric flask. Add distilled water to the 100 mL level. Stir until dissolved.

Calcium chloride, 2.0%

Place 2.0 g of calcium chloride in a graduated cylinder or volumetric flask. Add distilled water to the 100 mL level. Stir until dissolved.

Calcium hydroxide solution (limewater)

Add an excess of calcium hydroxide to 1 L of distilled water. Stopper the bottle and shake thoroughly. Allow the solution to stand for 24 hours. Pour the supernatant fluid through a filter. Store the clear filtrate in a stoppered container.

Epsom salt solution, 0.1%

Place 0.5 g of Epsom salt in a graduated cylinder or volumetric flask. Add distilled water to the 500 mL level. Stir until dissolved.

Glucose solutions

1. *1.0% solution.* Place 1 g of glucose in a graduated cylinder or volumetric flask. Add distilled water to the 100 mL level. Stir until dissolved.
2. *10% solution.* Place 10 g of glucose in a graduated cylinder or volumetric flask. Add distilled water to the 100 mL level. Stir until dissolved.

Iodine-potassium-iodide (IKI solution)

Add 20 g of potassium iodide to 1 L of distilled water, and stir until dissolved. Then add 4.0 g of iodine, and stir again until dissolved. Solution should be stored in a dark stoppered bottle.

Methylene blue

Dissolve 0.3 g of methylene blue powder in 30 mL of 95% ethyl alcohol. In a separate container, dissolve 0.01 g of potassium hydroxide in 100 mL of distilled water. Mix the two solutions. (Prepared solution is available from various suppliers.)

Monosodium glutamate (MSG) solution, 1%

Place 1 g of monosodium glutamate in a graduated cylinder or volumetric flask. Add distilled water to the 100 mL level. Stir until dissolved.

Physiological saline solution

Place 0.9 g of sodium chloride in a graduated cylinder or volumetric flask. Add distilled water to the 100 mL level. Stir until dissolved.

Potassium chloride, 5%

Place 5.0 g of potassium chloride in a graduated cylinder or volumetric flask. Add distilled water to the 100 mL level. Stir until dissolved.

Quinine sulfate, 0.5%

Place 0.5 g of quinine sulfate in a graduated cylinder or volumetric flask. Add distilled water to the 100 mL level. Stir until dissolved.

Ringer's solution (frog)

Dissolve the following salts in 1 L of distilled water:

6.50 g sodium chloride
0.20 g sodium bicarbonate
0.14 g potassium chloride
0.12 g calcium chloride

Sodium chloride solutions

1. *0.9% solution.* Place 0.9 g of sodium chloride in a graduated cylinder or volumetric flask. Add distilled water to the 100 mL level. Stir until dissolved.
2. *1.0% solution.* Place 1.0 g of sodium chloride in a graduated cylinder or volumetric flask. Add distilled water to the 100 mL level. Stir until dissolved.
3. *3.0% solution.* Place 3.0 g of sodium chloride in a graduated cylinder or volumetric flask. Add distilled water to the 100 mL level. Stir until dissolved.
4. *5.0% solution.* Place 5.0 g of sodium chloride in a graduated cylinder or volumetric flask. Add distilled water to the 100 mL level. Stir until dissolved.

Starch solutions

1. *0.5% solution.* Add 5 g of cornstarch to 1 L of distilled water. Heat until the mixture boils. Cool the liquid, and pour it through a filter. Store the filtrate in a refrigerator.
2. *1.0% solution.* Add 10 g of cornstarch to 1 L of distilled water. Heat until the mixture boils. Cool the liquid, and pour it through a filter. Store the filtrate in a refrigerator.
3. *10% solution.* Add 100 g of cornstarch to 1 L of distilled water. Heat until the mixture boils. Cool the liquid, and pour it through a filter. Store the filtrate in a refrigerator.

Sucrose, 5% solution

Place 5.0 g of sucrose in a graduated cylinder or volumetric flask. Add distilled water to the 100 mL level. Stir until dissolved.

Wright's stain

Prepared solution is available from various suppliers.

Appendix 2

Assessments of Laboratory Reports

Many assessment models can be used for laboratory reports. A rubric, which can be used for performance assessments, contains a description of the elements (requirements or criteria) of success to various degrees. The term *rubric* originated from *rubrica terra,* which is Latin for the application of red earth to indicate anything of importance. A rubric used for assessment contains elements for judging student performance, with points awarded for varying degrees of success in meeting the learning outcomes. The content and the quality level necessary to attain certain points are indicated in the rubric. It is effective if the assessment tool is shared with the students before the laboratory exercise is performed.

Following are two sample rubrics that could easily be modified to meet the needs of a specific course. Some of the elements for these sample rubrics may not be necessary for every laboratory exercise. The generalized rubric needs to contain the possible assessment points that correspond to learning outcomes for a specific course. The point value for each element may vary. The specific rubric example contains performance levels for laboratory reports. The elements and the point values could easily be altered to meet the value placed on laboratory reports for a specific course.

Assessment: Generalized Laboratory Report Rubric

Element	Assessment Points Possible	Assessment Points Earned
1. Figures are completely and accurately labeled.		
2. Sketches are accurate, contain proper labels, and are of sufficient detail.		
3. Colored pencils were used extensively to differentiate structures on illustrations.		
4. Matching and fill-in-the-blank answers are completed and accurate.		
5. Short-answer/discussion questions contain complete, thorough, and accurate answers. Some elaboration is evident for some answers.		
6. Data collected are complete, accurately displayed, and contain a valid explanation.		

Total Points: Possible __________ Earned ____________________

Assessment: Specific Laboratory Report Rubric

Element	Excellent Performance (4 points)	Proficient Performance (3 points)	Marginal Performance (2 points)	Novice Performance (1 point)	Points Earned
Figure labels	Labels completed with ≥ 90% accuracy.	Labels completed with 80%–89% accuracy.	Labels completed with 70%–79% accuracy.	Labels < 70% accurate.	
Sketches	Accurate use of scale, details illustrated, and all structures labeled accurately.	Minor errors in sketches. Missing or inaccurate labels on one or more structures.	Sketch is not realistic. Missing or inaccurate labels on two or more structures.	Several missing or inaccurate labels.	
Matching and fill-in-the-blanks	All completed and accurate.	One to two errors or omissions.	Three to four errors or omissions.	Five or more errors or omissions.	
Short-answer and discussion questions	Answers are complete, valid, and contain some elaboration. No misinterpretations are noted.	Answers are generally complete and valid. Only minor inaccuracies were noted. Minimal elaboration exists.	Marginal answers to the questions and contains inaccurate information.	Many answers are incorrect or fail to address the topic. There may be misinterpretations.	
Data collection and analysis	Data are complete and displayed with a valid interpretation.	Only minor data missing or a slight misinterpretation exists.	Some omissions. Not displayed or interpreted accurately.	Data are incomplete or show serious misinterpretations.	

Total Points Earned ________

Credits

Photo Credits

Lab Exercise 1
Figures 1.2–1.3: © The McGraw-Hill Companies/J. Womack Photography.

Lab Exercise 2
Figures 2.3–2.8a–c: © The McGraw-Hill Companies/J. Womack Photography.

Lab Exercise 4
Figures 4.1–4.6: © The McGraw-Hill Companies/J. Womack Photography.

Lab Exercise 5
Figure 5.3: © Ed Reschke; 5.4a–b: Richard Rodewald/Biological Photo Service.

Lab Exercise 6
Figure 6.1: © The McGraw-Hill Companies/J. Womack Photography.

Lab Exercise 7
Figures 7.1(all)–7.3: © Ed Reschke; 7.4a–b: Courtesy of the March of Dimes Birth Defects Foundation; 7.5a–d: © Ed Reschke.

Lab Exercise 8
Figure 8.1a–b: © Ed Reschke; 8.1c: © Manfred Kage/Peter Arnold; 8.1d: © Ed Reschke; 8.1e: © Fred Hossler/Visuals Unlimited; 8.1f: © Ed Reschke.

Lab Exercise 9
Figure 9.1a–c: © Ed Reschke; 9.1d: © John D. Cunningham/Visuals Unlimited; 9.1e: © Ed Reschke/Peter Arnold; 9.1f–g: © Ed Reschke; 9.1h: © The McGraw-Hill Companies/Al Telser, photographer; 9.1i: © Ed Reschke/Peter Arnold; 9.1j: © The McGraw-Hill Companies/Al Telser, photographer.

Lab Exercise 10
Figure 10.1a–b: © Ed Reschke; 10.1c: © Manfred Kage/Peter Arnold; 10.1d: © Ed Reschke.

Lab Exercise 11
Figure 11.2: © The McGraw-Hill Companies/Al Telser, photographer; 11.4a: © Victor B. Eichler, Ph.D; 11.4b: © The McGraw-Hill Companies/Carol D. Jacobson, Ph.D. Dept. of Veterinary Anatomy, Iowa State University; 11.4c: © The McGraw-Hill Companies/Al Telser, photographer; 11.4d: © Per H. Kjeldsen.

Lab Exercise 12
Figure 12.3: © Victor B. Eichler; 12.4a–12.5: © The McGraw-Hill Companies/J. Womack Photography.

Lab Exercise 13
Figure 13.2: © The McGraw-Hill Companies/J. Womack Photography.

Lab Exercise 14
Figures 14.6–14.12: © The McGraw-Hill Companies/J. Womack Photography.

Lab Exercise 15
Figures 15.2–15.3b: © The McGraw-Hill Companies/J. Womack Photography; 15.6: © Dr. Kent M. Van De Graaff.

Lab Exercise 16
Figure 16.7: © Martin M. Rotker; 16.8: Courtesy of Eastman Kodak; 16.9: © Martin M. Rotker; 16.10: © The McGraw-Hill Companies/J. Womack Photography.

Lab Exercise 17
Figures 17.7–17.9: © Martin M. Rotker; 17.10: © The McGraw-Hill Companies/J. Womack Photography.

Lab Exercise 18
Figures 18.2a–18.3e: © The McGraw-Hill Companies/Womack Photography Ltd.

Lab Exercise 20
Figure 20.4a–c: © The McGraw-Hill Companies/J. Womack Photography.

Lab Exercise 21
Figure 21.5a–c: © The McGraw-Hill Companies/J. Womack Photography.

Lab Exercise 23
Figure 23.7a–b: © The McGraw-Hill Companies/J. Womack Photography.

Lab Exercise 24
Figure 24.1a: © The McGraw-Hill Companies/Joe DeGrandis, Photographer; 24.1b–c: © The McGraw-Hill Companies/Eric Wise, photographer; 24.2: Suza Scalora/PhotoDisc/Getty Images; 24.3a: © The McGraw-Hill Companies/Joe DeGrandis, Photographer; 24.3b: Suza Scalora/PhotoDisc/Getty Images; 24.4–24.5a–b: © The McGraw-Hill Companies/Joe DeGrandis, Photographer; 24.6: © The McGraw-Hill Companies/Eric Wise, photographer; 24.7–24.8c: © The McGraw-Hill Companies/Joe DeGrandis, Photographer; 24.9a–b: © The McGraw-Hill Companies/Eric Wise, photographer.

Lab Exercise 25
Figures 25.3–25.7: © Ed Reschke.

Lab Exercise 26
Figure 26.4: © Per H. Kjeldsen.

Lab Exercise 27
Figure 27.2a–e: © The McGraw-Hill Companies/J. Womack Photography.

Lab Exercise 28
Figure 28.5: © Martin Rotker/Photo Researchers, Inc.

Lab Exercise 29
Figures 29.2–29.5: © The McGraw-Hill Companies/J. Womack Photography.

Lab Exercise 30
Figures 30.1–30.2: © Ed Reschke.

Lab Exercise 31
Figure 31.2: © Dwight Kuhn; 31.4: © Ed Reschke.

Lab Exercise 32
Figure 32.4: © The McGraw-Hill Companies/J. Womack Photography; 32.5: © Biophoto Associates/Photo Researchers, Inc. 32.6a–32.7: © The McGraw-Hill Companies/J. Womack Photography; 32.8: © John D. Cunningham/Visuals Unlimited.

Lab Exercise 33
Figure 33.2: © John D. Cunningham/Visuals Unlimited; 33.3: © The McGraw-Hill Companies/J. Womack Photography.

Lab Exercise 34
Figure 34.4: © Per H. Kjeldsen, University of Michigan, Ann Arbor; 34.5: © The McGraw-Hill Companies/J. Womack Photography; 34.7: Mediscan Corbis; 34.9–34.10: © The McGraw-Hill Companies/J. Womack Photography; 34.11a–b: © Carroll Weiss/Camera M.D. Studios.

Lab Exercise 35
Figure 35.3: © The McGraw-Hill Companies/J. Womack Photography; 35.4: Ishihara's Tests for Colour Blindness published by Kanehara & Co., Ltd., Tokyo, Japan.

Lab Exercise 36
Figure 36.2: © Biophoto Associates/Photo Researchers, Inc.; 36.3: © Ed Reschke; 36.4: © Biophoto Associates/Photo Researchers, Inc.; 36.5–36.6: © Ed Reschke; 36.7: © The McGraw-Hill Companies/J. Womack Photography; 36.8: © The McGraw-Hill Companies/Al Telser, photographer; 36.9: © Dr. F.C. Skvara/Visuals Unlimited.

Lab Exercise 37
Figure 37.1(all): © The McGraw-Hill Companies/Al Telser, photographer.

Lab Exercise 38
Figures 38.4–38.6: © The McGraw-Hill Companies/J. Womack Photography.

Lab Exercise 40
Figure 40.2: © Biophoto Associates/Photo Researchers, Inc.

Lab Exercise 42
Figure 42.3a: © The McGraw-Hill Companies/J. Womack Photography; 42.3b: © John Watney/Photo Researchers, Inc.; 42.4: © John D. Cunningham/Visuals Unlimited; 42.5: © Biophoto Associates/Photo Researchers, Inc.

Lab Exercise 43
Figure 43.4: © Biophoto Associates/Photo Researchers, Inc.; 43.6–43.9: © Ed Reschke; 43.12: © The McGraw-Hill Companies/J. Womack Photography; 43.13: © Dennis Strete; 43.14: © Ed Reschke/Peter Arnold.

Lab Exercise 45
Figure 45.5: © Ed Reschke; 45.6: © Biophoto Associates/Photo Researchers, Inc.; 45.7: Victor B. Eichler, Ph.D.; 45.8: © Dwight Kuhn.

Lab Exercise 46
Figures 46.1–46.3: © The McGraw-Hill Companies/J. Womack Photography.

Lab Exercise 48
Figure 48.3: © The McGraw-Hill Companies/J. Womack Photography; 48.5a: © Biophoto Associates/Photo Researchers, Inc.; 48.5b: © Manfred Kage/Peter Arnold.

Lab Exercise 50

Figure 50.3: © Biophoto Associates/Photo Researchers, Inc.; 50.4–50.5: © Ed Reschke; 50.6: Michael Peres.

Lab Exercise 51

Figure 51.6: © Ed Reschke/Peter Arnold; 51.7: © Ed Reschke; 51.8–51.9: © 1989 Michael Peres.

Lab Exercise 53

Figure 53.1a–j: © The McGraw-Hill Companies/J. Womack Photography.

Lab Exercise 54

Figures 54.2–54.22: Courtesy of Dr. Sheril Burton; 54.23: © The McGraw-Hill Companies/Photo Ralph W. Stevens III, Ph.d.

Lab Exercise 55

Figures 55.2–55.9: Courtesy of Dr. Sheril Burton.

Lab Exercise 56

Figure 56.4: Courtesy of Dr. Sheril Burton.

Lab Exercise 57

Figure 57.1: Courtesy of Dr. Sheril Burton.

Lab Exercise 58

Figure 58.1: Courtesy of Dr. Sheril Burton; 58.3: © The McGraw-Hill Companies/Photo Ralph W. Stevens III, Ph.d.

Lab Exercise 59

Figures 59.2–59.4: © The McGraw-Hill Companies/Photo Ralph W. Stevens III, Ph.d.

Index

Note: Page numbers followed by *f* refer to figures only; page numbers followed by *t* refer to tables only.

A

C

D

E

F

I

J

K

L

M

N

O

P

Q

R

S

T

U

V

W

X

Y

Z

Notes

Notes

Notes

Notes

Notes